Entropy and Diversity

The global biodiversity crisis is one of humanity's most urgent problems, but even quantifying biological diversity is a difficult mathematical and conceptual challenge. This book brings new mathematical rigour to the ongoing debate. It was born of research in category theory, is given strength by information theory, and is fed by the ancient field of functional equations. It applies the power of the axiomatic method to a biological problem of pressing concern, but it also presents new theorems that stand up as mathematics in their own right, independently of any application. The question 'what is diversity?' has surprising mathematical depth, and this book covers a wide breadth of mathematics, from functional equations to geometric measure theory, from probability theory to number theory. Despite this range, the mathematical prerequisites are few: the main narrative thread of this book requires no more than an undergraduate course in analysis.

TOM LEINSTER is Professor of Category Theory at the University of Edinburgh, a member of the University of Glasgow's Boyd Orr Centre for Population and Ecosystem Health, and co-author of a highly cited *Ecology* article on measuring biodiversity. He was awarded the 2019 Chauvenet Prize for mathematical writing.

Entropy and Diversity

The Axiomatic Approach

TOM LEINSTER
University of Edinburgh

CAMBRIDGE
UNIVERSITY PRESS

University Printing House, Cambridge CB2 8BS, United Kingdom

One Liberty Plaza, 20th Floor, New York, NY 10006, USA

477 Williamstown Road, Port Melbourne, VIC 3207, Australia

314–321, 3rd Floor, Plot 3, Splendor Forum, Jasola District Centre,
New Delhi – 110025, India

79 Anson Road, #06–04/06, Singapore 079906

Cambridge University Press is part of the University of Cambridge.

It furthers the University's mission by disseminating knowledge in the pursuit of education, learning, and research at the highest international levels of excellence.

www.cambridge.org
Information on this title: www.cambridge.org/9781108832700
DOI: 10.1017/9781108963558

First published 2021

A catalogue record for this publication is available from the British Library.

ISBN 978-1-108-83270-0 Hardback
ISBN 978-1-108-96557-6 Paperback

Much of this book was written in Catalonia during the years 2017 to 2019.

I dedicate it to all those who defend democracy.

Contents

Acknowledgements

Many people have given me encouragement and the benefit of their insight, wisdom, and expertise. I am especially grateful to John Baez, Jim Borger, Tony Carbery, Josep Elgueta, José Figueroa-O'Farrill, Tobias Fritz, Herbert Gangl, Heiko Gimperlein, Dan Haydon, Richard Hepworth, André Joyal, Joachim Kock, Barbara Mable, Louise Matthews, Richard Reeve, Emily Roff, Mike Shulman, Zoran Škoda, Todd Trimble, Simon Willerton, and Xīlíng Zhāng. I also thank Roger Astley, Clare Dennison and Anna Scriven at CUP, as well as the copyeditor, Siriol Jones.

My heartfelt thanks go to Christina Cobbold and Mark Meckes, not only for many thought-provoking mathematical conversations over the years, but also for taking the time to read drafts of this text, which their perceptive and knowledgeable comments helped to improve.

I thank all of those involved in the original functional equations seminar course in Edinburgh for their good humour and friendly participation. Interactions during those seminars were formative for my understanding. So too were many conversations at *The n-Category Café*, a research blog that has been invaluable for my learning of new mathematics.

I owe a great deal to the Boyd Orr Centre for Population and Ecosystem Health, an interdisciplinary research centre at the University of Glasgow, of which I have been happy to have been a member since 2010. It is a model of the interdisciplinary spirit: welcoming, collaborative, informal, and scientifically ambitious. Warm thanks to all who foster that positive atmosphere.

This work was supported at different times by an EPSRC Advanced Research Fellowship, a BBSRC FLIP award (BB/P004210/1), and a Leverhulme Trust Research Fellowship. Critical research advances were made during a 2012 programme on the Mathematics of Biodiversity at the Centre de Recerca Matemàtica (CRM) in Barcelona, which was supported by the CRM, a Spanish government grant, and a BBSRC International Workshop Grant; I was also supported personally there by the Carnegie Trust for the Universities of Scotland. Finally, I thank the Department of Mathematics at the Universitat Autònoma de Barcelona for their hospitality during part of the writing of this book.

Some parts of Section 6.3 first appeared in Leinster and Meckes [221], and are reproduced with the second author's permission.

Note to the Reader

This book began life as a seminar course on functional equations at the University of Edinburgh in 2017, motivated by recent research on the quantification of biological diversity. The course attracted not only mathematicians in subjects from stochastic analysis to algebraic topology, but also participants from physics and biology. In response, I did everything I could to minimize the mathematical prerequisites.

I have tried here to retain the broad accessibility of the course. At the same time, I have not censored myself from including the many fruitful connections with more advanced parts of mathematics.

These two opposing forces have been reconciled by confining the more advanced material to separate chapters or sections that can easily be omitted. Chapter 9 requires some probability theory, Chapter 11 some abstract algebra, and Chapter 12 some category theory, while Sections 3.4, 6.4 and 6.5 also call on parts of geometry, analysis and statistics. However, the core narrative thread requires no more mathematics than a first course in rigorous (ε-δ) analysis. Readers with this background are promised that they are equipped to follow all the main ideas and results. The parts just listed, and any remarks that refer to more specialized knowledge, can safely be omitted.

Moreover, those who regard themselves as wholly 'pure' mathematicians will find no barriers here. Although much of this book is about the diversity of ecological systems, no knowledge of ecology is needed. Similarly, the information theory that we use is introduced from the ground up.

In the middle parts of the book, many conditions on means and diversity measures are defined: homogeneity, consistency, symmetry, etc. Appendix B contains a summary of this terminology for easy reference. There is also an index of notation.

Interdependence of Chapters

A dotted line indicates that one chapter is helpful, but not essential, for another.

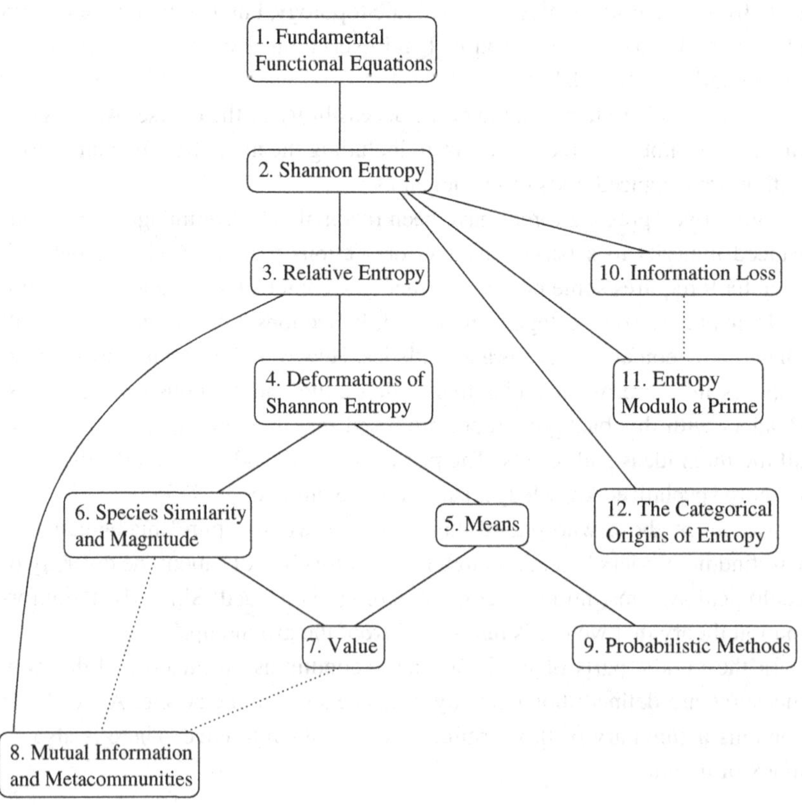

Introduction

This book was born of research in category theory, brought to life by the ongoing vigorous debate on how to quantify biological diversity, given strength by information theory, and fed by the ancient field of functional equations. It applies the power of the axiomatic method to a biological problem of pressing concern, but it also presents new advances in 'pure' mathematics that stand in their own right, independently of any application.

The starting point is the connection between diversity and entropy. We will discover:

- how Shannon entropy, originally defined for communications engineering, can also be understood through biological diversity (Chapter 2);
- how deformations of Shannon entropy express a spectrum of viewpoints on the meaning of biodiversity (Chapter 4);
- how these deformations *provably* provide the only reasonable abundance-based measures of diversity (Chapter 7);
- how to derive such results from characterization theorems for the power means, of which we prove several, some new (Chapters 5 and 9).

Complementing the classical techniques of these proofs is a large-scale categorical programme, which has produced both new mathematics and new measures of diversity now used in scientific applications. For example, we will find:

- that many invariants of size from across the breadth of mathematics (including cardinality, volume, surface area, fractional dimension, and both topological and algebraic notions of Euler characteristic) arise from one single invariant, defined in the wide generality of enriched categories (Chapter 6);
- a way of measuring diversity that reflects not only the varying abundances of

species (as is traditional), but also the varying similarities between them, or, more generally, any notion of the values of the species (Chapters 6 and 7);
- that these diversity measures belong to the extended family of measures of size (Chapter 6);
- a 'best of all possible worlds', an abundance distribution on any given set of species that maximizes diversity from an infinite number of viewpoints simultaneously (Chapter 6);
- an extension of Shannon entropy from its classical context of finite sets to distributions on a metric space or a graph (Chapter 6), obtained by translating the similarity-sensitive diversity measures into the language of entropy.

Shannon entropy is a fundamental concept of information theory, but information theory contains many riches besides. We will mine them, discovering:

- how the concept of relative entropy not only touches subjects from Bayesian inference to coding theory to Riemannian geometry, but also provides a way of quantifying local diversity within a larger context (Chapter 3);
- quantitative methods for identifying particularly unusual or atypical parts of an ecological community (Chapter 8, drawing on work of Reeve et al. [293]).

The main narrative thread is modest in its mathematical prerequisites. But we also take advantage of some more specialized bodies of knowledge (large deviation theory, the theory of operads, and the theory of finite fields), establishing:

- how probability theory can be used to solve functional equations (Chapter 9, following work of Aubrun and Nechita [20]);
- a streamlined characterization of information loss, as a natural consequence of categorical and operadic thinking (Chapters 10 and 12);
- that the concept of entropy is (provably) inescapable even in the pure-mathematical heartlands of category theory, algebra and topology, quite separately from its importance in scientific applications (Chapter 12);
- the right definition of entropy for probability distributions whose 'probabilities' are elements of the ring $\mathbb{Z}/p\mathbb{Z}$ of integers modulo a prime p (Chapter 11, drawing on work of Kontsevich [195]).

The question of how to quantify diversity is far more mathematically profound than is generally appreciated. This book makes the case that the theory of diversity measurement is fertile soil for new mathematics, just as much as the neighbouring but far more thoroughly worked field of information theory.

* * *

What *is* the problem of quantifying diversity? Briefly, it is to take a biological community and extract from it a numerical measure of its 'diversity' (whatever that should mean). This task is certainly beset with practical problems: for instance, field ecologists recording woodland animals will probably observe the noisy, the brightly coloured and the gregarious more frequently than the quiet, the camouflaged and the shy. There are also statistical difficulties: if a survey of one community finds 10 different species in a sample of 50 individuals, and a survey of another finds 18 different species in a sample of 100, which is more diverse?

However, we will not be concerned with either the practical or the statistical difficulties. Instead, we will focus on a fundamental conceptual problem: in an ideal world where we have complete, perfect data, how can we quantify diversity in a meaningful and logical way?

In both the news media and the scientific literature, the most common meaning given to the word 'diversity' (or 'biodiversity') is simply the number of species present. Certainly this is an important quantity. However, it is not always very informative. For instance, the number of species of great ape on the planet is 8 (Example 4.3.8), but 99.99% of all great apes belong to just one species: us. In terms of global ecology, it is arguably more accurate to say that there is effectively only one species of great ape.

An example illustrates the spectrum of possible interpretations of the concept of diversity. Consider two bird communities:

In community A, there are four species, but the majority of individuals belong to a single dominant species. Community B contains the first three species in equal abundance, but the fourth is absent. Which community, A or B, is more diverse?

One viewpoint is that the presence of *species* is what matters. Rare species count for as much as common ones: every species is precious. From this viewpoint, community A is more diverse, simply because more species are present. The abundances of species are irrelevant; presence or absence is all that matters.

But there is an opposing viewpoint that prioritizes the balance of *communities*. Common species are important; they are the ones that exert the most influence on the community. Community A has a single very common species, which has largely outcompeted the others, whereas community B has three common species, evenly balanced. From this viewpoint, community B is more diverse.

These two viewpoints are the two ends of a continuum. More precisely, there is a continuous one-parameter family $(D_q)_{q\in[0,\infty]}$ of diversity measures encoding this spectrum of viewpoints. Low values of q attach high importance to rare species; for example, D_0 measures community A as more diverse than community B. When q is high, D_q is most strongly influenced by the balance of more common species; thus, D_∞ judges B to be more diverse. No single viewpoint is right or wrong. Different scientists adopt different viewpoints (that is, different values of q) for different purposes, as the literature amply attests (Examples 4.3.5).

Long ago, it was realized that the concept of diversity is closely related to the concept of entropy. Entropy appears in dozens of guises across dozens of branches of science, of which thermodynamics is probably the most famous. (The introduction to Chapter 2 gives a long but highly incomplete list.) The most simple incarnation is Shannon entropy, which is a real number associated with any probability distribution on a finite set. It is, in fact, the logarithm of the diversity measure D_1. Most often, Shannon entropy is explained and understood through the theory of coding; indeed, we provide such an explanation here. But the diversity interpretation provides a new perspective.

For example, the diversity measures D_q, known in ecology as the Hill numbers, are the exponentials of what information theorists know as the Rényi entropies. From the very beginning of information theory, an important role has been played by characterization theorems: results stating that any measure (of information, say) satisfying a list of desirable properties must be of a particular form (a scalar multiple of Shannon entropy, say). But what counts as a desirable property depends on one's perspective. We will prove that the Hill numbers D_q are, in a precise sense, the only measures of diversity with certain natural properties (Theorem 7.4.3). This theorem translates into a new characterization of the Rényi entropies, but it is not one that necessarily would have been thought of from a purely information-theoretic perspective.

However, something is missing. In the real world, diversity is understood as involving not only the number and abundances of the species, but also how *dif-*

ferent they are. (For example, this affects conservation policy; see the OECD quotation on p. 169.) We describe the remedy in Chapter 6, defining a family of diversity measures that take account of the varying similarity between species, while still incorporating the spectrum of viewpoints discussed above. This definition unifies into one family a large number of the diversity measures proposed and used in the ecological and genetics literature.

This family of diversity measures first appeared in a paper in *Ecology* [220], but it can also be understood and motivated from a purely mathematical perspective. The classical Rényi entropies are a family of real numbers assigned to any probability distribution on a finite *set*. By factoring in the differences or distances between points (species), we extend this to a family of real numbers assigned to any probability distribution on a finite *metric space*. In the extreme case where $d(x, y) = \infty$ for all distinct points x and y, we recover the Rényi entropies. In this way, the similarity-sensitive diversity measures extend the definition of Rényi entropy from sets to metric spaces.

Different values of the viewpoint parameter $q \in [0, \infty]$ produce different judgements on which of two distributions is the more diverse. But it turns out that for any metric space (or in biological terms, any set of species), there is a single distribution that maximizes diversity from all viewpoints simultaneously. For a generic finite metric space, this maximizing distribution is unique. Thus, almost every finite metric space carries a canonical probability distribution (not usually uniform). The maximum diversity itself is also independent of q, and is therefore a numerical invariant of metric spaces. This invariant has geometric significance in its own right (Section 6.5).

We go further. One might wish to evaluate an ecological community in a way that takes into account some notion of the values of the species (such as phylogenetic distinctiveness). Again, there is a sensible family of measures that does this job, extending not only the similarity-sensitive diversity measures just described, but also further measures already existing in the ecological literature. The word 'sensible' can be made precise: as soon as we subject an abstract measure of the value of a community to some basic logical requirements, it is forced to belong to a certain one-parameter family (σ_q) (Theorem 7.3.4), which are essentially the Rényi *relative* entropies.

Information theory also helps us to analyse the diversity of metacommunities, that is, ecological communities made up of a number of smaller communities such as geographical regions. The established notions of relative entropy, conditional entropy and mutual information provide meaningful measures of the structure of a metacommunity (Chapter 8). But we will do more than simply translate information theory into ecological language. For example, the new characterization of the Rényi entropies mentioned above is a byproduct of the characterization theorem for measures of ecological value. In this way, the theory of diversity gives back to information theory.

* * *

The scientific importance of biological diversity goes far beyond the obvious setting of conservation of animals and plants. Certainly such conservation efforts are important, and the need for meaningful measures of diversity is well appreciated in that context. For example, Vane-Wright et al. [342] wrote thirty years ago of the 'agony of choice' in conservation of flora and fauna, and emphasized how crucial it is to use the right diversity measures.

But most life is microscopic. Nee [262] argued in 2004 that

> [w]e are still at the very beginning of a golden age of biodiversity discovery, driven largely by the advances in molecular biology and a new open-mindedness about where life might be found,

and that

> all of the marvels in biodiversity's new bestiary are invisible.

Even excluding exotic new discoveries of microscopic life, two recent lines of research illustrate important uses of diversity measures at the microbial level.

First, the extensive use of antimicrobial drugs on animals unfortunate enough to be born into the modern meat industry is commonly held to be a cause of antimicrobial resistance in pathogens affecting humans. However, a 2012 study of Mather et al. [246] suggests that the causality may be more complex. By analysing the diversity of antimicrobial resistance in *Salmonella* taken from animal populations on the one hand, and from human populations on the other, the authors concluded that the animal population is 'unlikely to be the major source of resistance' for humans, and that 'current policy emphasis on restricting antimicrobial use in domestic animals may be overly simplistic'. The diversity measures used in this analysis were the Hill numbers D_q mentioned above and central to this book.

Second, the increasing problem of obesity in humans has prompted research into causes and treatments, and there is evidence of a negative correlation between obesity and diversity of the gut microbiome (Turnbaugh et al. [335, 336]). Almost all traditional measures of diversity rely on a division of organisms into species or other taxonomic groups, but in this case, only a fraction of the microbial species concerned have been isolated and classified taxonomically. Researchers in this field therefore use DNA sequence data, applying sophisticated but somewhat arbitrary clustering algorithms to create artificial species-like groups ('operational taxonomic units'). On the other hand, the similarity-sensitive diversity measures mentioned above and introduced in Chapter 6 can be applied directly to the sequence data, bypassing the clustering step and producing a measure of genetic diversity. A test case was carried out in Leinster and Cobbold [220] (Example 4), with results that supported the conclusions of Turnbaugh et al.

Despite the wide variety of uses of diversity measures in biology, none of the mathematics presented in this text is intrinsically biological. Indeed, the mathematics of diversity was being developed as early as 1912 by the economist Corrado Gini [118] (best known for the Gini coefficient of disparity of wealth), and by the statistician Udny Yule in the 1940s for the analysis of lexical diversity in literature [361]. Some of the diversity measures most common in ecology have recently been used to analyse the ethnic and sociological diversity of judges (Barton and Moran [30]), and the similarity-sensitive diversity measures that are the subject of Chapter 6 have been used not only in multiple ecological contexts (as listed after Example 6.1.8), but also in non-biological applications such as computer network security (Wang et al. [347]).

In mathematical terms, simple diversity measures such as the Hill numbers are invariants of a probability distribution on a finite set. The similarity-sensitive diversity measures are defined for any probability distribution on a finite set with an assigned degree of similarity between each pair of points. (This includes any finite metric space or graph.) The value measures are defined for any finite set equipped with a probability distribution and an assignment of a nonnegative value to each element. The metacommunity measures are defined for any probability distribution on the cartesian product of a pair of finite sets. Much of this text is written using ecological terminology, but the mathematics is entirely general.

* * *

This work grew out of a general category-theoretic study of size. In many parts of mathematics, there is a canonical notion of the size of the objects of study: sets have cardinality, vector spaces have dimension, subsets of Euclidean space have volume, topological spaces have Euler characteristic, and so on. Typically, such measures of size satisfy analogues of the elementary inclusion-exclusion and multiplicativity formulas for counting finite sets:

$$|X \cup Y| = |X| + |Y| - |X \cap Y|,$$
$$|X \times Y| = |X| \cdot |Y|.$$

(The interpretation of Euler characteristic as the topological analogue of cardinality is not as well known as it should be; this is an insight of Schanuel on which we elaborate in Section 6.4.) From a categorical perspective, it is natural to seek a single invariant unifying all of these measures of size.

Some unification is achieved by defining a notion of size for categories themselves, called *magnitude* or Euler characteristic. (Finiteness hypotheses are required, but will not be mentioned in this overview.) This definition already brings together several established invariants of size [210]: cardinality of sets, and the various notions of Euler characteristic for partially ordered sets, topological spaces, and even orbifolds (whose Euler characteristics are in general not integers). The theory of magnitude of categories is closely related to the theory of Möbius–Rota inversion for partially ordered sets [301, 215].

But the decisive, unifying step is the generalization of the definition of magnitude from categories to the wider class of *enriched* categories [216], which includes not only categories themselves, but also metric spaces, graphs, and the additive categories that are a staple of homological algebra.

The definition of the magnitude of an enriched category unifies still more established invariants of size. For example, in the representation theory of associative algebras, one frequently considers the indecomposable projective modules, which form an additive category. The magnitude of that additive category turns out to be the Euler form of a certain canonical module, defined as an alternating sum of dimensions of Ext groups (equation (6.20)). Magnitude for enriched categories can also be realized as the Euler characteristic of a certain Hochschild-like homology theory of enriched categories, in the same sense that the Jones polynomial for knots is the Euler characteristic of Khovanov homology [189]. This was established in recent work led by Shulman [224], building on the case of magnitude homology for graphs previously developed by Hepworth and Willerton [144].

Since any metric space can be regarded as an enriched category, the general definition of the magnitude of an enriched category gives, in particular, a definition of the magnitude $|X| \in \mathbb{R}$ of a metric space X. Unlike the other special cases just mentioned, this invariant is essentially new.

Recent, increasingly sophisticated, work in analysis has connected magnitude with classical invariants of geometric measure. For example, for a compact subset $X \subseteq \mathbb{R}^n$ satisfying certain regularity conditions, if one is given the magnitude of all of the rescalings tX of X (for $t > 0$), then one can recover:

- the Minkowski dimension of X (one of the principal notions of fractional dimension), a result proved by Meckes using results in potential theory (Theorem 6.5.9);
- the volume of X, a result proved by Barceló and Carbery using PDE methods (Theorem 6.5.6);

- the surface area of X, a result proved by Gimperlein and Goffeng using global analysis (or more specifically, tools for computing heat trace asymptotics; see Theorem 6.5.8).

Gimperlein and Goffeng also proved an asymptotic inclusion-exclusion principle:

$$|t(X \cup Y)| + |t(X \cap Y)| - |tX| - |tY| \to 0$$

as $t \to \infty$, for sufficiently regular $X, Y \subseteq \mathbb{R}^n$ (Section 6.5). This is another manifestation of the cardinality-like nature of magnitude.

We have seen that every finite metric space X has an unambiguous maximum diversity $D_{\max}(X) \in \mathbb{R}$, defined in terms of the similarity-sensitive diversity measures (p. 5). We have also seen that X has a magnitude $|X| \in \mathbb{R}$. These two real numbers are not in general equal (ultimately because probabilities or species abundances are forbidden to be negative), but they are closely related. Indeed, $D_{\max}(X)$ is always equal to the magnitude of some *subspace* of X, and in important families of cases is equal to the magnitude of X itself. So, magnitude is closely related to maximum diversity. Indeed, this relationship was exploited by Meckes to prove the result on Minkowski dimension.

There is a historical surprise. Although this author arrived at the definition of the magnitude of a metric space by the route of enriched category theory, it had already arisen in earlier work on the quantification of biodiversity. In 1994, the environmental scientists Andrew Solow and Stephen Polasky carried out a probabilistic analysis of the benefits of high biodiversity ([319], Section 4), and isolated a particular quantity that they called the 'effective number of species'. They did not investigate it mathematically, merely remarking mildly that it 'has some appealing properties'. It is exactly our magnitude.

* * *

Ecologists began to propose quantitative definitions of biological diversity in the mid-twentieth century [314, 351], setting in motion more than sixty years of heated debate, with dozens of further proposed diversity measures, hundreds of scholarly papers, at least one book devoted to the subject [240], and consequently, for some, despair (expressed as early as 1971 in a famously titled paper of Hurlbert [150]). Meanwhile, parallel debates were taking place in genetics and other disciplines.

The connections between diversity measurement on the one hand, and information theory and category theory on the other, are fruitful for both mathematics and biology. But any measure of biological diversity must be justifiable in purely biological terms, rather than by borrowing authority from information theory, category theory, or any other field. The ecologist E. C. Pielou warned against attaching ecological significance to diversity measures for anything other than ecological reasons:

> It should not be (but it is) necessary to emphasize that the object of calculating indices of diversity is to solve, not to create, problems. The indices are merely numbers, useful in some circumstances but not in all. [...] Indices should be calculated for the light (not the shadow) they cast on genuine ecological problems.

(Pielou [283], p. 293).

In a series of incisive papers beginning in 2006, the conservationist and botanist Lou Jost insisted that whatever diversity measures one uses, they must exhibit *logical behaviour* [166, 167, 168, 169]. For example, Shannon entropy is commonly used as a diversity measure by practising ecologists, and it does behave logically if one is only using it to ask whether one community is more or less diverse than another. But as Jost observed, any attempt to reason about percentage changes in diversity using Shannon entropy runs into logical absurdities: Examples 2.4.7 and 2.4.11 describe the plague that exterminates 90% of species but only causes a 17% drop in 'diversity', and the oil drilling that simultaneously destroys *and* preserves 83% of the 'diversity' of an ecosystem. It is, in fact, the *exponential* of Shannon entropy that should be used for this purpose.

In this sense, origin stories are irrelevant. Inventing new diversity measures is easy, and it is nearly as easy to tell a story of how a new measure fits with some intuitive idea of diversity, or to justify it in terms of its importance in some related discipline. But if a measure does not pass basic logical tests (as in Section 4.4), it is useless or worse.

Jost noted that all of the Hill numbers D_q do behave logically. Again, we go further: Theorem 7.4.3 states that the Hill numbers are in fact the *only* measures of diversity satisfying certain logically fundamental properties. (At least, this is so for the simple model of a community in terms of species abundances only.) This is the ideal of the axiomatic approach: to prove results stating that if one wishes to have a measure with such-and-such properties, then it can only be one of *these* measures.

Mathematically, such results belong to the field of functional equations. We review a small corner of this vast and classical theory, beginning with the fact that the only measurable functions $f\colon \mathbb{R} \to \mathbb{R}$ satisfying the Cauchy functional equation $f(x + y) = f(x) + f(y)$ are the linear mappings $x \mapsto cx$. Building on classical results, we obtain new axiomatic characterizations of a variety of measures of diversity, entropy and value. We also explain a new method, pioneered by Aubrun and Nechita in 2011 [20], for solving functional equations by harnessing the power of probability theory. This produces new characterizations of the ℓ^p norms and the power means.

Characterization theorems for the power means are, in fact, the engine of this book (Chapter 5). By definition, the power mean of order t of real numbers $x_1, \ldots, x_n$, weighted by a probability distribution $(p_1, \ldots, p_n)$, is

$$M_t(\mathbf{p}, \mathbf{x}) = \left(\sum_{i=1}^{n} p_i x_i^t \right)^{1/t}.$$

The power means $(M_t)_{t \in \mathbb{R}}$ form a one-parameter family of operations, and the central place that they occupy in this text is explained by their relationship with several other important one-parameter families: the Hill numbers, the Rényi entropies, the q-logarithms, the q-logarithmic entropies (also known as Tsallis entropies), the value measures of Chapter 7, and the ℓ^p-norms. We will prove characterization theorems for all of these families, in each case finding a short list of properties that determines them uniquely.

* * *

Much of this text can be described as 'mathematical anthropology'. The mathematical anthropologist begins by observing that some group of scientists attaches great importance to a particular object or concept: homotopy theorists talk a lot about simplicial sets, harmonic analysts constantly use the Fourier transform, ecologists often count the number of species present in a community, and so on. The next step is to ask: why do they attach such importance to that particular thing, not something slightly different? Is it the *only* object that enjoys the useful properties that it enjoys? If not, why do they use the object they use, and not some other object with those properties? And if it *is* the only object with those properties, can we prove it? For example, 2008 work of Alesker, Artstein-Avidan and Milman [7] proved that the Fourier transform is, in fact, the only transform that enjoys its familiar properties.

This is the animating spirit of the field of functional equations. But there is another field that has been enormously successful in mathematical anthropology: category theory. There, objects of mathematical interest are typically characterized by universal properties. For instance, the tensor product $M \otimes N$ of modules M and N is the universal module equipped with a bilinear map $M \times N \to M \otimes N$; the Hilbert space completion $\hat{X}$ of an inner product space X is the universal Hilbert space equipped with an isometry $X \to \hat{X}$; the real interval $[0, 1]$ is the universal bipointed topological space equipped with a map $[0, 1] \to [0, 1] \vee [0, 1]$ (Theorem 2.2 of Leinster [212] and Theorem 2.5 of

Leinster [209], building on results of Freyd [111]). Any universal property involves uniqueness at two levels: the literal uniqueness of a connecting *map*, and the fact that the universal property characterizes the *object* possessing it uniquely up to isomorphism. Thus, category theory is a potent tool for proving characterization theorems.

We demonstrate this with a categorically motivated characterization theorem for entropy (Baez, Fritz and Leinster [25]). Briefly put, the probability distributions on finite sets form an operad, we construct a certain universal category acted on by that operad, and this leads naturally to the concept of Shannon entropy. The categorical approach amounts to a shift of emphasis from the entropy of a probability space (an object) to the amount of information lost by a deterministic process (a map).

The moral of this result is that entropy is not just something for applied scientists. It emerges inevitably from a general categorical machine, given as its inputs nothing more obscure than the real line and the standard topological simplices. In other words, even in algebra and topology, entropy is inescapable.

To demonstrate the strength of the axiomatic approach, we finish by applying it to an entity of purely mathematical interest: entropy modulo a prime number. The topic was first introduced as a curiosity by Kontsevich, as a byproduct of work on polylogarithms [195]. Just as any real probability distribution $\boldsymbol{\pi} = (\pi_1, \ldots, \pi_n)$ has a Shannon entropy $H_{\mathbb{R}}(\boldsymbol{\pi}) \in \mathbb{R}$, one can define, for any prime p and 'probabilities' $\pi_1, \ldots, \pi_n \in \mathbb{Z}/p\mathbb{Z}$, a kind of entropy $H_p(\boldsymbol{\pi}) \in \mathbb{Z}/p\mathbb{Z}$. The functional forms are quite different:

$$\begin{aligned} H_{\mathbb{R}}(\pi_1, \ldots, \pi_n) &= -\sum_{1 \le i \le n} \pi_i \log \pi_i && \in \mathbb{R}, \\ H_p(\pi_1, \ldots, \pi_n) &= -\sum_{\substack{0 \le r_1, \ldots, r_n < p \\ r_1 + \cdots + r_n = p}} \frac{\pi_1^{r_1} \cdots \pi_n^{r_n}}{r_1! \cdots r_n!} && \in \mathbb{Z}/p\mathbb{Z}. \end{aligned}$$

One would probably not guess that the second formula is the correct mod p analogue of the first. However, the definition is fully justified by a characterization theorem strictly analogous to the one that characterizes real Shannon entropy. And from the categorical perspective, there is a strictly analogous characterization of information loss mod p. In short, the apparatus developed for the real field can be successfully applied to the field of integers modulo a prime.

* * *

Finally, this book aims to challenge outdated conceptions of what applied mathematics can look like. Too often, 'applied mathematics' is subconsciously understood to mean 'methods of analysis applied to problems of physics'. (Or, worse, 'applied' is taken to be a euphemism for 'unrigorous'.) Those applications are certainly enormously important. However, this excessively narrow interpretation ignores the glittering array of applications of other parts of mathematics to other kinds of problem. It is mere historical accident that a researcher using PDEs in the study of fluids is usually called an applied mathematician, but one applying category theory to the design of programming languages is not.

Mathematicians are coming to appreciate that applications of their subject to biology are enormously fruitful and, with the revolution in the availability of genetic data, will only grow. Mackey and Maini asked and answered the question 'What has mathematics done for biology?' [239], quoting the evolutionary biologist and slime mould specialist John Bonner on the 'rocking back and forth between the reality of experimental facts and the dream world of hypotheses'. They reviewed some major contributions, including striking success stories in ecology, epidemiology, developmental biology, physiology, and neuro-oncology. But still, most of the work cited there (and most of mathematical biology as a whole) uses parts of mathematics traditionally thought of as 'applied', such as differential equations, dynamical systems, and stochastic analysis.

The reality is that many parts of mathematics conventionally called 'pure' are now being successfully applied in diverse contexts, both biological and otherwise. Knot theory has solved longstanding problems in genetic recombination (Buck and Flapan [52, 53]). Group theory has illuminated virus structure (Twarock, Valiunas and Zappa [338]). Topological data analysis, founded on the theory of persistent homology and calling on the power of algebraic topology, succeeded in identifying a hitherto unknown subtype of breast cancer with a 100% survival rate (Nicolau, Levine and Carlsson [263]; see Lesnick [227] for an expository account). Order theory, topos theory and classical logic have all been employed in the quest for improved ways of specifying, modelling and designing concurrent systems (Nygaard and Winskel [267]; Joyal, Nielsen and Winskel [172]; Hennessy and Milner [142]). And, famously, number theory is used to both provide and undermine security of communications on the internet (Hales [135]). All of these are real applications of mathematics. None is 'applied mathematics' as traditionally construed.

But applications are not the only product of applied mathematics. It also *nourishes* the core of mathematics, providing new questions, answers, and perspectives. Mathematics applied to physics has done this from Archimedes to Newton to Witten. Reed [291] lists dozens of ways in which mathematics applied to biology is doing it now. The developments surveyed in this book provide further evidence that a body of mathematics can simultaneously be entirely rigorous, be applied effectively to another branch of science, use parts of mathematics that do not fit the narrow stereotype of 'applied mathematics', and produce new results that are significant and satisfying from a purely mathematical aesthetic.

1

Fundamental Functional Equations

Throughout this book, we will make contact with the venerable subject of functional equations. A functional equation is an equation in an unknown function satisfied at all values of its arguments; or more generally, it is an equation relating several functions to each other in this way.

To set the scene, we give some brief indicative examples. Viewing sequences as functions on the set of positive integers, the Fibonacci sequence $(F_n)_{n\geq 1}$ satisfies the functional equation

$$F_{n+2} = F_n + F_{n+1}$$

($n \geq 1$). Together with the boundary conditions $F_1 = F_2 = 1$, this functional equation uniquely characterizes the sequence. But more typically, one is concerned with functions of *continuous* variables. For instance, one might notice that the function

$$\begin{array}{rccc} f\colon & \mathbb{R} \cup \{\infty\} & \to & \mathbb{R} \cup \{\infty\} \\ & x & \mapsto & \dfrac{1}{1-x} \end{array}$$

satisfies the functional equation

$$f(f(f(x))) = x \qquad (1.1)$$

($x \in \mathbb{R} \cup \{\infty\}$). The natural question, then, is whether f is the *only* function satisfying equation (1.1) for all x. In this case, it is not. (This can be shown by constructing an explicit counterexample or via the theory of Möbius transformations.) So, it is then natural to seek the whole set of solutions f, perhaps restricting the search to just those functions that are continuous, differentiable, etc.

A more sophisticated example is the functional equation

$$\zeta(1-s) = \frac{2^{1-s}}{\pi^s} \cos\left(\frac{\pi s}{2}\right) \Gamma(s) \zeta(s)$$

($s \in \mathbb{C}$) satisfied by the Riemann zeta function ζ (Theorem 12.7 of Apostol [16], for instance). Here Γ is Euler's gamma function. This functional equation, proved by Riemann himself, is a fundamental property of the zeta function.

In this chapter, we solve three classical, fundamental, functional equations. The first is Cauchy's equation on a function $f : \mathbb{R} \to \mathbb{R}$:

$$f(x+y) = f(x) + f(y)$$

($x, y \in \mathbb{R}$) (Section 1.1). Once we have solved this, we will easily be able to deduce the solutions of related equations such as

$$f(xy) = f(x) + f(y) \tag{1.2}$$

($x, y \in (0, \infty)$).

The second is the functional equation

$$f(mn) = f(m) + f(n)$$

($m, n \geq 1$) on a *sequence* $(f(n))_{n\geq 1}$. Despite the resemblance to equation (1.2), the shift from continuous to discrete makes it necessary to develop quite different techniques (Section 1.2).

Third and finally, we solve the functional equation

$$f(xy) = f(x) + g(x)f(y)$$

in two unknown functions $f, g : (0, \infty) \to \mathbb{R}$. The nontrivial, measurable solutions f turn out to be the constant multiples of the so-called q-logarithms (Section 1.3), a one-parameter family of functions of which the ordinary logarithm is just the best-known member.

1.1 Cauchy's Equation

A function $f : \mathbb{R} \to \mathbb{R}$ is **additive** if

$$f(x+y) = f(x) + f(y) \tag{1.3}$$

for all $x, y \in \mathbb{R}$. This is **Cauchy's functional equation**, some of whose long history is recounted in Section 2.1 of Aczél [2]. Let us say that f is **linear** if there exists $c \in \mathbb{R}$ such that

$$f(x) = cx$$

for all $x \in \mathbb{R}$. Putting $x = 1$ shows that if such a constant c exists then it must be equal to $f(1)$.

Evidently any linear function is additive. The question is to what extent the converse holds. If we are willing to assume that f is differentiable then the converse is very easy.

Proposition 1.1.1 *Every differentiable additive function $\mathbb{R} \to \mathbb{R}$ is linear.*

Proof Let $f\colon \mathbb{R} \to \mathbb{R}$ be a differentiable additive function. Differentiating equation (1.3) with respect to y gives

$$f'(x + y) = f'(y)$$

for all $x, y \in \mathbb{R}$. Taking $y = 0$ then shows that f' is constant. Hence there are constants $c, d \in \mathbb{R}$ such that $f(x) = cx + d$ for all $x \in \mathbb{R}$. Substituting this expression back into equation (1.3) gives $d = 0$. □

However, differentiability is a stronger condition than we will want to assume for our later purposes. It is, in fact, unnecessarily strong. In the rest of this section, we prove that additivity implies linearity under a succession of ever weaker regularity conditions, starting with continuity and finishing with mere measurability.

We begin with a lemma that needs no regularity conditions at all.

Lemma 1.1.2 *Let $f\colon \mathbb{R} \to \mathbb{R}$ be an additive function. Then $f(qx) = qf(x)$ for all $q \in \mathbb{Q}$ and $x \in \mathbb{R}$.*

Proof First, $f(0 + 0) = f(0) + f(0)$, so $f(0) = 0$. Then, for all $x \in \mathbb{R}$,

$$0 = f(0) = f(-x + x) = f(-x) + f(x),$$

so $f(-x) = -f(x)$.

Let $x \in \mathbb{R}$. By induction,

$$f(nx) = nf(x) \tag{1.4}$$

for all integers $n > 0$, and we have just shown that equation (1.4) also holds when $n = 0$. Moreover, when $n < 0$,

$$f(nx) = f(-(-n)x) = -f((-n)x) = -(-n)f(x) = nf(x),$$

using equation (1.4) for positive integers. Hence (1.4) holds for all integers n.

Now let $x \in \mathbb{R}$ and $q \in \mathbb{Q}$. Write $q = m/n$, where $m, n \in \mathbb{Z}$ with $n \neq 0$. Then by two applications of equation (1.4),

$$f(qx) = \tfrac{1}{n}f(nqx) = \tfrac{1}{n}f(mx) = \tfrac{m}{n}f(x) = qf(x),$$

as required. □

Remark 1.1.3 The same argument proves that any additive function between vector spaces over $\mathbb{Q}$ is linear over $\mathbb{Q}$. In the case of functions $\mathbb{R} \to \mathbb{R}$, our question is whether (or under what conditions) $\mathbb{Q}$-linearity implies $\mathbb{R}$-linearity, which here we are just calling 'linearity'.

Lemma 1.1.2 enables us to improve Proposition 1.1.1, relaxing differentiability to continuity. The following result was known to Cauchy himself (cited in Hardy, Littlewood and Pólya [137], proof of Theorem 84).

Proposition 1.1.4 *Every continuous additive function $\mathbb{R} \to \mathbb{R}$ is linear.*

Proof Let $f: \mathbb{R} \to \mathbb{R}$ be a continuous additive function, and write $c = f(1)$. By Lemma 1.1.2, $f(q) = cq$ for all $q \in \mathbb{Q}$. Thus, the two functions f and $x \mapsto cx$ are equal when restricted to $\mathbb{Q}$. But both are continuous, so they are equal on all of $\mathbb{R}$. □

It is now straightforward to relax continuity of f to a much weaker condition.

Proposition 1.1.5 *Every additive function $\mathbb{R} \to \mathbb{R}$ that is continuous at one or more point is linear.*

In other words, every additive function is linear unless, perhaps, it is discontinuous everywhere.

Proof Let $f: \mathbb{R} \to \mathbb{R}$ be an additive function continuous at a point $x \in \mathbb{R}$. By Proposition 1.1.4, it is enough to show that f is continuous. Let $y, t \in \mathbb{R}$: then by additivity,

$$f(y + t) - f(y) = f(t) = f(x + t) - f(x) \to 0$$

as $t \to 0$, as required. □

Next we show that mere measurability suffices: every measurable additive function is linear.

Remark 1.1.6 Readers unfamiliar with measure theory may wish to read the rest of this remark then resume at Corollary 1.1.11. Measurability is an extremely weak condition. In the usual logical framework for mathematics, there do exist nonmeasurable functions and nonlinear additive functions (Remark 1.1.9). However, every function for which anyone has ever written down an explicit formula, or ever will, is measurable (by Remark 1.1.10). So it is not too dangerous to assume that every function is measurable and, therefore, that every additive function is linear.

There are several proofs that every measurable additive function is linear. The first was published by Maurice Fréchet in his 1913 paper 'Pri la funkcia ekvacio $f(x + y) = f(x) + f(y)$' [110]. (Fréchet wrote many papers in Esperanto, and served three years as the president of the Internacia Scienca Asocio Esperantista.) Here we give the proof by Banach [27]. It is based on a standard measure-theoretic result of Lusin [235], which makes precise Littlewood's maxim that every measurable function is 'nearly continuous' [233].

Write λ for Lebesgue measure on $\mathbb{R}$.

Theorem 1.1.7 (Lusin) *Let $a \leq b$ be real numbers, and let $f: [a, b] \to \mathbb{R}$ be a measurable function. Then for all $\varepsilon > 0$, there exists a closed subset $V \subseteq [a, b]$ such that $f|_V$ is continuous and $\lambda([a, b] \setminus V) < \varepsilon$.*

Proof See Theorem 7.5.2 of Dudley [85], for instance. □

Following Banach, we deduce:

Theorem 1.1.8 *Every measurable additive function $\mathbb{R} \to \mathbb{R}$ is linear.*

Proof Let $f: \mathbb{R} \to \mathbb{R}$ be a measurable additive function. By Lusin's theorem, we can choose a closed set $V \subseteq [0, 1]$ such that $f|_V$ is continuous and $\lambda(V) > 2/3$. Since V is compact, $f|_V$ is uniformly continuous.

By Proposition 1.1.5, it is enough to prove that f is continuous at 0. Let $\varepsilon > 0$. We have to show that $|f(x)| < \varepsilon$ for all x in some neighbourhood of 0.

By uniform continuity, we can choose $\delta > 0$ such that for $v, v' \in V$,

$$|v - v'| < \delta \implies |f(v) - f(v')| < \varepsilon.$$

I claim that $|f(x)| < \varepsilon$ for all $x \in \mathbb{R}$ such that $|x| < \min\{\delta, 1/3\}$. Indeed, take such an x. Then, writing $V - x = \{v - x : v \in V\}$, the inclusion-exclusion property of Lebesgue measure λ gives

$$\lambda(V \cap (V - x)) = \lambda(V) + \lambda(V - x) - \lambda(V \cup (V - x)).$$

Consider the right-hand side. For the first two terms, we have $\lambda(V) > 2/3$ and so $\lambda(V - x) > 2/3$. For the last, if $x \geq 0$ then $V \cup (V - x) \subseteq [-1/3, 1]$, if $x \leq 0$ then $V \cup (V - x) \subseteq [0, 4/3]$, and in either case, $\lambda(V \cup (V - x)) \leq 4/3$. Hence

$$\lambda(V \cap (V - x)) > 2/3 + 2/3 - 4/3 = 0.$$

In particular, $V \cap (V - x)$ is nonempty, so we can choose an element y. Then $y, x + y \in V$ with $|y - (x + y)| = |x| < \delta$, so $|f(y) - f(x + y)| < \varepsilon$ by definition of δ. But since f is additive, this means that $|f(x)| < \varepsilon$, as required. □

The regularity condition can be weakened still further; see Reem [292] for a recent survey. However, measurability is as weak a condition as we will need.

Remark 1.1.9 Assuming the axiom of choice, there do exist additive functions $\mathbb{R} \to \mathbb{R}$ that are not linear. To see this, first note that the real line $\mathbb{R}$ is a vector space over $\mathbb{Q}$ in the evident way. Choose a basis B for $\mathbb{R}$ over $\mathbb{Q}$. Choose an element b of B, and let $\phi: B \to \mathbb{R}$ be the function taking value 1 at b and 0 elsewhere. By the universal property of bases, ϕ extends uniquely to a $\mathbb{Q}$-linear map $f: \mathbb{R} \to \mathbb{R}$.

Certainly f is additive. On the other hand, we can show that f is not $\mathbb{R}$-linear (that is, not 'linear' in the terminology of this section). Indeed, any $\mathbb{R}$-linear function $\mathbb{R} \to \mathbb{R}$ either is identically zero or vanishes nowhere except at 0. Now f is not identically zero, since $f(b) = \phi(b) = 1$. But also, for any $b' \neq b$ in B, we have $f(b') = \phi(b') = 0$ with $b' \neq 0$, so f vanishes at some point other than 0. Hence f is a nonlinear, additive function $\mathbb{R} \to \mathbb{R}$.

Remark 1.1.10 It is consistent with the Zermelo–Fraenkel axioms of set theory (that is, ZFC without the axiom of choice) that all functions $\mathbb{R} \to \mathbb{R}$ are measurable. This is a 1970 theorem of Solovay [318]. If all functions $\mathbb{R} \to \mathbb{R}$ are measurable then by Theorem 1.1.8, all additive functions are linear.

On the other hand, the axiom of choice is also consistent with ZF. If the axiom of choice holds then by Remark 1.1.9, not all additive functions are linear.

Hence, starting from ZF, one may consistently assume *either* that every additive function is linear *or* that not every additive function is linear.

Theorem 1.1.8 classifies the measurable functions that convert addition into addition. One can easily adapt it to classify the functions that convert addition into multiplication, multiplication into multiplication, and so on:

Corollary 1.1.11 *i. Let $f: \mathbb{R} \to (0, \infty)$ be a measurable function. The following are equivalent:*

a. $f(x + y) = f(x)f(y)$ for all $x, y \in \mathbb{R}$;

b. there exists $c \in \mathbb{R}$ such that $f(x) = e^{cx}$ for all $x \in \mathbb{R}$.

ii. Let $f: (0, \infty) \to \mathbb{R}$ be a measurable function. The following are equivalent:

a. $f(xy) = f(x) + f(y)$ for all $x, y \in (0, \infty)$;

b. there exists $c \in \mathbb{R}$ such that $f(x) = c \log x$ for all $x \in (0, \infty)$.

iii. Let $f: (0, \infty) \to (0, \infty)$ be a measurable function. The following are equivalent:

a. $f(xy) = f(x)f(y)$ for all $x, y \in (0, \infty)$;

b. there exists $c \in \mathbb{R}$ such that $f(x) = x^c$ for all $x \in (0, \infty)$.

Proof For (i), evidently (b) implies (a). Assuming (a), define $g\colon \mathbb{R} \to \mathbb{R}$ by $g(x) = \log f(x)$. Then g is measurable and additive, so by Theorem 1.1.8, there is some constant $c \in \mathbb{R}$ such that $g(x) = cx$ for all $x \in \mathbb{R}$. It follows that $f(x) = e^{cx}$ for all $x \in \mathbb{R}$, as required.

Parts (ii) and (iii) are proved similarly, putting $g(x) = f(e^x)$ and $g(x) = \log f(e^x)$. □

Remark 1.1.12 In this book, the notation log means the natural logarithm ln = $\log_e$. However, the choice of base for logarithms is usually unimportant, as it is in Corollary 1.1.11(ii): changing the base amounts to multiplying the logarithm by a positive constant, which is in any case absorbed by the free choice of the constant c.

Theorem 1.1.8 also allows us to classify the additive functions that are defined on only half of the real line.

Corollary 1.1.13 *Let $f\colon [0,\infty) \to \mathbb{R}$ be a measurable function satisfying $f(x+y) = f(x)+f(y)$ for all $x, y \in [0,\infty)$. Then there exists $c \in \mathbb{R}$ such that $f(x) = cx$ for all $x \in [0,\infty)$.*

Proof First we extend $f\colon [0,\infty) \to \mathbb{R}$ to a measurable additive function $g\colon \mathbb{R} \to \mathbb{R}$. By the hypothesis on f, for all $a^+, a^-, b^+, b^- \in [0,\infty)$,

$$a^+ - a^- = b^+ - b^- \implies f(a^+) - f(a^-) = f(b^+) - f(b^-).$$

We can, therefore, consistently define a function $g\colon \mathbb{R} \to \mathbb{R}$ by

$$g(a^+ - a^-) = f(a^+) - f(a^-)$$

($a^+, a^- \in [0,\infty)$). To prove that g is additive, let $x, y \in \mathbb{R}$, and choose $a^\pm, b^\pm \in [0,\infty)$ such that

$$x = a^+ - a^-, \qquad y = b^+ - b^-.$$

Then

$$x + y = (a^+ + b^+) - (a^- + b^-)$$

with $a^+ + b^+, a^- + b^- \in [0,\infty)$. Hence

$$\begin{aligned} g(x+y) &= f(a^+ + b^+) - f(a^- + b^-) \\ &= f(a^+) + f(b^+) - f(a^-) - f(b^-) \\ &= f(a^+) - f(a^-) + f(b^+) - f(b^-) \\ &= g(x) + g(y), \end{aligned}$$

as required. To prove that g is measurable, note that

$$g(x) = \begin{cases} f(x) & \text{if } x \geq 0, \\ -f(-x) & \text{if } x \leq 0 \end{cases}$$

($x \in \mathbb{R}$), as if $x \geq 0$ then we can take $a^+ = x$ and $a^- = 0$ in the definition of g, and similarly for $x \leq 0$. Since f is measurable, so is g.

By Theorem 1.1.8, there exists a constant c such that $g(x) = cx$ for all $x \in \mathbb{R}$. It follows that $f(x) = cx$ for all $x \in [0, \infty)$. □

The techniques and results of this section can be assembled in several ways to derive variant theorems. Rather than attempting to catalogue all the possibilities, we illustrate the point with two particular variants needed later.

Corollary 1.1.14 *Let $f: (0, 1] \to \mathbb{R}$ be a measurable function. The following are equivalent:*

i. $f(xy) = f(x) + f(y)$ for all $x, y \in (0, 1]$;
ii. there exists a constant $c \in \mathbb{R}$ such that $f(x) = c \log x$ for all $x \in (0, 1]$.

Proof Trivially, (ii) implies (i). Now assuming (i), define $g: [0, \infty) \to \mathbb{R}$ by $g(u) = f(e^{-u})$. Then g is measurable and $g(u + v) = g(u) + g(v)$ for all $u, v \in [0, \infty)$, so by Corollary 1.1.13, $g(u) = bu$ for some real constant b. It follows that $f(x) = -b \log x$ for all $x \in (0, 1]$, as required. □

The moral of Corollary 1.1.14 is that for the Cauchy-like functional equation $f(xy) = f(x) + f(y)$, there is no substantial difference between solving it on the domain $(0, \infty)$ and solving it on the domain $(0, 1]$ (or $[1, \infty)$, similarly). But matters become very different when we seek solutions on the discrete domain $\{1, 2, 3, \ldots\}$, as we will discover in the next section.

Remark 1.1.15 In this text, we always use the terms 'increasing' and 'decreasing' in their non-strict senses. Thus, a function $f: S \to \mathbb{R}$ on a subset $S \subseteq \mathbb{R}$ is **increasing** if

$$x \leq y \implies f(x) \leq f(y)$$

($x, y \in S$), and **decreasing** if $-f$ is increasing. It is **strictly** increasing or decreasing if $x < y$ implies $f(x) < f(y)$ or $f(x) > f(y)$, respectively. The same terminology applies to sequences.

Corollary 1.1.16 *Let $f: (0, 1) \to (0, \infty)$ be an increasing function. The following are equivalent:*

i. $f(xy) = f(x)f(y)$ for all $x, y \in (0, 1)$;
ii. there exists a constant $c \in [0, \infty)$ such that $f(x) = x^c$ for all $x \in (0, 1)$.

Proof Trivially, (ii) implies (i). Assuming (i), define $g\colon (0,\infty) \to \mathbb{R}$ by $g(u) = -\log f(e^{-u})$. Then $g(u+v) = g(u) + g(v)$ for all $u, v \in (0,\infty)$, and g is also increasing.

By the same argument as in the proof of Lemma 1.1.2, $g(qu) = qg(u)$ for all $q, u \in (0,\infty)$ with q rational. Define $\widetilde{g}\colon (0,\infty) \to \mathbb{R}$ by $\widetilde{g}(u) = g(1)u$. Then $g(q) = \widetilde{g}(q)$ for all $q \in (0,\infty) \cap \mathbb{Q}$. Since g is increasing and $\widetilde{g}$ is either increasing or decreasing (depending on the sign of $g(1)$), it follows that $\widetilde{g}$ is increasing. But now $g, \widetilde{g}\colon (0,\infty) \to \mathbb{R}$ are increasing functions that are equal on the positive rationals, so $g = \widetilde{g}$. Hence $f(x) = x^{g(1)}$ for all $x \in (0,1)$. □

1.2 Logarithmic Sequences

A sequence $f(1), f(2), \ldots$ of real numbers is **logarithmic** if

$$f(mn) = f(m) + f(n) \tag{1.5}$$

for all $m, n \geq 1$. Certainly the sequence $(c\log n)_{n\geq 1}$ is logarithmic, for any real constant c. But in contrast to the situation for functions $f\colon (0,\infty) \to \mathbb{R}$ satisfying $f(xy) = f(x) + f(y)$ (Corollary 1.1.11(ii)), it is easy to write down logarithmic sequences that are not of this simple form. Indeed, we can choose $f(p)$ arbitrarily for each prime p, and these choices uniquely determine a logarithmic sequence, generally not of the form $(c\log n)$.

However, there are reasonable conditions on a logarithmic sequence $(f(n))$ guaranteeing that it is of the form $(c\log n)$. One such condition is that f is increasing:

$$f(1) \leq f(2) \leq \cdots.$$

An alternative condition is that

$$\lim_{n\to\infty} (f(n+1) - f(n)) = 0.$$

We will prove a single theorem implying both of these results. But a direct proof of the result on increasing sequences is short enough to be worth giving separately, even though it is not logically necessary.

Theorem 1.2.1 (Erdős) *Let $(f(n))_{n\geq 1}$ be an increasing sequence of real numbers. The following are equivalent:*

i. f is logarithmic;
ii. there exists a constant $c \geq 0$ such that $f(n) = c\log n$ for all $n \geq 1$.

This was first proved by Erdős [92]. In fact, he showed more: as is customary in number theory, he only required equation (1.5) to hold when m and n are relatively prime. But since we will not need the extra precision of that result, we will not prove it.

The argument presented here follows Khinchin ([188], p. 11).

Proof Certainly (ii) implies (i). Now assume (i). By the logarithmic property,

$$f(1) = f(1 \cdot 1) = f(1) + f(1),$$

so $f(1) = 0$. Since f is increasing, $f(n) \geq 0$ for all n. If $f(n) = 0$ for all n then (ii) holds with $c = 0$. Assuming otherwise, we can choose some $N > 1$ such that $f(N) > 0$.

Let $n \geq 1$. For each integer $r \geq 1$, there is an integer $\ell_r \geq 1$ such that

$$N^{\ell_r} \leq n^r \leq N^{\ell_r+1}$$

(since $N > 1$). As f is increasing and logarithmic,

$$\ell_r f(N) \leq r f(n) \leq (\ell_r + 1) f(N),$$

which since $f(N) > 0$ implies that

$$\frac{\ell_r}{r} \leq \frac{f(n)}{f(N)} \leq \frac{\ell_r + 1}{r}. \tag{1.6}$$

As log is also increasing and logarithmic, the same argument gives

$$\frac{\ell_r}{r} \leq \frac{\log n}{\log N} \leq \frac{\ell_r + 1}{r}. \tag{1.7}$$

Inequalities (1.6) and (1.7) together imply that

$$\left| \frac{f(n)}{f(N)} - \frac{\log n}{\log N} \right| \leq \frac{1}{r}.$$

But this conclusion holds for all $r \geq 1$, so

$$\frac{f(n)}{f(N)} = \frac{\log n}{\log N}.$$

Hence $f(n) = c \log n$, where $c = f(N)/\log N$. And since this is true for all $n \geq 1$, we have proved (ii). □

We now prove the unified theorem promised above. Before stating it, let us recall the concept of **limit inferior**. Given a real sequence $(g(n))_{n\geq 1}$, define

$$h(n) = \inf\{g(n), g(n+1), \ldots\} \in [-\infty, \infty)$$

($n \geq 1$). The sequence $(h(n))_{n\geq 1}$ is increasing and therefore has a limit (perhaps $\pm\infty$), written as

$$\liminf_{n\to\infty} g(n) = \lim_{n\to\infty} h(n) \in [-\infty, \infty].$$

If the ordinary limit $\lim_{n\to\infty} g(n)$ exists then $\liminf_{n\to\infty} g(n) = \lim_{n\to\infty} g(n)$. However, the limit inferior exists whether or not the limit does. For instance, the sequence $1, -1, 1, -1, \ldots$ has a limit inferior of -1, but no limit.

If $(f(n))$ is a sequence that either is increasing or satisfies $f(n+1)-f(n) \to 0$ as $n \to \infty$, then

$$\liminf_{n\to\infty}(f(n+1) - f(n)) \geq 0.$$

The following theorem therefore implies both of the results mentioned above.

Theorem 1.2.2 (Erdős, Kátai, Máté) *Let $(f(n))_{n\geq 1}$ be a sequence of real numbers such that*

$$\liminf_{n\to\infty}(f(n+1) - f(n)) \geq 0.$$

The following are equivalent:

i. f is logarithmic;
ii. there exists a constant c such that $f(n) = c \log n$ for all $n \geq 1$.

This result was stated without proof by Erdős in 1957 [93], then proved independently by Kátai [183] and by Máté [245], both in 1967. Again, the logarithmic condition can be relaxed by only requiring that (1.5) holds when m and n are relatively prime, but again, we have no need for this extra precision.

The proof below follows Aczél and Daróczy's adaptation of Kátai's argument (Theorem 0.4.3 of [3]). The strategy is to put $c = \liminf_{n\to\infty} f(n)/\log n$ and show that $f(N)/\log N = c$ for all N.

Proof It is trivial that (ii) implies (i). Now assume (i). I claim that for all $N \geq 2$,

$$\liminf_{n\to\infty} \frac{f(n)}{\log n} = \frac{f(N)}{\log N}. \tag{1.8}$$

Let $N \geq 2$. First we show that the left-hand side of (1.8) is less than or equal to the right. For each $r \geq 1$, the logarithmic property of f implies that

$$\frac{f(N^r)}{\log(N^r)} = \frac{rf(N)}{r\log N} = \frac{f(N)}{\log N}.$$

Since $N^r \to \infty$ as $r \to \infty$, it follows from the definition of limit inferior that

$$\liminf_{n\to\infty} \frac{f(n)}{\log n} \leq \frac{f(N)}{\log N}.$$

Now we prove the opposite inequality,

$$\liminf_{n\to\infty} \frac{f(n)}{\log n} \geq \frac{f(N)}{\log N}. \tag{1.9}$$

Let $\varepsilon > 0$. By hypothesis, we can choose $k \geq 1$ such that for all $n \geq N^k$,

$$f(n+1) - f(n) \geq -\varepsilon. \tag{1.10}$$

Any integer $n \geq N^k$ has a base N expansion

$$n = c_\ell N^\ell + \cdots + c_1 N + c_0$$

with $c_0, \ldots, c_\ell \in \{0, \ldots, N-1\}$, $c_\ell \neq 0$, and $\ell \geq k$. Then

$$\begin{aligned} f(n) &\geq f(c_\ell N^\ell + \cdots + c_1 N) - c_0\varepsilon && (1.11)\\ &\geq f(c_\ell N^\ell + \cdots + c_1 N) - N\varepsilon && (1.12)\\ &= f(c_\ell N^{\ell-1} + \cdots + c_1) + f(N) - N\varepsilon, && (1.13) \end{aligned}$$

where inequality (1.11) follows from (1.10) using induction and the fact that $\ell \geq k$, inequality (1.12) holds because $c_0 \leq N$, and equation (1.13) follows from the logarithmic property of f. As long as $\ell - 1 \geq k$, we can apply the same argument again with $c_\ell N^{\ell-1} + \cdots + c_1$ in place of $n = c_\ell N^\ell + \cdots + c_0$, giving

$$f(c_\ell N^{\ell-1} + \cdots + c_1) \geq f(c_\ell N^{\ell-2} + \cdots + c_2) + f(N) - N\varepsilon$$

and so

$$f(n) \geq f(c_\ell N^{\ell-2} + \cdots + c_2) + 2(f(N) - N\varepsilon).$$

Repeated application of this argument gives

$$f(n) \geq f(c_\ell N^{k-1} + \cdots + c_{\ell-k+1}) + (\ell - k + 1)(f(N) - N\varepsilon).$$

Hence, writing $A = \min\{f(1), f(2), \ldots, f(N^k)\}$,

$$f(n) \geq A + (\ell - k + 1)(f(N) - N\varepsilon). \tag{1.14}$$

In (1.14), the only term on the right-hand side that depends on n is ℓ, which is equal to $\lfloor \log_N n \rfloor$, and $\lfloor \log_N n \rfloor / \log_N n \to 1$ as $n \to \infty$. Hence

$$\begin{aligned} \liminf_{n\to\infty} \frac{f(n)}{\log_N n} &\geq \liminf_{n\to\infty} \left\{ \frac{A}{\log_N n} + \left(\frac{\lfloor \log_N n \rfloor}{\log_N n} + \frac{-k+1}{\log_N n} \right)(f(N) - N\varepsilon) \right\} \\ &= f(N) - N\varepsilon. \end{aligned}$$

This holds for all $\varepsilon > 0$, so

$$\liminf_{n\to\infty} \frac{f(n)}{\log_N n} \geq f(N).$$

Since $\log_N n = (\log n)/(\log N)$, this proves the claimed inequality (1.9) and, therefore, equation (1.8).

Putting $c = \liminf_{n\to\infty} f(n)/\log n \in \mathbb{R}$, we have $f(N) = c \log N$ for all $N \geq 2$. Finally, the logarithmic property of f implies that $f(1) = 0$, so $f(1) = c \log 1$ too. □

Corollary 1.2.3 *Let $(f(n))_{n\geq 1}$ be a sequence such that*

$$\lim_{n\to\infty} (f(n+1) - f(n)) = 0. \tag{1.15}$$

The following are equivalent:

i. f is logarithmic;
ii. there exists a constant c such that $f(n) = c \log n$ for all $n \geq 1$. □

To apply this corollary, we will need to be able to verify the limit condition (1.15). The following improvement lemma will be useful.

Lemma 1.2.4 *Let $(a_n)_{n\geq 1}$ be a real sequence such that $a_{n+1} - \frac{n}{n+1} a_n \to 0$ as $n \to \infty$. Then $a_{n+1} - a_n \to 0$ as $n \to \infty$.*

Our proof of Lemma 1.2.4 follows that of Feinstein [99] (pp. 6–7), and uses the following standard result.

Proposition 1.2.5 (Cesàro) *Let $(x_n)_{n\geq 1}$ be a real sequence, and for $n \geq 1$, write*

$$\overline{x}_n = \tfrac{1}{n}(x_1 + \cdots + x_n).$$

Suppose that $\lim_{n\to\infty} x_n$ exists. Then $\lim_{n\to\infty} \overline{x}_n$ exists and is equal to $\lim_{n\to\infty} x_n$.

Proof This can be found in introductory analysis texts such as Apostol [15] (Theorem 12-48). □

Proof of Lemma 1.2.4 It is enough to prove that $a_n/(n+1) \to 0$ as $n \to \infty$. Write $b_1 = a_1$ and $b_n = a_n - \frac{n-1}{n} a_{n-1}$ for $n \geq 2$; then by hypothesis, $b_n \to 0$ as $n \to \infty$. We have $na_n = nb_n + (n-1)a_{n-1}$ for all $n \geq 2$, so

$$na_n = nb_n + (n-1)b_{n-1} + \cdots + 1b_1$$

for all $n \geq 1$. Dividing through by $n(n+1)$ gives

$$\begin{aligned} \frac{a_n}{n+1} &= \frac{1}{2} \cdot \frac{1}{\frac{1}{2}n(n+1)} (b_1 + b_2 + b_2 + b_3 + b_3 + b_3 + \cdots + \underbrace{b_n + \cdots + b_n}_{n}) \\ &= \frac{1}{2} \cdot M_1(b_1, b_2, b_2, b_3, b_3, b_3, \ldots, \underbrace{b_n, \ldots, b_n}_{n}), \end{aligned} \tag{1.16}$$

where M_1 denotes the arithmetic mean. Since $b_n \to 0$ as $n \to \infty$, the sequence

$$b_1, b_2, b_2, b_3, b_3, b_3, \ldots, \underbrace{b_n, \ldots, b_n}_{n}, \ldots$$

also converges to 0. Proposition 1.2.5 applied to this sequence then implies that

$$M_1(b_1, b_2, b_2, b_3, b_3, b_3, \ldots, \underbrace{b_n, \ldots, b_n}_{n}) \to 0 \text{ as } n \to \infty.$$

But by equation (1.16), this means that $a_n/(n+1) \to 0$ as $n \to \infty$, completing the proof. □

Remark 1.2.6 Lemma 1.2.4 can also be deduced from the Stolz–Cesàro theorem (Section 3.1.7 of Mureşan [258], for instance). This is a discrete analogue of l'Hôpital's rule, and states that given a real sequence (x_n) and a strictly increasing sequence (y_n) diverging to ∞, if

$$\frac{x_{n+1} - x_n}{y_{n+1} - y_n} \to \ell$$

as $n \to \infty$ then $x_n/y_n \to \ell$ as $n \to \infty$. Lemma 1.2.4 follows by taking $x_n = na_n$ and $y_n = \frac{1}{2}n(n+1)$. (I thank Xīlíng Zhāng for this observation.)

1.3 The q-Logarithm

The q-logarithms ($q \in \mathbb{R}$) form a continuous one-parameter family of functions that include the ordinary natural logarithm as the case $q = 1$. They can be regarded as deformations of the natural logarithm. We will show that as a family, they are characterized by a single functional equation.

For $q \in \mathbb{R}$, the **q-logarithm** is the function

$$\ln_q : (0, \infty) \to \mathbb{R}$$

defined by

$$\ln_q(x) = \int_1^x t^{-q}\, dt$$

($x \in (0, \infty)$). Thus,

$$\ln_1(x) = \log(x)$$

and for $q \neq 1$,

$$\ln_q(x) = \frac{x^{1-q} - 1}{1 - q}. \tag{1.17}$$

Then $\ln_q(x) \to \ln_1(x)$ as $q \to 1$, by l'Hôpital's rule.

Let $q \in \mathbb{R}$. The q-logarithm shares with the natural logarithm the property that

$$\ln_q(1) = 0.$$

However, in general

$$\ln_q(xy) \neq \ln_q(x) + \ln_q(y).$$

One can see this without calculation: for by Corollary 1.1.11(ii), the only measurable functions that transform multiplication into addition are the multiples of the natural logarithm. There is nevertheless a simple formula for $\ln_q(xy)$ in terms of $\ln_q(x)$ and $\ln_q(y)$:

$$\ln_q(xy) = \ln_q(x) + \ln_q(y) + (1 - q)\ln_q(x)\ln_q(y).$$

Later, we will use a second formula for $\ln_q(xy)$:

$$\ln_q(xy) = \ln_q(x) + x^{1-q}\ln_q(y). \tag{1.18}$$

Similarly, in general

$$\ln_q(1/x) \neq -\ln_q(x),$$

but instead we have the following three formulas for $\ln_q(1/x)$:

$$\begin{aligned}\ln_q(1/x) &= \frac{-\ln_q(x)}{1 + (1-q)\ln_q(x)}\\ &= -x^{q-1}\ln_q(x)\\ &= -\ln_{2-q}(x).\end{aligned} \tag{1.19}$$

By (1.19), replacing $\ln_q$ by the function $x \mapsto -\ln_q(1/x)$ defines an involution $\ln_q \leftrightarrow \ln_{2-q}$ of the family of q-logarithms, with a fixed point at the classical logarithm $\ln_1$. Finally, there is a quotient formula

$$\ln_q(x/y) = y^{q-1}(\ln_q(x) - \ln_q(y)), \tag{1.20}$$

obtained from equation (1.18) by substituting y for x and x/y for y.

Remark 1.3.1 The history of the q-logarithms as an *explicit* object of study goes back at least as far as a 1964 paper of Box and Cox in statistics (Section 3 of [49]). The notation $\ln_q$ appeared in a 1994 article of Tsallis [332], and the name 'q-logarithm' has been used since at least the late 1990s (as in Borges [45]).

But there is more than one system of q-analogues of the classical notions of calculus. For instance, there is the system developed by the early twentieth-century clergyman F. H. Jackson [155] (a modern account of which can be found in Kac and Cheung [175]). In particular, this has given rise to a different notion of q-logarithm, as developed in Chung, Chung, Nam and Kang [70]. Ernst [94] gives a full historical treatment of the various branches of q-calculus. In any case, none of the developments just mentioned use the q-logarithms considered here.

We now prove that the q-logarithms are characterized by a simple functional equation. The proof is essentially the argument behind Theorem 84 in the classic text of Hardy, Littlewood and Pólya [137].

Theorem 1.3.2 *Let $f: (0, \infty) \to \mathbb{R}$ be a measurable function. The following are equivalent:*

i. *there exists a function $g: (0, \infty) \to \mathbb{R}$ such that for all $x, y \in (0, \infty)$,*

$$f(xy) = f(x) + g(x)f(y); \tag{1.21}$$

ii. *$f = c \ln_q$ for some $c, q \in \mathbb{R}$, or f is constant.*

Proof First suppose that (ii) holds. If $f = c \ln_q$ for some $c, q \in \mathbb{R}$ then equation (1.21) holds with $g(x) = x^{1-q}$, by equation (1.18). Otherwise, f is constant, so (1.21) holds with $g \equiv 0$.

Now assume (i). Since $f(xy) = f(yx)$, equation (1.21) implies that

$$f(x) + g(x)f(y) = f(y) + g(y)f(x),$$

or equivalently

$$f(x)(1 - g(y)) = f(y)(1 - g(x)), \tag{1.22}$$

for all $x, y \in (0, \infty)$. If $f \equiv 0$ then f is constant and (ii) holds. Assuming otherwise, we can choose $y_0 \in (0, \infty)$ such that $f(y_0) \neq 0$. Taking $y = y_0$ in (1.22) and putting $a = (1 - g(y_0))/f(y_0)$ gives

$$g(x) = 1 - af(x) \tag{1.23}$$

($x \in \mathbb{R}$). Since f is measurable, so is g. There are now two cases: $a = 0$ and $a \neq 0$.

If $a = 0$ then $g \equiv 1$, so the original functional equation (1.21) states that $f(xy) = f(x) + f(y)$. Since f is measurable, Corollary 1.1.11(ii) implies that $f = c \log = c \ln_1$ for some $c \in \mathbb{R}$.

If $a \neq 0$ then equation (1.23) can be rewritten as

$$f(x) = \tfrac{1}{a}(1 - g(x)) \tag{1.24}$$

($x \in (0, \infty)$). Substituting this into the original functional equation (1.21) gives

$$g(xy) = g(x)g(y) \tag{1.25}$$

($x, y \in (0, \infty)$). In particular, $g(x) = g(\sqrt{x})^2 \geq 0$ for all $x \in (0, \infty)$. There are now two subcases: g either sometimes vanishes or never vanishes.

If $g(x_0) = 0$ for some $x_0 \in (0, \infty)$ then

$$g(x) = g(x_0)g(x/x_0) = 0$$

for all $x \in (0, \infty)$, so $g \equiv 0$. Hence by equation (1.24), f is constant.

Otherwise, $g(x) > 0$ for all $x \in (0, \infty)$. Since g is measurable and satisfies the multiplicativity condition (1.25), Corollary 1.1.11(iii) implies that there is some constant $t \in \mathbb{R}$ such that $g(x) = x^t$ for all $x \in (0, \infty)$. We have assumed that $f \not\equiv 0$, so $g \not\equiv 1$ (by equation (1.24)), so $t \neq 0$. Hence

$$f(x) = \tfrac{1}{a}(1 - x^t) = \tfrac{-t}{a} \ln_{1-t}(x)$$

for all $x \in (0, \infty)$, completing the proof. □

2

Shannon Entropy

> My greatest concern was what to call it. I thought of calling it 'information', but the word was overly used, so I decided to call it 'uncertainty'. When I discussed it with John von Neumann, he had a better idea. Von Neumann told me, 'You should call it entropy, for two reasons. In the first place your uncertainty function has been used in statistical mechanics under that name, so it already has a name. In the second place, and more important, no one knows what entropy really is, so in a debate you will always have the advantage.' – Claude Shannon (quoted in [328], p. 180).

Entropy appears in almost every branch of science. The most casual literature search quickly brings up works on entropy in thermodynamics [101], quantum physics [279], communications engineering [309, 316], information theory [238], statistical inference [156, 157], machine learning and artificial intelligence [48, 81, 290], malware detection [35], macroecology [138], the quantification of biological diversity [240], biochemistry [243], water network engineering [128], the theory of algorithms and complexity [114], ergodic theory and dynamical systems [270, 84], algebraic dynamics [95], combinatorial dynamics [9], topological dynamics [4], and climate science [139]. (The references given are a random sample.) The word 'entropy' has many meanings, all related, and is applied in more ways still.

This chapter is an introduction to the simplest kind of entropy: the Shannon entropy of a probability distribution on a finite set. There are several ways of interpreting Shannon entropy, and we develop two in depth. The first is through coding theory (Section 2.3), an interpretation that is very standard in the mathematical literature and goes back to Shannon's seminal paper of 1948 [309]. The second is through the theory of diversity (Section 2.4). This is much less well known, and is one of the main themes of this book.

The single most important property of Shannon entropy is the chain rule, which is a formula for the entropy of a composite distribution. The major theoretical goal of this chapter is to prove that Shannon entropy is essentially the *only* quantity that satisfies the chain rule. To that end, we begin by reviewing probability distributions and composition of distributions (Section 2.1). The chain rule itself is derived in Section 2.2, along with other basic properties of Shannon entropy, and is explained in terms of coding and diversity in the next two sections. In the final section, we prove the unique characterization of Shannon entropy by the chain rule.

2.1 Probability Distributions on Finite Sets

Let $n \geq 1$. A **probability distribution** on the finite set $\{1, \ldots, n\}$ is an n-tuple $\mathbf{p} = (p_1, \ldots, p_n)$ of real numbers $p_i \geq 0$ such that $\sum p_i = 1$.

Of the various interpretations of probability distributions, one will be especially important for us.

Example 2.1.1 Consider an ecological community of living organisms classified into n species. Let p_i be the relative abundance of the ith species, where 'relative' means that the abundances have been normalized so that $\sum p_i = 1$. Then the probability distribution $\mathbf{p} = (p_1, \ldots, p_n)$ is a model of the community, albeit a very crude one.

Some remarks are in order. First, the distinction between species is inexact and sometimes arbitrary. Mayden [249] lists 24 inequivalent ways of defining 'species' (further discussed in Hey [145]). The difficulty is most acute for microbes, many of which have not been classified into species at all. In practice, for microbes, scientists sequence the DNA of their sample and use software that applies a clustering algorithm, thus automatically creating 'species' according to a pre-chosen (and somewhat arbitrary) level of genetic similarity. We will find a way through this difficulty in Chapter 6.

Second, the meaning of 'abundance' is completely flexible. In some contexts, it may be appropriate to simply count individuals. But when the organisms are of very different sizes, it may be better to interpret the abundance of a species as the total mass of the members of that species. Or, for plants, the area of land covered by a species may be a more appropriate measure than the number of individuals.

Third, as emphasized in the Introduction (p. 7), nothing that we will say about 'communities' or 'species' is actually specific to ecology: mathematically speaking, it is entirely general.

For $n \geq 1$, write

$$\Delta_n = \{\text{probability distributions on } \{1, \ldots, n\}\}.$$

Occasionally we will want to include the case $n = 0$, and we put $\Delta_0 = \emptyset$. The **support** of $\mathbf{p} \in \Delta_n$ is

$$\text{supp}(\mathbf{p}) = \{i \in \{1, \ldots, n\} : p_i > 0\}.$$

We say that $\mathbf{p} \in \Delta_n$ has **full support** if $\text{supp}(\mathbf{p}) = \{1, \ldots, n\}$, and write

$$\Delta_n^\circ = \{\mathbf{p} \in \Delta_n : p_i > 0 \text{ for all } i\}$$

for the set of probability distributions of full support. Finally,

$$\mathbf{u}_n = (1/n, \ldots, 1/n)$$

denotes the **uniform distribution** on n elements. Geometrically, Δ_n is the standard $(n-1)$-dimensional simplex, Δ_n° is its interior, and $\mathbf{u}_n$ is its centre.

Example 2.1.2 Consider a community consisting of species numbered $1, \ldots, n$, with relative abundance distribution $\mathbf{p} \in \Delta_n$. Then $\text{supp}(\mathbf{p})$ is the set of species that are actually present in the community, and $\mathbf{p} \in \Delta_n^\circ$ if and only if every species is present. (A typical situation in which some species are absent is a longitudinal study: if the same site is surveyed every year over several years, it may be that in some years, not every species is present.) The uniform distribution $\mathbf{u}_n$ represents the situation in which all species are equally common.

We now define a fundamental operation: composition of probability distributions (Figure 2.1).

Definition 2.1.3 Let $n, k_1, \ldots, k_n \geq 1$ and let

$$\mathbf{w} \in \Delta_n, \ \mathbf{p}^1 \in \Delta_{k_1}, \ \ldots, \ \mathbf{p}^n \in \Delta_{k_n}.$$

Write $\mathbf{p}^i = (p^i_1, \ldots, p^i_{k_i})$. The **composite distribution** is

$$\mathbf{w} \circ (\mathbf{p}^1, \ldots, \mathbf{p}^n) = (w_1 p^1_1, \ldots, w_1 p^1_{k_1}, \ \ldots, \ w_n p^n_1, \ldots, w_n p^n_{k_n}) \in \Delta_{k_1 + \cdots + k_n}.$$

Example 2.1.4 Flip a coin. If it comes up heads, roll a die. If it comes up tails, draw from a pack of cards. Thus, the final outcome of the process is either a number between 1 and 6 or a playing card. There are, therefore, $6 + 52 = 58$ possible final outcomes.

Assuming that the coin toss, die roll, and card draw are all fair, the probabilities of the 58 possible outcomes are as shown in Figure 2.2. That is, the final

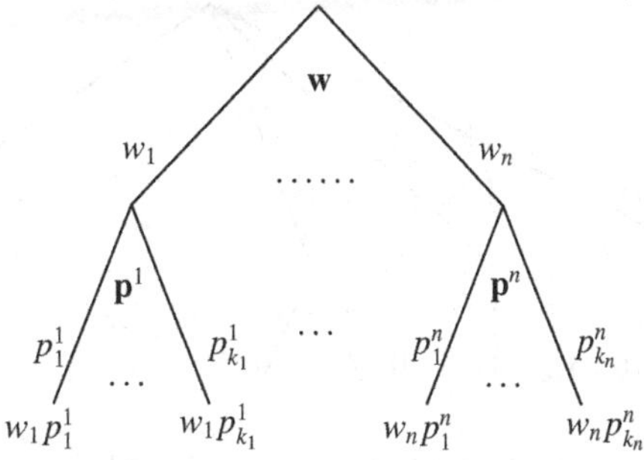

Figure 2.1 Composition of probability distributions.

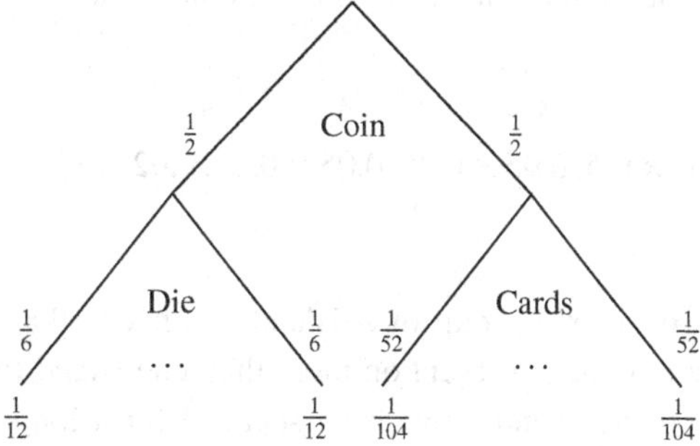

Figure 2.2 The composite distribution of Example 2.1.4.

outcome has probability distribution

$$\mathbf{u}_2 \circ (\mathbf{u}_6, \mathbf{u}_{52}) = \Big(\underbrace{\tfrac{1}{12}, \ldots, \tfrac{1}{12}}_{6}, \underbrace{\tfrac{1}{104}, \ldots, \tfrac{1}{104}}_{52}\Big).$$

Example 2.1.5 The French language is written with the same letters as English, but some are sometimes decorated by an accent (diacritical mark). For instance, the letter a appears in the three forms a (no accent), à and â, the letter b appears only as b, and the letter c appears in the two forms c and ç. Let us make the conventions that a **letter** is one of a, b, ..., z and a **symbol** is a letter together with, optionally, an accent. Thus, the symbols are a, à, â, b, c, ç, ...

Let $\mathbf{w} \in \Delta_{26}$ denote the frequency distribution of the letters as used in written French. For the sake of argument, let us suppose that $w_1, w_2, w_3, \ldots, w_{26}$ have the values shown in Figure 2.3. Suppose also that the letter a appears without accent 50% of the time, as à 25% of the time, and as â 25% of the time, again as in the figure. Write $\mathbf{p}^1 = (0.5, 0.25, 0.25)$, and similarly for $\mathbf{p}^2, \ldots, \mathbf{p}^{26}$. Then

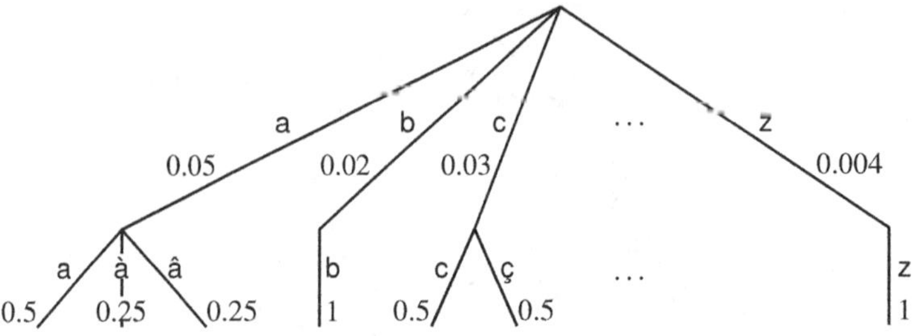

Figure 2.3 The composite distribution for French symbols (Example 2.1.5).

the frequency distribution of the symbols is the composite

$$\begin{aligned}&\mathbf{w} \circ (\mathbf{p}^1, \ldots, \mathbf{p}^{26})\\&\qquad = (0.05 \times 0.5, 0.05 \times 0.25, 0.05 \times 0.25, 0.02 \times 1, \ldots, 0.004 \times 1).\end{aligned}$$

Example 2.1.6 Consider a group of n islands. Suppose that among all the species living there, none is present on more than one island (as may in principle be the case if the islands have been separate for a long enough period of evolutionary time). Write k_i for the number of species on the ith island, and $\mathbf{p}^i \in \Delta_{k_i}$ for their relative abundance distribution. Also write $\mathbf{w} \in \Delta_n$ for the relative sizes of the n islands, where 'size' means the total abundance of organisms on each island. Then the composite

$$\mathbf{w} \circ (\mathbf{p}^1, \ldots, \mathbf{p}^n) \in \Delta_{k_1 + \cdots + k_n}$$

is the relative abundance distribution for the whole island group, with the species on the first island listed first, then the species on the second island, and so on.

Example 2.1.7 Recall that in the standard taxonomic system, the next level up from species is genus (plural: genera). Take an ecological community of n genera, with relative abundances $\mathbf{w} = (w_1, \ldots, w_n)$. Let $\mathbf{p}^i$ be the relative abundance distribution of the species within the ith genus. Then the relative abundance distribution of the species in the community is the composite $\mathbf{w} \circ (\mathbf{p}^1, \ldots, \mathbf{p}^n)$.

Remark 2.1.8 Composition of probability distributions satisfies an associative law: for each $n, k_i, \ell_{ij} \geq 1$ and $\mathbf{w} \in \Delta_n$, $\mathbf{p}^i \in \Delta_{k_i}$, $\mathbf{r}^{ij} \in \Delta_{\ell_{ij}}$,

$$\begin{aligned}\Big(\mathbf{w} \circ (\mathbf{p}^1, \ldots, \mathbf{p}^n)\Big) \circ (\mathbf{r}^{11}, \ldots, \mathbf{r}^{1k_1}, \ \ldots, \ \mathbf{r}^{n1}, \ldots, \mathbf{r}^{nk_n}) \\ = \mathbf{w} \circ \Big(\mathbf{p}^1 \circ (\mathbf{r}^{11}, \ldots, \mathbf{r}^{1k_1}), \ \ldots, \ \mathbf{p}^n \circ (\mathbf{r}^{n1}, \ldots, \mathbf{r}^{nk_n})\Big).\end{aligned}$$

The unique distribution $\mathbf{u}_1$ on the one-element set acts as an identity for composition:

$$\mathbf{p} \circ (\underbrace{\mathbf{u}_1, \ldots, \mathbf{u}_1}_{n}) = \mathbf{p} = \mathbf{u}_1 \circ (\mathbf{p})$$

for all $n \geq 1$ and $\mathbf{p} \in \Delta_n$.

These equations are straightforward to check. In the language of abstract algebra, they state that the sequence of sets $(\Delta_n)_{n\geq 0}$, equipped with the operation of composition and the trivial distribution $\mathbf{u}_1$, is an operad. We explain and exploit this observation in Chapter 12.

Now consider the *decomposition* problem: given $\mathbf{r} \in \Delta_k$ and positive integers $n, k_1, \ldots, k_n$ such that $\sum k_i = k$, do there exist distributions $\mathbf{w} \in \Delta_n$ and $\mathbf{p}^i \in \Delta_{k_i}$ such that

$$\mathbf{w} \circ (\mathbf{p}^1, \ldots, \mathbf{p}^n) = \mathbf{r}? \tag{2.1}$$

The answer is yes. In fact, $\mathbf{w}$ and $\mathbf{p}^1, \ldots, \mathbf{p}^n$ are very nearly uniquely determined, ambiguity only arising if some of the probabilities r_i are zero. The exact situation is as follows.

Lemma 2.1.9 *Let $k \geq 1$ and $\mathbf{r} \in \Delta_k$. Let $n, k_1, \ldots, k_n$ be positive integers such that $k_1 + \cdots + k_n = k$. Then there exist*

$$\mathbf{w} \in \Delta_n, \ \mathbf{p}^1 \in \Delta_{k_1}, \ \ldots, \ \mathbf{p}^n \in \Delta_{k_n}$$

such that equation (2.1) *holds. Moreover,* $\mathbf{w}, \mathbf{p}^1, \ldots, \mathbf{p}^n$ *satisfy* (2.1) *if and only if*

$$w_i = r_{k_1+\cdots+k_{i-1}+1} + \cdots + r_{k_1+\cdots+k_{i-1}+k_i} \tag{2.2}$$

for each $i \in \{1, \ldots, n\}$ and

$$\mathbf{p}^i = \frac{1}{w_i}(r_{k_1+\cdots+k_{i-1}+1}, \ldots, r_{k_1+\cdots+k_{i-1}+k_i}) \tag{2.3}$$

for each $i \in$ supp($\mathbf{w}$). *In particular, equation* (2.1) *determines* $\mathbf{w}$ *uniquely.*

Proof Define $\mathbf{w}$ by equation (2.2), define $\mathbf{p}^i$ by equation (2.3) for each $i \in$ supp($\mathbf{w}$), and for $i \notin$ supp($\mathbf{w}$), let $\mathbf{p}^i$ be any element of Δ_{k_i}. It is then trivial to verify equation (2.1).

Conversely, suppose that $\mathbf{w}, \mathbf{p}^1, \ldots, \mathbf{p}^n$ are distributions satisfying (2.1). Write $\mathbf{p}^i = (p^i_1, \ldots, p^i_{k_i})$. We have

$$w_1 = w_1(p^1_1 + \cdots + p^1_{k_1}) = w_1 p^1_1 + \cdots + w_1 p^1_{k_1} = r_1 + \cdots r_{k_1},$$

since $\mathbf{p}^1 \in \Delta_1$. A similar argument holds for $w_2, \ldots, w_n$, giving equation (2.2), and equation (2.3) then follows. □

Some further terminology illuminates this result, and will be useful throughout.

Definition 2.1.10 Let $k, n \geq 1$, let

$$\pi\colon \{1, \ldots, k\} \to \{1, \ldots, n\}$$

be a map of sets, and let $\mathbf{r} \in \Delta_k$. The **pushforward** of $\mathbf{r}$ along π is the distribution $\pi\mathbf{r} \in \Delta_n$ with ith coordinate

$$(\pi\mathbf{r})_i = \sum_{j\colon \pi(j)=i} r_j$$

$(i \in \{1, \ldots, n\})$.

In the situation of Lemma 2.1.9, consider the function

$$\pi\colon \{1, \ldots, k\} \to \{1, \ldots, n\}$$

that maps the first k_1 elements of $\{1, \ldots, k\}$ to 1, the next k_2 elements to 2, and so on. Then part of the statement of the lemma is that equation (2.1) determines $\mathbf{w}$ uniquely as $\mathbf{w} = \pi\mathbf{r}$.

Remark 2.1.11 Definition 2.1.10 is a special case of the general measure-theoretic notion of the pushforward $\pi_*\mu$ of a measure μ along a measurable map π. (We omit the star.) Our statements about composition and decomposition on finite sets are trivial cases of a general measure-theoretic theory of integration and disintegration. For a summary of disintegration, see Section 3.2 of Dahlqvist, Danos, Garnier and Kammar [77], or for a more comprehensive account, see around Theorem III.71 of Dellacherie and Meyer [80].

An important special case of composition is the **tensor product**. Given $\mathbf{w} \in \Delta_n$ and $\mathbf{p} \in \Delta_k$, define

$$\begin{aligned} \mathbf{w} \otimes \mathbf{p} &= \mathbf{w} \circ \underbrace{(\mathbf{p}, \ldots, \mathbf{p})}_{n} \\ &= (w_1 p_1, \ldots, w_1 p_k, \; \ldots, \; w_n p_1, \ldots, w_n p_k) \\ &\in \Delta_{nk}. \end{aligned}$$

Probabilistically, $\mathbf{w} \otimes \mathbf{p}$ is the joint distribution of two independent random variables with distributions $\mathbf{w}$ and $\mathbf{p}$ respectively.

Example 2.1.12 Consider a large ecological community – a **metacommunity** – divided into N subcommunities of relative sizes $w_1, \ldots, w_N$. Write S for the number of species in the metacommunity, and $p_1, \ldots, p_S$ for their relative abundances across the whole metacommunity. There is an $S \times N$ matrix representing how the organisms are distributed across the S species and N communities, with the ith row summing to p_i and the jth column summing to w_j.

If the metacommunity is homogeneous in the sense that the species distributions in all the subcommunities are identical, then the (i, j)-entry of this matrix is $w_j p_i$. In that case, when the SN entries of the matrix are expressed as an SN-dimensional vector (concatenating the columns in order), that vector is exactly $\mathbf{w} \otimes \mathbf{p}$.

The tensor product of distributions has the usual algebraic properties of a product: it satisfies the associativity and identity laws

$$(\mathbf{w} \otimes \mathbf{p}) \otimes \mathbf{r} = \mathbf{w} \otimes (\mathbf{p} \otimes \mathbf{r}), \quad \mathbf{p} \otimes \mathbf{u}_1 = \mathbf{p} = \mathbf{u}_1 \otimes \mathbf{p}.$$

These follow from the equations in Remark 2.1.8. For $\mathbf{p} \in \Delta_n$ and $d \geq 1$, we write

$$\mathbf{p}^{\otimes d} = \underbrace{\mathbf{p} \otimes \cdots \otimes \mathbf{p}}_{d} \in \Delta_{n^d},$$

interpreted as $\mathbf{u}_1 \in \Delta_1$ if $d = 0$.

2.2 Definition and Properties of Shannon Entropy

Let $\mathbf{p} = (p_1, \ldots, p_n)$ be a probability distribution on n elements. The **Shannon entropy** of $\mathbf{p}$ is

$$H(\mathbf{p}) = - \sum_{i \in \operatorname{supp}(\mathbf{p})} p_i \log p_i = \sum_{i \in \operatorname{supp}(\mathbf{p})} p_i \log \frac{1}{p_i}.$$

Equivalently, instead of restricting the sum to just those i such that $p_i > 0$, one can let i run over all of $\{1, \ldots, n\}$, with the conventions that

$$0 \log 0 = 0 = 0 \log \frac{1}{0}.$$

These conventions are justified by the facts that

$$\lim_{p \to 0+} p \log p = 0 = \lim_{p \to 0+} p \log \frac{1}{p}.$$

Remark 2.2.1 Although we take log to denote the *natural* logarithm (Remark 1.1.12), changing the base of the logarithm simply multiplies H by a constant factor, and in this sense is unimportant. In information and coding theory, where one is typically concerned with strings of binary digits, it is normal to take entropy to base 2. We write base 2 entropy as $H^{(2)}$; thus, $H^{(2)}(\mathbf{p}) = H(\mathbf{p})/\log 2$.

Much of this chapter is devoted to explaining and interpreting Shannon entropy, but we can immediately give several interpretations in brief.

Uniformity For distributions $\mathbf{p}$ on a fixed number of elements, the entropy of $\mathbf{p}$ is greatest when $\mathbf{p}$ is uniform, and least when $\mathbf{p}$ is concentrated on a single element (Figure 2.4 and Lemma 2.2.4 below).

Information Regard $\log(1/p_i)$ as the amount of information gained by observing an event of probability p_i. For a near-inevitable event such as the sun rising, $p_i \approx 1$ and so $\log(1/p_i) \approx 0$: knowing that the sun rose this morning tells us nothing that we could not have predicted with very high confidence beforehand. The entropy $H(\mathbf{p})$ is the average amount of information gained per observation. We develop this interpretation in the pages that follow.

Expected surprise Similarly, $\log(1/p_i)$ can be regarded as our surprise at observing an event of probability p_i, and then $H(\mathbf{p})$ is the expected surprise. We return to this viewpoint in Section 4.1.

Genericity In thermodynamics, a system in a state of high entropy is disordered, or generic. For instance, it is the usual state of a box of gas that every cubic centimetre contains about the same number of molecules; this is a high-entropy, generic, state. If, by some unlikely chance, all the molecules were concentrated into one cubic centimetre, this would be a low-entropy and very non-generic state.

The logarithm of diversity Let $\mathbf{p}$ be a probability distribution modelling an ecological community, as in Example 2.1.1. In Section 2.4, we will see that $\exp(H(\mathbf{p}))$ is a sensible measure of the diversity of a community. In later chapters, we will meet other types of entropy and show that their exponentials are also meaningful measures of diversity.

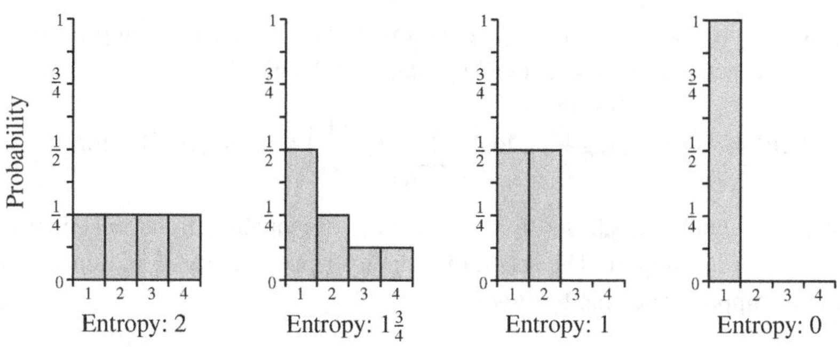

Figure 2.4 Four probability distributions on $\{1, 2, 3, 4\}$, and their entropies to base 2.

Example 2.2.2 Figure 2.4 shows the base 2 entropies $H^{(2)}(\mathbf{p})$ of four distributions $\mathbf{p} \in \Delta_4$. For instance, the second is computed as

$$H^{(2)}\left(\tfrac{1}{2}, \tfrac{1}{4}, \tfrac{1}{8}, \tfrac{1}{8}\right) = \tfrac{1}{2}\log_2 2 + \tfrac{1}{4}\log_2 4 + \tfrac{1}{8}\log_2 8 + \tfrac{1}{8}\log_2 8 = 1\tfrac{3}{4}.$$

These examples illustrate the interpretation of entropy as uniformity. The highest entropy belongs to the first, uniform, distribution. Each of the four distributions on $\{1, 2, 3, 4\}$ is less uniform than its predecessor, and, correspondingly, has lower entropy.

We now set out the basic properties of entropy. Here and later, we will repeatedly use the following elementary fact about logarithms.

Lemma 2.2.3 *Let* $\mathbf{p} \in \Delta_n$ *and* $x_1, \ldots, x_n \in (0, \infty)$. *Then*

$$\log\left(\sum_{i=1}^{n} p_i x_i\right) \geq \sum_{i=1}^{n} p_i \log x_i,$$

with equality if and only if $x_i = x_j$ *for all* $i, j \in \mathrm{supp}(\mathbf{p})$.

Proof The function $\log\colon (0, \infty) \to \mathbb{R}$ is strictly concave, since $\frac{d^2}{dx^2}\log x = -1/x^2 < 0$. The result follows. □

We now show that among all probability distributions on a finite set, entropy is maximized by the uniform distribution and minimized by any distribution of the form $(0, \ldots, 0, 1, 0, \ldots, 0)$.

Lemma 2.2.4 *Let* $n \geq 1$.

i. $H(\mathbf{p}) \geq 0$ *for all* $\mathbf{p} \in \Delta_n$, *with equality if and only if* $p_i = 1$ *for some* $i \in \{1, \ldots, n\}$.
ii. $H(\mathbf{p}) \leq \log n$ *for all* $\mathbf{p} \in \Delta_n$, *with equality if and only if* $\mathbf{p} = \mathbf{u}_n$.

Proof Part (i) follows from the fact that $\log(1/p_i) \geq 0$ for all $i \in \operatorname{supp}(\mathbf{p})$, with equality if and only if $p_i = 1$. For (ii), Lemma 2.2.3 gives

$$H(\mathbf{p}) = \sum_{i \in \operatorname{supp}(\mathbf{p})} p_i \log \frac{1}{p_i} \leq \log\left(\sum_{i \in \operatorname{supp}(\mathbf{p})} p_i \cdot \frac{1}{p_i} \right) = \log|\operatorname{supp}(\mathbf{p})| \leq \log n.$$

Again by Lemma 2.2.3, the first inequality is an equality if and only if $\mathbf{p}$ is uniform on its support. The second inequality is an equality if and only if $\mathbf{p}$ has full support. The result follows. □

It is often useful to express entropy in terms of the function

$$\partial\colon [0,1] \to \mathbb{R}$$

defined by

$$\partial(x) = \begin{cases} -x \log x & \text{if } x > 0, \\ 0 & \text{if } x = 0. \end{cases} \tag{2.4}$$

Thus,

$$H(\mathbf{p}) = \sum_{i=1}^{n} \partial(p_i) \tag{2.5}$$

for all $n \geq 1$ and $\mathbf{p} \in \Delta_n$.

Lemma 2.2.5 *For each $n \geq 1$, the entropy function $H\colon \Delta_n \to \mathbb{R}$ is continuous.*

Proof This follows from equation (2.5) and the elementary fact that ∂ is continuous. □

The operator ∂ is a nonlinear derivation:

Lemma 2.2.6 $\partial(xy) = \partial(x)y + x\partial(y)$ *for all $x, y \in [0,1]$.* □

Remark 2.2.7 Up to a constant factor, ∂ is the only measurable function $d\colon [0,1] \to \mathbb{R}$ satisfying $d(xy) = d(x)y + xd(y)$ for all x, y. Indeed, taking $x = y = 0$ forces $d(0) = 0$, and the result follows by applying Corollary 1.1.14 to the function $x \mapsto d(x)/x$ on $(0,1]$.

We now use Lemma 2.2.6 to prove the most important algebraic property of Shannon entropy.

Proposition 2.2.8 (Chain rule) *Let $\mathbf{w} \in \Delta_n$ and $\mathbf{p}^1 \in \Delta_{k_1}, \ldots, \mathbf{p}^n \in \Delta_{k_n}$. Then*

$$H(\mathbf{w} \circ (\mathbf{p}^1, \ldots, \mathbf{p}^n)) = H(\mathbf{w}) + \sum_{i=1}^{n} w_i H(\mathbf{p}^i).$$

Proof Writing $\mathbf{p}^i = (p^i_1, \ldots, p^i_{k_i})$ and using Lemma 2.2.6, we have

$$\begin{aligned} H(\mathbf{w} \circ (\mathbf{p}^1, \ldots, \mathbf{p}^n)) &= \sum_{i=1}^{n} \sum_{j=1}^{k_i} \partial(w_i p^i_j) \\ &= \sum_i \sum_j (\partial(w_i) p^i_j + w_i \partial(p^i_j)) \\ &= \sum_i \partial(w_i) + \sum_i w_i \sum_j \partial(p^i_j) \\ &= H(\mathbf{w}) + \sum_i w_i H(\mathbf{p}^i), \end{aligned}$$

as required. □

Example 2.2.9 Consider again the coin-die-card process of Example 2.1.4. How much information do we expect to gain from observing the final outcome of the process?

Let us measure information by base 2 entropy, in bits. The information gained is as follows.

- Whether the final outcome is a number between 1 and 6 or a card tells us whether the coin came up heads or tails. This gives us $H^{(2)}(\mathbf{u}_2) = 1$ bit of information.
- With probability 1/2, the outcome is the result of a die roll, which would give us $H^{(2)}(\mathbf{u}_6) = \log_2 6$ bits of information.
- With probability 1/2, the outcome is the result of a card draw, which would give us $H^{(2)}(\mathbf{u}_{52}) = \log_2 52$ bits of information.

Hence in total, the expected information gained from observing the outcome of the composite process is

$$H^{(2)}(\mathbf{u}_2) + \tfrac{1}{2} H^{(2)}(\mathbf{u}_6) + \tfrac{1}{2} H^{(2)}(\mathbf{u}_{52}) = 1 + \tfrac{1}{2} \log_2 6 + \tfrac{1}{2} \log_2 52$$

bits. If we have reasoned correctly, this should be equal to the entropy of the composite process, which is

$$H^{(2)}(\mathbf{u}_2 \circ (\mathbf{u}_6, \mathbf{u}_{52})) = H^{(2)}\Big(\underbrace{\tfrac{1}{12}, \ldots, \tfrac{1}{12}}_{6}, \underbrace{\tfrac{1}{104}, \ldots, \tfrac{1}{104}}_{52}\Big)$$

bits. The chain rule guarantees that these two numbers are, indeed, equal.

Corollary 2.2.10 *For all* $\mathbf{w} \in \Delta_n$ *and* $\mathbf{p} \in \Delta_k$,

$$H(\mathbf{w} \otimes \mathbf{p}) = H(\mathbf{w}) + H(\mathbf{p}). \tag{2.6}$$

Proof Take $\mathbf{p}^1 = \cdots = \mathbf{p}^n = \mathbf{p}$ in the chain rule. □

In other words, H has the logarithmic property of converting products into sums. Indeed, in the special case $\mathbf{w} = \mathbf{u}_n$ and $\mathbf{p} = \mathbf{u}_k$, we have $\mathbf{w} \otimes \mathbf{p} = \mathbf{u}_{nk}$, so equation (2.6) is precisely the characteristic property of the logarithm,

$$\log(nk) = \log n + \log k.$$

In the general case, equation (2.6) states that the amount of information gained by observing the outcome of a pair of *independent* events is equal to the information gained from the first plus the information gained from the second.

Remark 2.2.11 With the understanding that H is symmetric in its arguments, the chain rule as stated in Proposition 2.2.8 is equivalent to the superficially less general statement that

$$H(pw_1, (1-p)w_1, w_2, \dots, w_n) = H(\mathbf{w}) + w_1 H(p, 1-p) \tag{2.7}$$

for all $p \in [0, 1]$ and $\mathbf{w} \in \Delta_n$. This is the special case $k_1 = 2$, $k_2 = \cdots = k_n = 1$ of Proposition 2.2.8, and is sometimes known as the **recursivity** of entropy (Definition 1.2.8 of Aczél and Daróczy [3]) or the **grouping rule** (Problem 4 of Chapter 2 of Cover and Thomas [71]).

The general chain rule of Proposition 2.2.8 is also equivalent to a different special case:

$$H(wp_1, \dots, wp_k, (1-w)r_1, \dots, (1-w)r_\ell) = \\ H(w, 1-w) + wH(\mathbf{p}) + (1-w)H(\mathbf{r})$$

for all $w \in [0, 1]$, $\mathbf{p} \in \Delta_k$, and $\mathbf{r} \in \Delta_\ell$. This is the special case $n = 2$ of Proposition 2.2.8.

Both equivalences are routine inductions, carried out in Appendix A.1.

2.3 Entropy in Terms of Coding

The theory of coding provides a very concrete way of understanding the concept of information. The fundamental concepts and theorems of coding theory were set out in Shannon's original 1948 paper [309], with rigour and detail added soon afterwards by researchers such as Khinchin [188] and Feinstein [99]. This section presents parts of that early work, and in particular, Shannon's source coding theorem.

The source coding theorem can be described informally as follows. Take an alphabet of symbols, say the English letters a to z, which occur with known frequencies $p_1, \dots, p_{26}$. We want to design a scheme that encodes each letter as a finite sequence of 0s and 1s. Using this system, any message in English can also be encoded as a sequence of 0s and 1s, by concatenating the codes for the letters in the message. Of course, we want our coding scheme to have the property that the encoded message can be decoded unambiguously, and it is also natural to want it to use as few bits as possible. Roughly speaking, the theorem is that in the most efficient coding scheme, the number of bits needed per symbol is the base 2 entropy of the frequency distribution $\mathbf{p}$.

We now give a more precise account. In this section, entropy will always be taken to base 2. Details of everything that follows can be found in introductions to information theory such as Cover and Thomas ([71], Chapter 5), MacKay ([238], Chapter 4), and Jones and Jones [164].

Take an alphabet of n symbols, with frequency distribution $\mathbf{p} \in \Delta_n$; thus, in messages written using this alphabet, we expect the symbols to be used in proportions $p_1, \ldots, p_n$. A **code** is an assignment to each $i \in \{1, \ldots, n\}$ of a finite sequence of bits (a **code word**). The ith code word is, then, an element of the set $\{0, 1\}^{L_i}$ for some integer $L_i \geq 0$, and L_i is called the **word length** of the ith symbol. The expected word length of a symbol in our alphabet is

$$\sum_{i=1}^{n} p_i L_i.$$

We seek a code that minimizes the average word length, subject to the natural constraint of unambiguous decodability (made precise shortly).

Example 2.3.1 Take an alphabet of four symbols a, b, c, d, with frequency distribution $\mathbf{p} = (1/2, 1/4, 1/8, 1/8)$. How should we encode our symbols as strings of bits, in a way that uses as few bits as possible?

The basic principle is that common symbols should have short code words. (The same principle guided the design of Morse code, where the most common letter, e, is encoded as a single dot, and uncommon letters such as z use four dots or dashes.) So let us encode as follows:

$$\mathsf{a}: 0, \quad \mathsf{b}: 10, \quad \mathsf{c}: 110, \quad \mathsf{d}: 111.$$

For instance, 11110011010 represents dbacb. The average word length is

$$\tfrac{1}{2} \cdot 1 + \tfrac{1}{4} \cdot 2 + \tfrac{1}{8} \cdot 3 + \tfrac{1}{8} \cdot 3 = 1\tfrac{3}{4}.$$

This is more efficient than the most naive coding system, which would simply assign the four two-bit strings 00, 01, 10, 11 to the four symbols, for an average word length of 2.

A code is **instantaneous** if none of the code words is a prefix (initial segment) of any other. Thus, if $\delta_1 \cdots \delta_\ell$ and $\varepsilon_1 \cdots \varepsilon_m$ are code words in an instantaneous code, with $\ell \leq m$, then $(\delta_1, \ldots, \delta_\ell) \neq (\varepsilon_1, \ldots, \varepsilon_\ell)$. This is the non-ambiguity condition, guaranteeing that any string of bits produced by the system can only be decoded in one possible way.

Example 2.3.2 The code of Example 2.3.1 is instantaneous. But if we changed the code word for b to 11, the code would no longer be instantaneous, since 11 is a prefix of the code words for both c and d. Messages in this new code are not uniquely decodable; for instance, the string 110 could be decoded as either c or ba.

The average word length $1\frac{3}{4}$ of the code in Example 2.3.1 happens to be equal to the entropy of the frequency distribution of the symbols, calculated in Example 2.2.2. In fact, it is not possible to find an instantaneous code whose average word length is any shorter. This is an instance of part (ii) of the following result.

Proposition 2.3.3 *Let $n, L_1, \ldots, L_n \geq 1$, and suppose that there exists an instantaneous code on the alphabet $\{1, \ldots, n\}$ with word lengths $L_1, \ldots, L_n$. Then:*

i. $\sum_{i=1}^{n} (1/2)^{L_i} \leq 1$;

ii. $\sum_{i=1}^{n} p_i L_i \geq H^{(2)}(\mathbf{p})$ *for all* $\mathbf{p} \in \Delta_n$.

Part (i), together with part (i) of Proposition 2.3.4 below, is known as **Kraft's inequality** (Theorem 5.2.1 of Cover and Thomas [71], for instance).

Proof To prove (i), we consider binary expansions $0.b_1 b_2 \ldots$ of elements of $[0, 1)$, where $b_i \in \{0, 1\}$. We make the convention that if $x \in [0, 1)$ has two binary expansions, one ending with an infinite sequence of 0s and the other with an infinite sequence of 1s, we choose the former. In this way, each $x \in [0, 1)$ determines an infinite sequence of bits $b_1, b_2, \ldots$

Take an instantaneous code with word lengths $L_1, \ldots, L_n$. For $i \in \{1, \ldots, n\}$, write

$$J_i = \{x \in [0, 1) : \text{the binary expansion of } x \text{ begins with the } i\text{th code word}\}.$$

Then J_i is a half-open interval of length $(1/2)^{L_i}$. Since the code is instantaneous, the intervals $J_1, \ldots, J_n$ are disjoint. But since they are all subsets of $[0, 1)$, their total length is at most 1, giving the desired inequality.

For (ii), let $\mathbf{p} \in \Delta_n$. By Lemma 2.2.3 and part (i),

$$\begin{aligned}
H^{(2)}(\mathbf{p}) - \sum_{i=1}^{n} p_i L_i &= \sum_{i \in \mathrm{supp}(\mathbf{p})} p_i \big(\log_2(1/p_i) + \log_2((1/2)^{L_i})\big) \\
&= \sum_{i \in \mathrm{supp}(\mathbf{p})} p_i \log_2 \frac{(1/2)^{L_i}}{p_i} \\
&\leq \log_2 \Bigg(\sum_{i \in \mathrm{supp}(\mathbf{p})} p_i \cdot \frac{(1/2)^{L_i}}{p_i} \Bigg) \\
&\leq \log_2 \sum_{i=1}^{n} (1/2)^{L_i} \\
&\leq \log_2 1 = 0,
\end{aligned}$$

as required. □

The frequency distribution of Example 2.3.1 has the exceptional property that all the frequencies are powers of $1/2$. In such cases, it is always possible to find an instantaneous code in which the ith symbol is encoded in $\log_2(1/p_i)$ bits, so that the average word length is exactly the entropy. In the general case, this is not quite possible; but it is nearly possible, as follows.

Proposition 2.3.4 *Let* $\mathbf{p} \in \Delta_n$. *Then:*

i. *there is an instantaneous code with word lengths* $\lceil \log_2(1/p_1) \rceil, \ldots, \lceil \log_2(1/p_n) \rceil$;
ii. *any such code has expected word length strictly less then* $H^{(2)}(\mathbf{p}) + 1$.

Here $\lceil x \rceil$ denotes the smallest integer greater than or equal to x. Codes with the property in (i) are called **Shannon codes**.

Proof For (i), suppose without loss of generality that $p_1 \geq \cdots \geq p_n$. For each $i \in \{1, \ldots, n\}$, put

$$L_i = \lceil \log_2(1/p_i) \rceil, \quad q_i = (1/2)^{L_i}.$$

In other words, q_i is maximal among all powers of $1/2$ less than or equal to p_i. Now, $q_1, \ldots, q_i$ are all integer multiples of $(1/2)^{L_i}$, so $q_1 + \cdots + q_{i-1}$ and $q_1 + \cdots + q_i$ are integer multiples of $(1/2)^{L_i}$ too. It follows that the binary expansions of the elements of the interval

$$J_i = [q_1 + \cdots + q_{i-1}, q_1 + \cdots + q_{i-1} + q_i)$$

all begin with the same L_i bits, and, moreover, that no other element of $[0, 1)$ begins with this bit-sequence. (Here we use the same convention on binary expansions as in the proof of Proposition 2.3.3.) Take the ith code word to be this bit-sequence. Since the intervals $J_1, \ldots, J_n$ are disjoint, none of the code words is a prefix of any other; that is, the code is instantaneous.

For (ii), take a code as in (i), again writing $L_i = \lceil \log_2(1/p_i) \rceil$. We have

$$L_i < \log_2(1/p_i) + 1$$

for each $i \in \{1, \ldots, n\}$, so

$$\sum_{i=1}^{n} p_i L_i = \sum_{i \in \text{supp}(\mathbf{p})} p_i L_i < \sum_{i \in \text{supp}(\mathbf{p})} p_i(\log_2(1/p_i) + 1) = H^{(2)}(\mathbf{p}) + 1,$$

as required. □

Example 2.3.5 Take the alphabet consisting of a, b, c, d with frequencies $\mathbf{p} = (0.4, 0.3, 0.2, 0.1)$. Following the construction in the proof of Proposition 2.3.4, we round each frequency down to the next power of 1/2, giving

$$(q_1, q_2, q_3, q_4) = \left(\tfrac{1}{4}, \tfrac{1}{4}, \tfrac{1}{8}, \tfrac{1}{16}\right) = \left(\left(\tfrac{1}{2}\right)^2, \left(\tfrac{1}{2}\right)^2, \left(\tfrac{1}{2}\right)^3, \left(\tfrac{1}{2}\right)^4\right).$$

Thus, $(L_1, L_2, L_3, L_4) = (2, 2, 3, 4)$ and the intervals J_i are as follows, in binary notation:

$$\begin{aligned} J_1 &= \left[0, \tfrac{1}{4}\right) = [0.00, 0.01), \\ J_2 &= \left[\tfrac{1}{4}, \tfrac{1}{2}\right) = [0.01, 0.10), \\ J_3 &= \left[\tfrac{1}{2}, \tfrac{5}{8}\right) = [0.100, 0.101), \\ J_4 &= \left[\tfrac{5}{8}, \tfrac{11}{16}\right) = [0.1010, 0.1011). \end{aligned}$$

We therefore encode as follows:

$$\mathsf{a}: 00, \quad \mathsf{b}: 01, \quad \mathsf{c}: 100, \quad \mathsf{d}: 1010.$$

Short calculations show that

$$\sum_{i=1}^{4} p_i L_i = 2.4 < 2.846\ldots = H^{(2)}(\mathbf{p}) + 1,$$

as the proof of Proposition 2.3.4 guarantees.

This is not the most efficient code. For instance, we could have encoded d as 101 for a smaller average word length. There are in fact algorithms that construct for each $\mathbf{p}$ a code with the least possible average word length, such as that of Huffman [148]. But we will not need such precision here.

Example 2.3.6 Similarly, the code in Example 2.3.1 is the one constructed by the algorithm in the proof of Proposition 2.3.4.

Remark 2.3.7 The bound $H^{(2)}(\mathbf{p})+1$ in Proposition 2.3.4 cannot be improved to $H^{(2)}(\mathbf{p}) + c$ for any constant $c < 1$. For instance, if a two-symbol alphabet has frequency distribution $\mathbf{p} = (0.99, 0.01)$ then $H^{(2)}(\mathbf{p}) \approx H^{(2)}(1, 0) = 0$ (since $H^{(2)}$ is continuous), but clearly the average word length cannot be reduced to below 1.

We now state a version of Shannon's source coding theorem.

Theorem 2.3.8 (Shannon) *For an alphabet with frequency distribution* $\mathbf{p} = (p_1, \ldots, p_n)$,

$$H^{(2)}(\mathbf{p}) \leq \inf \sum_{i=1}^{n} p_i L_i < H^{(2)}(\mathbf{p}) + 1,$$

where the infimum is over all instantaneous codes on n elements, with L_i denoting the ith word length.

Proof This is immediate from Propositions 2.3.3 and 2.3.4. □

A crucial further insight of Shannon was that the upper bound $H^{(2)}(\mathbf{p}) + 1$ can be reduced to $H^{(2)}(\mathbf{p})+\varepsilon$, for any $\varepsilon > 0$, as long as we are willing to encode symbols in blocks rather than one at a time. Informally, this works as follows.

For an alphabet with n symbols, there are n^{10} blocks of 10 symbols. Writing $\mathbf{p} \in \Delta_n$ for the frequency distribution of the original alphabet and assuming that successive symbols in messages are distributed independently, the frequency distribution of the n^{10} blocks is $\mathbf{p}^{\otimes 10}$.

Now treat each 10-symbol block as a unit, and consider ways of encoding each block as a sequence of bits. By Proposition 2.3.4, we can find an instantaneous code for the blocks that uses an average of less than $H^{(2)}(\mathbf{p}^{\otimes 10}) + 1$ bits per block. But $H^{(2)}(\mathbf{p}^{\otimes 10}) = 10H^{(2)}(\mathbf{p})$ by Corollary 2.2.10, so the average number of bits per letter is less than

$$\tfrac{1}{10}\left(H^{(2)}(\mathbf{p}^{\otimes 10}) + 1\right) = H^{(2)}(\mathbf{p}) + \tfrac{1}{10}.$$

In this way, by encoding symbols in large blocks rather than individually, we can make the average number of bits per letter as close as we please to the lower bound of $H^{(2)}(\mathbf{p})$.

(In applications, successive symbols are often not independent. For instance, in English, the letter pair ch is more frequent than hc. But it will follow from Remark 8.1.13 that even if they are not independent, the actual frequency distribution of the n^{10} blocks has entropy at most $H(\mathbf{p}^{\otimes 10})$. For that reason, the argument above is valid even without the assumption of independence.)

Example 2.3.9 Take a two-symbol alphabet a, b with frequency distribution $\mathbf{p} = (0.6, 0.4)$. Then $H^{(2)}(\mathbf{p}) = 0.9709\ldots$ We compute the average number of bits per letter when encoding in larger and larger blocks, following the code construction in the proof of Proposition 2.3.4.

- First encode one symbol at a time. We round each p_i down to the next power of $1/2$, giving $((1/2)^1, (1/2)^2)$. Hence the average number of bits per symbol is

$$0.6 \times 1 + 0.4 \times 2 = 1.4.$$

- Now encode symbols in blocks of two. The frequency distribution of aa, ab, ba, bb is $(0.36, 0.24, 0.24, 0.16)$ (assuming that successive symbols are distributed independently). Following the same algorithm, we round down to $((1/2)^2, (1/2)^3, (1/2)^3, (1/2)^3)$ and obtain an average of

$$0.36 \times 2 + 0.24 \times 3 + 0.24 \times 3 + 0.16 \times 3 = 2.64$$

  bits per two-symbol block, or equivalently an average of 1.32 bits per symbol. This is an improvement on the original code.
- Similarly, encoding in three-symbol blocks gives an average of $1.117\ldots$ bits per symbol, which is closer still to the ideal of $H^{(2)}(\mathbf{p}) \approx 0.971$ bits per symbol.

None of these three codes is as efficient as the naive code that assigns the code words 0 to a and 1 to b, which has an average word length of 1. But we can improve on that by encoding in large enough blocks. For instance, since

$$0.971 + \tfrac{1}{35} < 1,$$

we can attain an average word length of less than 1 by coding blocks of 35 symbols at a time.

Example 2.3.10 In written English, the base 2 entropy of the frequency distribution of the 26 letters of the alphabet is approximately 4.1 (Section 2 of Shannon [310]). Thus, by using sufficiently large blocks, one can encode English using about four bits per letter. (It is as if English had only $2^{4.1} \approx 17$ letters, used with equal frequency.) This is without taking advantage of the fact that ch occurs more often than hc, for instance. Using the non-independence of neighbouring letters would enable us to reduce the number of bits still further, as detailed by Shannon [310] and later researchers.

A convenient fiction when reasoning about entropy is that for every probability distribution $\mathbf{p}$, there is an instantaneous code with average word length $H^{(2)}(\mathbf{p})$. This is not true unless all the nonzero frequencies happen to be powers

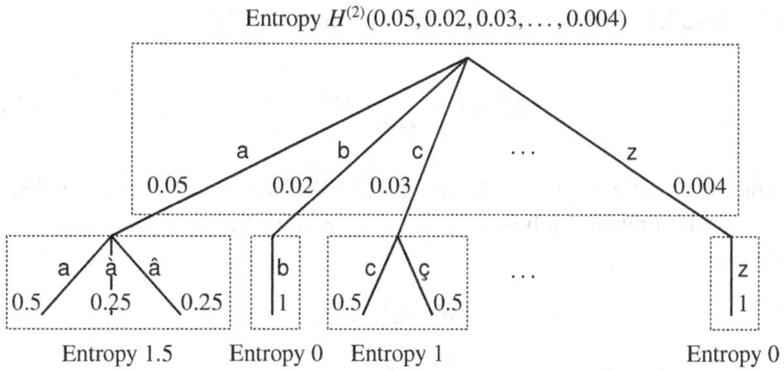

Figure 2.5 The entropy of the French language (Example 2.3.11).

of $\frac{1}{2}$, but it is approximately true in the sense just described: we can come arbitrarily close by encoding in sufficiently large blocks. Let us call this (usually nonexistent) code an **ideal code** for $\mathbf{p}$.

Ideal codes provide a way to understand the chain rule (Proposition 2.2.8), as follows.

Example 2.3.11 Consider again the French language (Example 2.1.5), which is written with symbols such as à made up of a letter (in this case, a) and an accent (in this case, `). Figure 2.5 shows a hypothetical frequency distribution $\mathbf{w}$ of the letters, hypothetical frequency distributions

$$\mathbf{p}^1 \in \Delta_3, \ \mathbf{p}^2 \in \Delta_1, \ \ldots, \ \mathbf{p}^{26} \in \Delta_1$$

of the accents on each letter, and the base 2 entropy of each of the distributions $\mathbf{w}, \mathbf{p}^1, \ldots, \mathbf{p}^{26}$.

To transmit a French symbol (such as à), we need to transmit both its base letter (a) and its accent (`). Using ideal codes, the average number of bits needed per symbol is as follows. For the base letter, we need $H^{(2)}(\mathbf{w})$ bits. The number of bits needed for the accent depends on which letter it decorates:

- with probability w_1, the letter is a, and then the average number of bits needed for the accent is $H^{(2)}(\mathbf{p}^1)$;
- with probability w_2, the letter is b, and then the average number of bits needed for the accent is $H^{(2)}(\mathbf{p}^2)$;

and so on. Hence the average number of bits needed to encode the accent is $\sum_{i=1}^{26} w_i H^{(2)}(\mathbf{p}^i)$. The average number of bits needed per symbol is the number for the base letter plus the number for the accent, which is

$$H^{(2)}(\mathbf{w}) + \sum_{i=1}^{26} w_i H^{(2)}(\mathbf{p}^i). \tag{2.8}$$

On the other hand, we saw in Example 2.1.5 that the overall frequency distribution of the French symbols a, à, â, b, ..., z is $\mathbf{w} \circ (\mathbf{p}^1, \ldots, \mathbf{p}^n)$, whose ideal code uses

$$H^{(2)}(\mathbf{w} \circ (\mathbf{p}^1, \ldots, \mathbf{p}^n)) \tag{2.9}$$

bits per symbol. If we have reasoned correctly then the expressions (2.8) and (2.9) should be equal. The chain rule states that, indeed, they are.

2.4 Entropy in Terms of Diversity

Entropies of various kinds have been used to measure biological diversity for almost as long as diversity measures have been considered. For instance, among all the measures of diversity used by ecologists, one of the most common is the Shannon entropy $H(\mathbf{p})$. Here $\mathbf{p} = (p_1, \ldots, p_n)$ is the relative abundance distribution of the community concerned, as in Example 2.1.1. For reasons that will be explained, when it comes to measuring diversity, it is better to use the *exponential* of entropy than entropy itself.

Let us begin by considering intuitively what it means for a community of n species to be diverse, for a fixed value of n. As described in the Introduction, there is a spectrum of viewpoints on what the word 'diversity' should mean. Loosely, though, diversity is low when most of the population is concentrated into one or two very common species, and high when the population is spread evenly across all species. Another way to say this is that diversity is low when an individual chosen at random usually belongs to a common species, and high when an individual chosen at random usually belongs to a rare species. So, the diversity of a community can be understood as the average rarity of an individual belonging to it.

Since p_i represents the relative abundance of the ith species, $1/p_i$ is a measure of its rarity or specialness. We want to take the average rarity, and for now we will use the geometric mean as our notion of average. (Later, we will use different notions of average. The most important are the power means, which are introduced in Section 4.2 and include the geometric mean.) Thus, one reasonable measure of the diversity of a community is the geometric mean of the species rarities $1/p_1, \ldots, 1/p_n$, weighted by the species sizes $p_1, \ldots, p_n$:

$$\left(\frac{1}{p_1}\right)^{p_1} \cdots \left(\frac{1}{p_n}\right)^{p_n}.$$

We therefore make the following definition.

Definition 2.4.1 Let $n \geq 1$ and $\mathbf{p} \in \Delta_n$. The **diversity of order** 1 of $\mathbf{p}$ is

$$D(\mathbf{p}) = \frac{1}{p_1^{p_1} p_2^{p_2} \cdots p_n^{p_n}},$$

with the convention that $0^0 = 1$.

Equivalently,

$$D(\mathbf{p}) = \prod_{i \in \text{supp}(\mathbf{p})} p_i^{-p_i} = e^{H(\mathbf{p})}.$$

In short: diversity is the exponential of entropy.

Remarks 2.4.2 i. The meaning of 'order 1' will be revealed in Section 4.3. It is related to the different possible notions of average. In this section, 'diversity' will always mean diversity of order 1.

ii. No choice of base is involved in the definition of D, in contrast to the situation for H (Remark 2.2.1). For instance, $D(\mathbf{p})$ is equal to both $e^{H(\mathbf{p})}$ and $2^{H^{(2)}(\mathbf{p})}$.

Crucially, the word 'diversity' refers only to the *relative*, not absolute, abundances. If half of a forest burns down, or if a patient loses 90% of their gut bacteria, then it may be an ecological or medical disaster; but assuming that the system is well-mixed, the diversity does not change. In the language of physics, diversity is an intensive quantity (like density or temperature) rather than an extensive quantity (like mass or heat), meaning that it is independent of the system's size.

Lemma 2.2.4 immediately implies the following.

Lemma 2.4.3 *Let* $n \geq 1$.

i. $D(\mathbf{p}) \geq 1$ *for all* $\mathbf{p} \in \Delta_n$, *with equality if and only if* $p_i = 1$ *for some* $i \in \{1, \ldots, n\}$.

ii. $D(\mathbf{p}) \leq n$ *for all* $\mathbf{p} \in \Delta_n$, *with equality if and only if* $\mathbf{p} = \mathbf{u}_n$. □

Similarly, the continuity of entropy (Lemma 2.2.5) implies:

Lemma 2.4.4 *For each* $n \geq 1$, *the diversity function* $D\colon \Delta_n \to \mathbb{R}$ *of order* 1 *is continuous.* □

Evidently

$$D(\mathbf{u}_n) = n$$

for all $n \geq 1$. This is a very important property for a diversity measure, and we adopt the standard terminology for it.

Definition 2.4.5 Let $(E\colon \Delta_n \to (0, \infty))_{n\geq 1}$ be a sequence of functions. Then E is an **effective number** if $E(\mathbf{u}_n) = n$ for all $n \geq 1$.

Thus, D is an effective number. When the species are all present in equal quantities, we think of the community as containing n fully present species and assign it a diversity value of n. On the other hand, if one species accounts for nearly 100% of the community and all the others are very rare, then the diversity value is barely more than 1 (by Lemmas 2.4.3(i) and 2.4.4). Effectively, there is barely more than one species present.

For instance, if a community has a diversity of 18.2, then the community is slightly more diverse than a community of 18 equally abundant species. There are 'effectively' slightly more than 18 balanced species.

Examples 2.4.6 For the four distributions on $\{1, 2, 3, 4\}$ in Example 2.2.2, the diversities are

$$2^2 = 4, \quad 2^{7/4} \approx 3.364, \quad 2^1 = 2, \quad 2^0 = 1,$$

respectively. In particular, the community represented by the second distribution is judged by D to be somewhat more diverse than a community of three species in equal proportions, but less diverse than a balanced community of four species.

Despite the popularity of Shannon entropy as a measure of biological diversity, many ecologists have argued that it should be rejected in favour of its exponential, including MacArthur [237] in 1965, Buzas and Gibson [56] in 1969, and Whittaker [352] in 1972. More recently and more generally, Jost [166, 167, 169] has argued convincingly that when measuring diversity, we should only use effective numbers. (That principle appears to be gaining acceptance, judging by the editorial [91] of Ellison.) The following example is adapted from Jost [167].

Example 2.4.7 Suppose that a plague strikes a continent of a million equally common species, rendering 90% of the species extinct and leaving the remaining 10% untouched. How do H and D respond to this catastrophe?

The Shannon entropy H drops by just

$$1 - \frac{\log(10^5)}{\log(10^6)} = \frac{1}{6} \approx 17\%,$$

suggesting a change of considerably smaller magnitude than the one that actually occurred. For comparison, if a community of four equally common species loses only *one* of its species, the rest remaining unchanged, this causes a drop in Shannon entropy of

$$1 - \frac{\log 3}{\log 4} \approx 21\%.$$

So, if we judge by percentage change in Shannon entropy, losing 25% of four species destroys a greater proportion of the diversity than losing 90% of a million species. Shannon entropy drops *more* in the situation where the species loss is *less*. So as an indicator of change in diversity, percentage change in Shannon entropy is plainly unsuitable.

However, the effect of the plague on the diversity D is to make it drop by 90% (from 10^6 to 10^5), because D is an effective number. And for the same reason, in the four-species example, D drops by 25% (from 4 to 3). This is intuitively reasonable behaviour, faithfully reflecting the scale of the change.

In information and coding theory, the logarithmic measure H is the more useful form, corresponding as it does to the number of bits per symbol in an ideal code. But for species diversity, it is the number of species (not its logarithm) with which we reason most naturally.

We now consider the chain rule in terms of diversity. Taking exponentials in Proposition 2.2.8 gives the following.

Corollary 2.4.8 *Let $n, k_1, \ldots, k_n \geq 1$. Then*

$$D(\mathbf{w} \circ (\mathbf{p}^1, \ldots, \mathbf{p}^n)) = D(\mathbf{w}) \cdot \prod_{i=1}^{n} D(\mathbf{p}^i)^{w_i}$$

for all $\mathbf{w} \in \Delta_n$ and $\mathbf{p}^i \in \Delta_{k_i}$. □

The second factor on the right-hand side is the geometric mean of the diversities $D(\mathbf{p}^1), \ldots, D(\mathbf{p}^n)$, weighted by $w_1, \ldots, w_n$.

The most important aspect of this result is not the specific formula, but the fact that the diversity of the composite distribution depends only on $\mathbf{w}$ and $D(\mathbf{p}^1), \ldots, D(\mathbf{p}^n)$, *not* on $\mathbf{p}^1, \ldots, \mathbf{p}^n$ themselves. This can be understood in either of the following ways.

Example 2.4.9 As in Example 2.1.6, consider a group of n islands of relative sizes $w_1, \ldots, w_n$, with no species shared between islands. Let d_i denote $D(\mathbf{p}^i)$, the diversity of the ith island. Then the diversity of the whole island group is

$$D(\mathbf{w}) \cdot d_1^{w_1} \cdots d_n^{w_n}. \tag{2.10}$$

Thus, the diversity of the whole island group is determined by the diversities and relative sizes of the islands. It can be computed without reference to the population distributions on each island.

Example 2.4.10 As in Example 2.1.7, consider a community of n genera, with the ith genus divided into k_i species. Let $\mathbf{w}$ denote the genus distribution and d_i the diversity of the species in the ith genus. Then the species diversity of the whole community is again given by (2.10). For instance, if there are 2 equally abundant genera, with the first genus consisting of 45 species of equal abundance and the second consisting of 5 species of equal abundance, then the diversity of the whole community is

$$D(\mathbf{u}_2 \circ (\mathbf{u}_{45}, \mathbf{u}_5)) = D(\mathbf{u}_2) \cdot D(\mathbf{u}_{45})^{1/2} D(\mathbf{u}_5)^{1/2} = 2\sqrt{45}\sqrt{5} = 30.$$

In other words, the whole community of $45+5 = 50$ species, which has relative abundance distribution

$$\Big(\underbrace{\tfrac{1}{90}, \ldots, \tfrac{1}{90}}_{45}, \underbrace{\tfrac{1}{10}, \ldots, \tfrac{1}{10}}_{5}\Big),$$

has the same diversity as a community of 30 species of equal abundance.

Different chain rules will appear in Sections 4.3 and 6.2, where we consider diversity of orders other than 1. But all share the crucial property that $D(\mathbf{w} \circ (\mathbf{p}^1, \ldots, \mathbf{p}^n))$ depends only on $\mathbf{w}$ and $D(\mathbf{p}^1), \ldots, D(\mathbf{p}^n)$.

We refer to this property of D as **modularity**. The word is used here in the sense of modular software design, buildings or furniture (as opposed to modular arithmetic or modules over a ring, say). In this metaphor, the islands of Example 2.4.9 or the genera of Example 2.4.10 are the 'modules': when it comes to computing the diversity of the whole assemblage, they are black boxes whose internal features we do not need to know.

The logarithmic property of H (Corollary 2.2.10) translates into a multiplicative property of D:

$$D(\mathbf{w} \otimes \mathbf{p}) = D(\mathbf{w}) \cdot D(\mathbf{p}) \tag{2.11}$$

($n, k \geq 1$, $\mathbf{w} \in \Delta_n$, $\mathbf{p} \in \Delta_k$). An important special case is the **replication principle**:

$$D(\mathbf{u}_n \otimes \mathbf{p}) = nD(\mathbf{p})$$

($n, k \geq 1$, $\mathbf{p} \in \Delta_k$). In the language of Example 2.4.9, this principle states that given n islands of equal size and the same species distributions, but with no actual shared species, the diversity of the whole island group is n times the diversity of any individual island.

Another argument of Jost (adapted from [169] and [171]) makes a compelling case for the importance of the replication principle:

Example 2.4.11 An oil company is planning to carry out work on a group of islands that will destroy all wildlife on half of the islands. Environmentalists are bringing a legal case to stop them. What would be the impact of the work on biodiversity?

Suppose that there are 16 equally sized islands in the group, that there are no species shared between islands, and that each island has diversity 4. Then before the oil work, the diversity of the island group is

$$16 \times 4 = 64.$$

Afterwards, similarly, it will be 32. Thus, the diversity is reduced by 50%. This is intuitively reasonable, and is a consequence of the replication principle for D.

However, one of the most popular measures of diversity in ecology is Shannon entropy ('many long-term investigations have chosen it as their benchmark of biological diversity', Magurran [240], p. 101). The oil company's lawyers can therefore argue as follows. Before the works, the 'diversity' (Shannon entropy) is $\log 64$, and afterwards, it will be $\log 32$. Thus, the proportion of diversity preserved is

$$\frac{\log 32}{\log 64} = \frac{5}{6} \approx 83\%.$$

On the other hand, the environmentalists' lawyers can argue that the islands whose wildlife is to be exterminated have a diversity of $\log 32$, out of a total of $\log 64$, so the proportion of diversity destroyed will be

$$\frac{\log 32}{\log 64} = \frac{5}{6} \approx 83\%.$$

So the oil company can truthfully claim that by the scientifically accepted measure, 83% of the diversity will be preserved, while the environmentalists can just as legitimately claim that 83% of the diversity will be lost. They cannot both be right, and, of course, both are wrong: by any reasonable measure, 50% of the diversity is preserved and 50% is lost. The reason for the contradictory and illogical conclusions is that Shannon entropy does not satisfy the replication principle.

Although this is an idealized hypothetical example, it is not hard to see how a choice of diversity measure, far from being some obscure theoretical issue, could have genuine environmental consequences.

Although the diversity measure D does satisfy the replication principle, and in that sense behaves logically, it has a glaring deficiency: it takes no notice of the varying similarities between species. A forest consisting of ten equally abundant species of larch is intuitively less diverse than a forest of ten equally abundant but highly varied tree species. However, the measure D gives the same diversity to both. The same criticism can be levelled at most of the diversity measures used in ecology, and a remedy is presented in Chapter 6.

2.5 The Chain Rule Characterizes Entropy

There are many characterizations of Shannon entropy, beginning with one in the original paper by Shannon himself ([309], Theorem 2). Here, we prove a variant of one of the best-known such theorems, due to Dmitry Faddeev [96].

Theorem 2.5.1 (Faddeev) *Let $(I\colon \Delta_n \to \mathbb{R})_{n\geq 1}$ be a sequence of functions. The following are equivalent:*

i. *the functions I are continuous and satisfy the chain rule*

$$I(\mathbf{w} \circ (\mathbf{p}^1, \ldots, \mathbf{p}^n)) = I(\mathbf{w}) + \sum_{i=1}^{n} w_i I(\mathbf{p}^i)$$

$(n, k_1, \ldots, k_n \geq 1,\ \mathbf{w} \in \Delta_n,\ \mathbf{p}^i \in \Delta_{k_i})$;

ii. *$I = cH$ for some $c \in \mathbb{R}$.*

In other words, up to a constant factor, entropy is uniquely characterized by the chain rule and continuity. We already know that (ii) implies (i); the challenge is to show that (i) implies (ii).

Remarks 2.5.2 i. As noted in Remark 2.2.1, the appearance of the constant factor should not be a surprise. We could eliminate it by adding the axiom that $I(\mathbf{u}_2) = \log 2$, for instance.

ii. The theorem that Faddeev proved in [96] was slightly different. He assumed that I was symmetric, that is, unchanged when the arguments $p_1, \ldots, p_n$ are permuted, but he assumed only the superficially simpler form of the chain rule stated as equation (2.7) (Remark 2.2.11). As noted in that remark, if we assume symmetry then the two forms of the chain rule are equivalent via a straightforward induction (Appendix A.1). On the other hand, Theorem 2.5.1 tells us that if we assume the chain rule in its general form then we do not need symmetry. This is not an obvious consequence of Faddeev's original theorem.

iii. If we assume symmetry, the hypotheses of Faddeev's original theorem can be weakened in a different direction, replacing continuity by measurability. This is a 1964 theorem of Lee [206]. We return to Lee's theorem at the end of Chapter 11, but omit the proof.

iv. It is not possible to prove a Faddeev-type theorem with no regularity conditions at all (unless one drops the axiom of choice). Indeed, let $f\colon \mathbb{R} \to \mathbb{R}$ be an additive nonlinear function, as in Remark 1.1.9. Then the assignment

$$\mathbf{p} \mapsto - \sum_{i \in \operatorname{supp}(\mathbf{p})} p_i f(\log p_i)$$

satisfies the chain rule but is not a scalar multiple of Shannon entropy.

The remainder of this section is devoted to the proof of Theorem 2.5.1. ***For the rest of this section***, let $(I\colon \Delta_n \to \mathbb{R})_{n \geq 1}$ be a sequence of continuous functions satisfying the chain rule.

The strategy of the proof is to show that I is proportional to H on successively larger classes of probability distributions. First we prove it for the uniform distributions $\mathbf{u}_n$, using the results on logarithmic sequences in Section 1.2. This forms the bulk of the proof. It is then relatively easy to extend the result to distributions $\mathbf{p}$ for which each p_i is a positive rational number, and from there, by continuity, to all distributions.

We begin by studying the real sequence $(I(\mathbf{u}_n))_{n \geq 1}$.

Lemma 2.5.3 *i.* $I(\mathbf{u}_{mn}) = I(\mathbf{u}_m) + I(\mathbf{u}_n)$ *for all* $m, n \geq 1$.
ii. $I(\mathbf{u}_1) = 0$.

Proof By the chain rule, I has the logarithmic property

$$I(\mathbf{w} \otimes \mathbf{p}) = I(\mathbf{w} \circ (\mathbf{p}, \ldots, \mathbf{p})) = I(\mathbf{w}) + I(\mathbf{p})$$

($\mathbf{w} \in \Delta_m$, $\mathbf{p} \in \Delta_n$). In particular, for all $m, n \geq 1$,

$$I(\mathbf{u}_{mn}) = I(\mathbf{u}_m \otimes \mathbf{u}_n) = I(\mathbf{u}_m) + I(\mathbf{u}_n),$$

proving (i). For (ii), take $m = n = 1$ in (i). □

As we saw in Section 1.2, the property $I(\mathbf{u}_{mn}) = I(\mathbf{u}_m) + I(\mathbf{u}_n)$ alone does not tell us very much about the sequence $(I(\mathbf{u}_n))$. To take advantage of the results in that section, we will need to prove some analytic condition on the sequence. Specifically, we will show that $I(\mathbf{u}_{n+1}) - I(\mathbf{u}_n) \to 0$ as $n \to \infty$, then apply Corollary 1.2.3.

Lemma 2.5.4 $I(1,0) = 0.$

Proof We compute $I(1,0,0)$ in two ways. On the one hand, using the chain rule,

$$I(1,0,0) = I\big((1,0) \circ ((1,0), \mathbf{u}_1)\big) = I(1,0) + 1 \cdot I(1,0) + 0 \cdot I(\mathbf{u}_1) = 2I(1,0).$$

On the other, using the chain rule again and the fact that $I(\mathbf{u}_1) = 0$,

$$I(1,0,0) = I\big((1,0) \circ (\mathbf{u}_1, (1,0))\big) = I(1,0) + 1 \cdot I(\mathbf{u}_1) + 0 \cdot I(1,0) = I(1,0).$$

Hence $I(1,0) = 0$. □

Lemma 2.5.5 *$I(\mathbf{u}_{n+1}) - \frac{n}{n+1} I(\mathbf{u}_n) \to 0$ as $n \to \infty$.*

Proof We have

$$\mathbf{u}_{n+1} = \left(\frac{n}{n+1}, \frac{1}{n+1}\right) \circ (\mathbf{u}_n, \mathbf{u}_1),$$

so by the chain rule and the fact that $I(\mathbf{u}_1) = 0$,

$$I(\mathbf{u}_{n+1}) = I\left(\frac{n}{n+1}, \frac{1}{n+1}\right) + \frac{n}{n+1} I(\mathbf{u}_n).$$

Hence

$$I(\mathbf{u}_{n+1}) - \frac{n}{n+1} I(\mathbf{u}_n) = I\left(\frac{n}{n+1}, \frac{1}{n+1}\right) \to I(1,0) = 0$$

as $n \to \infty$, by continuity and Lemma 2.5.4. □

Now we can use the results of Section 1.2.

Lemma 2.5.6 *There exists a constant $c \in \mathbb{R}$ such that $I(\mathbf{u}_n) = cH(\mathbf{u}_n)$ for all $n \geq 1$.*

Proof By Lemma 2.5.3(i), the sequence $(I(\mathbf{u}_n))$ is logarithmic. By Lemmas 2.5.5 and 1.2.4, $\lim_{n\to\infty}(I(\mathbf{u}_{n+1}) - I(\mathbf{u}_n)) = 0$. Hence by Corollary 1.2.3, there is some $c \in \mathbb{R}$ such that for all $n \geq 1$,

$$I(\mathbf{u}_n) = c \log n = cH(\mathbf{u}_n).$$ □

We now move to the second phase of the proof of Theorem 2.5.1. Let c be the constant of Lemma 2.5.6 (which is uniquely determined).

Lemma 2.5.7 *Let $\mathbf{p} \in \Delta_n$ with $p_1, \ldots, p_n$ rational and nonzero. Then $I(\mathbf{p}) = cH(\mathbf{p})$.*

Proof We can write

$$\mathbf{p} = \left(\frac{k_1}{k}, \ldots, \frac{k_n}{k}\right)$$

for some positive integers $k_1, \ldots, k_n$, where $k = k_1 + \cdots + k_n$. Then

$$\mathbf{p} \circ (\mathbf{u}_{k_1}, \ldots, \mathbf{u}_{k_n}) = \mathbf{u}_k.$$

Since I satisfies the chain rule and $I(\mathbf{u}_r) = cH(\mathbf{u}_r)$ for all $r \geq 1$, we have

$$I(\mathbf{p}) + \sum_{i=1}^{n} p_i \cdot cH(\mathbf{u}_{k_i}) = cH(\mathbf{u}_k).$$

But since cH satisfies the chain rule too, we also have

$$cH(\mathbf{p}) + \sum_{i=1}^{n} p_i \cdot cH(\mathbf{u}_{k_i}) = cH(\mathbf{u}_k).$$

The result follows. □

The third and final phase of the proof is trivial: since the probability distributions with positive rational probabilities are dense in the space Δ_n of all probability distributions, and since I and cH are continuous functions agreeing on this dense set, they are equal everywhere. This proves Theorem 2.5.1.

Like any result on entropy, Faddeev's theorem can be translated into diversity terms. In the following corollary, we eliminate the arbitrary constant factor by requiring that E be an effective number.

Corollary 2.5.8 *Let* $(E\colon \Delta_n \to (0, \infty))_{n \geq 1}$ *be a sequence of functions. The following are equivalent:*

i. *the functions E are continuous and satisfy the chain rule*

$$E(\mathbf{w} \circ (\mathbf{p}^1, \ldots, \mathbf{p}^n)) = E(\mathbf{w}) \cdot \prod_{i=1}^{n} E(\mathbf{p}^i)^{w_i} \tag{2.12}$$

($n, k_1, \ldots, k_n \geq 1$, $\mathbf{w} \in \Delta_n$, $\mathbf{p}^i \in \Delta_{k_i}$), and E is an effective number;

ii. $E = D$.

Proof By Faddeev's theorem applied to $\log E$, the sequences of continuous functions E satisfying the diversity chain rule (2.12) are exactly the real powers D^c ($c \in \mathbb{R}$). But the effective number property (or indeed, the single equation $E(\mathbf{u}_2) = 2$) then forces $c = 1$. □

3

Relative Entropy

The notion of relative entropy allows us to compare two probability distributions on the same space. More specifically, for each pair of probability distributions $\mathbf{p}, \mathbf{r}$ on the same finite set, there is defined a real number $H(\mathbf{p} \parallel \mathbf{r}) \geq 0$, the entropy of $\mathbf{p}$ relative to $\mathbf{r}$. It is zero just when $\mathbf{p} = \mathbf{r}$. It extends the definition of Shannon entropy, in the sense that the Shannon entropy of a single distribution $\mathbf{p}$ on $\{1, \ldots, n\}$ is a function of $H(\mathbf{p} \parallel \mathbf{u}_n)$, the entropy of $\mathbf{p}$ relative to the uniform distribution.

Relative entropy goes by a remarkable number of names, attesting to its wide variety of interpretations and uses. It is also known as Kullback–Leibler information (as in [305], for instance), Kullback–Leibler distance [71], Kullback–Leibler divergence [173], directed divergence [198], information divergence [129], information deficiency [46], amount of information [294], discrimination information [199], relative information [324], gain of information or information gain ([295], Section IX.4), discrimination distance [180], and error [186], among others. This chapter provides multiple explanations and applications of relative entropy, as well as a theorem pinpointing what makes relative entropy uniquely useful.

Our first explanation of relative entropy is in terms of coding (Section 3.2). As we saw in Section 2.3, the Shannon entropy of $\mathbf{p}$ gives the average number of bits per symbol needed to encode an alphabet with frequency distribution $\mathbf{p}$ in a coding system optimized for that purpose. In a similar sense, $H(\mathbf{p} \parallel \mathbf{r})$ measures the *extra* number of bits per symbol needed to encode an alphabet with frequencies $\mathbf{p}$ using a coding system that was optimized for the frequency distribution $\mathbf{r}$. In other words, it is the penalty for using the wrong system.

The exponential of relative entropy is called relative diversity (Section 3.3). Often we have a preconceived idea of what an ordinary or default distribution of species is, and we judge how unusual a community is relative to that expectation. For instance, if we were assessing the diversity of flowering plants in a particular region of the island of Tasmania, we would naturally judge it by the standards of Tasmania as a whole. The relative diversity $\exp(H(\mathbf{p} \,\|\, \mathbf{r}))$ reflects the unusualness of a community with distribution $\mathbf{p}$ relative to a reference distribution $\mathbf{r}$.

Section 3.4 gives short accounts of roles played by relative entropy in three other subjects. In measure theory, we find that the definition of relative entropy generalizes easily from finite sets to arbitrary measurable spaces, while ordinary Shannon entropy does not. The slogan is: *all entropy is relative*. In geometry, although $H(- \,\|\, -)$ does not define a distance function on the set Δ_n of distributions, it turns out that infinitesimally, it behaves like the square of a distance. We can extend this infinitesimal metric to a global metric in the manner of Riemannian geometry. In statistics, the second argument $\mathbf{r}$ of $H(\mathbf{p} \,\|\, \mathbf{r})$ should be thought of as a prior, and maximizing likelihood can be reinterpreted as minimizing relative entropy. The concept of relative entropy also gives rise to the notions of Fisher information and the Jeffreys prior, an objective prior distribution in the sense of Bayesian statistics.

We finish the chapter with a characterization theorem for relative entropy (Section 3.5), which first appeared in [218]. Just as for Faddeev's characterization of ordinary entropy, the main characterizing property is a chain rule. And just as for ordinary entropy, many characterization theorems for relative entropy have previously been proved; but the one presented here appears to be the simplest yet.

3.1 Definition and Properties of Relative Entropy

This short section presents the definition and basic properties of relative entropy, without motivation for now. The later sections provide multiple interpretations of, and justifications for, the definition.

Definition 3.1.1 Let $n \geq 1$ and $\mathbf{p}, \mathbf{r} \in \Delta_n$. The **entropy of p relative to r** is

$$H(\mathbf{p} \,\|\, \mathbf{r}) = \sum_{i \in \mathrm{supp}(\mathbf{p})} p_i \log \frac{p_i}{r_i}. \tag{3.1}$$

If there is some i such that $p_i > 0 = r_i$ then $H(\mathbf{p} \,\|\, \mathbf{r})$ is defined to be ∞.

In the literature, relative entropy is more often denoted by $D(\mathbf{p} \,\|\, \mathbf{r})$, but in this text, we reserve the letter D for measures of diversity.

Example 3.1.2 Let $\mathbf{p} \in \Delta_n$. Then

$$\begin{aligned} H(\mathbf{p} \,\|\, \mathbf{u}_n) &= \sum_{i \in \mathrm{supp}(\mathbf{p})} p_i \log(np_i) \\ &= \log n - H(\mathbf{p}) \\ &= H(\mathbf{u}_n) - H(\mathbf{p}). \end{aligned}$$

Thus, ordinary entropy is essentially a special case of relative entropy.

Example 3.1.3 As well as sometimes taking the value ∞, relative entropy can take arbitrarily large finite values (even for fixed n). For instance, for $t \in (0, 1)$,

$$H(\mathbf{u}_2 \,\|\, (t, 1-t)) = \frac{1}{2} \log \frac{1}{2t} + \frac{1}{2} \log \frac{1}{2(1-t)} \to \infty$$

as $t \to 0$.

Unless $\mathbf{p} = \mathbf{r}$, there are some values of i for which $p_i > r_i$ and others for which $p_i < r_i$. Hence, some of the summands in (3.1) are positive and others are negative. Nevertheless:

Lemma 3.1.4 *$H(\mathbf{p} \,\|\, \mathbf{r}) \geq 0$, with equality if and only if $\mathbf{p} = \mathbf{r}$.*

Proof If $p_i > 0 = r_i$ for some i then $H(\mathbf{p} \,\|\, \mathbf{r}) = \infty$. Suppose otherwise, so that $\mathrm{supp}(\mathbf{p}) \subseteq \mathrm{supp}(\mathbf{r})$. Using Lemma 2.2.3,

$$\begin{aligned} H(\mathbf{p} \,\|\, \mathbf{r}) &= - \sum_{i \in \mathrm{supp}(\mathbf{p})} p_i \log \frac{r_i}{p_i} \\ &\geq - \log \left(\sum_{i \in \mathrm{supp}(\mathbf{p})} p_i \frac{r_i}{p_i} \right) \\ &\geq - \log \left(\sum_{i \in \mathrm{supp}(\mathbf{r})} r_i \right) \\ &= - \log 1 = 0, \end{aligned}$$

with equality in the first inequality if and only if $r_i/p_i = r_j/p_j$ for all $i, j \in \mathrm{supp}(\mathbf{p})$. Equality holds in the second inequality if and only if $\mathrm{supp}(\mathbf{p}) = \mathrm{supp}(\mathbf{r})$. Hence for equality to hold throughout, there must be some constant α such that $r_i = \alpha p_i$ for all $i \in \mathrm{supp}(\mathbf{p}) = \mathrm{supp}(\mathbf{r})$. But since $\sum_{i \in \mathrm{supp}(\mathbf{p})} p_i = 1 = \sum_{i \in \mathrm{supp}(\mathbf{r})} r_i$, this forces $\alpha = 1$ and so $\mathbf{p} = \mathbf{r}$. □

Lemma 3.1.4 suggests that *very* roughly speaking, $H(\mathbf{p} \,\|\, \mathbf{r})$ can be understood as a kind of distance between $\mathbf{p}$ and $\mathbf{r}$. However, relative entropy does not satisfy the triangle inequality (Example 3.4.2). Nor is it symmetric: for as Examples 3.1.2 and 3.1.3 show, $H(\mathbf{p} \,\|\, \mathbf{u}_2) \leq \log 2$ for all $\mathbf{p} \in \Delta_2$, whereas

$H(\mathbf{u}_2 \,\|\, \mathbf{p})$ can be arbitrarily large. We will return to the interpretation of relative entropy as a measure of distance in Section 3.4.

We now list some of the basic properties of relative entropy. Matters are simplified if we restrict to just those pairs $(\mathbf{p}, \mathbf{r})$ such that $H(\mathbf{p} \,\|\, \mathbf{r}) < \infty$. For $n \geq 1$, write

$$\begin{aligned} A_n &= \{(\mathbf{p}, \mathbf{r}) \in \Delta_n \times \Delta_n : r_i = 0 \implies p_i = 0\} \\ &= \{(\mathbf{p}, \mathbf{r}) \in \Delta_n \times \Delta_n : \operatorname{supp}(\mathbf{p}) \subseteq \operatorname{supp}(\mathbf{r})\}. \end{aligned}$$

Then

$$H(\mathbf{p} \,\|\, \mathbf{r}) < \infty \iff (\mathbf{p}, \mathbf{r}) \in A_n.$$

So for each $n \geq 1$, we have the function

$$\begin{array}{rccc} H(- \,\|\, -)\colon & A_n & \to & \mathbb{R} \\ & (\mathbf{p}, \mathbf{r}) & \mapsto & H(\mathbf{p} \,\|\, \mathbf{r}). \end{array}$$

This sequence of functions has the following properties, among others.

Measurability in the second argument For each fixed $\mathbf{p} \in \Delta_n$, the function

$$\begin{array}{ccc} \{\mathbf{r} \in \Delta_n : (\mathbf{p}, \mathbf{r}) \in A_n\} & \to & \mathbb{R} \\ \mathbf{r} & \mapsto & H(\mathbf{p} \,\|\, \mathbf{r}) \end{array}$$

is measurable. Indeed, the function $H(- \,\|\, -)\colon A_n \to \mathbb{R}$ is continuous, but for the unique characterization of relative entropy proved in Section 3.5, measurability in the second argument is all we will need.

Permutation-invariance The relative entropy $H(\mathbf{p} \,\|\, \mathbf{r})$ is unchanged if the same permutation is applied to the indices of both $\mathbf{p}$ and $\mathbf{r}$. That is,

$$H(\mathbf{p} \,\|\, \mathbf{r}) = H(\mathbf{p}\sigma \,\|\, \mathbf{r}\sigma)$$

for all $(\mathbf{p}, \mathbf{r}) \in A_n$ and permutations σ of $\{1, \ldots, n\}$, where

$$\mathbf{p}\sigma = (p_{\sigma(1)}, \ldots, p_{\sigma(n)}) \tag{3.2}$$

and similarly $\mathbf{r}\sigma$.

Vanishing $H(\mathbf{p} \,\|\, \mathbf{p}) = 0$ for all $\mathbf{p} \in \Delta_n$.

Chain rule Let $n, k_1, \ldots, k_n \geq 1$ and

$$(\mathbf{w}, \widetilde{\mathbf{w}}) \in A_n, \ (\mathbf{p}^1, \widetilde{\mathbf{p}}^1) \in A_{k_1}, \ \ldots, \ (\mathbf{p}^n, \widetilde{\mathbf{p}}^n) \in A_{k_n}.$$

Then

$$H(\mathbf{w} \circ (\mathbf{p}^1, \ldots, \mathbf{p}^n) \,\|\, \widetilde{\mathbf{w}} \circ (\widetilde{\mathbf{p}}^1, \ldots, \widetilde{\mathbf{p}}^n)) = H(\mathbf{w} \,\|\, \widetilde{\mathbf{w}}) + \sum_{i=1}^n w_i H(\mathbf{p}^i \,\|\, \widetilde{\mathbf{p}}^i). \tag{3.3}$$

This is a straightforward check, similar to Proposition 2.2.8. Note that

$$(\mathbf{w} \circ (\mathbf{p}^1, \dots, \mathbf{p}^n), \widetilde{\mathbf{w}} \circ (\widetilde{\mathbf{p}}^1, \dots, \widetilde{\mathbf{p}}^n)) \in A_{k_1 + \cdots + k_n},$$

so the relative entropy of this pair is guaranteed to be finite.

As a special case, relative entropy has a logarithmic property:

$$H(\mathbf{w} \otimes \mathbf{p} \,\|\, \widetilde{\mathbf{w}} \otimes \widetilde{\mathbf{p}}) = H(\mathbf{w} \,\|\, \widetilde{\mathbf{w}}) + H(\mathbf{p} \,\|\, \widetilde{\mathbf{p}}) \tag{3.4}$$

for all $(\mathbf{w}, \widetilde{\mathbf{w}}) \in A_n$ and $(\mathbf{p}, \widetilde{\mathbf{p}}) \in A_k$. This follows from the chain rule by taking $k_i = k$, $\mathbf{p}^i = \mathbf{p}$ and $\widetilde{\mathbf{p}^i} = \widetilde{\mathbf{p}}$ for all $i \in \{1, \dots, n\}$.

Just as for ordinary entropy, different choices of the base of the logarithm in the definition of relative entropy only change it by a constant factor. We will see in Section 3.5 that up to a constant factor, the four properties just listed characterize relative entropy uniquely.

3.2 Relative Entropy in Terms of Coding

We have already interpreted Shannon entropy in terms of coding (Section 2.3). Here we do the same for relative entropy.

To help our understanding, let us regard a probability distribution $\mathbf{p} \in \Delta_n$ as the frequency distribution of the n symbols in some human language, which we call **language p**. We make use of the convenient fiction introduced on p. 50, imagining that there exists an ideal code for language $\mathbf{p}$: a code whose average word length is exactly $H^{(2)}(\mathbf{p})$. We will suppose that the encoding is performed by a machine, called **machine p**. Although most distributions $\mathbf{p}$ have no ideal code, one can come arbitrarily close (as in Section 2.3), and this justifies the use of ideal codes as an explanatory device.

For $\mathbf{p} \in \Delta_n$, the ordinary base 2 entropy

$$H^{(2)}(\mathbf{p}) = \sum_{i \in \operatorname{supp}(\mathbf{p})} p_i \log_2 \frac{1}{p_i}$$

satisfies

$$H^{(2)}(\mathbf{p}) = \text{number of bits per symbol to encode language } \mathbf{p} \text{ using machine } \mathbf{p}.$$

Now let $\mathbf{p}, \mathbf{r} \in \Delta_n$, with $\mathbf{p}$ and $\mathbf{r}$ viewed as the frequency distributions of two languages on the same set of symbols. Write

$$H^{(2)}(\mathbf{p} \,\|\, \mathbf{r}) = \sum_{i \in \operatorname{supp}(\mathbf{p})} p_i \log_2 \frac{p_i}{r_i} = \frac{H(\mathbf{p} \,\|\, \mathbf{r})}{\log 2}$$

for the base 2 relative entropy. We will interpret $H^{(2)}(\mathbf{p}\|\mathbf{r})$ in terms of languages **p** and **r** and machines **p** and **r**.

To do this, first consider the quantity

$$H^{(2)\times}(\mathbf{p} \,\|\, \mathbf{r}) = \sum_{i \in \operatorname{supp}(\mathbf{p})} p_i \log_2 \frac{1}{r_i}.$$

Here $\log_2(1/r_i)$ is the number of bits that machine **r** uses to encode the ith symbol. (Of course, this is not usually an integer, but recall the comments on ideal codes on p. 51.) Hence

$$H^{(2)\times}(\mathbf{p} \,\|\, \mathbf{r}) = \text{bits per symbol to encode language } \mathbf{p} \text{ using machine } \mathbf{r}.$$

This quantity $H^{(2)\times}(\mathbf{p} \,\|\, \mathbf{r})$, or its base e analogue

$$H^{\times}(\mathbf{p} \,\|\, \mathbf{r}) = \sum_{i \in \operatorname{supp}(\mathbf{p})} p_i \log \frac{1}{r_i} = H^{(2)\times}(\mathbf{p} \,\|\, \mathbf{r}) \cdot \log 2, \tag{3.5}$$

is the **cross entropy** of **p** with respect to **r**.

The relative, cross and ordinary entropies are related by the equation

$$H(\mathbf{p} \,\|\, \mathbf{r}) = H^{\times}(\mathbf{p} \,\|\, \mathbf{r}) - H(\mathbf{p}). \tag{3.6}$$

Hence

$$\begin{aligned} H^{(2)}(\mathbf{p} \,\|\, \mathbf{r}) &= H^{(2)\times}(\mathbf{p} \,\|\, \mathbf{r}) - H^{(2)}(\mathbf{p}) \\ &= \text{bits per symbol to encode language } \mathbf{p} \text{ using machine } \mathbf{r} \\ &\quad - \text{bits per symbol to encode language } \mathbf{p} \text{ using machine } \mathbf{p}. \end{aligned}$$

So, for the task of encoding language **p**, the relative entropy $H^{(2)}(\mathbf{p} \,\|\, \mathbf{r})$ is the number of *extra* bits needed if one uses machine **r** instead of machine **p**. Machine **p** is ideal for the job: it is optimized for exactly this purpose. Relative entropy is, then, the penalty for using the wrong machine.

This provides an intuitive explanation of why $H(\mathbf{p}\|\mathbf{r})$ is always nonnegative and why $H(\mathbf{p} \,\|\, \mathbf{p}) = 0$. It also suggests why relative entropy can be arbitrarily large, as in the following example.

Examples 3.2.1 i. Consider an alphabet with $n = 2$ symbols. Suppose that language **p** uses the two symbols with equal frequency, and that in language **r** the frequency distribution is $(2^{-1000}, 1 - 2^{-1000})$. Then machine **r** encodes the first symbol with a word of 1000 bits. Since language **p** uses this symbol half the time, the average word length when encoding language **p** using machine **r** is at least 500 bits. This is drastically worse than when language **p** is encoded using the most suitable machine, machine **p**, which has an average word length of just 1 bit. So the relative entropy $H^{(2)}(\mathbf{p} \,\|\, \mathbf{r})$ is at least 499.

ii. The same example provides intuition for the fact that

$$H(\mathbf{p} \,\|\, \mathbf{r}) \neq H(\mathbf{r} \,\|\, \mathbf{p}).$$

Machine $\mathbf{p}$ encodes the two symbols of the alphabet as the binary words 0 and 1, of length 1 each. Hence the average number of bits used when encoding language $\mathbf{r}$ (or indeed, any other language) in machine $\mathbf{p}$ is 1. So $H^{(2)}(\mathbf{r} \,\|\, \mathbf{p})$ is less than 1, and is therefore much smaller than the value of $H^{(2)}(\mathbf{p} \,\|\, \mathbf{r})$ derived in (i).

Remark 3.2.2 The name 'cross entropy' has a tangled history. It was introduced by Jack Good in 1956 ([121], Section 6), who defined it as in equation (3.5) above and gave it its name. But later, Good used 'cross entropy' as a synonym for relative entropy ([122], p. 913), and others have done the same (Shore and Johnson [313], for instance). Nowadays the term is often used in the context of the cross entropy method in operational research [79]. In the broadest terms, this involves fixing a distribution $\mathbf{p}$ and minimizing $H(\mathbf{p} \,\|\, \mathbf{r})$, or equivalently $H^{\times}(\mathbf{p} \,\|\, \mathbf{r})$, among all $\mathbf{r}$ subject to certain constraints. It makes no difference which one minimizes, by equation (3.6). From that point of view, the concepts are essentially interchangeable, which has not helped to clarify the terminological situation either.

This text uses the term with its original meaning, in part because relative entropy already has an overabundance of synonyms.

The chain rule for relative entropy (equation (3.3)) can also be explained in terms of coding, as in the following example.

Example 3.2.3 In Example 2.3.11, we interpreted the chain rule for ordinary Shannon entropy in terms of letters and their accents in French. There are many dialects of French, using the same letters and accents but slightly different vocabulary, hence slightly different frequency distributions of both letters and accents. Here we consider Swiss and Canadian French, which for brevity we call just 'Swiss' and 'Canadian'.

Define distributions $\mathbf{w}, \widetilde{\mathbf{w}}, \mathbf{p}^i, \widetilde{\mathbf{p}}^i$ as follows:

$$\mathbf{w} \in \Delta_{26} : \text{frequency distribution of letters in Swiss}$$
$$\widetilde{\mathbf{w}} \in \Delta_{26} : \text{frequency distribution of letters in Canadian}$$

and then

$$\begin{aligned}
\mathbf{p}^1 &\in \Delta_3 : \text{frequency distribution of accents on } \mathsf{a} \text{ in Swiss}\\
\widetilde{\mathbf{p}}^1 &\in \Delta_3 : \text{frequency distribution of accents on } \mathsf{a} \text{ in Canadian}\\
&\vdots\\
\mathbf{p}^{26} &\in \Delta_1 : \text{frequency distribution of accents on } \mathsf{z} \text{ in Swiss}\\
\widetilde{\mathbf{p}}^{26} &\in \Delta_1 : \text{frequency distribution of accents on } \mathsf{z} \text{ in Canadian.}
\end{aligned}$$

So, recalling the convention that a 'symbol' is a letter plus a (possibly nonexistent) accent,

$$\begin{aligned}
\mathbf{w} \circ (\mathbf{p}^1, \ldots, \mathbf{p}^{26}) &= \text{frequency distribution of symbols in Swiss}\\
\widetilde{\mathbf{w}} \circ (\widetilde{\mathbf{p}}^1, \ldots, \widetilde{\mathbf{p}}^{26}) &= \text{frequency distribution of symbols in Canadian.}
\end{aligned}$$

Now suppose that we encode Swiss using the Canadian machine. How much extra does this cost (in bits/symbol) compared to encoding Swiss using the Swiss machine?

Since every symbol consists of a letter with an accent, we expect to have:

$$\begin{aligned}
&\text{mean extra cost per symbol} =\\
&\qquad\qquad \text{mean extra cost per letter} + \text{mean extra cost per accent.} \qquad (3.7)
\end{aligned}$$

The mean extra cost per symbol is

$$H^{(2)}(\mathbf{w} \circ (\mathbf{p}^1, \ldots, \mathbf{p}^{26}) \,\|\, \widetilde{\mathbf{w}} \circ (\widetilde{\mathbf{p}}^1, \ldots, \widetilde{\mathbf{p}}^{26})).$$

The mean extra cost per letter is

$$H^{(2)}(\mathbf{w} \,\|\, \widetilde{\mathbf{w}}).$$

The mean extra cost per accent is computed by conditioning on the letter that it decorates. Since it is Swiss rather than Canadian that we are encoding, the probability of the ith letter occurring is w_i, so the mean extra cost per accent is

$$\sum_{i=1}^{26} w_i H^{(2)}(\mathbf{p}^i \,\|\, \widetilde{\mathbf{p}}^i).$$

Hence the hoped for equation (3.7) predicts that

$$H^{(2)}(\mathbf{w} \circ (\mathbf{p}^1, \ldots, \mathbf{p}^{26}) \,\|\, \widetilde{\mathbf{w}} \circ (\widetilde{\mathbf{p}}^1, \ldots, \widetilde{\mathbf{p}}^{26})) = H^{(2)}(\mathbf{w} \,\|\, \widetilde{\mathbf{w}}) + \sum_{i=1}^{26} w_i H^{(2)}(\mathbf{p}^i \,\|\, \widetilde{\mathbf{p}}^i).$$

This is indeed true. It is exactly the chain rule of Section 3.1.

3.3 Relative Entropy in Terms of Diversity

In Section 2.4, we interpreted the exponential of Shannon entropy as the diversity of a biological community. Here we interpret the exponential of relative entropy as a measure of how diverse or atypical one community is when seen from the perspective of another. This interpretation elaborates on ideas of Reeve et al. [293].

As in Section 2.4, we consider communities of individuals drawn from n species, whose relative abundances define a probability distribution on the set $\{1, \ldots, n\}$.

Definition 3.3.1 Let $n \geq 1$ and $\mathbf{p}, \mathbf{r} \in \Delta_n$. The **diversity of p relative to r (of order** 1) is

$$D(\mathbf{p} \,\|\, \mathbf{r}) = e^{H(\mathbf{p}\|\mathbf{r})} = \prod_{i \in \mathrm{supp}(\mathbf{p})} \left(\frac{p_i}{r_i}\right)^{p_i} \in [1, \infty].$$

(We repeat the warning that although in the literature, the notation $D(\mathbf{p} \,\|\, \mathbf{r})$ is often used to mean relative *entropy*, we reserve the letter D for diversity.)

By Lemma 3.1.4, $D(\mathbf{p} \,\|\, \mathbf{r}) \geq 1$, with equality if and only if $\mathbf{p} = \mathbf{r}$.

It is helpful to regard $\mathbf{r}$ as the distribution of a reference community (a community that one considers to be normal or the default) and $\mathbf{p}$ as the distribution of the community in which we are primarily interested. As we will see, $D(\mathbf{p}\|\mathbf{r})$ measures how exotic or unusual this other community is from the viewpoint of the reference community.

To explain this, it is helpful to begin with another quantity.

Definition 3.3.2 Let $n \geq 1$ and $\mathbf{p}, \mathbf{r} \in \Delta_n$. The **cross diversity of p with respect to r (of order** 1) is

$$D^{\times}(\mathbf{p} \,\|\, \mathbf{r}) = e^{H^{\times}(\mathbf{p}\|\mathbf{r})} = \prod_{i \in \mathrm{supp}(\mathbf{p})} \left(\frac{1}{r_i}\right)^{p_i} \in [1, \infty].$$

In Section 2.4, the ordinary diversity of $\mathbf{p}$,

$$D(\mathbf{p}) = \prod_{i \in \mathrm{supp}(\mathbf{p})} \left(\frac{1}{p_i}\right)^{p_i},$$

was interpreted as follows: $1/p_i$ is the rarity of the ith species within the community, and $D(\mathbf{p})$ is therefore the average rarity of individuals in the community. (In this case, 'average' means geometric mean.) Cross diversity can be understood in a similar way. If we use the second community $\mathbf{r}$ as our reference point – the community by which others are to be judged – then we naturally take the rarity or specialness of the ith species to be $1/r_i$ rather than

$1/p_i$. Thus, $D^{\times}(\mathbf{p} \,\|\, \mathbf{r})$ is the average rarity of individuals in the first community, seen from the viewpoint of the second.

Since

$$D(\mathbf{p} \,\|\, \mathbf{r}) = \frac{D^{\times}(\mathbf{p} \,\|\, \mathbf{r})}{D(\mathbf{p})}, \tag{3.8}$$

the relative diversity measures how much *more* diverse the first community looks from the viewpoint of the second than from the viewpoint of itself. Some examples illuminate this interpretation.

Example 3.3.3 We have $D(\mathbf{p} \,\|\, \mathbf{p}) = 1$, which is the minimal possible value of relative diversity: any community perceives itself as completely normal.

Example 3.3.4 Let $\mathbf{p}$ and $\mathbf{r}$ be the relative abundance distributions of reptiles in Portugal and Russia, respectively. Geckos are commonplace in Portugal but rare in Russia. Hence from the Russian viewpoint, the ecology of Portugal seems exotic or atypical, in this respect at least.

Mathematically, there are several values of i (corresponding to species of gecko) such that r_i is small but p_i is not. This means that the cross diversity contains some large factors, $(1/r_i)^{p_i}$, and the relative diversity also contains some large factors, $(p_i/r_i)^{p_i}$. Thus, both the cross diversity $D^{\times}(\mathbf{p} \,\|\, \mathbf{r})$ and the relative diversity $D(\mathbf{p} \,\|\, \mathbf{r})$ are large, regardless of the diversity $D(\mathbf{p})$ of reptiles in Portugal.

Example 3.3.5 Taking the previous example to the extreme, if one or more species is present in the test community $\mathbf{p}$ but absent in the reference community $\mathbf{r}$ then $D(\mathbf{p} \,\|\, \mathbf{r}) = \infty$.

Example 3.3.6 Suppose now that we judge communities from the reference point of a community with a uniform distribution. (This is in some sense the canonical choice of reference, and is the one produced by the maximum entropy method of statistics [156, 51].) The cross diversity $D^{\times}(\mathbf{p} \,\|\, \mathbf{u}_n)$ is equal to n, regardless of $\mathbf{p}$. Hence equation (3.8) gives

$$D(\mathbf{p} \,\|\, \mathbf{u}_n) = \frac{n}{D(\mathbf{p})}. \tag{3.9}$$

This is also the exponential of the equation

$$H(\mathbf{p} \,\|\, \mathbf{u}_n) = \log n - H(\mathbf{p})$$

derived in Example 3.1.2.

Equation (3.9) implies that for a fixed number of species, the diversity of a community relative to the uniform distribution is inversely proportional to the intrinsic diversity of the community itself. From the viewpoint of a community in which all n species are balanced equally, any variation from this balance looks unusual – and the more unbalanced, the more unusual.

As an illustration of the general point, house sparrows are common throughout Britain, but a region of the country in which the *only* birds were house sparrows would be highly unusual. Correspondingly, the relative diversity $D(\mathbf{p} \,\|\, \mathbf{r})$ of that region relative to the country would be high, even though the intrinsic diversity $D(\mathbf{p})$ of the region would take the minimum possible value, 1.

By equation (3.9) and Lemma 2.4.3, $D(\mathbf{p} \,\|\, \mathbf{u}_n)$ takes its minimal value, 1, when $\mathbf{p} = \mathbf{u}_n$. It takes its maximal value, n, when

$$\mathbf{p} = (0, \ldots, 0, 1, 0, \ldots, 0).$$

That is, from the viewpoint of a completely balanced community, the most unusual possible community is one consisting of a single species.

Often, we want to assess a community from the viewpoint of a larger community that *contains* it. For instance, we are more likely to study the diversity of plankton in the eastern Mediterranean Sea with reference to the Mediterranean as a whole than with reference to the Arctic Ocean.

Consider, then, an ecological community with relative abundance distribution $\mathbf{r} \in \Delta_n$, and a subcommunity consisting of some of its organisms. Write π_i for the proportion of the community consisting of individuals that belong to both the subcommunity and the ith species. Then $0 \leq \pi_i \leq r_i$. The proportion of the whole community made up by the subcommunity is $w = \sum \pi_i \leq 1$, and the relative abundance distribution of the subcommunity is

$$\mathbf{p} = (\pi_1/w, \ldots, \pi_n/w) \in \Delta_n.$$

The inequality $\pi_i \leq r_i$ gives $wp_i \leq r_i$, or equivalently,

$$\frac{p_i}{r_i} \leq \frac{1}{w}, \tag{3.10}$$

for all $i \in \operatorname{supp}(\mathbf{r})$. Hence

$$D(\mathbf{p} \,\|\, \mathbf{r}) = \prod_{i \in \operatorname{supp}(\mathbf{p})} \left(\frac{p_i}{r_i}\right)^{p_i} \leq \prod_{i \in \operatorname{supp}(\mathbf{p})} \left(\frac{1}{w}\right)^{p_i} = \frac{1}{w},$$

giving

$$1 \leq D(\mathbf{p} \,\|\, \mathbf{r}) \leq \frac{1}{w}. \tag{3.11}$$

We now consider cases in which these bounds are attained.

Examples 3.3.7 i. By Lemma 3.1.4, $D(\mathbf{p} \,\|\, \mathbf{r})$ attains its lower bound of 1 precisely when $\mathbf{p} = \mathbf{r}$. This means that the subcommunity is exactly typical or representative of the larger community; it is minimally unusual.

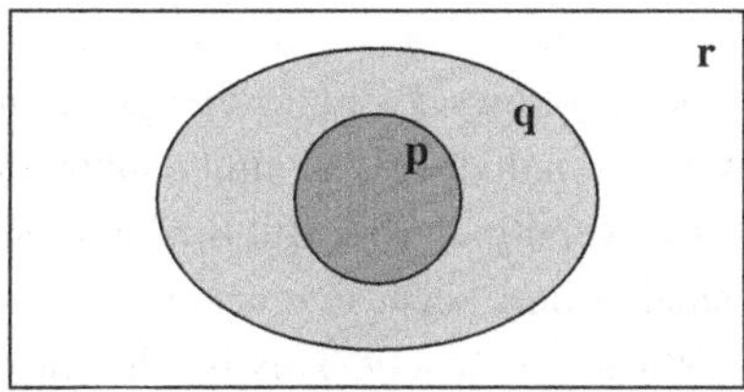

Figure 3.1 The three communities of Example 3.3.8.

ii. The maximum $D(\mathbf{p} \,\|\, \mathbf{r}) = 1/w$ in (3.11) is attained when $r_i = wp_i$ for all $i \in \operatorname{supp}(\mathbf{p})$. In the notation above, this is equivalent to $\pi_i = r_i$ for all $i \in \operatorname{supp}(\mathbf{p})$. In other words, the subcommunity is **isolated**: the species occurring in the subcommunity occur nowhere else in the community.

   If the isolated subcommunity is very small then its species distribution appears highly unusual from the viewpoint of the whole community, and correspondingly $D(\mathbf{p} \,\|\, \mathbf{r}) = 1/w$ is large. But if, say, the isolated subcommunity makes up 90% of the whole community, then from the viewpoint of the whole, the ecology of the subcommunity looks very typical. So, it is intuitively reasonable that $D(\mathbf{p} \,\|\, \mathbf{r}) = 1/0.9$ is close to the minimal possible value of 1 .

The difference between relative diversity and cross diversity is illustrated by the case of a uniform reference community (Example 3.3.6) and by the following example.

Example 3.3.8 Consider a community with species distribution $\mathbf{r}$, containing a subcommunity with species distribution $\mathbf{q}$, which in turn contains a subcommunity with species distribution $\mathbf{p}$ (Figure 3.1). Suppose that the two subcommunities consist only of species that are rare in the whole community, with $r_i = 1/100$ for all $i \in \operatorname{supp}(\mathbf{q})$. The larger subcommunity consists of 50 such species, and the smaller of just one.

For the smaller subcommunity,

$$D(\mathbf{p}) = 1, \quad D^{\times}(\mathbf{p} \,\|\, \mathbf{r}) = 100, \quad D(\mathbf{p} \,\|\, \mathbf{r}) = 100.$$

Indeed, $D(\mathbf{p}) = 1$ since the subcommunity contains just one species. For the cross diversity, $1/r_i = 100$ for all $i \in \operatorname{supp}(\mathbf{p})$, and $D^{\times}(\mathbf{p} \,\|\, \mathbf{r})$ is the geometric mean of $1/r_i$ over $i \in \operatorname{supp}(\mathbf{p})$, so $D^{\times}(\mathbf{p}\|\mathbf{r}) = 100$. Then $D(\mathbf{p}\|\mathbf{r}) = 100/1 = 100$.

For the larger subcommunity,

$$D(\mathbf{q}) = 50, \quad D^{\times}(\mathbf{q} \,\|\, \mathbf{r}) = 100, \quad D(\mathbf{q} \,\|\, \mathbf{r}) = 2,$$

by a similar argument.

This can be understood as follows. From the viewpoint of the whole community, the average rarity of the individuals in either subcommunity is 100. This is why both have a cross diversity of 100. But the larger subcommunity looks less unusual than the smaller one, because it occupies more of the community and therefore resembles it more closely. This is why the relative diversity of the larger subcommunity is lower.

Remark 3.3.9 In ecology, there are concepts of alpha-, beta- and gamma-diversity. The quantities $D(\mathbf{p})$, $D(\mathbf{p} \parallel \mathbf{r})$ and $D^{\times}(\mathbf{p} \parallel \mathbf{r})$ are, respectively, kinds of alpha-, beta- and gamma-diversities, and equation (3.8) is a version of the equation $\beta = \gamma/\alpha$ that appears in the ecological literature (beginning with Whittaker [351], p. 321).

However, $D(\mathbf{p})$, $D(\mathbf{p} \parallel \mathbf{r})$ and $D^{\times}(\mathbf{p} \parallel \mathbf{r})$ are somewhat different from alpha-, beta- and gamma-diversity as usually construed. In the traditional ecological framework, a large community is divided into a number of subcommunities, alpha-diversity is some kind of average of the intrinsic diversities of the subcommunities, beta-diversity is a measure of the variation between the subcommunities, and gamma-diversity is, simply, the diversity of the whole. Here, non-traditionally, our beta-diversity (the relative diversity $D(\mathbf{p} \parallel \mathbf{r})$) and our gamma-diversity (the cross diversity $D^{\times}(\mathbf{p} \parallel \mathbf{r})$) express properties of an *individual* subcommunity with reference to the larger community. This is one of the innovations introduced in recent work of Reeve et al. [293], explored in depth in Chapter 8.

3.4 Relative Entropy in Measure Theory, Geometry and Statistics

Here we give brief interpretations of relative entropy as seen from specific standpoints in these three subjects.

Measure Theory

Let us attempt to generalize the notion of Shannon entropy from probability distributions on a finite set to probability measures on an arbitrary measurable space Ω. Starting from the definition

$$H(\mathbf{p}) = -\sum_{i \in \mathrm{supp}(\mathbf{p})} p_i \log p_i$$

for finite sets, and reasoning purely formally, one might try to define the entropy of a probability measure ν on Ω as

$$H(\nu) = -\int_{\Omega} (\log \nu)\, d\nu.$$

But this makes no sense, since there is no such function as '$\log \nu$'.

However, *relative* entropy generalizes easily. Indeed, given probability measures ν and μ on Ω, the **entropy of ν relative to μ** is defined as

$$H(\nu \,\|\, \mu) = \int_{\Omega} \log\left(\frac{d\nu}{d\mu}\right) d\nu \in [0, \infty], \tag{3.12}$$

where $d\nu/d\mu$ is the Radon–Nikodym derivative. (If ν is not absolutely continuous with respect to μ then $d\nu/d\mu$ sometimes takes the value ∞; but as in the finite case, we allow ∞ as a relative entropy.)

Examples 3.4.1 i. Fix a measure λ on Ω, and take measures ν and μ on Ω with densities p and r with respect to λ. Thus, $d\nu = p\, d\lambda$, $d\mu = r\, d\lambda$, and $d\nu/d\mu = p/r$. It follows that

$$H(\nu \,\|\, \mu) = \int_{\operatorname{supp}(p)} p \log\left(\frac{p}{r}\right) d\lambda.$$

Provided that the choice of reference measure λ is understood, $H(\nu \,\|\, \mu)$ can be written as $H(p \,\|\, r)$.

ii. In particular, when Ω is a finite set with counting measure λ, we recover Definition 3.1.1.

The measure-theoretic viewpoint also explains some earlier notation. On p. 65, we introduced the set A_n of pairs $(\mathbf{p}, \mathbf{r})$ of probability distributions on $\{1, \ldots, n\}$ such that $H(\mathbf{p} \,\|\, \mathbf{r}) < \infty$. Regarding $\mathbf{p}$ and $\mathbf{r}$ as measures on $\{1, \ldots, n\}$, the set A_n consists of exactly the pairs such that $\mathbf{p}$ is absolutely continuous with respect to $\mathbf{r}$.

The slogan

all entropy is relative

is partly justified by the fact just established: relative entropy makes sense in a wide measure-theoretic context in a way that ordinary entropy does not. A different justification is given in Section 8.5.

There is, nevertheless, a useful concept of the entropy of a probability distribution on Euclidean space. Indeed, the **differential entropy** of a probability density function f on $\mathbb{R}^n$ is defined as

$$H(f) = -\int_{\operatorname{supp}(f)} f(x) \log f(x)\, dx,$$

and it is a fundamental fact that among all density functions with a given mean and variance, the one with the maximal entropy is the normal distribution. (This fact is closely related to the central limit theorem, as explained in Johnson [162], for instance.) However, $H(f)$ is still a kind of relative entropy, since the integration takes place with respect to Lebesgue measure λ. Writing $\nu = f\,d\lambda$ for the probability measure corresponding to the density f, we have $f = d\nu/d\lambda$, hence

$$H(f) = -\int \log\left(\frac{d\nu}{d\lambda}\right) d\nu.$$

Formally, the right-hand side is the negative of the expression for $H(\nu\|\lambda)$ given by equation (3.12), even though λ is not a *probability* measure.

Geometry

We have tentatively evoked the idea that $H(\mathbf{p} \,\|\, \mathbf{r})$ is some kind of measure of distance, difference or divergence between the probability distributions $\mathbf{p}$ and $\mathbf{r}$, and it is true that $H(\mathbf{p}\|\mathbf{r}) \geq 0$ with equality if and only if $\mathbf{p} = \mathbf{r}$. However, we have also seen that relative entropy does not have one of the standard properties of a distance function:

$$H(\mathbf{p} \,\|\, \mathbf{r}) \neq H(\mathbf{r} \,\|\, \mathbf{p}).$$

If that were the only problem, it would not be so bad: for as Lawvere, Gromov, and others have argued ([204], pp. 138–139 and [131], p. xv), and as anyone who has walked up and down a hill already knows, there are useful notions of distance that are not symmetric. A more serious problem is that relative entropy fails the triangle inequality, as the following example shows.

Example 3.4.2 Define $\mathbf{p}, \mathbf{q}, \mathbf{r} \in \Delta_2$ by

$$\mathbf{p} = (0.9, 0.1), \quad \mathbf{q} = (0.2, 0.8), \quad \mathbf{r} = (0.1, 0.9).$$

Then

$$H(\mathbf{p} \,\|\, \mathbf{q}) + H(\mathbf{q} \,\|\, \mathbf{r}) = 1.190\ldots < 1.757\ldots = H(\mathbf{p} \,\|\, \mathbf{r}).$$

So, relative entropy only crudely resembles a distance function or metric in the sense of metric spaces.

However, it is a highly significant fact that the *square root* of relative entropy is an *infinitesimal* distance on the set of probability distributions. We explain this twice: first informally, then in the language of Riemannian geometry.

Informally, let $\mathbf{p} \in \Delta_n^\circ$, and consider the relative entropy

$$H(\mathbf{p} + \mathbf{t} \,\|\, \mathbf{p})$$

for $\mathbf{t} \in \mathbb{R}^n$ close to $\mathbf{0}$ such that $\sum t_i = 0$. (Then $\mathbf{p} + \mathbf{t} \in \Delta_n^\circ$.) We can expand $H(\mathbf{p}+\mathbf{t} \| \mathbf{p})$ as a Taylor series in $t_1, \ldots, t_n$. Since $H(\mathbf{p}+\mathbf{t} \| \mathbf{p})$ attains its minimum of 0 at $\mathbf{t} = \mathbf{0}$, the constant term in the Taylor expansion is 0 and the terms in $t_1, \ldots, t_n$ also vanish. A straightforward calculation shows that, in fact,

$$H(\mathbf{p} + \mathbf{t} \| \mathbf{p}) = \sum_{i=1}^{n} \frac{1}{2p_i} t_i^2 + \text{higher order terms.}$$

Thus, up to a different scale factor $1/2p_i$ in each coordinate, relative entropy locally resembles the square of Euclidean distance. The same is true with the arguments reversed:

$$H(\mathbf{p} \| \mathbf{p} + \mathbf{t}) = \sum_{i=1}^{n} \frac{1}{2p_i} t_i^2 + \text{higher order terms.}$$

So although $H(-\|-)$ is not symmetric in its two arguments, it is infinitesimally so, to second order.

These formulas suggest that we regard the square root of relative entropy, rather than relative entropy itself, as a metric. But again, it is not a metric in the sense of metric spaces, because it fails the triangle inequality. The same $\mathbf{p}$, $\mathbf{q}$ and $\mathbf{r}$ as in Example 3.4.2 provide a counterexample:

$$\sqrt{H(\mathbf{p} \| \mathbf{q})} + \sqrt{H(\mathbf{q} \| \mathbf{r})} = 1.281\ldots < 1.325\ldots = \sqrt{H(\mathbf{p} \| \mathbf{r})}.$$

Nevertheless, $\sqrt{H(- \| -)}$ can successfully be used as an *infinitesimal* metric. Still speaking informally, the process is as follows.

Suppose that we are given a set $X \subseteq \mathbb{R}^n$ and a nonnegative real-valued function δ defined on all pairs of points of X that are sufficiently close together. Then under suitable hypotheses on δ, we can define a metric d on X. First, define the length of any path γ in X by finite approximations: plot a large number of close together points $\mathbf{x}_0, \ldots, \mathbf{x}_m$ along γ, use $\sum_{r=1}^{m} \delta(\mathbf{x}_{r-1}, \mathbf{x}_r)$ as an approximation to the length of γ, then pass to the limit. The distance $d(\mathbf{x}, \mathbf{y}) \in [0, \infty]$ between two points $\mathbf{x}, \mathbf{y} \in X$ is defined as the length of a shortest path between $\mathbf{x}$ and $\mathbf{y}$. This d is a metric in the sense of metric spaces.

Applied when $X = \Delta_n^\circ$ and $\delta = \sqrt{H(- \| -)}$, this process gives a new metric d on the simplex. 'Have you ever seen anything like that?' asked Gromov ([132], Section 2). As it turns out, d is not so exotic. Let

$$S^{n-1} = \left\{ \mathbf{x} \in \mathbb{R}^n : \sum x_i^2 = 1 \right\}$$

denote the unit $(n-1)$-sphere. It carries the geodesic metric $d_{S^{n-1}}$, in which $d_{S^{n-1}}(\mathbf{x}, \mathbf{y})$ is the length of a shortest path between $\mathbf{x}$ and $\mathbf{y}$ on the sphere (an arc of a great circle). Any distribution $\mathbf{p} \in \Delta_n^\circ$ has a corresponding point $\sqrt{\mathbf{p}} =$

$(\sqrt{p_1}, \ldots, \sqrt{p_n})$ on S^{n-1}. And as we will see, the metric d on Δ_n° satisfies

$$d(\mathbf{p}, \mathbf{r}) = \sqrt{2} d_{S^{n-1}}(\sqrt{\mathbf{p}}, \sqrt{\mathbf{r}})$$

($\mathbf{p}, \mathbf{r} \in \Delta_n^\circ$). So when the simplex is equipped with this distance d, it is isometric to a subset of the sphere of radius $\sqrt{2}$. With different constant factors, $d(\mathbf{p}, \mathbf{r})$ is known as the Fisher distance or Bhattacharyya angle between $\mathbf{p}$ and $\mathbf{r}$, as detailed below.

We now sketch the precise development. The story told here is the beginning of the subject of information geometry, and we refer to the literature in that subject for details of what follows. The books by Ay, Jost, Lê and Schwachhöfer [22] and Amari [12] are comprehensive modern introductions to information geometry. Other important sources are the 1964 paper and 1972 book of Čencov [62, 63], the book of Amari and Nagaoka [13], the 1983 paper of Amari [11], and the 1987 articles of Lauritzen [202] and Rao [288]. The idea of converting an infinitesimal distance-like function on a manifold into a genuine distance function is developed systematically in Eguchi's theory of contrast functions [86, 87], a summary of which can be found in Section 3.2 of [13].

Let $M = (M, g)$ be a Riemannian manifold, and write d for its geodesic distance function. (We temporarily adopt the Riemannian geometers' practice of using **metric** to mean a Riemannian metric, and **distance** for a metric in the sense of metric spaces.) For each point $p \in M$, we have the function

$$d(-, p)^2 \colon M \to \mathbb{R}.$$

It takes its minimum value, 0, at p, and is smooth on a neighbourhood of p. We can therefore take its Hessian (with respect to the Levi-Civita connection) at any point x near p, giving a bilinear form

$$\mathrm{Hess}_x(d(-, p)^2)$$

on the tangent space $T_x M$. In particular, we can take $x = p$, giving a bilinear form on $T_p M$. But of course, we already have another bilinear form on $T_p M$, the Riemannian metric g_p at p. And up to a constant factor, the two forms are equal:

$$g_p = \frac{1}{2} \mathrm{Hess}_p(d(-, p)^2). \tag{3.13}$$

This equation expresses the Riemannian metric in terms of the geodesic distance (together with the connection). That is, it expresses infinitesimal distance in terms of global distance.

(Equation (3.13) is proved by an elementary calculation, although it is not often stated directly in the literature. It can be derived from more sophisticated results such as Theorem 6.6.1 of Jost [165], by taking the limit as $x \to p$ there, or equation (5) in Supplement A of Pennec [278].)

The idea now is that given any manifold M with connection and any function $\delta\colon M \times M \to \mathbb{R}$ with primitive distance-like properties, we can define a Riemannian metric g on M by

$$g_p = \frac{1}{2}\mathrm{Hess}_p(\delta(-, p)^2) \tag{3.14}$$

($p \in M$). Then, in turn, g gives rise to a geodesic distance function d on M. So, starting from a distance-like function δ, we will have derived a *genuine* distance function d. By equations (3.13) and (3.14), d and δ are equal infinitesimally to second order, and d is entirely determined by the second-order infinitesimal behaviour of δ.

We apply this procedure to the open simplex Δ_n°, taking δ to be the square root of relative entropy. Each of the tangent spaces of Δ_n° is naturally identified with

$$T_n = \left\{\mathbf{t} \in \mathbb{R}^n : \sum_{i=1}^n t_i = 0\right\},$$

so Δ_n° carries a canonical connection. For each $\mathbf{p} \in \Delta_n^\circ$, we define a bilinear form g on $T_{\mathbf{p}}\Delta_n^\circ = T_n$ by

$$g(\mathbf{t}, \mathbf{u}) = \frac{1}{2}\mathrm{Hess}_{\mathbf{p}}(H(- \parallel \mathbf{p}))$$

($\mathbf{t}, \mathbf{u} \in T_n$). By a straightforward calculation, this reduces to

$$g(\mathbf{t}, \mathbf{u}) = \sum_{i=1}^n \frac{1}{2p_i} t_i u_i. \tag{3.15}$$

This is a Riemannian metric on Δ_n°. Without the factor of $1/2$, it is called the **Fisher metric**, $(\mathbf{t}, \mathbf{u}) \mapsto \sum t_i u_i / p_i$.

Now write

$$S_+^{n-1} = S^{n-1} \cap (0, \infty)^n$$

for the positive orthant of the unit $(n-1)$-sphere S^{n-1}. There is a diffeomorphism of smooth manifolds

$$\sqrt{\ }: \Delta_n^\circ \to S_+^{n-1}$$

defined by taking square roots in each coordinate. Transferring the standard Riemannian structure on S_+^{n-1} across this diffeomorphism gives a Riemannian structure on Δ_n°. Explicitly, since $\frac{d}{dx}\sqrt{x} = 1/(2\sqrt{x})$, the induced inner product $\langle -, - \rangle$ on the tangent space T_n at $\mathbf{p} \in \Delta_n^\circ$ is given by

$$\langle \mathbf{t}, \mathbf{u} \rangle = \sum_{i=1}^{n} \frac{t_i}{2\sqrt{p_i}} \frac{u_i}{2\sqrt{p_i}} = \sum_{i=1}^{n} \frac{1}{4p_i} t_i u_i \tag{3.16}$$

(as in Proposition 2.1 of Ay, Jost, Lê and Schwachhöfer [22]). Equations (3.15) and (3.16) together give $g(\mathbf{t}, \mathbf{u}) = 2\langle \mathbf{t}, \mathbf{u} \rangle$. Thus, the Riemannian manifold (Δ_n°, g) is isometric to $\sqrt{2}S_+^{n-1}$, the positive orthant of the $(n-1)$-sphere of radius $\sqrt{2}$.

Like any Riemannian metric, g induces a distance function. The isometry just established makes it easy to compute. Indeed, we already know the geodesic distance on S_+^{n-1} induced by its Riemannian structure; it is given by

$$d_{S^{n-1}}(\mathbf{x}, \mathbf{y}) = \cos^{-1}(\mathbf{x} \cdot \mathbf{y}) \in [0, \pi/2]$$

$(\mathbf{x}, \mathbf{y} \in S_+^{n-1})$, where $\cdot$ denotes the standard inner product on $\mathbb{R}^n$. But by the previous paragraph, the geodesic distance d induced by the Riemannian metric g on Δ_n° is given by

$$d(\mathbf{p}, \mathbf{r}) = \sqrt{2} d_{S^{n-1}}(\sqrt{\mathbf{p}}, \sqrt{\mathbf{r}})$$

$(\mathbf{p}, \mathbf{r} \in \Delta_n^\circ)$. Hence

$$d(\mathbf{p}, \mathbf{r}) = \sqrt{2} \cos^{-1}\left(\sum_{i=1}^{n} \sqrt{p_i r_i} \right) \in [0, \pi/\sqrt{2}].$$

With different normalizations, this distance function has established names: the **Fisher distance** and the **Bhattacharyya angle** [38] between $\mathbf{p}$ and $\mathbf{r}$ are, respectively,

$$2\cos^{-1}\left(\sum_{i=1}^{n} \sqrt{p_i r_i} \right), \quad \cos^{-1}\left(\sum_{i=1}^{n} \sqrt{p_i r_i} \right).$$

The Fisher distance is the geodesic distance induced by the Fisher metric $(\mathbf{t}, \mathbf{u}) \mapsto \sum t_i u_i / p_i$, and makes Δ_n° isometric to the positive orthant of a sphere of radius 2. The Bhattacharyya angle has the advantage that when it is used as a distance function, Δ_n° is isometric to a subset of the *unit* sphere.

In summary, relative entropy produces a notion of distance between two probability distributions on a finite set, obeying the axioms of a metric space. If the square root of relative entropy is regarded as an infinitesimal metric, then its global counterpart is (up to a constant) the Fisher distance.

Further development of these ideas leads to the notion of a statistical manifold. Loosely, this is a Riemannian manifold whose points are to be thought of as probability distributions (on some usually infinite space). We refer to the original paper of Lauritzen [202] and, again, information geometry texts such as [22] and [12].

Statistics

Cross entropy and relative entropy arise naturally from elementary statistical considerations, as follows.

Suppose that we make k observations of elements drawn (by any method) from $\{1, \ldots, n\}$, with outcomes

$$x_1, \ldots, x_k \in \{1, \ldots, n\}.$$

The **empirical distribution** $\hat{\mathbf{p}} = (\hat{p}_1, \ldots, \hat{p}_n) \in \Delta_n$ of the observations is given by

$$\hat{p}_i = \frac{\left|\{j \in \{1, \ldots, k\} : x_j = i\}\right|}{k},$$

or equivalently, $\hat{\mathbf{p}} = \frac{1}{k}\sum_{j=1}^{k} \delta_{x_j}$, where δ_x denotes the point mass at x. For example, if $n = 4$, $k = 3$ and $(x_1, x_2, x_3) = (4, 1, 4)$, then $\hat{\mathbf{p}} = (1/3, 0, 0, 2/3)$.

Now let $\mathbf{p} \in \Delta_n$, and suppose that k elements of $\{1, \ldots, n\}$ are drawn independently at random according to $\mathbf{p}$. The probability $\mathbb{P}(x_1, \ldots, x_k)$ of observing $x_1, \ldots, x_k$ in that order is, in fact, a function of the cross diversity or cross entropy of $\hat{\mathbf{p}}$ with respect to $\mathbf{p}$. Indeed,

$$\begin{aligned}\mathbb{P}(x_1, \ldots, x_k) &= \prod_{j=1}^{k} p_{x_j} = \prod_{i=1}^{n} p_i^{|\{j : x_j = i\}|} = \prod_{i=1}^{n} p_i^{k\hat{p}_i} \\ &= D^{\times}(\hat{\mathbf{p}} \,\|\, \mathbf{p})^{-k} \\ &= \exp(-kH^{\times}(\hat{\mathbf{p}} \,\|\, \mathbf{p})).\end{aligned}$$

Example 3.4.3 Let $\mathbf{p}$ be a probability distribution on $\{1, \ldots, n\}$ with rational probabilities:

$$\mathbf{p} = (k_1/k, \ldots, k_n/k)$$

($k_i \geq 0$, $k = \sum k_i$). Make k observations using this distribution. What is the probability that the results observed are, in order,

$$\underbrace{1, \ldots, 1}_{k_1}, \ \ldots, \ \underbrace{n, \ldots, n}_{k_n}\,?$$

The empirical distribution of those observations is just $\mathbf{p}$, so the answer is

$$D^{\times}(\mathbf{p} \,\|\, \mathbf{p})^{-k} = D(\mathbf{p})^{-k} = e^{-kH(\mathbf{p})}.$$

So, when k is fixed, the probability of obtaining these observations is a decreasing function of the entropy of $\mathbf{p}$. For instance, take $k = n$. At one extreme, if $p_i = 1$ for some i, then the probability of the observed results being $i, \ldots, i$ is maximal (with value 1) and the entropy is minimal (with value 0). At the other extreme, if $\mathbf{p} = \mathbf{u}_n$, then the probability of the results being $1, \ldots, n$ is small ($1/n^n$), corresponding to the fact that $\mathbf{p}$ has the maximal possible entropy.

A standard situation in statistics is that we are in the presence of a probability distribution that is unknown, but which we are willing to assume is a member of a specific family $(\mathbf{p}_\theta)_{\theta\in\Theta}$. We make some observations drawn from the distribution, then we attempt to make inferences about the value of the unknown parameter θ.

(In our current setting, Θ is any set and each $\mathbf{p}_\theta$ is a distribution on $\{1, \ldots, n\}$. But usually in statistics, Θ is a subset of $\mathbb{R}^n$ and the set on which the distributions are defined is infinite. For instance, one may be interested in the family of all normal distributions on $\mathbb{R}$, parametrized by pairs (μ, σ) where $\mu \in \mathbb{R}$ is the mean and $\sigma \in \mathbb{R}^+$ is the standard deviation.)

How to make such inferences is one of the central questions of statistics. The simplest way is the **maximum likelihood** method, as follows. Write

$$\mathbb{P}(x_1, \ldots, x_k \mid \theta)$$

for the probability of observing $x_1, \ldots, x_k$ when drawing from the distribution $\mathbf{p}_\theta$. The maximum likelihood method is this: given observations $x_1, \ldots, x_k$, choose the value of θ that maximizes $\mathbb{P}(x_1, \ldots, x_k \mid \theta)$.

We have already shown that

$$\mathbb{P}(x_1, \ldots, x_k \mid \theta) = \exp(-kH^\times(\hat{\mathbf{p}} \parallel \mathbf{p}_\theta)),$$

so it follows from equation (3.6) that

$$\mathbb{P}(x_1, \ldots, x_k \mid \theta) = \exp\Big(-k(H(\hat{\mathbf{p}} \parallel \mathbf{p}_\theta) + H(\hat{\mathbf{p}}))\Big).$$

The term $H(\hat{\mathbf{p}})$ is fixed, in the sense of depending only on the observed data and not on the unknown θ. The right-hand side is a decreasing function of $H(\hat{\mathbf{p}} \| \mathbf{p}_\theta)$. Thus, the maximum likelihood method amounts to choosing θ to minimize the relative entropy $H(\hat{\mathbf{p}} \| \mathbf{p}_\theta)$. Regarding $H(\hat{\mathbf{p}} \| \mathbf{p})$ as a kind of difference or distance between $\hat{\mathbf{p}}$ and $\mathbf{p}$ (with the caveats above), this means choosing θ so that $\mathbf{p}_\theta$ is as close as possible to the observed distribution $\hat{\mathbf{p}}$, as in Figure 3.2.

Further details and context can be found in Csiszár and Shields [76]. The method of minimizing relative entropy has uniquely good properties, as was proved by Shore and Johnson [313] in a slightly different context to the one described here.

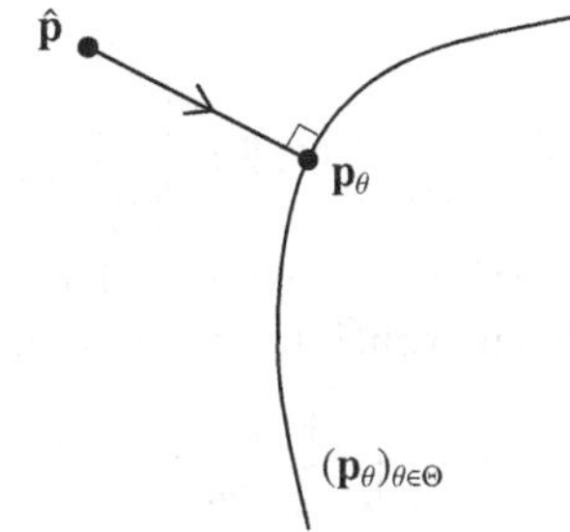

Figure 3.2 Maximum likelihood and minimum relative entropy.

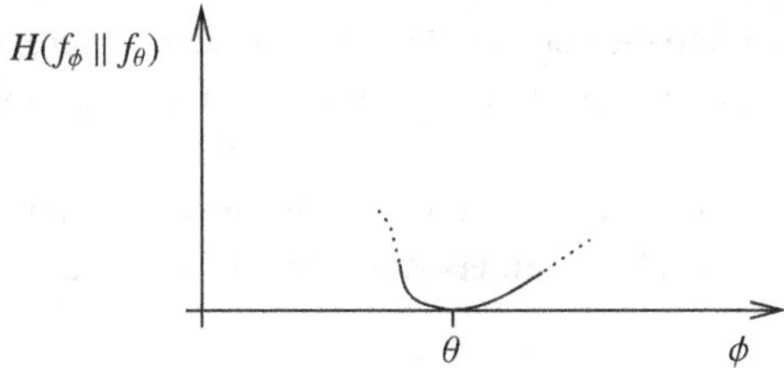

Figure 3.3 Relative entropy for a parametrized family of probability distributions. The Fisher information $I(\theta)$ is the second derivative of the graph at θ, that is, the curvature there.

Measure Theory, Geometry and Statistics

The connection between maximum likelihood and relative entropy involves relative entropies $H(\hat{\mathbf{p}} \,\|\, \mathbf{p}_\theta)$ in which the arguments, $\hat{\mathbf{p}}$ and $\mathbf{p}_\theta$, need not be close together in Δ_n. Nevertheless, we saw in the discussion of the Fisher metric that the behaviour of $H(\mathbf{p} \,\|\, \mathbf{r})$ when $\mathbf{p}$ and $\mathbf{r}$ are close is especially significant. More exactly, it is its infinitesimal behaviour to second order that matters. What follows is a brief further exploration of this second-order behaviour, from a statistical perspective.

Let Ω be a measure space and let $(f_\theta)_{\theta\in\Theta}$ be a smooth family of probability density functions on Ω, indexed over some real interval Θ. Fix $\theta \in \Theta$. The relative entropy

$$H(f_\phi \,\|\, f_\theta) = \int_\Omega f_\phi(x) \log \frac{f_\phi(x)}{f_\theta(x)} \, dx$$

($\phi \in \Theta$), defined as in Example 3.4.1(i), attains its minimum value 0 at $\phi = \theta$ (Figure 3.3). Thus, the function $\phi \mapsto H(f_\phi \,\|\, f_\theta)$ has both value 0 and first derivative 0 at $\phi = \theta$. The second derivative measures how fast the distribution f_ϕ changes as ϕ varies near θ. It is called the **Fisher information** $I(\theta)$ of our family at θ:

$$I(\theta) = \frac{\partial^2}{\partial \phi^2} H(f_\phi \parallel f_\theta)\Big|_{\phi=\theta}. \tag{3.17}$$

Substituting the definition of $H(f_\phi \parallel f_\theta)$ into (3.17) and performing some elementary calculations leads to an explicit formula for the Fisher information:

$$I(\theta) = \int_\Omega \frac{1}{f_\theta}\left(\frac{\partial f_\theta}{\partial \theta}\right)^2.$$

Detailed discussions of Fisher information can be found in texts such as Amari and Nagaoka [13] (Section 2.2), where the definition is given for families of distributions parametrized by *several* real variables $\theta_1, \ldots, \theta_n$, and Fisher information is put into the context of the Fisher metric. Here, we simply describe two uses of Fisher information in statistics, remaining in the single-parameter case.

The first is the Cramér–Rao bound. Suppose that we have an unbiased estimator $\hat{\theta}$ of the parameter θ. The **Cramér–Rao bound** for $\hat{\theta}$ is a lower bound on its variance:

$$\mathrm{Var}(\hat{\theta}) \geq \frac{1}{I(\theta)} \tag{3.18}$$

(Cramér [72], Rao [285]).

This statement can be understood as follows. Let θ denote the true but unknown value of our parameter, which we are trying to infer from the data. If the Fisher information $I(\theta)$ at θ is small, then f_ϕ changes only slowly when ϕ is near θ. Different parameter values near θ produce similar distributions, so it is difficult to infer the parameter value from observations with any degree of accuracy. The Cramér–Rao bound (3.18) formalizes this intuition: since $1/I(\theta)$ is in this case large, any unbiased estimator of θ must be imprecise, in the sense of having large variance. In contrast, if f_ϕ varies rapidly near $\phi = \theta$ then inferring θ from the data is easier, and it may be possible to find a more precise unbiased estimator.

A second use of Fisher information is in the definition of the Jeffreys prior. A fundamental challenge in Bayesian statistics is how to choose a prior distribution on the parameter space Θ. In particular, one can ask for a universal method that takes as its input a family $(f_\theta)_{\theta \in \Theta}$ of probability distributions and produces as its output a canonical distribution on Θ, intended to be used as a prior. In 1939, the statistician Harold Jeffreys proposed using as a prior the density function

$$\theta \mapsto \sqrt{I(\theta)},$$

normalized (if possible) to integrate to 1 [158, 159]. This is the **Jeffreys prior**.

The Jeffreys prior has the crucial property of invariance under reparametrization. For example, suppose that one person works with the family $(f_\sigma)_{0\leq\sigma\leq 10}$ of normal distributions on $\mathbb{R}$ with mean 0 and standard deviation between 0 and 10, while another works with the family $(g_V)_{0\leq V\leq 100}$ of normal distributions with mean 0 and variance between 0 and 100. The difference between the two families is obviously cosmetic, and if calculations based on the different parametrizations resulted in different outcomes, something would be seriously wrong.

But the Jeffreys prior behaves correctly. The first person can calculate the Jeffreys prior of their family to produce a probability density function on $[0, 10]$, hence a probability measure ν_1 on $[0, 10]$. The second person, similarly, obtains a probability measure ν_2 on $[0, 100]$. The invariance property is that when ν_1 is pushed forward along the squaring map $[0, 10] \to [0, 100]$, the resulting measure on $[0, 100]$ is equal to ν_2. In other words, the choice of parametrization makes no difference to the Jeffreys prior.

This is a very important logical property, and not all systems for assigning a prior possess it. For instance, suppose that we simply assign the uniform prior to any family (Bernoulli and Laplace's principle of insufficient reason, discussed in Section 3 of Kass and Wasserman [182]). Then invariance fails: in the example above, a probability of $1/2$ is assigned to the standard deviation being less than 5, but a probability of $1/4$ to the variance being less than 25. This is a fatal flaw.

A careful account of the Jeffreys prior, with historical and mathematical context, can be found in Section 4.7 of Robert, Chopin and Rousseau [298]. This includes the full multi-parameter definition, extending the single-parameter version to which we have confined ourselves here.

3.5 Characterization of Relative Entropy

Here we show that relative entropy is uniquely characterized by the four properties listed in Section 3.1, proving:

Theorem 3.5.1 *Let $(I(- \parallel -)\colon A_n \to \mathbb{R})_{n\geq 1}$ be a sequence of functions. The following are equivalent:*

i. $I(- \parallel -)$ is measurable in the second argument, permutation-invariant, and satisfies the vanishing and chain rules (equation (3.3), with I in place of H);

ii. $I(- \parallel -) = cH(- \parallel -)$ for some $c \in \mathbb{R}$.

Just as ordinary Shannon entropy has been the subject of many characterization theorems, so too has relative entropy. Theorem 3.5.1 and its proof first appeared in [218] (Theorem II.1), and was strongly influenced by a categorical characterization of relative entropy by Baez and Fritz [24], which in turn built on work of Petz [281]. It is also very close to a result of Kannappan and Ng, although the proof is entirely different. Historical commentary can be found in Remark 3.5.7.

We now embark on the proof of Theorem 3.5.1.

The four conditions in part (i) are satisfied by $H(- \parallel -)$ (as observed in Section 3.1), hence by $cH(- \parallel -)$ for any scalar c. Thus, (ii) implies (i).

For the rest of this section, let $I(- \parallel -)$ be a sequence of functions satisfying (i). We have to prove that $I(- \parallel -)$ is a scalar multiple of $H(- \parallel -)$.

Define a function $L\colon (0, 1] \to \mathbb{R}$ by

$$L(\alpha) = I((1, 0) \parallel (\alpha, 1 - \alpha)).$$

(Since $\alpha > 0$, we have $((1, 0), (\alpha, 1 - \alpha)) \in A_2$, so $L(\alpha) \in \mathbb{R}$ is well defined.) The idea is that if $I(- \parallel -) = H(- \parallel -)$ then $L = -\log$. We will show that in any case, L is a scalar multiple of log.

Lemma 3.5.2 *Let* $(\mathbf{p}, \mathbf{r}) \in A_n$ *with* $p_{k+1} = \cdots = p_n = 0$, *where* $1 \leq k \leq n$. *Then* $r_1 + \cdots + r_k > 0$ *and*

$$I(\mathbf{p} \parallel \mathbf{r}) = L(r_1 + \cdots + r_k) + I(\mathbf{p}' \parallel \mathbf{r}'),$$

where

$$\mathbf{p}' = (p_1, \ldots, p_k), \qquad \mathbf{r}' = \frac{(r_1, \ldots, r_k)}{r_1 + \cdots + r_k}.$$

Proof The case $k = n$ reduces to the statement that $L(1) = 0$, which follows from the vanishing property. Suppose, then, that $k < n$.

Since $\mathbf{p}$ is a probability distribution with $p_i = 0$ for all $i > k$, there is some $i \leq k$ such that $p_i > 0$, and then $r_i > 0$ since $(\mathbf{p}, \mathbf{r}) \in A_n$. Hence $r_1 + \cdots + r_k > 0$. Let $\mathbf{r}'' \in \Delta_{n-k}$ be the normalization of $(r_{k+1}, \ldots, r_n)$ if $r_{k+1} + \cdots + r_n > 0$, or choose $\mathbf{r}''$ arbitrarily in Δ_{n-k} otherwise. (The set Δ_{n-k} is nonempty since $k < n$.) Then by definition of composition,

$$\begin{aligned} \mathbf{p} &= (1, 0) \circ (\mathbf{p}', \mathbf{r}''), \\ \mathbf{r} &= (r_1 + \cdots + r_k,\ r_{k+1} + \cdots + r_n) \circ (\mathbf{r}', \mathbf{r}''). \end{aligned}$$

Hence by the chain rule,

$$I(\mathbf{p} \parallel \mathbf{r}) = L(r_1 + \cdots + r_k) + 1 \cdot I(\mathbf{p}' \parallel \mathbf{r}') + 0 \cdot I(\mathbf{r}'' \parallel \mathbf{r}''),$$

and the result follows. □

Lemma 3.5.3 *$L(\alpha\beta) = L(\alpha) + L(\beta)$ for all $\alpha, \beta \in (0, 1]$.*

Proof By the chain rule, $I(-\,\|\,-)$ has the logarithmic property stated at the end of Section 3.1 (equation (3.4), with I in place of H). Hence

$$I((1,0) \otimes (1,0) \,\|\, (\alpha, 1-\alpha) \otimes (\beta, 1-\beta)) = L(\alpha) + L(\beta).$$

But also

$$\begin{aligned} &I((1,0) \otimes (1,0) \,\|\, (\alpha, 1-\alpha) \otimes (\beta, 1-\beta)) \\ &\quad = I\Big((1,0,0,0) \,\Big\|\, (\alpha\beta, \alpha(1-\beta), (1-\alpha)\beta, (1-\alpha)(1-\beta))\Big) \\ &\quad = L(\alpha\beta) + I(\mathbf{u}_1 \,\|\, \mathbf{u}_1) \\ &\quad = L(\alpha\beta), \end{aligned}$$

by Lemma 3.5.2 (with $k = 1$) and the vanishing property. □

We can now deduce:

Lemma 3.5.4 *There is some $c \in \mathbb{R}$ such that $L(\alpha) = -c \log \alpha$ for all $\alpha \in (0, 1]$.*

Proof By hypothesis, L is measurable, so this follows from Lemma 3.5.3 and Corollary 1.1.14. □

Our next lemma is an adaptation of the most ingenious part of Baez and Fritz's argument (Lemma 4.2 of [24]).

Lemma 3.5.5 *Let $n \geq 1$ and $(\mathbf{p}, \mathbf{r}) \in A_n$. Suppose that $\mathbf{p}$ has full support. Then $I(\mathbf{p} \,\|\, \mathbf{r}) = cH(\mathbf{p} \,\|\, \mathbf{r})$.*

Proof Since $(\mathbf{p}, \mathbf{r}) \in A_n$, the distribution $\mathbf{r}$ also has full support. We can therefore choose some $\alpha \in (0, 1]$ such that $r_i - \alpha p_i \geq 0$ for all i.

We will compute the number

$$x = I((p_1, \ldots, p_n, \underbrace{0, \ldots, 0}_{n}) \,\|\, (\alpha p_1, \ldots, \alpha p_n, r_1 - \alpha p_1, \ldots, r_n - \alpha p_n))$$

in two ways. (The pair of distributions on the right-hand side belongs to A_{2n}, so x is well defined.) First, by Lemma 3.5.2 and the vanishing property,

$$x = L(\alpha) + I(\mathbf{p} \,\|\, \mathbf{p}) = -c \log \alpha.$$

Second, by permutation-invariance and then the chain rule,

$$\begin{aligned} x &= I((p_1, 0, \dots, p_n, 0) \,\|\, (\alpha p_1, r_1 - \alpha p_1, \dots, \alpha p_n, r_n - \alpha p_n)) \\ &= I\Big(\mathbf{p} \circ ((1,0), \dots, (1,0)) \,\Big\|\, \mathbf{r} \circ \Big(\big(\alpha \tfrac{p_1}{r_1}, 1 - \alpha \tfrac{p_1}{r_1}\big), \dots, \big(\alpha \tfrac{p_n}{r_n}, 1 - \alpha \tfrac{p_n}{r_n}\big)\Big)\Big) \\ &= I(\mathbf{p} \,\|\, \mathbf{r}) + \sum_{i=1}^{n} p_i L\big(\alpha \tfrac{p_i}{r_i}\big) \\ &= I(\mathbf{p} \,\|\, \mathbf{r}) - c \log \alpha - cH(\mathbf{p} \,\|\, \mathbf{r}). \end{aligned}$$

Comparing the two expressions for x gives the result. □

We have now proved that $I(\mathbf{p} \,\|\, \mathbf{r}) = cH(\mathbf{p} \,\|\, \mathbf{r})$ when $\mathbf{p}$ has full support. It only remains to prove it for arbitrary $\mathbf{p}$.

Proof of Theorem 3.5.1 Let $(\mathbf{p}, \mathbf{r}) \in A_n$. By permutation-invariance, we can assume that

$$p_1, \dots, p_k > 0, \quad p_{k+1} = \cdots = p_n = 0,$$

where $1 \le k \le n$. Writing $R = r_1 + \cdots + r_k$,

$$I(\mathbf{p} \,\|\, \mathbf{r}) = L(R) + I((p_1, \dots, p_k) \,\|\, \tfrac{1}{R}(r_1, \dots, r_k))$$

by Lemma 3.5.2. Hence by Lemmas 3.5.4 and 3.5.5,

$$I(\mathbf{p} \,\|\, \mathbf{r}) = -c \log R + cH((p_1, \dots, p_k) \,\|\, \tfrac{1}{R}(r_1, \dots, r_k)).$$

But by the same argument applied to cH in place of I (or by direct calculation), we also have

$$cH(\mathbf{p} \,\|\, \mathbf{r}) = -c \log R + cH((p_1, \dots, p_k) \,\|\, \tfrac{1}{R}(r_1, \dots, r_k)).$$

The result follows. □

Remarks 3.5.6 i. Cross entropy satisfies all the properties listed in Theorem 3.5.1(i) except for vanishing, which it does not satisfy. Hence the vanishing axiom cannot be dropped from the theorem.

ii. The chain rule can equivalently be replaced by a special case:

$$\begin{aligned} I\big((pw_1, (1-p)w_1, w_2, \dots, w_n) \,\big\|\, (\widetilde{p}\widetilde{w}_1, (1-\widetilde{p})\widetilde{w}_1, \widetilde{w}_2, \dots, \widetilde{w}_n)\big) \\ = I(\mathbf{w} \,\|\, \widetilde{\mathbf{w}}) + w_1 I((p, 1-p) \,\|\, (\widetilde{p}, 1-\widetilde{p})) \end{aligned}$$

for all $(\mathbf{w}, \widetilde{\mathbf{w}}) \in A_n$ and $((p, 1-p), (\widetilde{p}, 1-\widetilde{p})) \in A_2$. Alternatively, it can be replaced by a different special case:

$$\begin{aligned} &I(w\mathbf{p} \oplus (1-w)\mathbf{r} \,\|\, \widetilde{w}\widetilde{\mathbf{p}} \oplus (1-\widetilde{w})\widetilde{\mathbf{r}}) \\ &\qquad = I((w, 1-w) \,\|\, (\widetilde{w}, 1-\widetilde{w})) + wI(\mathbf{p} \,\|\, \widetilde{\mathbf{p}}) + (1-w)I(\mathbf{r} \,\|\, \widetilde{\mathbf{r}}) \end{aligned}$$

for all $(\mathbf{p}, \widetilde{\mathbf{p}}) \in A_k$, $(\mathbf{r}, \widetilde{\mathbf{r}}) \in A_\ell$, and $((w, 1-w), (\widetilde{w}, 1-\widetilde{w})) \in A_2$. Here we have used the notation

$$w\mathbf{p} \oplus (1-w)\mathbf{r} = (wp_1, \ldots, wp_k, (1-w)r_1, \ldots, (1-w)r_\ell).$$

Both special cases are equivalent to the general case by elementary inductions, as in Remark 2.2.11 and Appendix A.1.

Remark 3.5.7 The first characterization of relative entropy appears to have been proved by Rényi in 1961 ([294], Theorem 4). It relied on $H(\mathbf{p} \,\|\, \mathbf{r})$ being defined not only for probability distributions $\mathbf{p}$ and $\mathbf{r}$, but also for all 'generalized' distributions (in which the requirement that $\sum p_i = \sum r_i = 1$ is weakened to $\sum p_i, \sum r_i \leq 1$). The result does not translate easily into a characterization of relative entropy for ordinary probability distributions only.

Among the theorems characterizing relative entropy for ordinary probability distributions, one of the first was that of Hobson [147] in 1969. His hypotheses were stronger than those of Theorem 3.5.1, for the same conclusion. In common with Theorem 3.5.1, he assumed permutation-invariance, vanishing, and the chain rule (in the second of the two equivalent forms given in Remark 3.5.6(ii)). But he also assumed continuity in both variables (instead of just measurability in one) and a monotonicity hypothesis unlike anything in Theorem 3.5.1.

In 1973, Kannappan and Ng [177] proved a result very close to Theorem 3.5.1. They did not explicitly *state* that result in their paper, but the closing remarks in another paper by the same authors [178] and the approach of a contemporaneous paper by Kannappan and Rathie [179] suggest the intent. The result resembling Theorem 3.5.1 was stated explicitly in a 2008 article of Csiszár ([75], Section 2.1), who attributed it to Kannappan and Ng.

There are some small differences between the hypotheses of Kannappan and Ng's theorem and those of Theorem 3.5.1. They assumed measurability in both variables, whereas we only assumed measurability in the second (and actually only used that $I((1,0) \,\|\, -)$ is measurable). On the other hand, they only needed the vanishing condition for $\mathbf{u}_2$, whereas we needed it for all $\mathbf{p}$. Like many authors on functional equations in information theory, they used the chain rule in the first of the equivalent forms in Remark 3.5.6(ii), under the name of recursivity.

The proofs, however, are completely different. Theirs was a tour de force of functional equations, putting at its heart the so-called fundamental equation of information theory (equation (11.17)), and involving the solution of such functional equations as

$$f(x) + (1 - x)g\left(\frac{y}{1 - x}\right) = h(y) + (1 - y)j\left(\frac{x}{1 - y}\right)$$

in four unknown functions. The proof above bypasses these considerations entirely.

4

Deformations of Shannon Entropy

Shannon entropy is fundamental, but it is not the only useful or natural notion of entropy, even in the context of a single probability distribution on a finite set. In this chapter, we meet two one-parameter families of entropies that both include Shannon entropy as a member (Figure 4.1). Both are indexed by a real parameter q, and both have Shannon entropy as the case $q = 1$. Moving the value of q away from 1 can be thought of as deforming Shannon entropy. As in other mathematical contexts where the word 'deformation' is used, the undeformed object (Shannon entropy) has uniquely good properties that are lost after deformation, but the deformed objects nevertheless retain some of the original object's features.

We begin with the q-logarithmic entropies $(S_q)_{q\in\mathbb{R}}$, often called 'Tsallis entropies' (a misattribution detailed in Remark 4.1.4). The q-logarithmic entropies have been used as measures of biological diversity, but should probably not be, as we will see (Examples 4.1.3).

Perhaps surprisingly, it is *easier* to uniquely characterize the entropy S_q for

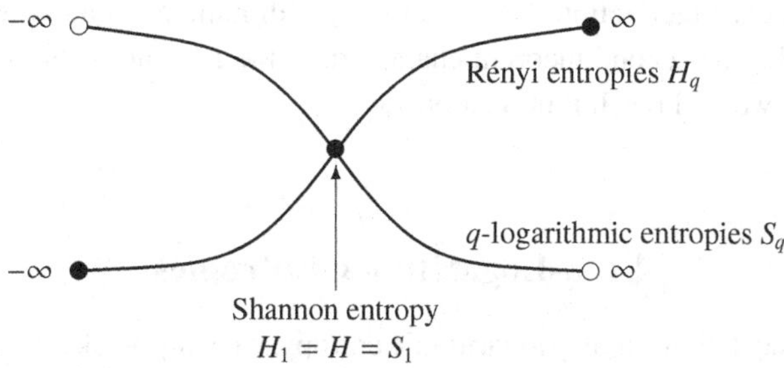

Figure 4.1 Two families of deformations of Shannon entropy: the Rényi entropies $(H_q)_{q\in[-\infty,\infty]}$ and the q-logarithmic entropies $(S_q)_{q\in(-\infty,\infty)}$.

$q \neq 1$ than it is in the Shannon case $S_1 = H$. Moreover, the characterization theorem that we prove does not require any regularity conditions at all, not even measurability. The same goes for the q-logarithmic relative entropy, which we also introduce and characterize.

After some necessary preliminaries on the classical topic of power means (Section 4.2), we introduce the other main family of deformations of Shannon entropy: the Rényi entropies $(H_q)_{q\in[-\infty,\infty]}$ (Section 4.3). The q-logarithmic and Rényi entropies have exactly the same content: for each finite value of q, there is a simple formula for $S_q(\mathbf{p})$ in terms of $H_q(\mathbf{p})$, and vice versa. But they have different and complementary algebraic properties. For instance, the q-logarithmic entropies satisfy a simple chain rule similar to that for Shannon entropy, whereas the chain rule for the Rényi entropies is more cumbersome. On the other hand, the Rényi entropies have the same log-like property as Shannon entropy,

$$H_q(\mathbf{p} \otimes \mathbf{r}) = H_q(\mathbf{p}) + H_q(\mathbf{r}),$$

but the q-logarithmic entropies do not.

The exponential of Rényi entropy, $D_q(\mathbf{p}) = \exp(H_q(\mathbf{p}))$, is known in ecology as the Hill number of order q. The Hill numbers are the most important measures of biological diversity (at least, if we are using the crude model of a community as a probability distribution on the set of species). Different values of q reflect different aspects of a community's composition, and graphing $D_q(\mathbf{p})$ against q enables one to read off meaningful features of the community. In Sections 4.3 and 4.4, we illustrate this point and establish the properties that make the Hill numbers so suitable as measures of diversity.

We finish by showing that the Hill number of a given order q is uniquely characterized by certain properties (Section 4.5). The same is therefore true of the Rényi entropies (since one is the exponential of the other), although the properties appear more natural when stated for the Hill numbers. This is the first of two characterization theorems for the Hill numbers that we will prove in this book. The second theorem characterizes the Hill numbers of *unknown* orders, and we will reach it in Section 7.4.

4.1 q-Logarithmic Entropies

To obtain the definition of q-logarithmic entropy, we simply take the definition of Shannon entropy and replace the logarithm by the q-logarithm $\ln_q$ defined in Section 1.3.

Definition 4.1.1 Let $q \in \mathbb{R}$ and $n \geq 1$. The **q-logarithmic entropy**

$$S_q \colon \Delta_n \to \mathbb{R}$$

is defined by

$$S_q(\mathbf{p}) = \sum_{i \in \mathrm{supp}(\mathbf{p})} p_i \ln_q\left(\frac{1}{p_i}\right).$$

Thus, $S_1(\mathbf{p})$ is the Shannon entropy $H(\mathbf{p})$, and for $q \neq 1$,

$$S_q(\mathbf{p}) = \frac{1}{1-q}\left(\sum_{i \in \mathrm{supp}(\mathbf{p})} p_i^q - 1\right). \tag{4.1}$$

Remark 4.1.2 We chose to generalize the expression $\sum p_i \log(1/p_i)$ for Shannon entropy, but we could instead have used $-\sum p_i \log p_i$. Since $\ln_q(1/x) \neq -\ln_q(x)$, this would have given a different result. But by equation (1.19),

$$-\sum_{i \in \mathrm{supp}(\mathbf{p})} p_i \ln_q p_i = S_{2-q}(\mathbf{p}),$$

so this different choice only amounts to a different parametrization.

The q-logarithmic entropy $S_q(\mathbf{p})$ can be interpreted as expected surprise. Let $s \colon [0,1] \to \mathbb{R} \cup \{\infty\}$ be a decreasing function such that $s(1) = 0$, thought of as assigning to each probability p the degree of surprise $s(p)$ that one would experience on witnessing an event with that probability. Then our expected surprise at an event drawn from a probability distribution $\mathbf{p} = (p_1, \ldots, p_n)$ is

$$\sum_{i \in \mathrm{supp}(\mathbf{p})} p_i \cdot s(p_i).$$

Expected surprise is a measure of uncertainty. If $\mathbf{p} = (1, 0, \ldots, 0)$ then the expected surprise is 0: the process of drawing from $\mathbf{p}$ is completely predictable. If $\mathbf{p} = \mathbf{u}_n$ then the expected surprise is $s(1/n)$, which is an increasing function of n: the greater the number of possibilities, the less predictable the outcome.

(Informally, the concept of expected surprise is familiar: someone who lives in a stable environment will expect that most days, something may mildly surprise them but nothing will astonish them. The less stable the environment, the greater the expected surprise.)

In these terms, $S_q(\mathbf{p})$ is the expected surprise at an event drawn from the distribution $\mathbf{p}$ when we use $p \mapsto \ln_q(1/p)$ as our surprise function. Figure 4.2 shows the surprise functions for $q = 0, 1, 2, 3$. For a general $q > 0$, we have

$$0 \leq S_q(\mathbf{p}) \leq \ln_q(\mathbf{u}_n)$$

for all $\mathbf{p} \in \Delta_n$, with $S_q(\mathbf{p}) = 0$ if and only if $\mathbf{p} = (0, \ldots, 0, 1, 0, \ldots, 0)$ and

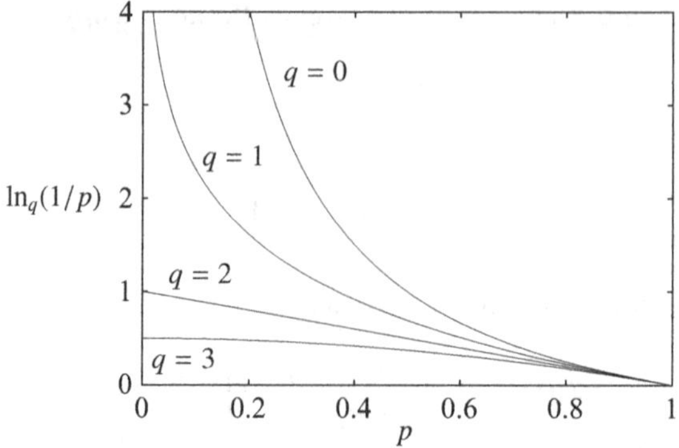

Figure 4.2 The functions $p \mapsto \ln_q(1/p)$ for several values of q.

$S_q(\mathbf{p}) = \ln_q(n)$ if and only if $\mathbf{p} = \mathbf{u}_n$. This will follow from the corresponding properties of the Hill number D_q, once we have established the relationship between the q-logarithmic entropies and the Hill numbers (Remark 4.4.4(i)).

Examples 4.1.3 In these examples, we regard $\mathbf{p} = (p_1, \ldots, p_n) \in \Delta_n$ as the relative abundance distribution of n species making up a biological community. Sometimes $S_q(\mathbf{p})$ has been advocated as a measure of diversity (as in Patil and Taillie [273], Keylock [187], and Ricotta and Szeidl [296]), but this is problematic, as now explained.

i. $S_0(\mathbf{p}) = |\mathrm{supp}(\mathbf{p})| - 1$. That is, the 0-logarithmic entropy is one less than the number of species present.
ii. $S_1(\mathbf{p}) = H(\mathbf{p})$. The plague and oil company arguments of Examples 2.4.7 and 2.4.11 show why S_1 should not be used as a diversity measure. More generally, S_q should not be used as a diversity measure either, for any value of q, since it is not an effective number:

$$S_q(\mathbf{u}_n) = \ln_q(n) \neq n.$$

However, we will see in Section 4.3 that S_q can be transformed into a well-behaved diversity measure, and that the result is the Hill number of order q.
iii. The 2-logarithmic entropy of $\mathbf{p}$ is

$$S_2(\mathbf{p}) = 1 - \sum_{i=1}^{n} p_i^2 = \sum_{i,j:\ i \neq j} p_i p_j.$$

This is the probability that two individuals chosen at random are of different species. In ecology, $S_2(\mathbf{p})$ is associated with the names of Edward H. Simpson, who introduced $S_2(\mathbf{p})$ as an index of diversity in 1949 [314], and Corrado Gini, who used $S_2(\mathbf{p})$ in a wide-ranging 1912 monograph on economics, statistics and demography [118]. It is such a natural quantity that it has been used in many different fields; it also has the advantage that it admits an unbiased estimator. These points are discussed in the 1982 note of Good [123], who wrote that 'any statistician of this century who wanted a measure of homogeneity would have taken about two seconds to suggest $\sum p_i^2$'.

Despite all this, $S_2(\mathbf{p})$ has the defect of not being an effective number. Again, Section 4.3 describes the remedy.

Remark 4.1.4 The q-logarithmic entropies have been discovered and rediscovered repeatedly. They seem to have first appeared in a 1967 paper on information and classification by Havrda and Charvát [141], in a form adapted to base 2 logarithms:

$$S_q^{(2)}(\mathbf{p}) = \frac{1}{2^{1-q} - 1}\left(\sum p_i^q - 1\right).$$

The constant factor is chosen so that $S_q^{(2)}(\mathbf{p})$ converges to the base 2 Shannon entropy $H^{(2)}(\mathbf{p})$ as $q \to 1$, and so that $S_q^{(2)}(\mathbf{u}_2) = 1$ for all q.

Further work on the entropies $S_q^{(2)}$ was carried out in 1968 by Vajda [340] (with reference to Havrda and Charvát). They were rediscovered in 1970 by Daróczy [78] (without reference to Havrda and Charvát), and were the subject of Section 6.3 of the 1975 book [3] by Aczél and Daróczy (with reference to all of the above).

The base e entropies S_q themselves seem to have appeared first in Section 3.2 of a 1982 paper [273] by Patil and Taillie (with reference to Aczél and Daróczy but none of the others), where S_q was proposed as an index of biological diversity.

In physics, meanwhile, the q-logarithmic entropies appeared in a 1971 article of Lindhard and Nielsen [229] (according to Csiszar [75], Section 2.4). They also made a brief appearance in a review article on entropy in physics by Wehrl ([350], p. 247). Finally, they were rediscovered again in a 1988 paper on statistical physics by Tsallis [331] (with reference to none of the above).

Despite the twenty years of active life that the q-logarithmic entropies had already enjoyed, it is after Tsallis that they are most commonly named. The term 'q-logarithmic entropy' is new, but has the benefits of being descriptive and of not perpetuating a misattribution.

The chief advantage of the q-logarithmic entropies over the Rényi entropies (introduced in Section 4.3) is that they satisfy a simple chain rule:

$$S_q(\mathbf{w} \circ (\mathbf{p}^1, \ldots, \mathbf{p}^n)) = S_q(\mathbf{w}) + \sum_{i \in \operatorname{supp}(\mathbf{w})} w_i^q S_q(\mathbf{p}^i) \tag{4.2}$$

($q \in \mathbb{R}$, $\mathbf{w} \in \Delta_n$, $\mathbf{p}^i \in \Delta_{k_i}$). This is easily checked directly. Alternatively, one can imitate the proof of the chain rule for Shannon entropy (Proposition 2.2.8), replacing ∂ by $\partial_q \colon x \mapsto x \ln_q(1/x)$ and showing that

$$\partial_q(xy) = \partial_q(x)y + x^q \partial_q(y).$$

We will also prove a more general chain rule as Proposition 6.2.13.

The special case $\mathbf{p}^1 = \cdots = \mathbf{p}^n = \mathbf{p}$ gives

$$S_q(\mathbf{w} \otimes \mathbf{p}) = S_q(\mathbf{w}) + \left(\sum_{i \in \operatorname{supp}(\mathbf{w})} w_i^q \right) S_q(\mathbf{p}) \tag{4.3}$$

($q \in \mathbb{R}$, $\mathbf{w} \in \Delta_n$, $\mathbf{p} \in \Delta_k$). In particular, the symmetry present in the case $q = 1$,

$$H(\mathbf{w} \otimes \mathbf{p}) = H(\mathbf{w}) + H(\mathbf{p}),$$

disappears when we deform away from $q = 1$. This is the key to the characterization theorem that follows.

Before we state it, let us record one other property: S_q is **symmetric**, meaning that

$$S_q(\mathbf{p}) = S_q(\mathbf{p}\sigma) \tag{4.4}$$

for all $q \in \mathbb{R}$, $\mathbf{p} \in \Delta_n$, and permutations σ of $\{1, \ldots, n\}$.

Theorem 4.1.5 *Let $1 \neq q \in \mathbb{R}$. Let $(I \colon \Delta_n \to \mathbb{R})_{n \geq 1}$ be a sequence of functions. The following are equivalent:*

i. I is symmetric and satisfies

$$I(\mathbf{w} \otimes \mathbf{p}) = I(\mathbf{w}) + \left(\sum_{i \in \operatorname{supp}(\mathbf{w})} w_i^q \right) I(\mathbf{p})$$

for all $n, k \geq 1$, $\mathbf{w} \in \Delta_n$, and $\mathbf{p} \in \Delta_k$;

ii. $I = cS_q$ for some $c \in \mathbb{R}$.

This characterization of q-logarithmic entropy first appeared as Theorem III.1 of [218]. Notably, it needs no regularity conditions whatsoever. This is in contrast to the case $q = 1$ of Shannon entropy, where some form of regularity is indispensable (Remark 2.5.2(iv)).

Proof By the observations just made, (ii) implies (i). Now assume (i). By symmetry, $I(\mathbf{w} \otimes \mathbf{p}) = I(\mathbf{p} \otimes \mathbf{w})$, so

$$I(\mathbf{w}) + \left(\sum_{i \in \text{supp}(\mathbf{w})} w_i^q \right) I(\mathbf{p}) = I(\mathbf{p}) + \left(\sum_{i \in \text{supp}(\mathbf{p})} p_i^q \right) I(\mathbf{w}),$$

or equivalently,

$$\left(\sum_{i \in \text{supp}(\mathbf{w})} w_i^q - 1 \right) I(\mathbf{p}) = \left(\sum_{i \in \text{supp}(\mathbf{p})} p_i^q - 1 \right) I(\mathbf{w}),$$

for all $\mathbf{w} \in \Delta_n$ and $\mathbf{p} \in \Delta_k$. Take $\mathbf{w} = \mathbf{u}_2$: then

$$(2^{1-q} - 1) I(\mathbf{p}) = \left(\sum_{i \in \text{supp}(\mathbf{p})} p_i^q - 1 \right) I(\mathbf{u}_2)$$

for all $\mathbf{p} \in \Delta_k$. Since $q \neq 1$, we can define

$$c = \frac{1-q}{2^{1-q} - 1} I(\mathbf{u}_2),$$

and then $I = cS_q$. □

Remark 4.1.6 There have been several characterization theorems for the q-logarithmic entropies. One similar to Theorem 4.1.5 was published by Daróczy in 1970 [78], and also appears as Theorem 6.3.9 of the book of Aczél and Daróczy [3]. In one sense it is stronger than Theorem 4.1.5 (that is, has weaker hypotheses): where we have assumed that $I\colon \Delta_n \to \mathbb{R}$ is symmetric for all n, Daróczy assumed it only for $n = 3$. On the other hand, Daróczy's theorem essentially assumed the full q-chain rule for $I(\mathbf{w} \circ (\mathbf{p}^1, \ldots, \mathbf{p}^n))$ (equation (4.2)), rather than just the special case of $I(\mathbf{w} \otimes \mathbf{p})$ that we used.

The word 'essentially' here hides a historical wrinkle. In Remark 2.2.11, we noted that the chain rule for Shannon entropy is equivalent to the special case

$$H(pw_1, (1-p)w_1, w_2, \ldots, w_n) = H(\mathbf{w}) + w_1 H(p, 1-p),$$

by a simple inductive argument. Similarly, here, the q-chain rule of equation (4.2) is equivalent to the special case

$$S_q(pw_1, (1-p)w_1, w_2, \ldots, w_n) = S_q(\mathbf{w}) + w_1^q S_q(p, 1-p), \tag{4.5}$$

by the same simple inductive argument (given in Appendix A.1). So, it is reasonable to regard (4.2) and (4.5) as equivalent. But it was the special case (4.5), not the general case (4.2), that was a hypothesis in Daróczy's theorem.

The proof given by Daróczy was entirely different, involving a q-analogue of the 'fundamental equation of information theory' (equation (11.17)).

Other characterizations of S_q have been proved, but using stronger hypotheses than Theorem 4.1.5 to obtain the same conclusion (such as the theorem in Section 2 of Suyari [322], and Theorem V.2 of Furuichi [113]).

Just as ordinary entropy has a family of q-logarithmic deformations, so too does relative entropy.

Definition 4.1.7 Let $q \in \mathbb{R}$ and $\mathbf{p}, \mathbf{r} \in \Delta_n$. The **$q$-logarithmic entropy of p relative to r** is

$$S_q(\mathbf{p} \,\|\, \mathbf{r}) = -\sum_{i \in \mathrm{supp}(\mathbf{p})} p_i \ln_q \frac{r_i}{p_i} \in [0, \infty].$$

Explicitly, $S_1(\mathbf{p} \,\|\, \mathbf{r}) = H(\mathbf{p} \,\|\, \mathbf{r})$, and for $q \neq 1$,

$$S_q(\mathbf{p} \,\|\, \mathbf{r}) = \frac{1}{q-1}\left(\sum_{i \in \mathrm{supp}(\mathbf{p})} p_i^q r_i^{1-q} - 1 \right).$$

As for ordinary relative entropy $H(- \,\|\, -)$ (Section 3.1), we have

$$S_q(\mathbf{p} \,\|\, \mathbf{r}) < \infty \iff (\mathbf{p}, \mathbf{r}) \in A_n.$$

The definition of q-logarithmic relative entropy was given by Rathie and Kannappan in 1972 [289]. (They used a version adapted to base 2 logarithms, in the tradition of Havrda and Charvát described in Remark 4.1.4.) Their definition was taken up by Cressie and Read in 1984 ([73], Section 5), who used the base e version in statistical work on goodness-of-fit tests. It was rediscovered twice in physics in 1998, by Shiino [311] and Tsallis [333] independently.

Remark 4.1.8 As in the definition of non-relative q-logarithmic entropy (Remark 4.1.2), there is a choice in how to generalize the formula for ordinary relative entropy, given that $\ln_q(1/x) \neq -\ln_q(x)$. Again, making the other choice simply flips the parametrization:

$$\sum_{i \in \mathrm{supp}(\mathbf{p})} p_i \ln_q \frac{p_i}{r_i} = S_{2-q}(\mathbf{p} \,\|\, \mathbf{r}),$$

by equation (1.19). The choice made in Definition 4.1.7 has the advantage that, as in the case $q = 1$, the relative entropy $S_q(\mathbf{p} \,\|\, \mathbf{u}_n)$ is a function of $S_q(\mathbf{p})$ and n:

$$\begin{aligned} S_q(\mathbf{p} \,\|\, \mathbf{u}_n) &= n^{q-1}(\ln_q(n) - S_q(\mathbf{p})) \\ &= n^{q-1}(S_q(\mathbf{u}_n) - S_q(\mathbf{p})), \end{aligned}$$

as is easily checked.

Like its non-relative cousin, q-logarithmic relative entropy has an extremely simple characterization. It satisfies a chain rule

$$S_q(\mathbf{w} \circ (\mathbf{p}^1, \ldots, \mathbf{p}^n) \,\|\, \widetilde{\mathbf{w}} \circ (\widetilde{\mathbf{p}}^1, \ldots, \widetilde{\mathbf{p}}^n)) \\ = S_q(\mathbf{w} \,\|\, \widetilde{\mathbf{w}}) + \sum_{i \in \mathrm{supp}(\mathbf{w})} w_i^q \widetilde{w}_i^{1-q} S_q(\mathbf{p}^i \,\|\, \widetilde{\mathbf{p}}^i)$$

($\mathbf{w}, \widetilde{\mathbf{w}} \in \Delta_n$, $\mathbf{p}^i, \widetilde{\mathbf{p}}^i \in \Delta_{k_i}$), which specializes to a multiplication rule

$$S_q(\mathbf{w} \otimes \mathbf{p} \,\|\, \widetilde{\mathbf{w}} \otimes \widetilde{\mathbf{p}}) = S_q(\mathbf{w} \,\|\, \widetilde{\mathbf{w}}) + \Bigg(\sum_{i \in \mathrm{supp}(\mathbf{w})} w_i^q \widetilde{w}_i^{1-q} \Bigg) S_q(\mathbf{p} \,\|\, \widetilde{\mathbf{p}}) \tag{4.6}$$

($\mathbf{w}, \widetilde{\mathbf{w}} \in \Delta_n$, $\mathbf{p}, \widetilde{\mathbf{p}} \in \Delta_k$). Moreover, $S_q(- \,\|\, -)$ is permutation-invariant in the same sense as in the case $q = 1$ (Section 3.1). Equation (4.6) and permutation-invariance characterize $S_q(- \,\|\, -)$ uniquely up to a constant factor:

Theorem 4.1.9 *Let $1 \neq q \in \mathbb{R}$. Let $(I(- \,\|\, -) \colon A_n \to \mathbb{R})_{n \geq 1}$ be a sequence of functions. The following are equivalent:*

i. $I(- \,\|\, -)$ is permutation-invariant and satisfies the multiplication rule (4.6) *(with I in place of S_q);*
ii. $I(- \,\|\, -) = cS_q(- \,\|\, -)$ for some $c \in \mathbb{R}$.

This result first appeared as Theorem IV.1 of [218]. Compared with the characterization theorem for ordinary relative entropy (Theorem 3.5.1), it needs neither a regularity condition nor the vanishing axiom.

Proof The proof is very similar to that of Theorem 4.1.5. By the observations just made, (ii) implies (i). Now assume (i). By permutation-invariance,

$$I(\mathbf{p} \otimes \mathbf{r} \,\|\, \widetilde{\mathbf{p}} \otimes \widetilde{\mathbf{r}}) = I(\mathbf{r} \otimes \mathbf{p} \,\|\, \widetilde{\mathbf{r}} \otimes \widetilde{\mathbf{p}})$$

for all $(\mathbf{p}, \widetilde{\mathbf{p}}) \in A_n$ and $(\mathbf{r}, \widetilde{\mathbf{r}}) \in A_k$. So by the multiplication rule,

$$I(\mathbf{p} \,\|\, \widetilde{\mathbf{p}}) + \Big(\sum p_i^q \widetilde{p}_i^{1-q} \Big) I(\mathbf{r} \,\|\, \widetilde{\mathbf{r}}) = I(\mathbf{r} \,\|\, \widetilde{\mathbf{r}}) + \Big(\sum r_i^q \widetilde{r}_i^{1-q} \Big) I(\mathbf{p} \,\|\, \widetilde{\mathbf{p}}),$$

or equivalently,

$$\Big(\sum r_i^q \widetilde{r}_i^{1-q} - 1 \Big) I(\mathbf{p} \,\|\, \widetilde{\mathbf{p}}) = \Big(\sum p_i^q \widetilde{p}_i^{1-q} - 1 \Big) I(\mathbf{r} \,\|\, \widetilde{\mathbf{r}}).$$

Take $\mathbf{r} = (1, 0)$ and $\widetilde{\mathbf{r}} = \mathbf{u}_2$: then

$$(2^{q-1} - 1) I(\mathbf{p} \,\|\, \widetilde{\mathbf{p}}) = I((1, 0) \,\|\, \mathbf{u}_2) \Big(\sum p_i^q \widetilde{p}_i^{1-q} - 1 \Big)$$

for all $(\mathbf{p}, \widetilde{\mathbf{p}}) \in A_n$. Since $q \neq 1$, we can put

$$c = \frac{(q-1)I((1,0) \parallel \mathbf{u}_2)}{2^{q-1}-1},$$

and then $I(- \parallel -) = cS_q(- \parallel -)$. □

Remark 4.1.10 Other characterization theorems for q-logarithmic relative entropy have been proved. For example, Furuichi ([113], Section IV) obtained the same conclusion, but also assumed continuity and the full chain rule (or more precisely, an equivalent special case, as in Remark 4.1.6) instead of just the multiplication rule (4.6).

4.2 Power Means

We pause in our account of deformations of Shannon entropy to collect some basic facts about power means (also called generalized means). The reason for doing this now is that the language and theory of power means make possible a considerable streamlining of later material on Rényi entropies and diversity measures.

The reader not interested in means for their own sake may wish to read Definition 4.2.1 and then jump ahead to Section 4.3, referring back here only as necessary.

This section is essentially a list of properties satisfied by the power means, together with the terminology for those properties. A summary of the terminology can also be found in Appendix B. Means are a classical topic of analysis, and almost everything in this section can be found in Chapter II of Hardy, Littlewood and Pólya's book [137].

In what follows, n denotes a positive integer.

Definition 4.2.1 Let $t \in [-\infty, \infty]$, $\mathbf{p} \in \Delta_n$, and $\mathbf{x} \in [0, \infty)^n$. The **power mean of order t of $\mathbf{x}$, weighted by $\mathbf{p}$**, is defined for $0 < t < \infty$ by

$$M_t(\mathbf{p}, \mathbf{x}) = \left(\sum_{i \in \operatorname{supp}(\mathbf{p})} p_i x_i^t \right)^{1/t}, \tag{4.7}$$

for $-\infty < t < 0$ by

$$M_t(\mathbf{p}, \mathbf{x}) = \begin{cases} \left(\sum_{i \in \operatorname{supp}(\mathbf{p})} p_i x_i^t \right)^{1/t} & \text{if } x_i > 0 \text{ for all } i \in \operatorname{supp}(\mathbf{p}), \\ 0 & \text{otherwise,} \end{cases} \tag{4.8}$$

and for the remaining values of t by

$$M_{-\infty}(\mathbf{p},\mathbf{x}) = \min_{i\in\mathrm{supp}(\mathbf{p})} x_i,$$
$$M_0(\mathbf{p},\mathbf{x}) = \prod_{i\in\mathrm{supp}(\mathbf{p})} x_i^{p_i},$$
$$M_{\infty}(\mathbf{p},\mathbf{x}) = \max_{i\in\mathrm{supp}(\mathbf{p})} x_i.$$

The various exceptional cases in this definition are justified by continuity, as detailed after the following examples.

Examples 4.2.2 i. The mean of order 1 is the arithmetic mean $\sum p_i x_i$ of $\mathbf{x}$ weighted by $\mathbf{p}$.

ii. The mean of order 0 is the geometric mean of $\mathbf{x}$ weighted by $\mathbf{p}$.

iii. The mean of order -1 is the harmonic mean

$$\frac{1}{\frac{p_1}{x_1}+\cdots+\frac{p_n}{x_n}}$$

of $\mathbf{x}$ weighted by $\mathbf{p}$.

Example 4.2.3 Taking $\mathbf{p} = \mathbf{u}_n$ gives the **unweighted** (or uniformly weighted) power means $M_t(\mathbf{u}_n, \mathbf{x})$.

Example 4.2.4 For each $t \in [-\infty, \infty]$, the power mean M_t has at least the basic properties of an average:

$$M_t(\mathbf{p}, (x, \ldots, x)) = x$$

for all $\mathbf{p} \in \Delta_n$ and $x \in [0, \infty)$, and

$$\min_{i\in\mathrm{supp}(\mathbf{p})} x_i \leq M_t(\mathbf{p},\mathbf{x}) \leq \max_{i\in\mathrm{supp}(\mathbf{p})} x_i$$

for all $\mathbf{p} \in \Delta_n$ and $\mathbf{x} \in [0, \infty)^n$. The rest of this section is devoted to investigating the properties of power means in greater depth.

We now prove three statements on the continuity of power means $M_t(\mathbf{p}, \mathbf{x})$. The first is on continuity in $\mathbf{x}$.

Lemma 4.2.5 *Let $t \in [-\infty, \infty]$ and $\mathbf{p} \in \Delta_n$. Then the function*

$$M_t(\mathbf{p}, -)\colon [0, \infty)^n \to [0, \infty)$$

is continuous.

Proof Let $\mathbf{x} \in [0, \infty)^n$. From Definition 4.2.1, it is immediate that $M_t(\mathbf{p}, \mathbf{x})$ is continuous at $\mathbf{x}$ except perhaps in the case where $t \in (-\infty, 0)$ and $x_i = 0$ for some $i \in \mathrm{supp}(\mathbf{p})$. So, let $t \in (-\infty, 0)$ and suppose that, say, $x_1 = 0$ with $1 \in \mathrm{supp}(\mathbf{p})$. It suffices to show that $M_t(\mathbf{p}, \mathbf{y}) \to 0$ as $\mathbf{y} \to \mathbf{x}$ with $y_i > 0$ for all $i \in \mathrm{supp}(\mathbf{p})$. And indeed, for such $\mathbf{y}$,

$$\begin{aligned} \left|M_t(\mathbf{p}, \mathbf{y})\right| &= \Bigg(\sum_{i \in \mathrm{supp}(\mathbf{p})} p_i y_i^t \Bigg)^{1/t} \\ &\leq (p_1 y_1^t)^{1/t} \\ &= p_1^{1/t} y_1 \\ &\to p_1^{1/t} x_1 = 0 \end{aligned}$$

as $\mathbf{y} \to \mathbf{x}$, as required. □

The continuity properties of $M_t(\mathbf{p}, \mathbf{x})$ in $\mathbf{p}$ are more delicate. Indeed, the power means of order ≤ 0 are *not* continuous in $\mathbf{p}$: for when $t \leq 0$,

$$M_t((\varepsilon, 1 - \varepsilon), (0, 1)) = 0$$

for all $\varepsilon \in (0, 1]$, whereas

$$M_t((0, 1), (0, 1)) = 1.$$

Discontinuities do not only arise from zero values of x_i. For instance, $M_{-\infty}(\mathbf{p}, (1, 2))$ is not continuous in $\mathbf{p}$, since

$$M_{-\infty}((\varepsilon, 1 - \varepsilon), (1, 2)) = 1, \quad M_{-\infty}((0, 1), (1, 2)) = 2$$

for all $\varepsilon \in (0, 1]$. There is a similar counterexample for M_∞. But we do have the following.

Lemma 4.2.6 *i. For all $t \in [-\infty, \infty]$, the function*

$$M_t(-, -) \colon \Delta_n^\circ \times [0, \infty)^n \to [0, \infty)$$

is continuous.

ii. For all $t \in (-\infty, \infty)$, the function

$$M_t(-, -) \colon \Delta_n \times (0, \infty)^n \to (0, \infty)$$

is continuous.

iii. For all $t \in (0, \infty)$, the function

$$M_t(-, -) \colon \Delta_n \times [0, \infty)^n \to [0, \infty)$$

is continuous.

Proof Part (i) is immediate from the definition. For parts (ii) and (iii), just note that in the cases at hand, the formulas for M_t are unchanged if i is allowed to range over all of $\{1, \ldots, n\}$ instead of only $\text{supp}(\mathbf{p})$. □

Our third and final continuity lemma states that power means are continuous in their order.

Lemma 4.2.7 *Let* $\mathbf{p} \in \Delta_n$ *and* $\mathbf{x} \in [0, \infty)^n$. *Then* $M_t(\mathbf{p}, \mathbf{x})$ *is continuous in* $t \in [-\infty, \infty]$.

Proof This is clear except perhaps at $t = 0$ and $t = \pm\infty$.

For continuity at $t = 0$, first suppose that $x_i > 0$ for all $i \in \text{supp}(\mathbf{p})$. When t is finite and nonzero,

$$\log M_t(\mathbf{p}, \mathbf{x}) = \frac{\log(\sum_{i \in \text{supp}(\mathbf{p})} p_i x_i^t)}{t}. \tag{4.9}$$

As $t \to 0$,

$$\log\left(\sum_{i \in \text{supp}(\mathbf{p})} p_i x_i^t \right) \to \log\left(\sum_{i \in \text{supp}(\mathbf{p})} p_i \right) = 0,$$

so we can apply l'Hôpital's rule to equation (4.9), giving

$$\begin{aligned}
\lim_{t \to 0} \log M_t(\mathbf{p}, \mathbf{x}) &= \lim_{t \to 0} \frac{\sum p_i x_i^t \log x_i}{\sum p_i x_i^t} \\
&= \sum p_i \log x_i \\
&= \log M_0(\mathbf{p}, \mathbf{x}),
\end{aligned}$$

where all sums are over $i \in \text{supp}(\mathbf{p})$. Hence the map $t \mapsto M_t(\mathbf{p}, \mathbf{x})$ is continuous at $t = 0$.

Now suppose that $x_i = 0$ for some $i \in \text{supp}(\mathbf{p})$. By definition, $M_t(\mathbf{p}, \mathbf{x}) = 0$ for all $t \leq 0$, so it suffices to show that $M_t(\mathbf{p}, \mathbf{x}) \to 0$ as $t \to 0+$. For $t \in (0, \infty)$,

$$0 \leq M_t(\mathbf{p}, \mathbf{x}) = \left(\sum_{i \in \text{supp}(\mathbf{p})} p_i x_i^t \right)^{1/t} \leq M_\infty(\mathbf{p}, \mathbf{x}) \cdot \left(\sum_{i \in \text{supp}(\mathbf{p}) \cap \text{supp}(\mathbf{x})} p_i \right)^{1/t}.$$

But $\sum_{i \in \text{supp}(\mathbf{p}) \cap \text{supp}(\mathbf{x})} p_i < 1$, so our upper bound on $M_t(\mathbf{p}, \mathbf{x})$ converges to 0 as $t \to 0+$. Hence also $M_t(\mathbf{p}, \mathbf{x}) \to 0$ as $t \to 0+$, as required.

For continuity at $t = \infty$, suppose without loss of generality that $\max_{i \in \text{supp}(\mathbf{p})} x_i$ is achieved at $i = 1$. Then for $t \in (0, \infty)$,

$$M_t(\mathbf{p}, \mathbf{x}) \leq \left(\sum_{i \in \text{supp}(\mathbf{p})} p_i x_1^t \right)^{1/t} = x_1.$$

On the other hand,

$$M_t(\mathbf{p},\mathbf{x}) \geq (p_1 x_1^t)^{1/t} = p_1^{1/t} x_1 \to x_1$$

as $t \to \infty$. Hence

$$M_t(\mathbf{p},\mathbf{x}) \to x_1 = M_\infty(\mathbf{p},\mathbf{x})$$

as $t \to \infty$, as required. The proof for $M_{-\infty}$ is similar. □

We now come to the celebrated inequality of the arithmetic and geometric means:

$$\frac{1}{n}\sum_{i=1}^{n} x_i \geq \left(\prod_{i=1}^{n} x_i\right)^{1/n}$$

for all $x_1, \ldots, x_n \geq 0$. This is a very special case of the following classical and fundamental result (Theorem 9 of Hardy, Littlewood and Pólya [137], for instance). Recall from Remark 1.1.15 that we use the word 'increasing' in the non-strict sense.

Theorem 4.2.8 *Let* $\mathbf{p} \in \Delta_n$ *and* $\mathbf{x} \in [0,\infty)^n$, *with* $x_i > 0$ *for all* $i \in \operatorname{supp}(\mathbf{p})$. *Then the function*

$$\begin{array}{ccc} [-\infty,\infty] & \to & [0,\infty) \\ t & \mapsto & M_t(\mathbf{p},\mathbf{x}) \end{array}$$

is increasing. It is constant if $x_i = x_j$ *for all* $i, j \in \operatorname{supp}(\mathbf{p})$, *and strictly increasing otherwise.*

Proof If the coordinates x_i of $\mathbf{x}$ have the same value x for all $i \in \operatorname{supp}(\mathbf{p})$, then evidently $M_t(\mathbf{p},\mathbf{x}) = x$ for all $t \in [-\infty,\infty]$. Supposing otherwise, we have to prove that $M_t(\mathbf{p},\mathbf{x})$ is strictly increasing in $t \in [-\infty,\infty]$. We will prove that $\frac{d}{dt}\log M_t(\mathbf{p},\mathbf{x}) > 0$ for all $t \in (-\infty,0) \cup (0,\infty)$. Since $M_t(\mathbf{p},\mathbf{x})$ is continuous in $t \in [-\infty,\infty]$ (Lemma 4.2.7), this suffices. For real $t \neq 0$,

$$\begin{aligned}
\frac{d}{dt}\log M_t(\mathbf{p},\mathbf{x}) &= \frac{d}{dt}\left(\frac{\log \sum p_i x_i^t}{t}\right) \\
&= \frac{t(\sum p_i x_i^t \log x_i)/(\sum p_i x_i^t) - \log \sum p_i x_i^t}{t^2} \\
&= \frac{\sum p_i x_i^t \log x_i^t - (\sum p_i x_i^t)\log \sum p_i x_i^t}{t^2 \sum p_i x_i^t} \\
&= \frac{-\sum p_i \partial(x_i^t) + \partial(\sum p_i x_i^t)}{t^2 \sum p_i x_i^t},
\end{aligned} \tag{4.10}$$

where all sums are over $i \in \operatorname{supp}(\mathbf{p})$ and $\partial(x) = -x\log x$ (as in equation (2.4)).

But $\partial''(x) = -1/x < 0$ for all $x > 0$, so ∂ is strictly concave. Hence by equation (4.10),

$$\tfrac{d}{dt} \log M_t(\mathbf{p}, \mathbf{x}) \geq 0,$$

with equality if and only if $x_i^t = x_j^t$ for all $i, j \in \mathrm{supp}(\mathbf{p})$. But $t \neq 0$, so equality only holds if $x_i = x_j$ for all $i, j \in \mathrm{supp}(\mathbf{p})$, contrary to our earlier assumption. Hence the inequality is strict, as required. □

There is a simple duality law for power means:

$$M_{-t}(\mathbf{p}, \mathbf{x}) = \frac{1}{M_t(\mathbf{p}, 1/\mathbf{x})} \tag{4.11}$$

for all $t \in [-\infty, \infty]$, $\mathbf{p} \in \Delta_n$, and $\mathbf{x} \in (0, \infty)^n$. Here $1/\mathbf{x}$ denotes the vector $(1/x_1, \ldots, 1/x_n)$. For instance, in the case $t = 1$, the harmonic mean is the reciprocal of the arithmetic means of $1/x_1, \ldots, 1/x_n$.

Remark 4.2.9 Often in this text, we will want to perform coordinatewise algebraic operations on vectors. For instance, given $\mathbf{x}, \mathbf{y} \in \mathbb{R}^n$, we will use not only the (coordinatewise) sum and difference $\mathbf{x} + \mathbf{y}$ and $\mathbf{x} - \mathbf{y}$, but also the coordinatewise product and quotient

$$\mathbf{xy} = (x_1 y_1, \ldots, x_n y_n), \qquad \mathbf{x}/\mathbf{y} = (x_1/y_1, \ldots, x_n/y_n)$$

(with the usual caveats regarding $y_i = 0$ in the latter case). This is just the standard notation for the product and quotient of real-valued functions on a set S, applied to $S = \{1, \ldots, n\}$.

We now run through some basic properties satisfied by the power means

$$(M_t \colon \Delta_n \times [0, \infty)^n \to [0, \infty))_{n \geq 1}$$

of every order $t \in [-\infty, \infty]$. For later purposes, it is useful to set up the terminology in the generality of a sequence of functions

$$(M \colon \Delta_n \times I^n \to I)_{n \geq 1},$$

where I is an arbitrary real interval. The most important cases are $I = [0, \infty)$ and $I = (0, \infty)$.

Definition 4.2.10 Let I be a real interval and let $(M \colon \Delta_n \times I^n \to I)_{n \geq 1}$ be a sequence of functions.

i. M is **symmetric** if $M(\mathbf{p}, \mathbf{x}) = M(\mathbf{p}\sigma, \mathbf{x}\sigma)$ for all $n \geq 1$, $\mathbf{p} \in \Delta_n$, $\mathbf{x} \in I^n$, and permutations σ of $\{1, \ldots, n\}$, where $\mathbf{p}\sigma$ and $\mathbf{x}\sigma$ are defined as in equation (3.2).

ii. M is **absence-invariant** if whenever $\mathbf{p} \in \Delta_n$, $\mathbf{x} \in I^n$ and $1 \leq i \leq n$ with $p_i = 0$, then

$$M(\mathbf{p}, \mathbf{x}) = M((p_1, \ldots, p_{i-1}, p_{i+1}, \ldots, p_n), (x_1, \ldots, x_{i-1}, x_{i+1}, \ldots, x_n)).$$

iii. M has the **repetition** property if whenever $\mathbf{p} \in \Delta_n$, $\mathbf{x} \in I^n$ and $1 \leq i < n$ with $x_i = x_{i+1}$, then

$$\begin{aligned} &M(\mathbf{p}, \mathbf{x}) \\ &\quad = M((p_1, \ldots, p_{i-1}, p_i + p_{i+1}, p_{i+2}, \ldots, p_n), (x_1, \ldots, x_{i-1}, x_i, x_{i+2}, \ldots, x_n)). \end{aligned}$$

Absence-invariance states that M behaves logically with respect to elements x_i that are absent (have zero weight): such elements might as well be ignored.

Lemma 4.2.11 *Let $t \in [-\infty, \infty]$. Then M_t has the symmetry, absence-invariance and repetition properties.*

A direct proof of this lemma is, of course, elementary, but it is enlightening to derive all three properties from a single general law, as follows. Let

$$f\colon \{1, \ldots, m\} \to \{1, \ldots, n\}$$

be a map of finite sets. Any distribution $\mathbf{p} \in \Delta_m$ gives rise to a pushforward distribution $f\mathbf{p} \in \Delta_n$ (Definition 2.1.10). On the other hand, any vector $\mathbf{x} \in [0, \infty)^n$ can be pulled back along f to give a vector $\mathbf{x}f \in [0, \infty)^m$, where

$$(\mathbf{x}f)_i = x_{f(i)}$$

($i \in \{1, \ldots, m\}$).

Definition 4.2.12 Let I be a real interval. A sequence of functions $(M\colon \Delta_n \times I^n \to I)_{n \geq 1}$ is **natural** if

$$M(f\mathbf{p}, \mathbf{x}) = M(\mathbf{p}, \mathbf{x}f)$$

for all $m, n \geq 1$, $\mathbf{p} \in \Delta_m$, $\mathbf{x} \in I^n$, and maps of sets

$$f\colon \{1, \ldots, m\} \to \{1, \ldots, n\}.$$

Remark 4.2.13 If we write x_j as $\phi(j)$, so that ϕ is a function $\{1, \ldots, n\} \to [0, \infty)$, then $\mathbf{x}f = \phi \circ f$. If we also write $M(\mathbf{p}, -)$ as $\int - \, d\mathbf{p}$ then naturality states that

$$\int \phi \, d(f\mathbf{p}) = \int (\phi \circ f) \, d\mathbf{p},$$

the standard formula for integration under a change of variable. However, this notation is misleading: unlike an ordinary integral, $M(\mathbf{p}, \mathbf{x})$ need not be linear in $\mathbf{x}$ (and is not when $M = M_t$ for $t \neq 1$).

Lemma 4.2.14 (Naturality) *For each* $t \in [-\infty, \infty]$, *the power mean* M_t *on* $[0, \infty)$ *is natural.*

Proof Take $\mathbf{p}$, $\mathbf{x}$, and f as in Definition 4.2.12. We have to show that $M_t(f\mathbf{p}, \mathbf{x}) = M_t(\mathbf{p}, \mathbf{x}f)$. First suppose that $t \neq 0, \pm\infty$ and that $x_j > 0$ for all $j \in \{1, \ldots, n\}$. Then

$$\begin{aligned} M_t(f\mathbf{p}, \mathbf{x}) &= \left(\sum_{j \in \text{supp}(f\mathbf{p})} (f\mathbf{p})_j x_j^t \right)^{1/t} \\ &= \left(\sum_{j \in \text{supp}(f\mathbf{p})} \sum_{i \in f^{-1}(j)} p_i x_j^t \right)^{1/t} \\ &= \left(\sum_{i \in \text{supp}(\mathbf{p})} p_i x_{f(i)}^t \right)^{1/t} \\ &= M_t(\mathbf{p}, \mathbf{x}f), \end{aligned}$$

as required. The case where $x_j = 0$ for some values of j follows by continuity of $M_t(\mathbf{p}, \mathbf{x})$ in $\mathbf{x}$ (Lemma 4.2.5), and the result for $t = 0$ and $t = \pm\infty$ follows by continuity of M_t in t (Lemma 4.2.7). □

Proof of Lemma 4.2.11 We use the naturality of the power means for all three parts. Write $\mathbf{n} = \{1, \ldots, n\}$. Symmetry follows by taking f to be a bijection $\mathbf{n} \to \mathbf{n}$. Absence-invariance follows by taking f to be the order-preserving injection $\mathbf{n} - \mathbf{1} \to \mathbf{n}$ that omits i from its image. The repetition property follows by taking f to be the order-preserving surjection $\mathbf{n} \to \mathbf{n} - \mathbf{1}$ that identifies i with $i + 1$. □

Remark 4.2.15 The absence-invariance of the power means implies that $M_t(\mathbf{p}, \mathbf{x})$ is unaffected by the value of x_i for coordinates i such that $p_i = 0$. Indeed, writing $\text{supp}(\mathbf{p}) = \{i_1, \ldots, i_k\}$ with $i_1 < \cdots < i_k$, we have

$$M_t(\mathbf{p}, \mathbf{x}) = M_t((p_{i_1}, \ldots, p_{i_k}), (x_{i_1}, \ldots, x_{i_k}))$$

for all $\mathbf{x}$, by absence-invariance and induction. Hence

$$M_t(\mathbf{p}, \mathbf{x}) = M_t(\mathbf{p}, \mathbf{y})$$

whenever $\mathbf{x}, \mathbf{y} \in [0, \infty)^n$ with $x_i = y_i$ for all $i \in \text{supp}(\mathbf{p})$.

Because of this, the expression $M_t(\mathbf{p}, \mathbf{x})$ has a clear meaning even if x_i is *undefined* for some or all $i \notin \text{supp}(\mathbf{p})$. (We can arbitrarily put $x_i = 0$ or $x_i = 17$ for all such i; it makes no difference.) For example, the expression

$$M_t(\mathbf{p}, 1/\mathbf{p})$$

has a clear meaning for all $\mathbf{p} \in \Delta_n$, even if $p_i = 0$ for some i; writing $\text{supp}(\mathbf{p}) = \{i_1, \ldots, i_k\}$ as above, it is understood to mean

$$M_t((p_{i_1}, \ldots, p_{i_k}), (1/p_{i_1}, \ldots, 1/p_{i_k})).$$

We adopt the convention throughout this text that power means $M_t(\mathbf{p}, \mathbf{x})$ are valid expressions even if x_i is undefined for some $i \notin \text{supp}(\mathbf{p})$, and are to be interpreted as just described. This convention is strictly analogous to the standard interpretation of integral notation $\int f\, d\mu$, for a function f and a measure μ: the integral is unaffected by the value of f off the support of μ, and has an unambiguous meaning even if f is undefined there.

A minimal requirement on anything called a mean is that the mean of several copies of x should be x.

Definition 4.2.16 Let I be a real interval. A sequence of functions $(M\colon \Delta_n \times I^n \to I)_{n \geq 1}$ is **consistent** if

$$M(\mathbf{p}, (x, \ldots, x)) = x$$

for all $n \geq 1$, $\mathbf{p} \in \Delta_n$, and $x \in I$.

Lemma 4.2.17 *For each $t \in [-\infty, \infty]$, the power mean M_t is consistent.*

Proof Trivial. □

For $\mathbf{x}, \mathbf{y} \in \mathbb{R}^n$, write $\mathbf{x} \leq \mathbf{y}$ if $x_i \leq y_i$ for all $i \in \{1, \ldots, n\}$.

Definition 4.2.18 Let I be a real interval and let $(M\colon \Delta_n \times I^n \to I)_{n \geq 1}$ be a sequence of functions.

i. M is **increasing** if

$$\mathbf{x} \leq \mathbf{y} \implies M(\mathbf{p}, \mathbf{x}) \leq M(\mathbf{p}, \mathbf{y})$$

for all $n \geq 1$, $\mathbf{p} \in \Delta_n$, and $\mathbf{x}, \mathbf{y} \in I^n$.

ii. M is **strictly increasing** if

$$(\mathbf{x} \leq \mathbf{y} \text{ and } x_i < y_i \text{ for some } i \in \text{supp}(\mathbf{p})) \implies M(\mathbf{p}, \mathbf{x}) < M(\mathbf{p}, \mathbf{y})$$

for all $n \geq 1$, $\mathbf{p} \in \Delta_n$, and $\mathbf{x}, \mathbf{y} \in I^n$.

Whether the power mean M_t is *strictly* increasing depends on both the order t and whether the domain of definition is taken to be $[0, \infty)$ or $(0, \infty)$, as follows.

Lemma 4.2.19 *i. For all $t \in [-\infty, \infty]$, the power mean M_t on $[0, \infty)$ is increasing.*

ii. For all $t \in (-\infty, \infty)$, the power mean M_t on $(0, \infty)$ is strictly increasing.

iii. For all $t \in (0, \infty)$, the power mean M_t on $[0, \infty)$ is strictly increasing.

Proof Elementary. □

Remark 4.2.20 The careful statement of Lemma 4.2.19 is necessary because of various limiting counterexamples. The means $M_{\pm\infty}$ are not strictly increasing on $(0, \infty)$, since, for instance,

$$M_\infty(\mathbf{u}_2, (1, 3)) = 3 = M_\infty(\mathbf{u}_2, (2, 3)).$$

When $t \in [-\infty, 0]$, the mean M_t is not strictly increasing on $[0, \infty)$; for example,

$$M_t(\mathbf{u}_2, (0, 1)) = 0 = M_t(\mathbf{u}_2, (0, 2)).$$

Definition 4.2.21 Let I be a real interval closed under multiplication. A sequence of functions $(M\colon \Delta_n \times I^n \to I)_{n\geq 1}$ is **homogeneous** if

$$M(\mathbf{p}, c\mathbf{x}) = cM(\mathbf{p}, \mathbf{x})$$

for all $n \geq 1$, $\mathbf{p} \in \Delta_n$, $c \in I$, and $\mathbf{x} \in I^n$.

The hypothesis on I guarantees that $M(\mathbf{p}, c\mathbf{x})$ is defined.

Lemma 4.2.22 *For each $t \in [-\infty, \infty]$, the power mean M_t on $[0, \infty)$ is homogeneous.*

Proof Elementary. □

The most important algebraic property of the power means is a chain rule. Given vectors

$$\mathbf{x}^1 = (x^1_1, \dots, x^1_{k_1}) \in \mathbb{R}^{k_1}, \ \dots, \ \mathbf{x}^n = (x^n_1, \dots, x^n_{k_n}) \in \mathbb{R}^{k_n},$$

write

$$\mathbf{x}^1 \oplus \cdots \oplus \mathbf{x}^n = (x^1_1, \dots, x^1_{k_1}, \ \dots, \ x^n_1, \dots, x^n_{k_n}) \in \mathbb{R}^{k_1+\cdots+k_n}.$$

Definition 4.2.23 Let I be a real interval. A sequence of functions $(M\colon \Delta_n \times I^n \to I)_{n\geq 1}$ satisfies the **chain rule** if

$$M(\mathbf{w} \circ (\mathbf{p}^1, \dots, \mathbf{p}^n), \mathbf{x}^1 \oplus \cdots \oplus \mathbf{x}^n) = M\big(\mathbf{w}, (M(\mathbf{p}^1, \mathbf{x}^1), \dots, M(\mathbf{p}^n, \mathbf{x}^n))\big)$$

for all $\mathbf{w} \in \Delta_n$, $\mathbf{p}^i \in \Delta_{k_i}$, and $\mathbf{x}^i \in I^{k_i}$.

Proposition 4.2.24 (Chain rule) *For each $t \in [-\infty, \infty]$, the power mean M_t on $[0, \infty)$ satisfies the chain rule.*

Proof By the continuity of the power means in their second argument and in their order (Lemmas 4.2.5 and 4.2.7), it is enough to prove the equation in Definition 4.2.23 when $x^i_j > 0$ for all i, j and $0 \neq t \in \mathbb{R}$. Then

$$\begin{aligned} M_t(\mathbf{w} \circ (\mathbf{p}^1, \dots, \mathbf{p}^n), \mathbf{x}^1 \oplus \cdots \oplus \mathbf{x}^n) &= \left\{ \sum_{i=1}^{n} \sum_{j=1}^{k_i} w_i p^i_j (x^i_j)^t \right\}^{1/t} \\ &= \left\{ \sum_{i=1}^{n} w_i M_t(\mathbf{p}^i, \mathbf{x}^i)^t \right\}^{1/t} \\ &= M_t\Big(\mathbf{w}, (M_t(\mathbf{p}^1, \mathbf{x}^1), \dots, M_t(\mathbf{p}^n, \mathbf{x}^n))\Big), \end{aligned}$$

as required. □

An important consequence of the chain rule is that in order to calculate the mean of $\mathbf{x}^1 \oplus \cdots \oplus \mathbf{x}^n$ weighted by $\mathbf{w} \circ (\mathbf{p}^1, \dots, \mathbf{p}^n)$, we only need to know $\mathbf{w}$ and the means $M_t(\mathbf{p}^i, \mathbf{x}_i)$, not $\mathbf{p}^i$ and $\mathbf{x}^i$ themselves. We refer to this property as modularity, echoing the definition of modularity for diversity measures (p. 56). (Modularity of this kind has also been called **quasilinearity**, as in Section 6.21 of Hardy, Littlewood and Pólya [137].) Formally:

Definition 4.2.25 Let I be a real interval. A sequence of functions $(M\colon \Delta_n \times I^n \to I)_{n \geq 1}$ is **modular** if

$$\begin{aligned} &M(\mathbf{p}^i, \mathbf{x}^i) = M(\widetilde{\mathbf{p}}^i, \widetilde{\mathbf{x}}^i) \text{ for all } i \in \{1, \dots, n\} \\ \implies &M(\mathbf{w} \circ (\mathbf{p}^1, \dots, \mathbf{p}^n), \mathbf{x}^1 \oplus \cdots \oplus \mathbf{x}^n) = M(\mathbf{w} \circ (\widetilde{\mathbf{p}}^1, \dots, \widetilde{\mathbf{p}}^n), \widetilde{\mathbf{x}}^1 \oplus \cdots \oplus \widetilde{\mathbf{x}}^n) \end{aligned}$$

for all $n, k_1, \dots, k_n, \widetilde{k}_1, \dots, \widetilde{k}_n \geq 1$ and $\mathbf{w} \in \Delta_n$, $\mathbf{p}^i \in \Delta_{k_i}$, $\widetilde{\mathbf{p}}^i \in \Delta_{\widetilde{k}_i}$, $\mathbf{x}^i \in I^{k_i}$, $\widetilde{\mathbf{x}}^i \in I^{\widetilde{k}_i}$.

Corollary 4.2.26 *For each $t \in [-\infty, \infty]$, the power mean M_t on $[0, \infty)$ is modular.* □

As for diversity of order 1 (equation (2.11), p. 56), the chain rule also implies a multiplicativity property. For $\mathbf{x} \in \mathbb{R}^n$ and $\mathbf{y} \in \mathbb{R}^k$, write

$$\mathbf{x} \otimes \mathbf{y} = (x_1 y_1, \dots, x_1 y_k, \; \dots, \; x_n y_1, \dots, x_n y_k) \in \mathbb{R}^{nk}. \tag{4.12}$$

(To justify the notation: if the tensor product of vector spaces $\mathbb{R}^n \otimes \mathbb{R}^k$ is identified with $\mathbb{R}^{nk}$ in the standard way, then the vector usually written as $\mathbf{x} \otimes \mathbf{y} \in \mathbb{R}^n \otimes \mathbb{R}^k$ corresponds to what we are now writing as $\mathbf{x} \otimes \mathbf{y} \in \mathbb{R}^{nk}$.)

Definition 4.2.27 Let I be a real interval closed under multiplication. A sequence of functions $(M\colon \Delta_n \times I^n \to I)_{n \geq 1}$ is **multiplicative** if

$$M(\mathbf{p} \otimes \mathbf{p}', \mathbf{x} \otimes \mathbf{x}') = M(\mathbf{p}, \mathbf{x}) M(\mathbf{p}', \mathbf{x}')$$

for all $n, n' \geq 1$, $\mathbf{p} \in \Delta_n$, $\mathbf{p}' \in \Delta_{n'}$, $\mathbf{x} \in I^n$, and $\mathbf{x}' \in I^{n'}$.

Corollary 4.2.28 *For each $t \in [-\infty, \infty]$, the power mean M_t on $[0, \infty)$ is multiplicative.*

Proof We apply the chain rule (Proposition 4.2.24) to the composite distribution

$$\mathbf{p} \circ (\mathbf{p}', \ldots, \mathbf{p}') = \mathbf{p} \otimes \mathbf{p}'$$

and the vector

$$x_1\mathbf{x}' \oplus \cdots \oplus x_n\mathbf{x}' = \mathbf{x} \otimes \mathbf{x}'.$$

Doing this gives

$$M_t(\mathbf{p} \otimes \mathbf{p}', \mathbf{x} \otimes \mathbf{x}') = M_t\big(\mathbf{p}, (M_t(\mathbf{p}, x_1\mathbf{x}'), \ldots, M_t(\mathbf{p}, x_n\mathbf{x}'))\big).$$

Hence by two uses of homogeneity,

$$\begin{aligned} M_t(\mathbf{p} \otimes \mathbf{p}', \mathbf{x} \otimes \mathbf{x}') &= M_t\big(\mathbf{p}, (x_1 M_t(\mathbf{p}, \mathbf{x}'), \ldots, x_n M_t(\mathbf{p}, \mathbf{x}'))\big) \\ &= M_t(\mathbf{p}, \mathbf{x}) M_t(\mathbf{p}', \mathbf{x}'). \end{aligned}$$

□

The multiplicativity property is remarkably powerful, as we shall see in Chapter 9.

Finally, we record for later purposes a simple result connecting the power means with the q-logarithms.

Lemma 4.2.29 *Let $q \in [0, \infty)$, $\mathbf{p} \in \Delta_n$, and $\mathbf{x} \in [0, \infty)^n$, with $x_i > 0$ for all $i \in \operatorname{supp}(\mathbf{p})$. Then*

$$\ln_q M_{1-q}(\mathbf{p}, \mathbf{x}) = M_1(\mathbf{p}, \ln_q \mathbf{x}),$$

where $\ln_q \mathbf{x} = (\ln_q x_1, \ldots, \ln_q x_n)$.

Proof Trivial algebraic manipulation. □

4.3 Rényi Entropies and Hill Numbers

Historically, the first deformations of Shannon entropy were the Rényi entropies [294], defined as follows.

Definition 4.3.1 Let $q \in [-\infty, \infty]$, $n \geq 1$, and $\mathbf{p} \in \Delta_n$. The **Rényi entropy of order** q of $\mathbf{p}$ is

$$H_q(\mathbf{p}) = \log M_{1-q}(\mathbf{p}, 1/\mathbf{p}), \tag{4.13}$$

where $1/\mathbf{p} = (1/p_1, \ldots, 1/p_n)$.

Here we use the convention introduced in Remark 4.2.15, which covers the possibility that $1/p_i$ is undefined for some values of i.

Explicitly,

$$H_q(\mathbf{p}) = \frac{1}{1-q} \log \sum_{i \in \text{supp}(\mathbf{p})} p_i^q$$

for $q \neq 1, \pm\infty$, and

$$\begin{aligned} H_{-\infty}(\mathbf{p}) &= -\log \min_{i \in \text{supp}(\mathbf{p})} p_i, \\ H_1(\mathbf{p}) &= H(\mathbf{p}), \\ H_{\infty}(\mathbf{p}) &= -\log \max_{i \in \text{supp}(\mathbf{p})} p_i. \end{aligned}$$

By Lemma 4.2.7, $H_q(\mathbf{p})$ is continuous in q.

Rényi introduced these entropies in 1961 [294]. One of his purposes in doing so was to point out that Shannon entropy is far from the only useful quantity with the logarithmic property

$$H(\mathbf{p} \otimes \mathbf{r}) = H(\mathbf{p}) + H(\mathbf{r}) \tag{4.14}$$

($\mathbf{p} \in \Delta_n, \mathbf{r} \in \Delta_m$). Indeed, H_q has this same property for all $q \in [-\infty, \infty]$. This follows from the multiplicativity of the power means (Corollary 4.2.28), since

$$\begin{aligned} H_q(\mathbf{p} \otimes \mathbf{r}) &= \log M_{1-q}\left(\mathbf{p} \otimes \mathbf{r}, \tfrac{1}{\mathbf{p}} \otimes \tfrac{1}{\mathbf{r}}\right) \\ &= \log\left(M_{1-q}\left(\mathbf{p}, \tfrac{1}{\mathbf{p}}\right) M_{1-q}\left(\mathbf{r}, \tfrac{1}{\mathbf{r}}\right)\right) \\ &= H_q(\mathbf{p}) + H_q(\mathbf{r}). \end{aligned}$$

In this respect, the Rényi entropies resemble Shannon entropy more closely than the q-logarithmic entropies do. But there is a price to pay. Whereas the asymmetry of the multiplication formula for the q-logarithmic entropies (equation (4.3)) could be exploited to prove an extremely simple characterization theorem (Theorem 4.1.5), this avenue is not open to us for the Rényi entropies. We do prove a characterization theorem for the Rényi entropy of any given order (Section 4.5), but it is more involved.

The q-logarithmic and Rényi entropies each determine the other, since both are invertible functions of $\sum p_i^q$. Explicitly,

$$S_q(\mathbf{p}) = \frac{1}{1-q}\left(\exp((1-q)H_q(\mathbf{p})) - 1\right), \tag{4.15}$$

$$H_q(\mathbf{p}) = \frac{1}{1-q} \log((1-q)S_q(\mathbf{p}) + 1) \tag{4.16}$$

for real $q \neq 1$, and

$$S_1(\mathbf{p}) = H(\mathbf{p}) = H_1(\mathbf{p}). \tag{4.17}$$

Equations (4.15)–(4.17) can be written more compactly as

$$S_q(\mathbf{p}) = \ln_q(\exp H_q(\mathbf{p})), \tag{4.18}$$
$$H_q(\mathbf{p}) = \log(\exp_q S_q(\mathbf{p})) \tag{4.19}$$

($q \in \mathbb{R}$), where $\exp_q$ is the inverse function of $\ln_q$, given explicitly by

$$\exp_q(y) = \begin{cases} (1 + (1-q)y)^{1/(1-q)} & \text{if } q \neq 1, \\ \exp(y) & \text{if } q = 1. \end{cases}$$

The transformations relating $S_q(\mathbf{p})$ to $H_q(\mathbf{p})$ are strictly increasing, so maximizing or minimizing one is equivalent to maximizing or minimizing the other.

Remark 4.3.2 When $q = \pm\infty$, the Rényi entropy $H_q(\mathbf{p})$ is defined but the q-logarithmic entropy $S_q(\mathbf{p})$ is not. It is straightforward to check that

$$\lim_{q \to \infty} S_q(\mathbf{p}) = 0$$

for all $\mathbf{p}$, and

$$\lim_{q \to -\infty} S_q(\mathbf{p}) = \begin{cases} 0 & \text{if } p_i = 1 \text{ for some } i, \\ \infty & \text{otherwise.} \end{cases}$$

The only sensible way to define $S_\infty(\mathbf{p})$ and $S_{-\infty}(\mathbf{p})$ would be as these limits; but then the definitions would be trivial, would take infinite values in the latter case, and would break the result that $H_q(\mathbf{p})$ can be recovered from $S_q(\mathbf{p})$. We therefore leave $S_{\pm\infty}(\mathbf{p})$ undefined.

Remark 4.3.3 It is easy to manufacture other one-parameter families of entropies extending the Shannon entropy: simply take the formula

$$\frac{1}{1-q} \log \sum_{i \in \operatorname{supp}(\mathbf{p})} p_i^q$$

defining Rényi entropy for $q \neq 1$, and replace log by some other function λ. In order that the limit as $q \to 1$ is $H(\mathbf{p})$, the requirements on λ are that $\lambda(1) = 0$ and $\lambda'(1) = 1$. The simplest function λ with these properties is $\lambda(x) = x - 1$, the linear approximation to log at 1. Indeed, taking this simplest λ gives exactly the q-logarithmic entropy.

The *exponentials* of the Rényi entropies turn out to have slightly more convenient algebraic properties than the Rényi entropies themselves, and are important measures of biological diversity. We give the definition and examples here, and describe their properties in the next section.

Definition 4.3.4 Let $q \in [-\infty, \infty]$ and $\mathbf{p} \in \Delta_n$. The **Hill number of order** q of $\mathbf{p}$ is

$$D_q(\mathbf{p}) = \exp H_q(\mathbf{p}) = M_{1-q}(\mathbf{p}, 1/\mathbf{p}).$$

We also call this the **diversity of order** q of $\mathbf{p}$.

Thus, the Hill number D_q is related to the Rényi entropy H_q and q-logarithmic entropy S_q by

$$H_q = \log D_q, \qquad S_q = \ln_q D_q \tag{4.20}$$

(by definition and equation (4.18)). Explicitly,

$$D_q(\mathbf{p}) = \left(\sum_{i \in \operatorname{supp}(\mathbf{p})} p_i^q \right)^{1/(1-q)} \tag{4.21}$$

for $q \neq 1, \pm\infty$, and

$$D_{-\infty}(\mathbf{p}) = 1 \Big/ \min_{i \in \operatorname{supp}(\mathbf{p})} p_i, \tag{4.22}$$

$$D_1(\mathbf{p}) = \prod_{i \in \operatorname{supp}(\mathbf{p})} p_i^{-p_i} = D(\mathbf{p}), \tag{4.23}$$

$$D_\infty(\mathbf{p}) = 1 \Big/ \max_{i \in \operatorname{supp}(\mathbf{p})} p_i. \tag{4.24}$$

This definition of diversity of order q extends the earlier definition of diversity of order 1 (Definition 2.4.1), there written as D.

The quantities D_q are named after the ecologist Mark Hill [146], who introduced them in 1973 as measures of diversity (building on Rényi's work). In Section 7.4, we will prove a theorem pinpointing what makes the Hill numbers uniquely suitable as measures of diversity. For now, the following explanation can be given.

Let $\mathbf{p} = (p_1, \ldots, p_n)$ be the relative abundance distribution of a community. As in Section 2.4, $1/p_i$ measures the rarity or specialness of the ith species. There, we took the geometric mean $\prod (1/p_i)^{p_i}$ of the rarities as our measure of diversity. But we could just as reasonably use some other power mean $M_t(\mathbf{p}, 1/\mathbf{p})$. Reparametrizing as $q = 1-t$, this is exactly the Hill number $D_q(\mathbf{p})$.

The Hill numbers are effective numbers (Definition 2.4.5):

$$D_q(\mathbf{u}_n) = n \tag{4.25}$$

for all $n \geq 1$ and $q \in [-\infty, \infty]$. By equation (4.20), the quantities D_q, H_q and S_q are related to one another by increasing, invertible transformations. Thus, the Hill numbers are the result of taking either the Rényi entropies H_q or the q-logarithmic entropies S_q and converting them into effective numbers. In the terminology originating in economics (Bishop [39], p. 789) and now also used in ecology (Ellison [91], for instance), D_q is the **numbers equivalent** of both H_q and S_q.

Examples 4.3.5 i. The diversity or Hill number $D_0(\mathbf{p})$ of order 0 is simply $|\mathrm{supp}(\mathbf{p})|$, the number of species present. In ecology, this is called the **species richness**. It is the most common measure of diversity in both the popular media and the ecology literature, but makes no distinction between a rare species and a common species, and says nothing about the balance between the species present.

ii. We have already considered the diversity $D_1(\mathbf{p})$ of order 1 (Section 2.4), which is the exponential of Shannon entropy.

iii. The diversity of $\mathbf{p}$ of order 2 is

$$D_2(\mathbf{p}) = 1 \Big/ \sum_{i=1}^{n} p_i^2.$$

Being the reciprocal of a quadratic form, it is especially convenient mathematically. It also has an intuitive probabilistic interpretation: if we draw pairs of individuals at random from the community (with replacement), $D_2(\mathbf{p})$ is the expected number of trials needed in order to obtain a pair of the same species. Compare the probabilistic interpretation of $S_2(\mathbf{p})$ in Example 4.1.3(iii).

In ecology, $D_2(\mathbf{p})$ is called the **inverse Simpson concentration** [314].

iv. The diversity

$$D_\infty(\mathbf{p}) = 1 \Big/ \max_{i \in \mathrm{supp}(\mathbf{p})} p_i$$

of order ∞ is known as the **Berger–Parker index** [37]. It measures the extent to which the community is dominated by a single species. For instance, if one species has outcompeted the others and makes up nearly 100% of the community, then $D_\infty(\mathbf{p})$ is close to its minimum value of 1. At the opposite extreme, if $\mathbf{p} = \mathbf{u}_n$ then no species is dominant and $D_\infty(\mathbf{p})$ achieves its maximum value of n. (General statements on maximization and minimization of D_q will be made in Lemma 4.4.3.) So while diversity of order 0 gives rare species the same importance as any other, diversity of order ∞ ignores them altogether.

Example 4.3.6 Many of the diversity measures used in ecology are Hill numbers or transformations of them. Others can be expressed as combinations of several Hill numbers.

For instance, Hurlbert [150] and Smith and Grassle [317] studied the expected number $H_m^{\mathrm{HSG}}(\mathbf{p})$ of different species represented in a random sample (with replacement) of m individuals. Their measure turns out to be a combination of Hill numbers of integer orders:

$$H_m^{\mathrm{HSG}}(\mathbf{p}) = \sum_{q=1}^{m} (-1)^{q-1} \binom{m}{q} D_q(\mathbf{p})^{1-q}.$$

This was first proved as Proposition A8 in the appendix of Leinster and Cobbold [220], and the proof is also given in Appendix A.2 below.

Example 4.3.7 The reciprocals of the Hill numbers have been used in economics to measure concentration. One asks to what extent an industry or market is concentrated in the hands of a small number of large players. For example, if there are n competing companies in an industry, with market shares $p_1, \ldots, p_n$, then the concentration $1/D_q(\mathbf{p})$ is maximized when one company has a monopoly:

$$\mathbf{p} = (0, \ldots, 0, 1, 0, \ldots, 0).$$

See Hannah and Kay [136] or Chakravarty and Eichhorn [65], for instance.

The parameter q controls the sensitivity of the diversity measure D_q to rare species, with higher values of q corresponding to measures *less* sensitive to rare species. Thus, q is a 'viewpoint parameter', reflecting the importance that we wish to attach to rare species. For reasons to be explained, we usually restrict to parameter values $q \geq 0$.

With the multiplicity of diversity measures that exist in the literature, there is a risk of cherry-picking. Consciously or not, a scientist might choose the measure that best supports the desired conclusion. There is also a risk of attaching too much importance to a single number:

> The belief (or superstition) of some ecologists that a diversity index provides a basis (or talisman) for reaching a full understanding of community structure is totally unfounded

(Pielou [282], p. 19). Both problems are mitigated by systematically using *all* the diversity measures D_q $(0 \leq q \leq \infty)$. The graph of $D_q(\mathbf{p})$ against q is called the **diversity profile** of $\mathbf{p}$, and plotting it displays all viewpoints simultaneously.

Example 4.3.8 There are eight species of great ape in the world, but 99.99% of individual apes are humans. Figure 4.3 shows the absolute abundances of the eight species, their relative abundances p_i, and their diversity profile.

That there are eight extant species is conveyed by the value $D_0(\mathbf{p}) = 8$ of the profile at $q = 0$. However, this single statistic hides the fact that one of the species has all but totally outcompeted the others. For nearly any other value of the viewpoint parameter q, the diversity is almost exactly 1, reflecting the overwhelming dominance of a single species. For example, recall that $D_2(\mathbf{p})$ is the reciprocal of the probability that two individuals chosen at random belong to the same species (Example 4.3.5(iii)). In this case, the probability is very nearly 1, so $D_2(\mathbf{p})$ is only just greater than 1.

The very steep drop of the diversity profile at its left-hand end, from 8 to just above 1, indicates that seven of the eight species are exceptionally rare.

Example 4.3.9 Figure 4.4 shows the diversity profiles of the two bird communities of the Introduction (p. 3). From the viewpoint of low values of q, where rare species are given nearly as much importance as common species, community A is more diverse than community B. For instance, at $q = 0$, community A is more diverse than community B simply because it has more species. But from the viewpoint of high values, which give less importance to rare species, community B seems more diverse because it is better balanced. In the extreme, when $q = \infty$, we ignore all species except the most common, and the dominance of the first species in community A makes that community much less diverse than the well-balanced community B.

The flat profile of community B indicates the uniformity of the species present. Generally, we have seen in the last two examples that the shape of a diversity profile provides information on the community's structure. For more on the interpretation of diversity profiles, see Example 1, Example 2 and Figure 2 of Leinster and Cobbold [220].

Example 4.3.10 Diversity profiles arising from experimental data often cross one another (as in the previous example), indicating that different viewpoints on the importance of rare species lead to different judgements on which of the communities is more diverse. For example, Ellingsen tabulated $D_0(\mathbf{p})$, $D_1(\mathbf{p})$ and $D_2(\mathbf{p})$ for 16 distributions $\mathbf{p}$, corresponding to the populations of certain marine organisms at 16 sites on the Norwegian continental shelf (Table 1 of [90]). There are $\binom{16}{2} = 120$ pairs of sites, and it can be deduced from the data that for at least 53 of the 120 pairs, the profiles cross.

Typically, pairs of diversity profiles obtained from experimental data cross

Species	Absolute abundance	Relative abundance
Human	7 466 964 300	0.99989926
Bonobo	20 000	0.00000267
Chimpanzee	407 500	0.00005456
Eastern gorilla	4 700	0.00000063
Western gorilla	200 000	0.00002678
Bornean orangutan	104 700	0.00001040
Sumatran orangutan	14 600	0.00000196
Tapanuli orangutan	800	0.00000011

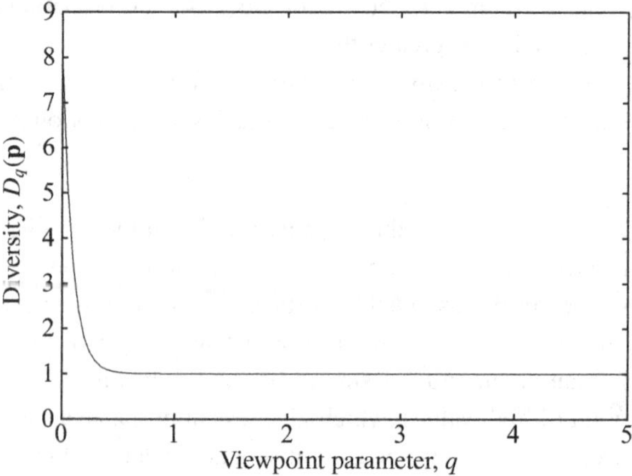

Figure 4.3 Abundances and species diversity profile of the estimated global distribution $\mathbf{p}$ of great apes (Hominidae). Population estimates are all for 2016, with human data from United Nations [339], Tapanuli orangutan data from Nater et al. [261], and all other data from the IUCN Red List of Threatened Species [14, 112, 149, 241, 284, 315].

at most once. But it can be shown that in principle, there is no upper bound on the number of times that a pair of diversity profiles can cross.

The ecological significance of the different judgements produced by different diversity measures is discussed in the highly readable 1974 paper of Peet [277]; see also Nagendra [259]. More specifically, diversity profiles of various types have long been discussed, beginning with Hill himself in 1973 [146], and continuing with Patil and Taillie [272, 273], Dennis and Patil [82], Tóthmérész [327], Patil [271], Mendes et al. [253], and others. In political science, $D_q(\mathbf{p})$ has been used as a measure of the effective number of parties in a parliamentary assembly, and diversity profiles have been used to

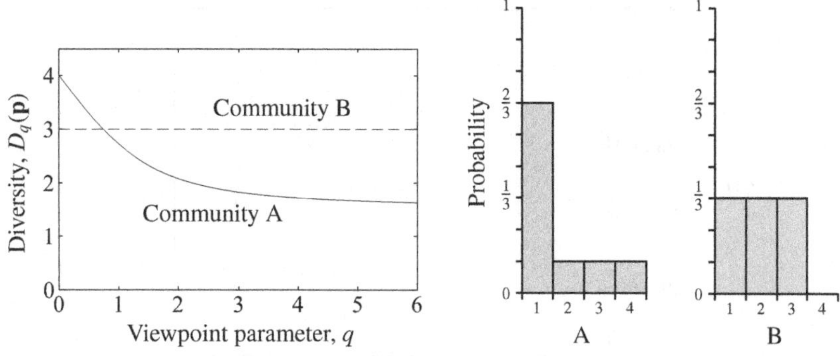

Figure 4.4 The diversity profiles of the two hypothetical bird communities in the Introduction (p. 3).

compare the political situations of different countries at different times (Laakso and Taagepera [200], especially equation [8] and Figure 1).

The next section establishes the mathematical properties of the Hill numbers and, therefore, of diversity profiles.

4.4 Properties of the Hill Numbers

Here we establish the main properties of the Hill numbers, using what we already know about properties of the power means. Of course, any statement about the Hill numbers can be translated into a statement about Rényi entropies, since one is the logarithm of the other. But here we work with the Hill numbers, interpreting them in terms of diversity.

We have already noted that for each $q \in [-\infty, \infty]$, the Hill number D_q is an effective number: $D_q(\mathbf{u}_n) = n$.

Diversity profiles are always decreasing. Intuitively, this is because diversity decreases as less importance is attached to rare species. The precise statement is as follows.

Proposition 4.4.1 *Let* $\mathbf{p} \in \Delta_n$*. Then* $D_q(\mathbf{p})$ *is a decreasing function of* $q \in [-\infty, \infty]$*. It is constant if* $\mathbf{p}$ *is uniform on its support, and strictly decreasing otherwise.*

Proof Since $D_q(\mathbf{p}) = M_{1-q}(\mathbf{p}, 1/\mathbf{p})$, this follows from Theorem 4.2.8. □

Figure 4.4 shows one strictly decreasing profile and one that is constant (being uniform on its support). Diversity profiles are always continuous, by Lemma 4.2.7.

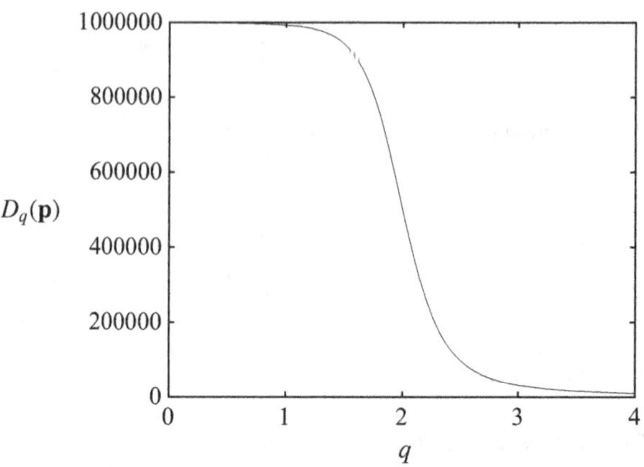

Figure 4.5 A non-convex diversity profile (Remark 4.4.2).

Remark 4.4.2 It is a curiosity that for most distributions $\mathbf{p}$ that arise experimentally, the diversity profile of $\mathbf{p}$ appears to be convex. (See the works cited at the end of Section 4.3, for example.) However, this is false for arbitrary $\mathbf{p}$. Figure 4.5 shows the diversity profile of the distribution

$$\mathbf{p} = (\underbrace{10^{-6}, \ldots, 10^{-6}}_{999\,000}, 10^{-3})$$

(adapted from an example of Willerton [354]), which is evidently not convex.

For each parameter value $q > 0$, the maximum and minimum values of the Hill number D_q, and the distributions at which they are attained, are exactly the same as for the diversity $D_1 = D$ of order 1 (Lemma 2.4.3):

Lemma 4.4.3 *Let $n \geq 1$ and $q \in [-\infty, \infty]$.*

i. *$D_q(\mathbf{p}) \geq 1$ for all $\mathbf{p} \in \Delta_n$, with equality if and only if $p_i = 1$ for some $i \in \{1, \ldots, n\}$.*
ii. *If $q > 0$ then $D_q(\mathbf{p}) \leq n$ for all $\mathbf{p} \in \Delta_n$, with equality if and only if $\mathbf{p} = \mathbf{u}_n$.*

Proof For (i), Proposition 4.4.1 implies that

$$D_q(\mathbf{p}) \geq D_\infty(\mathbf{p}) = 1 \Big/ \max_{i \in \operatorname{supp}(\mathbf{p})} p_i \geq 1.$$

If the second inequality is an equality then $p_i = 1$ for some i. Conversely, if $p_i = 1$ for some i then $D_q(\mathbf{p}) = 1$.

For (ii), Proposition 4.4.1 implies that

$$D_q(\mathbf{p}) \leq D_0(\mathbf{p}) = |\operatorname{supp}(\mathbf{p})| \leq n,$$

with equality in the first inequality if and only if $\mathbf{p}$ is uniform on its support. On the other hand, equality holds in the second inequality if and only if $\mathbf{p}$ has full support. Hence equality holds throughout if and only if $\mathbf{p} = \mathbf{u}_n$. □

Remarks 4.4.4 i. It follows that for $q > 0$, the Rényi entropy $H_q(\mathbf{p})$ is minimized exactly when $\mathbf{p}$ is of the form $(0, \ldots, 0, 1, 0, \ldots, 0)$, with value 0, and maximized exactly when $\mathbf{p} = \mathbf{u}_n$, with value $\log n$. Since the q-logarithmic entropy $S_q(\mathbf{p})$ is an increasing invertible transformation of $H_q(\mathbf{p})$ (equations (4.18) and (4.19)), it is minimized and maximized at these same distributions, with minimum 0 and maximum $S_q(\mathbf{u}_n) = \ln_q(n)$.

ii. The Hill numbers of negative orders are *not* maximized by the uniform distribution. Indeed, let $q < 0$, let $n \geq 2$, and take any non-uniform distribution $\mathbf{p} \in \Delta_n$ of full support. Then $D_0(\mathbf{p}) = |\mathrm{supp}(\mathbf{p})| = n$, and the diversity profile of $\mathbf{p}$ is strictly decreasing by Proposition 4.4.1, so

$$D_q(\mathbf{p}) > D_0(\mathbf{p}) = n = D_q(\mathbf{u}_n).$$

Whatever the word 'diverse' should mean, it is generally agreed that the most diverse abundance distribution on a given set of species should be the uniform distribution. (At least, this should be the case for the crude model of a community as a probability distribution, which we are using here. See also Section 6.3.) For this reason, the Hill numbers of negative orders are generally not used as measures of diversity.

On the other hand, the Hill numbers of negative orders measure *something*. For instance,

$$D_{-\infty}(\mathbf{p}) = 1 \Big/ \min_{i \in \mathrm{supp}(\mathbf{p})} p_i = \max_{i \in \mathrm{supp}(\mathbf{p})} (1/p_i)$$

measures the rarity of the rarest species, giving a high value to any community containing at least one species that is very rare. This is a meaningful quantity, even if it should not be called diversity.

We now show that the Hill number $D_q(\mathbf{p})$ of a given order q is very nearly continuous in $\mathbf{p} \in \Delta_n$, with the sole exception that D_q is discontinuous at the boundary of the simplex when $q \leq 0$. For instance, species richness D_0 is discontinuous: in terms of the number of species present, a relative abundance of 0.0001 is qualitatively different from a relative abundance of 0.

Definition 4.4.5 Let $(D\colon \Delta_n \to (0, \infty))_{n \geq 1}$ be a sequence of functions. Then D is **continuous** if the function $D\colon \Delta_n \to (0, \infty)$ is continuous for each $n \geq 1$, and **continuous in positive probabilities** if the restriction $D|_{\Delta_n^\circ}$ of D to the open simplex is continuous for each $n \geq 1$.

Continuity in positive probabilities means that small changes to the abundances of the species *present* cause only small changes in the perceived diversity. For example, D_0 is continuous in positive probabilities, even though it is not continuous.

Lemma 4.4.6 *i. For each $q \in [-\infty, \infty]$, the Hill number D_q is continuous in positive probabilities.*

ii. For each $q \in (0, \infty]$, the Hill number D_q is continuous.

Proof Part (i) is immediate from the explicit formulas for D_q (equations (4.21)–(4.24)), and part (ii) follows from the observation that when $q > 0$, the formulas for D_q are unchanged if we allow i to range over all of $\{1, \ldots, n\}$ instead of just $\operatorname{supp}(\mathbf{p})$. □

Next we establish the algebraic properties of the Hill numbers, beginning with the most elementary ones.

Definition 4.4.7 A sequence of functions $(D \colon \Delta_n \to (0, \infty))_{n \geq 1}$ is **absence-invariant** if whenever $\mathbf{p} \in \Delta_n$ and $1 \leq i \leq n$ with $p_i = 0$, then

$$D(\mathbf{p}) = D(p_1, \ldots, p_{i-1}, p_{i+1}, \ldots, p_n).$$

Absence-invariance means that as far as D is concerned, a species that is absent might as well not have been mentioned.

Recall from equation (4.4) that D is said to be symmetric if $D(\mathbf{p}\sigma) = D(\mathbf{p})$ for all $\mathbf{p} \in \Delta_n$ and permutations σ of $\{1, \ldots, n\}$. This means that the diversity is unaffected by the order in which the species happen to be listed.

Lemma 4.4.8 *For each $q \in [-\infty, \infty]$, the Hill number D_q of order q is symmetric and absence-invariant.*

Proof These statements follow from the symmetry and absence-invariance of the power means (Lemma 4.2.11). Alternatively, they can be deduced directly from the explicit formulas for D_q (equations (4.21)–(4.24)). □

Remark 4.4.9 By symmetry, $\mathbf{p}$ and $\mathbf{p}\sigma$ have the same diversity profile. In fact, the converse also holds: if $\mathbf{p}, \mathbf{r} \in \Delta_n$ have the same diversity profile then $\mathbf{p}$ and $\mathbf{r}$ must be the same up to a permutation. This is proved in Appendix A.3.

Thus, the diversity profile of a relative abundance distribution contains all the information about that distribution apart from which species is which, packaged in a way that displays meaningful information about the community's diversity.

We finally come to the chain rule. In Corollary 2.4.8, we treated the case $q = 1$, showing that

$$D_1(\mathbf{w} \circ (\mathbf{p}^1, \ldots, \mathbf{p}^n)) = D_1(\mathbf{w}) \cdot \prod_{i=1}^{n} D_1(\mathbf{p}^i)^{w_i}$$

for all $\mathbf{w} \in \Delta_n$ and $\mathbf{p}^i \in \Delta_{k_i}$. In Example 2.4.9, this formula was explained in terms of a group of n islands of relative sizes w_i and diversities $d_i = D_1(\mathbf{p}^i)$, with no shared species. We now give the chain rule for general q, in two different forms.

Proposition 4.4.10 (Chain rule, version 1) *Let $q \in [-\infty, \infty]$, $\mathbf{w} \in \Delta_n$, and $\mathbf{p}^1 \in \Delta_{k_1}, \ldots, \mathbf{p}^n \in \Delta_{k_n}$. Write $d_i = D_q(\mathbf{p}^i)$ and $\mathbf{d} = (d_1, \ldots, d_n)$. Then*

$$\begin{aligned} D_q(\mathbf{w} \circ (\mathbf{p}^1, \ldots, \mathbf{p}^n)) &= M_{1-q}(\mathbf{w}, \mathbf{d}/\mathbf{w}) \\ &= \begin{cases} \left(\sum w_i^q d_i^{1-q}\right)^{1/(1-q)} & \textit{if } q \neq 1, \pm\infty, \\ \max d_i/w_i & \textit{if } q = -\infty, \\ \prod (d_i/w_i)^{w_i} & \textit{if } q = 1, \\ \min d_i/w_i & \textit{if } q = \infty, \end{cases} \end{aligned}$$

where the sum, maximum, product, and minimum are over $i \in \mathrm{supp}(\mathbf{w})$.

Here $\mathbf{d}/\mathbf{w} = (d_1/w_1, \ldots, d_n/w_n)$, as in Remark 4.2.9.

Proof We have

$$\begin{aligned} D_q(\mathbf{w} \circ (\mathbf{p}^1, \ldots, \mathbf{p}^n)) &= M_{1-q}\left(\mathbf{w} \circ (\mathbf{p}^1, \ldots, \mathbf{p}^n), \tfrac{1}{w_1\mathbf{p}^1} \oplus \cdots \oplus \tfrac{1}{w_n\mathbf{p}^n}\right) \\ &= M_{1-q}\left(\mathbf{w}, \left(M_{1-q}\left(\mathbf{p}^1, \tfrac{1}{w_1\mathbf{p}^1}\right), \ldots, M_{1-q}\left(\mathbf{p}^n, \tfrac{1}{w_n\mathbf{p}^n}\right)\right)\right) \\ &= M_{1-q}(\mathbf{w}, (d_1/w_1, \ldots, d_n/w_n)), \end{aligned}$$

where the second equation follows from the chain rule for M_{1-q} (Proposition 4.2.24) and the last from the homogeneity of M_{1-q} (Lemma 4.2.22). This proves the first equality stated in the proposition, and the second follows from the explicit formulas for the power means. □

There is an alternative form of the chain rule, for which we will need some terminology. Given a probability distribution $\mathbf{w} \in \Delta_n$ and a real number q, the **escort distribution of order** q of $\mathbf{w}$ is the distribution $\mathbf{w}^{(q)} \in \Delta_n$ with ith coordinate

$$w_i^{(q)} = \begin{cases} w_i^q \Big/ \sum_{j \in \mathrm{supp}(\mathbf{w})} w_j^q & \text{if } i \in \mathrm{supp}(\mathbf{w}), \\ 0 & \text{otherwise.} \end{cases}$$

Lemma 4.4.11 *Let $q \in \mathbb{R}$, $\mathbf{w} \in \Delta_n$, and $\mathbf{d} \in [0,\infty)^n$. Then*

$$M_{1-q}(\mathbf{w}, \mathbf{d}/\mathbf{w}) = D_q(\mathbf{w}) \cdot M_{1-q}(\mathbf{w}^{(q)}, \mathbf{d}).$$

Proof For the case $q = 1$, note that

$$M_0(\mathbf{w}, \mathbf{xy}) = M_0(\mathbf{w}, \mathbf{x}) M_0(\mathbf{w}, \mathbf{y})$$

for all $\mathbf{x}, \mathbf{y} \in [0,\infty)^n$. It follows that

$$D_1(\mathbf{w}) \cdot M_0(\mathbf{w}^{(1)}, \mathbf{d}) = M_0(\mathbf{w}, 1/\mathbf{w}) \cdot M_0(\mathbf{w}, \mathbf{d}) = M_0(\mathbf{w}, \mathbf{d}/\mathbf{w}).$$

On the other hand, for $1 \neq q \in \mathbb{R}$,

$$\begin{aligned} M_{1-q}(\mathbf{w}, \mathbf{d}/\mathbf{w}) &= \left(\sum_{i \in \operatorname{supp}(\mathbf{w})} w_i^q d_i^{1-q} \right)^{1/(1-q)} \\ &= D_q(\mathbf{w}) \cdot \left(\frac{\sum_{i \in \operatorname{supp}(\mathbf{w})} w_i^q d_i^{1-q}}{\sum_{j \in \operatorname{supp}(\mathbf{w})} w_j^q} \right)^{1/(1-q)} \\ &= D_q(\mathbf{w}) \cdot M_{1-q}(\mathbf{w}^{(q)}, \mathbf{d}), \end{aligned}$$

as required. □

The last two results immediately imply the following.

Proposition 4.4.12 (Chain rule, version 2) *Let $q \in \mathbb{R}$, $\mathbf{w} \in \Delta_n$, and $\mathbf{p}^1 \in \Delta_{k_1}, \ldots, \mathbf{p}^n \in \Delta_{k_n}$. Write $d_i = D_q(\mathbf{p}^i)$ and $\mathbf{d} = (d_1, \ldots, d_n)$. Then*

$$D_q(\mathbf{w} \circ (\mathbf{p}^1, \ldots, \mathbf{p}^n)) = D_q(\mathbf{w}) \cdot M_{1-q}(\mathbf{w}^{(q)}, \mathbf{d}).$$ □

Remarks 4.4.13 Here we provide context for the notion of escort distribution.

i. The escort distributions of a distribution $\mathbf{w}$ form a one-parameter family

$$\left(\mathbf{w}^{(q)}\right)_{q \in \mathbb{R}}$$

of distributions, of which the original distribution $\mathbf{w}$ is the member corresponding to $q = 1$. The term 'escort distribution' is taken from thermodynamics (Chapter 9 of Beck and Schlögl [33]). There, one encounters expressions such as

$$\frac{(e^{-\beta E_1}, \ldots, e^{-\beta E_n})}{Z(\beta)},$$

where $Z(\beta) = e^{-\beta E_1} + \cdots + e^{-\beta E_n}$ is the partition function for energies E_i at inverse temperature β. Assuming without loss of generality that $\sum e^{-E_i} = 1$, the inverse temperature β plays the role of the parameter q.

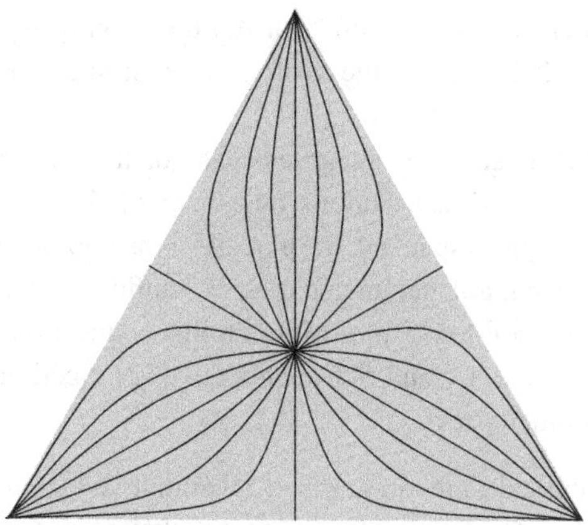

Figure 4.6 Twelve one-dimensional linear subspaces of the open simplex Δ_3° with the real vector space structure described in Remark 4.4.13(ii).

ii. The function $(q, \mathbf{w}) \mapsto \mathbf{w}^{(q)}$ is the scalar multiplication of a real vector space structure on the interior Δ_n° of the simplex. Addition is given by

$$(\mathbf{p}, \mathbf{r}) \mapsto \frac{(p_1 r_1, \ldots, p_n r_n)}{p_1 r_1 + \cdots + p_n r_n},$$

and the zero element is the uniform distribution $\mathbf{u}_n$. Figure 4.6 shows some one-dimensional linear subspaces of the two-dimensional vector space Δ_3°. This vector space structure was used in the field of statistical inference by Aitchison [5], and is sometimes named after him. It can be understood algebraically as follows.

Exponential and logarithm define a bijection between $\mathbb{R}$ and $(0, \infty)$. This induces a bijection between $\mathbb{R}^n$ and $(0, \infty)^n$, and transporting the vector space structure on $\mathbb{R}^n$ across this bijection gives a vector space structure on $(0, \infty)^n$. Explicitly, addition in the vector space $(0, \infty)^n$ is coordinatewise multiplication, the zero element is $(1, \ldots, 1)$, and scalar multiplication by $q \in \mathbb{R}$ raises each coordinate to the power of q.

Now take the linear subspace of $\mathbb{R}^n$ spanned by $(1, \ldots, 1)$. The corresponding subspace W of $(0, \infty)^n$ is $\{(\gamma, \ldots, \gamma) : \gamma \in (0, \infty)\}$, and we can form the quotient vector space $(0, \infty)^n / W$.

An element of this quotient is an equivalence class of vectors $\mathbf{y} \in (0, \infty)^n$, with $\mathbf{y}$ equivalent to $\mathbf{z}$ if and only if $\mathbf{y} = \gamma\mathbf{z}$ for some $\gamma > 0$. Geometrically, then, the equivalence classes are the rays through the origin in the positive orthant $(0, \infty)^n$. Each ray contains exactly one element of the open simplex

$$\Delta_n^\circ = \{\mathbf{y} \in (0, \infty)^n : y_1 + \cdots + y_n = 1\}.$$

This puts $(0, \infty)^n/W$ in bijection with Δ_n°, thus giving Δ_n° the structure of a vector space. It is exactly the vector space structure defined explicitly above.

iii. In statistical language, each linear subspace of the vector space Δ_n° is an exponential family of distributions on $\{1, \ldots, n\}$. For example, the one-dimensional subspace spanned by $\mathbf{p} \in \Delta_n^\circ$ is a one-parameter exponential family with natural parameter $q \in \mathbb{R}$, sufficient statistic $\log p_i$, and log-partition function $\log(\sum p_i^q)$. More on this connection can be found in Amari [10], Ay, Jost, Lê and Schwachhöfer ([22], Section 2.8), and other information geometry texts.

As already discussed in the case $q = 1$ (Example 2.4.9), the chain rule for the Hill numbers has the important consequence that when computing the total diversity of a group of islands with no shared species, the only information one needs is the diversities and relative sizes of the islands, not their internal make-up:

Definition 4.4.14 A sequence of functions $(D\colon \Delta_n \to (0, \infty))_{n \geq 1}$ is **modular** if

$$\begin{aligned} & D(\mathbf{p}^i) = D(\widetilde{\mathbf{p}}^i) \text{ for all } i \in \{1, \ldots, n\} \\ \implies & D(\mathbf{w} \circ (\mathbf{p}^1, \ldots, \mathbf{p}^n)) = D(\mathbf{w} \circ (\widetilde{\mathbf{p}}^1, \ldots, \widetilde{\mathbf{p}}^n)) \end{aligned}$$

for all $n, k_1, \ldots, k_n, \widetilde{k}_1, \ldots, \widetilde{k}_n \geq 1$ and $\mathbf{w} \in \Delta_n$, $\mathbf{p}^i \in \Delta_{k_i}$, $\widetilde{\mathbf{p}}^i \in \Delta_{\widetilde{k}_i}$.

In other words, D is modular if $D(\mathbf{w} \circ (\mathbf{p}^1, \ldots, \mathbf{p}^n))$ depends only on $\mathbf{w}$ and $D(\mathbf{p}^1), \ldots, D(\mathbf{p}^n)$.

Corollary 4.4.15 (Modularity) *For each $q \in [-\infty, \infty]$, the Hill number D_q is modular.* □

The chain rule has two further consequences.

Definition 4.4.16 A sequence of functions $(D\colon \Delta_n \to (0, \infty))_{n \geq 1}$ is **multiplicative** if

$$D(\mathbf{p} \otimes \mathbf{r}) = D(\mathbf{p}) D(\mathbf{r})$$

for all $m, n \geq 1$, $\mathbf{p} \in \Delta_m$, and $\mathbf{r} \in \Delta_n$.

Corollary 4.4.17 (Multiplicativity) *For each $q \in [-\infty, \infty]$, the Hill number D_q is multiplicative.*

Proof This follows from either the chain rule for the Hill numbers or the logarithmic property of the Rényi entropies (equation (4.14)). □

Definition 4.4.18 A sequence of functions $(D\colon \Delta_n \to (0,\infty))_{n\geq 1}$ satisfies the **replication principle** if

$$D(\mathbf{u}_n \otimes \mathbf{p}) = nD(\mathbf{p})$$

for all $n, k \geq 1$ and $\mathbf{p} \in \Delta_k$.

The oil company argument of Example 2.4.11 shows the fundamental importance of the replication principle. If n islands have the same relative abundance distribution $\mathbf{p}$, but on disjoint sets of species, the diversity of the whole system should be $nD(\mathbf{p})$.

Corollary 4.4.19 (Replication) *For each $q \in [-\infty, \infty]$, the Hill number D_q satisfies the replication principle.*

Proof This follows from multiplicativity and the fact that D_q is an effective number. □

Accompanying the Rényi entropies, there is also a notion of Rényi relative entropy (introduced in Section 3 of Rényi [294]). We defer discussion to Section 7.2.

4.5 Characterization of the Hill Number of a Given Order

In this book, we prove two characterization theorems for the Hill numbers. The first states that for each given q, the unique function satisfying certain conditions (which depend on q) is D_q. The second states that the only functions satisfying a different list of conditions (which make no mention of q) are those belonging to the family $(D_q)_{q\in[-\infty,\infty]}$. We prove the first characterization in this section, and the second in Section 7.4.

For the q-logarithmic entropies, we have already proved an analogue of the first result (Theorem 4.1.5). We will not prove an analogue of the second. However, there is a theorem of this type due to Forte and Ng, briefly discussed in Remark 7.4.15.

Here we build on work of Routledge [302] to characterize the Hill number D_q for each given $q \in (0, \infty)$. The restriction to positive q ensures that D_q is continuous on all of Δ_n (by Lemma 4.4.6(ii)).

Recall that D_q satisfies the chain rule

$$D_q(\mathbf{w} \circ (\mathbf{p}^1, \ldots, \mathbf{p}^n)) = D_q(\mathbf{w}) \cdot M_{1-q}\big(\mathbf{w}^{(q)}, (D_q(\mathbf{p}^1), \ldots, D_q(\mathbf{p}^n))\big), \qquad (4.26)$$

where $\mathbf{w} \in \Delta_n$ and $\mathbf{p}^i \in \Delta_{k_i}$ (Proposition 4.4.12). Let us reflect on equation (4.26), interpreting it in terms of the island scenario of Examples 2.1.6

and 2.4.9. Equation (4.26) can be interpreted as a decomposition of the diversity of the island group into two factors: the variation *between* the islands (given by $D_q(\mathbf{w})$), and the average variation or diversity *within* the islands (given by the second factor). Recall that in the island scenario, there is no overlap of species between islands, so the variation between the islands depends only on the variation in sizes.

Now, suppose that we want to list some properties that a reasonable diversity measure D ought to satisfy. One such property might be that D is decomposable in the sense just described: $D(\mathbf{w} \circ (\mathbf{p}^1, \ldots, \mathbf{p}^n))$ is equal to the variation $D(\mathbf{w})$ between islands multiplied by the average of the diversities $D(\mathbf{p}^1), \ldots, D(\mathbf{p}^n)$ within each island.

But what could 'average' reasonably mean? We have already seen that the power means have many good properties that we would expect of a notion of avarage, and we will see in Chapter 5 that in a certain precise sense, they are *uniquely* good. So, it is reasonable to take the 'average' to be some power mean M_t, and we can make the usual harmless reparametrization $t = 1 - q$.

This reasoning suggests that our hypothetical diversity measure D should satisfy something like equation (4.26), with D in place of D_q. Still, it does not explain why the average of the within-island diversities should be calculated using the weighting $\mathbf{w}^{(q)}$ on the islands, rather than some other weighting. All that seems clear is that the weighting should depend on the sizes of the islands only. If we write the weighting as $\theta(\mathbf{w})$, then our conclusion is that any reasonable diversity measure D ought to satisfy the equation

$$D(\mathbf{w} \circ (\mathbf{p}^1, \ldots, \mathbf{p}^n)) = D(\mathbf{w}) \cdot M_{1-q}\big(\theta(\mathbf{w}), (D(\mathbf{p}^1), \ldots, D(\mathbf{p}^n))\big)$$

for some q and some function $\theta \colon \Delta_n \to \Delta_n$. This explains the most substantial of the hypotheses in our main result:

Theorem 4.5.1 *Let $q \in (0, \infty)$. Let $(D \colon \Delta_n \to (0, \infty))_{n \geq 1}$ be a sequence of functions. The following are equivalent:*

i. the functions D are continuous, symmetric and effective numbers, and for each $n \geq 1$ there exists a function $\theta \colon \Delta_n \to \Delta_n$ such that

$$D(\mathbf{w} \circ (\mathbf{p}^1, \ldots, \mathbf{p}^n)) = D(\mathbf{w}) \cdot M_{1-q}\big(\theta(\mathbf{w}), (D(\mathbf{p}^1), \ldots, D(\mathbf{p}^n))\big)$$

for all $\mathbf{w} \in \Delta_n$, $k_1, \ldots, k_n \geq 1$ and $\mathbf{p}^i \in \Delta_{k_i}$;

ii. $D = D_q$.

Theorem 4.5.1 is a variation on a 1979 result of Routledge (Theorem 1 of the appendix to [302]).

The rest of this section is devoted to the proof. We already showed in Section 4.4 that (ii) implies (i). Conversely, and ***for the rest of this section***, take D and θ satisfying the conditions of (i). By the standard abuse of notation, we use the same letter θ for each of the functions $\theta\colon \Delta_1 \to \Delta_1$, $\theta\colon \Delta_2 \to \Delta_2$, etc. We have to prove that $D = D_q$.

For $\mathbf{p} \in \Delta_n$, write

$$\theta(\mathbf{p}) = (\theta_1(\mathbf{p}), \dots, \theta_n(\mathbf{p})).$$

Our first lemma shows how θ_1 can be expressed in terms of D. We temporarily adopt the notation

$$\mathbf{p}^\# = \mathbf{p} \circ (\mathbf{u}_2, \mathbf{u}_1, \dots, \mathbf{u}_1) = (\tfrac{1}{2}p_1, \tfrac{1}{2}p_1, p_2, \dots, p_n)$$

$(\mathbf{p} \in \Delta_n)$.

Lemma 4.5.2 *For all $n \geq 1$ and $\mathbf{p} \in \Delta_n$,*

$$\theta_1(\mathbf{p}) = \frac{1}{\ln_q 2} \cdot \ln_q \frac{D(\mathbf{p}^\#)}{D(\mathbf{p})}.$$

Proof By the main hypothesis on D and the effective number property,

$$D(\mathbf{p}^\#) = D(\mathbf{p}) \cdot M_{1-q}(\theta(\mathbf{p}), (2, 1, \dots, 1)).$$

Hence by Lemma 4.2.29,

$$\ln_q \frac{D(\mathbf{p}^\#)}{D(\mathbf{p})} = M_1(\theta(\mathbf{p}), (\ln_q 2, \ln_q 1, \dots, \ln_q 1)).$$

But $\ln_q 1 = 0$, so the right-hand side is just $\theta_1(\mathbf{p}) \ln_q 2$. □

We now use this lemma to compute the weighting $\theta(\mathbf{w} \circ (\mathbf{p}^1, \dots, \mathbf{p}^n))$ of a composite distribution.

Lemma 4.5.3 *Let $\mathbf{w} \in \Delta_n$ and $\mathbf{p}^1 \in \Delta_{k_1}, \dots, \mathbf{p}^n \in \Delta_{k_n}$. Then*

$$\theta_1(\mathbf{w} \circ (\mathbf{p}^1, \dots, \mathbf{p}^n)) = \frac{\theta_1(\mathbf{w}) D(\mathbf{p}^1)^{1-q}}{\sum_{i=1}^n \theta_i(\mathbf{w}) D(\mathbf{p}^i)^{1-q}} \theta_1(\mathbf{p}^1).$$

Proof Write $d_i = D(\mathbf{p}^i)$ and $d_1^\# = D\big(\mathbf{p}^{1\#}\big)$. We have

$$
\begin{aligned}
&\theta_1(\mathbf{w} \circ (\mathbf{p}^1, \ldots, \mathbf{p}^n)) \\
&= \frac{1}{\ln_q 2} \cdot \ln_q \frac{D\big((\mathbf{w} \circ (\mathbf{p}^1, \ldots, \mathbf{p}^n))^\#\big)}{D(\mathbf{w} \circ (\mathbf{p}^1, \ldots, \mathbf{p}^n))} && (4.27) \\
&= \frac{1}{\ln_q 2} \cdot \ln_q \frac{D\big(\mathbf{w} \circ \big(\mathbf{p}^{1\#}, \mathbf{p}^2, \ldots, \mathbf{p}^n\big)\big)}{D(\mathbf{w} \circ (\mathbf{p}^1, \mathbf{p}^2, \ldots, \mathbf{p}^n))} && (4.28) \\
&= \frac{1}{\ln_q 2} \cdot \ln_q \frac{M_{1-q}(\theta(\mathbf{w}), (d_1^\#, d_2, \ldots, d_n))}{M_{1-q}(\theta(\mathbf{w}), (d_1, d_2, \ldots, d_n))} && (4.29) \\
&= \frac{1}{\ln_q 2} \cdot \frac{\ln_q M_{1-q}(\theta(\mathbf{w}), (d_1^\#, d_2, \ldots, d_n)) - \ln_q M_{1-q}(\theta(\mathbf{w}), (d_1, d_2, \ldots, d_n))}{M_{1-q}(\theta(\mathbf{w}), (d_1, d_2, \ldots, d_n))^{1-q}} && (4.30) \\
&= \frac{1}{\ln_q 2} \cdot \frac{M_1(\theta(\mathbf{w}), (\ln_q d_1^\#, \ln_q d_2, \ldots)) - M_1(\theta(\mathbf{w}), (\ln_q d_1, \ln_q d_2, \ldots))}{\sum_{i=1}^n \theta_i(\mathbf{w}) d_i^{1-q}} && (4.31) \\
&= \frac{1}{\ln_q 2} \cdot \frac{\theta_1(\mathbf{w})(\ln_q d_1^\# - \ln_q d_1)}{\sum_{i=1}^n \theta_i(\mathbf{w}) d_i^{1-q}} && (4.32) \\
&= \frac{1}{\ln_q 2} \cdot \frac{\theta_1(\mathbf{w}) d_1^{1-q}}{\sum_{i=1}^n \theta_i(\mathbf{w}) d_i^{1-q}} \cdot \ln_q \frac{d_1^\#}{d_1} && (4.33) \\
&= \frac{\theta_1(\mathbf{w}) d_1^{1-q}}{\sum_{i=1}^n \theta_i(\mathbf{w}) d_i^{1-q}} \cdot \theta_1(\mathbf{p}^1), && (4.34)
\end{aligned}
$$

where equations (4.27) and (4.34) follow from Lemma 4.5.2, equation (4.28) from the definition of $^\#$, equation (4.29) from the main hypothesis on D, equations (4.30) and (4.33) from the quotient formula

$$
\ln_q \frac{x}{y} = \frac{\ln_q x - \ln_q y}{y^{1-q}}
$$

for the q-logarithm (equation (1.20)), equation (4.31) from Lemma 4.2.29 and the definition of M_{1-q}, and equation (4.32) from the definition of the arithmetic mean M_1. □

We now deduce that the weightings must be the q-escort distributions.

Lemma 4.5.4 $\theta(\mathbf{w}) = \mathbf{w}^{(q)}$ *for all* $n \geq 1$ *and* $\mathbf{w} \in \Delta_n$.

Proof Following a familiar pattern, we prove this first when $\mathbf{w}$ is uniform, then when the coordinates of $\mathbf{w}$ are positive and rational, and finally for arbitrary $\mathbf{w}$.

For the case $\mathbf{w} = \mathbf{u}_n$, we have to prove that $\theta(\mathbf{u}_n) = \mathbf{u}_n$. By Lemma 4.5.2,

$$\theta_1(\mathbf{u}_n) = \frac{1}{\ln_q 2} \ln_q \frac{D(\mathbf{u}_n \circ (\mathbf{u}_2, \mathbf{u}_1, \mathbf{u}_1, \dots, \mathbf{u}_1))}{D(\mathbf{u}_n)},$$

and by the same argument,

$$\theta_2(\mathbf{u}_n) = \frac{1}{\ln_q 2} \ln_q \frac{D(\mathbf{u}_n \circ (\mathbf{u}_1, \mathbf{u}_2, \mathbf{u}_1, \dots, \mathbf{u}_1))}{D(\mathbf{u}_n)}.$$

By symmetry of D, the right-hand sides of these two equations are equal. Hence $\theta_1(\mathbf{u}_n) = \theta_2(\mathbf{u}_n)$. Similarly, $\theta_i(\mathbf{u}_n) = \theta_j(\mathbf{u}_n)$ for all i, j, and so $\theta(\mathbf{u}_n) = \mathbf{u}_n$.

Now let $\mathbf{w} \in \Delta_n$ with

$$\mathbf{w} = (k_1/k, \dots, k_n/k)$$

for some positive integers k_i, where $k = \sum k_i$. We have

$$\mathbf{w} \circ (\mathbf{u}_{k_1}, \dots, \mathbf{u}_{k_n}) = \mathbf{u}_k. \tag{4.35}$$

Applying θ_1 to both sides gives

$$\frac{\theta_1(\mathbf{w}) k_1^{1-q}}{\sum \theta_i(\mathbf{w}) k_i^{1-q}} \frac{1}{k_1} = \frac{1}{k},$$

using Lemma 4.5.3, the effective number property of D, and the previous paragraph. This rearranges to

$$\theta_1(\mathbf{w}) = w_1^q \sum_{i=1}^{n} \theta_i(\mathbf{w}) w_i^{1-q}.$$

By the same argument,

$$\theta_j(\mathbf{w}) = w_j^q \sum_{i=1}^{n} \theta_i(\mathbf{w}) w_i^{1-q}$$

for all $j = 1, \dots, n$. The sum on the right-hand side is independent of j, so $\theta(\mathbf{w})$ is a probability distribution proportional to $(w_1^q, \dots, w_n^q)$, forcing $\theta(\mathbf{w}) = \mathbf{w}^{(q)}$.

Finally, we show that $\theta(\mathbf{w}) = \mathbf{w}^{(q)}$ for all $\mathbf{w} \in \Delta_n$. By Lemma 4.5.2 and the continuity hypothesis on D, the map θ_1 is continuous, and similarly for $\theta_2, \dots, \theta_n$. Hence $\theta \colon \Delta_n \to \Delta_n$ is continuous. So too is the map $\mathbf{w} \mapsto \mathbf{w}^{(q)}$. But by the previous paragraph, these last two maps are equal on the positive rational distributions, so they are equal everywhere. □

Proof of Theorem 4.5.1 First, consider distributions $\mathbf{w} = (k_1/k, \dots, k_n/k)$ with positive rational coordinates. Apply D to both sides of equation (4.35): then by Lemma 4.5.4 and the effective number hypothesis on D,

$$D(\mathbf{w}) \cdot M(\mathbf{w}^{(q)}, (k_1, \dots, k_n)) = k.$$

But we can also apply D_q to both sides of equation (4.35): then by the chain rule and the effective number property of D_q,

$$D_q(\mathbf{w}) \cdot M(\mathbf{w}^{(q)}, (k_1, \ldots, k_n)) = k.$$

Hence $D(\mathbf{w}) = D_q(\mathbf{w})$. And by the continuity hypothesis on D and the continuity property of D_q (Lemma 4.4.6(ii)), it follows that $D(\mathbf{w}) = D_q(\mathbf{w})$ for all $\mathbf{w} \in \Delta_n$. □

5

Means

The ideal of the axiomatic approach to diversity measurement is to be able to say 'any measure of diversity that satisfies properties X, Y and Z must be one of the following.' Our later theorems of this type will stand on the shoulders of characterization theorems for means.

The theory of means took shape in the first half of the twentieth century, with the 1930 papers of Kolmogorov [192, 194] and Nagumo [260] as well as Hardy, Littlewood and Pólya's seminal book *Inequalities* [137], first published in 1934. (Aczél [1] describes the early history.) But new results continue to be proved. The 2009 book by Grabisch, Marichal, Mesiar and Pap lists some modern developments ([126], Chapter 4), and most of the characterization theorems in this chapter also appear to be new.

The arguments that we will use are entirely elementary and require no specialist knowledge. Nevertheless, the reader could omit almost all of this chapter without affecting their ability to follow subsequent chapters. The only parts needed later are the statements of Theorems 5.5.10 and 5.5.11.

Compared to most of the literature on characterizations of means, the results and proofs in this chapter have a particular flavour. First, we are mainly interested in the *power* means, as opposed to the much larger class of quasiarithmetic means (defined below). That makes it reasonable to assume a homogeneity axiom, which in turn means that we can almost always do without continuity. (The absence of continuity hypotheses distinguishes our results from many others, such as those of Fodor and Marichal [107].) Second, we wish to include the end cases $M_\infty = \max$ and $M_{-\infty} = \min$ of the power means, and the fact that these means are not *strictly* increasing alters considerably the arguments that can be used.

A key role is played by what Tao calls the 'tensor power trick' ([325], Section 1.9), which can be described as follows. Take a set X and two functions $F, G\colon X \to \mathbb{R}^+$. Suppose we want to prove that $F \leq G$, but have only been

	Strictly increasing	Increasing
$(0, \infty)$	$t \in (-\infty, \infty)$ Theorem 5.3.2	$t \in [-\infty, \infty]$ Theorem 5.4.7
$[0, \infty)$	$t \in (0, \infty)$ Theorem 5.3.3	$t \in [-\infty, \infty]$ Theorem 5.4.9 (also assume continuous and nonzero)

Table 5.1 Summary of characterization theorems for symmetric, decomposable, homogeneous, unweighted means. For instance, the top-left entry indicates that the strictly increasing such means on $(0, \infty)$ are exactly the unweighted power means M_t of order $t \in (-\infty, \infty)$. Table 5.2 (p. 162) gives the corresponding results on weighted means.

able to find a constant C (perhaps large) such that $F \leq CG$. In general, there is nothing more to be said. However, suppose now that X can be equipped with a product that is preserved by both F and G. Let $x \in X$. Then for all $n \geq 1$,

$$F(x) = F(x^n)^{1/n} \leq (CG(x^n))^{1/n} = C^{1/n}G(x),$$

and letting $n \to \infty$ gives $F(x) \leq G(x)$, as desired.

Trivial as it may seem, the tensor power trick can be wielded to powerful effect. Typically X is taken to be a set of vectors or functions equipped with the tensor product. Tao [325] demonstrates the tensor power trick by using it to prove the Hausdorff–Young inequality, and notes that it also plays a part in Deligne's proof of the Weil conjectures. We will use it in the proof of the pivotal Lemma 5.4.3.

This chapter begins with the classical theory of quasiarithmetic means, which are just ordinary arithmetic means transported along a homeomorphism (Section 5.1). The bulk of the chapter (Sections 5.2–5.4) concerns general unweighted means, and culminates in the four characterization theorems shown in Table 5.1.

Finally, in Section 5.5, we develop a method for converting characterization theorems for *unweighted* means into characterization theorems for *weighted* means. This method is applied to the four theorems just mentioned. One of the resulting four characterizations of weighted means goes back to Hardy, Littlewood and Pólya in 1934, while the others may be new.

We will be defining a considerable amount of terminology for properties of means. Appendix B contains a summary for convenient reference. The word 'mean' in isolation will be used informally, without precise definition.

5.1 Quasiarithmetic Means

Let J be a real interval. The arithmetic mean defines a sequence of functions

$$(M_1 : \Delta_n \times J^n \to J)_{n \geq 1}.$$

For any other set I and bijection $\phi \colon I \to J$, we can transport the arithmetic mean on J along ϕ to obtain a kind of mean on I. We will focus on the case where I is also an interval and ϕ is a homeomorphism (that is, both ϕ and ϕ^{-1} are continuous), as follows.

Definition 5.1.1 Let $\phi \colon I \to J$ be a homeomorphism between real intervals. The **quasiarithmetic mean** on I generated by ϕ is the sequence of functions

$$(M_\phi \colon \Delta_n \times I^n \to I)_{n \geq 1}$$

defined by

$$M_\phi(\mathbf{p}, \mathbf{x}) = \phi^{-1}\left(\sum_{i=1}^{n} p_i \phi(x_i)\right)$$

($\mathbf{p} \in \Delta_n$, $\mathbf{x} \in I^n$).

The theory of quasiarithmetic means is classical, and most of the content of this section can be found, more or less explicitly, in Chapter III of Hardy, Littlewood and Pólya [137].

Remark 5.1.2 In the literature, the terms 'quasiarithmetic' and 'quasilinear' are both used, sometimes interchangeably, sometimes with the former reserved for the unweighted case, and sometimes with the latter meaning what we call modularity (Definition 4.2.25). 'Quasiarithmetic' has the advantage of evoking the fact that a quasiarithmetic mean is just an arithmetic mean disguised by a change of variable: the diagram

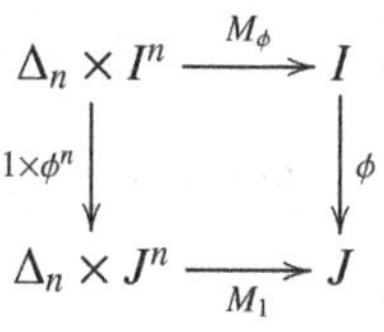

$$\begin{array}{ccc} \Delta_n \times I^n & \xrightarrow{M_\phi} & I \\ {\scriptstyle 1\times\phi^n}\downarrow & & \downarrow{\scriptstyle \phi} \\ \Delta_n \times J^n & \xrightarrow[M_1]{} & J \end{array}$$

commutes.

Example 5.1.3 For real $t \neq 0$, the power mean M_t on $(0, \infty)$ is the quasiarithmetic mean M_{ϕ_t} generated by the homeomorphism

$$\begin{array}{cccc} \phi_t \colon & (0,\infty) & \to & (0,\infty) \\ & t & \mapsto & x^t. \end{array}$$

The geometric mean M_0 on $(0, \infty)$ is the quasiarithmetic mean M_{ϕ_0} generated by the homeomorphism

$$\phi_0 = \log\colon (0, \infty) \to \mathbb{R}.$$

Thus, all the power means of *finite* order on $(0, \infty)$ are quasiarithmetic.

Example 5.1.4 The power means $M_{\pm\infty}$ on $(0, \infty)$ are not quasiarithmetic, as we will prove in Example 5.2.8(i).

Example 5.1.5 The quasiarithmetic mean on $\mathbb{R}$ generated by the homeomorphism $\exp\colon \mathbb{R} \to (0, \infty)$ is given by

$$M_{\exp}(\mathbf{p}, \mathbf{x}) = \log\left(\sum_{i=1}^{n} p_i e^{x_i}\right)$$

($\mathbf{p} \in \Delta_n$, $\mathbf{x} \in \mathbb{R}^n$). This is the **exponential mean**, whose special properties were established by Nagumo ([260], p. 78; or for a modern account, see Theorem 4.15(i) of Grabisch, Marichal, Mesiar and Pap [126]).

The rest of this section is dedicated to three questions.

First, when do two homeomorphisms out of an interval I generate the same quasiarithmetic mean on I?

Second, among all quasiarithmetic means on $(0, \infty)$, how can we pick out the power means M_t $(t \in \mathbb{R})$? In other words, what special properties do the power means possess?

Third (and imprecisely for now), given a mean on some large interval, if its restrictions to smaller intervals are quasiarithmetic, is it quasiarithmetic itself?

The answers to all three questions involve the notion of affine map.

Definition 5.1.6 Let I be a real interval. A function $\alpha\colon I \to \mathbb{R}$ is **affine** if

$$\alpha(px_1 + (1-p)x_2) = p\alpha(x_1) + (1-p)\alpha(x_2)$$

for all $x_1, x_2 \in I$ and $p \in [0, 1]$.

Lemma 5.1.7 *Let $\alpha\colon I \to J$ be a function between real intervals. The following are equivalent:*

i. *α is affine;*
ii. *$\alpha(\sum \lambda_i x_i) = \sum \lambda_i \alpha(x_i)$ for all $n \geq 1$, $x_1, \ldots, x_n \in I$ and $\lambda_1, \ldots, \lambda_n \in \mathbb{R}$ such that $\sum \lambda_i = 1$ and $\sum \lambda_i x_i \in I$;*
iii. *there exist constants $a, b \in \mathbb{R}$ such that $\alpha(x) = ax + b$ for all $x \in I$;*
iv. *α is continuous and $\alpha(\frac{1}{2}(x_1 + x_2)) = \frac{1}{2}(\alpha(x_1) + \alpha(x_2))$ for all $x_1, x_2 \in I$.*

Note that in (ii), some of the coefficients λ_i may be negative.

Proof See Appendix A.4. □

By part (iii), any affine map is either injective or constant. We will need the following elementary observation on extension of affine maps to larger domains.

Definition 5.1.8 A real interval is **trivial** if it has at most one element, and **nontrivial** otherwise.

Lemma 5.1.9 *Let $I \subseteq J$ be real intervals and let $\alpha\colon I \to \mathbb{R}$ be an affine map. Then:*

i. *there exists an affine map $\overline{\alpha}\colon J \to \mathbb{R}$ extending α;*
ii. *if α is injective then we may choose $\overline{\alpha}$ to be injective;*
iii. *if I is nontrivial then $\overline{\alpha}$ is uniquely determined by α.*

Proof Choose $a, b \in \mathbb{R}$ such that $\alpha(x) = ax + b$ for all $x \in I$. For (i), put $\overline{\alpha}(y) = ay + b$ for $y \in J$. For (ii), if α is injective then we can choose a to be nonzero, so $\overline{\alpha}$ is injective. Part (iii) is trivial. □

We are now ready to answer the first question: when are two quasiarithmetic means equal?

Proposition 5.1.10 *Let*

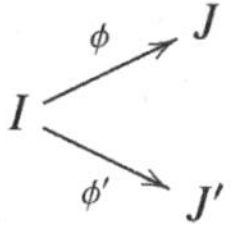

be homeomorphisms between real intervals. The following are equivalent:

i. $M_\phi = M_{\phi'}\colon \Delta_n \times I^n \to I$ *for all* $n \geq 1$;
ii. $M_\phi(\mathbf{u}_n, -) = M_{\phi'}(\mathbf{u}_n, -)\colon I^n \to I$ *for all* $n \geq 1$;
iii. *the map* $\phi' \circ \phi^{-1}\colon J \to J'$ *is affine.*

This is Theorem 83 of the book [137] by Hardy, Littlewood and Pólya, who attribute it to Jessen and Knopp.

Proof Trivially, (i) implies (ii).

Assuming (ii), we prove (iii). Write $\alpha = \phi' \circ \phi^{-1}$. We will prove that α is affine using Lemma 5.1.7(iv). Certainly α is continuous. Now let $y_1, y_2 \in J$. We have

$$M_\phi\big(\mathbf{u}_2, (\phi^{-1}(y_1), \phi^{-1}(y_2))\big) = M_{\phi'}\big(\mathbf{u}_2, (\phi^{-1}(y_1), \phi^{-1}(y_2))\big),$$

or explicitly,

$$\phi^{-1}(\tfrac{1}{2}y_1 + \tfrac{1}{2}y_2) = \phi'^{-1}(\tfrac{1}{2}\phi'\phi^{-1}(y_1) + \tfrac{1}{2}\phi'\phi^{-1}(y_2)).$$

But this can be rewritten as

$$\alpha(\tfrac{1}{2}(y_1 + y_2)) = \tfrac{1}{2}(\alpha(y_1) + \alpha(y_2)),$$

so condition (iv) of Lemma 5.1.7 holds and α is affine.

Finally, assuming (iii), we prove (i). Write α for the affine map $\phi' \circ \phi^{-1} : J \to J'$. Then $\phi' = \alpha \circ \phi$, so our task is to prove that

$$M_{\alpha\circ\phi}(\mathbf{p}, \mathbf{x}) = M_{\phi}(\mathbf{p}, \mathbf{x})$$

for all $n \geq 1$, $\mathbf{p} \in \Delta_n$, and $\mathbf{x} \in I^n$. And indeed,

$$\begin{aligned} M_{\alpha\circ\phi}(\mathbf{p}, \mathbf{x}) &= (\alpha \circ \phi)^{-1}\left(\sum_{i=1}^{n} p_i \alpha(\phi(x_i))\right) \\ &= \phi^{-1}\alpha^{-1}\alpha\left(\sum_{i=1}^{n} p_i \phi(x_i)\right) \\ &= M_{\phi}(\mathbf{p}, \mathbf{x}), \end{aligned}$$

using Lemma 5.1.7(ii) in the second equation. □

Example 5.1.11 This example concerns the quasiarithmetic mean $M_{\ln_q}$. Strictly speaking, $M_{\ln_q}$ is undefined, as the q-logarithm $\ln_q : (0, \infty) \to \mathbb{R}$ is not surjective (hence not a homeomorphism) unless $q = 1$. However, we can change the codomain to force it to be surjective; that is, we can consider the function

$$\begin{array}{ccc} (0, \infty) & \to & \ln_q(0, \infty) \\ x & \mapsto & \ln_q(x), \end{array}$$

where $\ln_q(0, \infty)$ is the image of $\ln_q$. This function, which by abuse of notation we also write as $\ln_q$, *is* a homeomorphism, and its codomain is a real interval. In this sense, we can speak of the quasiarithmetic mean $M_{\ln_q}$.

For $q \neq 1$, the function $\ln_q : (0, \infty) \to \ln_q(0, \infty)$ is the composite of homeomorphisms

$$\begin{array}{ccc} & \xrightarrow{\phi} & (0, \infty) \\ (0, \infty) & & \downarrow \alpha \\ & \xrightarrow{\ln_q} & \ln_q(0, \infty), \end{array}$$

where

$$\phi(x) = x^{1-q}, \quad \alpha(y) = \frac{y - 1}{1 - q}.$$

Here α is affine, so by Proposition 5.1.10 and Example 5.1.3,

$$M_{\ln_q} = M_{1-q}\colon \Delta_n \times (0,\infty)^n \to (0,\infty) \tag{5.1}$$

($n \geq 1$). This equation also holds for $q = 1$, by Example 5.1.3. Hence it holds for all real q.

Equation (5.1) can, of course, also be proved directly. It is equivalent to Lemma 4.2.29.

Next we answer the second question: among all quasiarithmetic means, what distinguishes the power means? The following result is Theorem 84 of Hardy, Littlewood and Pólya [137].

Theorem 5.1.12 *Let J be a real interval and let $\phi\colon (0,\infty) \to J$ be a homeomorphism. The following are equivalent:*

i. $M_\phi(\mathbf{u}_n, c\mathbf{x}) = cM_\phi(\mathbf{u}_n, \mathbf{x})$ *for all* $n \geq 1$, $\mathbf{x} \in (0,\infty)^n$, *and* $c \in (0,\infty)$;
ii. $M_\phi(\mathbf{p}, c\mathbf{x}) = cM_\phi(\mathbf{p}, \mathbf{x})$ *for all* $n \geq 1$, $\mathbf{p} \in \Delta_n$, $\mathbf{x} \in (0,\infty)^n$, *and* $c \in (0,\infty)$;
iii. $M_\phi = M_t$ *for some* $t \in \mathbb{R}$.

Proof Trivially, (iii) implies (ii) and (ii) implies (i).

Now assume (i); we prove (iii). By Proposition 5.1.10, we may assume that $\phi(1) = 0$: for if not, replace J by $J' = J - \phi(1)$ and ϕ by $\phi' = \phi - \phi(1)$, which is a homeomorphism $(0,\infty) \to J'$ satisfying $M_{\phi'} = M_\phi$ and $\phi'(1) = 0$.

For each $c > 0$, define $\phi_c\colon (0,\infty) \to J$ by $\phi_c(x) = \phi(cx)$. Then ϕ_c is a homeomorphism, and for all $\mathbf{x} \in (0,\infty)^n$,

$$\begin{aligned} M_{\phi_c}(\mathbf{u}_n, \mathbf{x}) &= \phi_c^{-1}\left(\sum_{i=1}^n \frac{1}{n}\phi_c(x_i)\right) \\ &= \frac{1}{c}\phi^{-1}\left(\sum_{i=1}^n \frac{1}{n}\phi(cx_i)\right) \\ &= \frac{1}{c}M_\phi(\mathbf{u}_n, c\mathbf{x}) \\ &= M_\phi(\mathbf{u}_n, \mathbf{x}), \end{aligned}$$

where the last step used the homogeneity hypothesis in (i). Hence by Proposition 5.1.10, there exist $\psi(c), \theta(c) \in \mathbb{R}$ such that $\phi_c = \psi(c)\phi + \theta(c)$.

We have now constructed functions $\psi, \theta\colon (0,\infty) \to \mathbb{R}$ such that

$$\phi(cx) = \psi(c)\phi(x) + \theta(c)$$

for all $c, x \in (0,\infty)$. Putting $x = 1$ and using $\phi(1) = 0$ gives $\theta = \phi$, so

$$\phi(cx) = \phi(c) + \psi(c)\phi(x)$$

for all $c, x \in (0, \infty)$. Since ϕ is measurable and not constant, the functional characterization of the q-logarithm (Theorem 1.3.2) implies that $\phi = A \ln_q$ for some $A, q \in \mathbb{R}$ with $A \neq 0$. Hence $M_\phi = M_{\ln_q}$ by Proposition 5.1.10. But $M_{\ln_q} = M_{1-q}$ by Example 5.1.11, so $M_\phi = M_{1-q}$, as required. □

We now answer the third and final question: loosely, given a mean on a large interval whose restriction to every small subinterval is quasiarithmetic, is the original mean also quasiarithmetic?

The most important means for us are the power means, which are defined on the unbounded interval $(0, \infty)$ or $[0, \infty)$. However, some results on means are most easily proved on closed bounded intervals. The following lemma allows us to leverage results on closed bounded intervals to prove results on arbitrary intervals. It states that whether a mean on an arbitrary interval is quasiarithmetic is entirely determined by its behaviour on closed bounded subintervals.

Our lemma concerns *unweighted* means. We will use the abbreviated notation

$$M_\phi(\mathbf{x}) = M_\phi(\mathbf{u}_n, \mathbf{x}) \tag{5.2}$$

for unweighted quasiarithmetic means, and we will say that a sequence of functions $(M\colon I^n \to I)_{n\geq 1}$ on a real interval I is a **quasiarithmetic mean** if there exist an interval J and a homeomorphism $\phi\colon I \to J$ such that M is the unweighted quasiarithmetic mean M_ϕ generated by ϕ.

Lemma 5.1.13 *Let I be a real interval and let $(M\colon I^n \to I)_{n\geq 1}$ be a sequence of functions. Suppose that M restricts to a quasiarithmetic mean on each nontrivial closed bounded subinterval of I. Then M is a quasiarithmetic mean.*

Proof If I is trivial then so is the result. Otherwise, we can write I as the union of an infinite nested sequence $I_1 \subseteq I_2 \subseteq \cdots$ of nontrivial closed bounded subintervals. By hypothesis, $M\colon I^n \to I$ restricts to a function $M|_{I_r}\colon I_r^n \to I_r$ for each $n, r \geq 1$, and the sequence of functions $(M|_{I_r}\colon I_r^n \to I_r)_{n\geq 1}$ is a quasiarithmetic mean for each $r \geq 1$.

We will construct, inductively, a nested sequence $J_1 \subseteq J_2 \subseteq \cdots$ of real intervals and a sequence of homeomorphisms $\phi_r\colon I_r \to J_r$, each satisfying $M_{\phi_r} = M|_{I_r}$ and each extending the last:

$$\begin{array}{ccccc} I_1 & \hookrightarrow & I_2 & \hookrightarrow & \cdots \\ \downarrow{\scriptstyle \phi_1} & & \downarrow{\scriptstyle \phi_2} & & \\ J_1 & \hookrightarrow & J_2 & \hookrightarrow & \cdots \end{array}$$

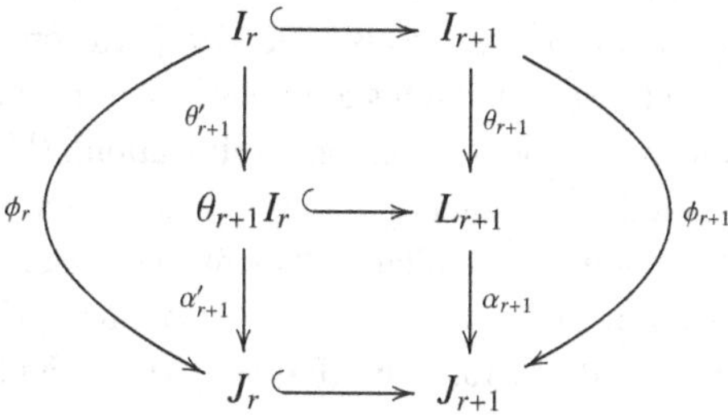

Figure 5.1 The inductive step in the proof of Lemma 5.1.13. The vertical and curved arrows are all homeomorphisms.

For the first step, since $M|_{I_1}$ is a quasiarithmetic mean, we can choose an interval J_1 and a homeomorphism $\phi_1 \colon I_1 \to J_1$ such that $M_{\phi_1} = M|_{I_1}$.

Now suppose inductively that J_r and ϕ_r have been defined for some $r \geq 1$, in such a way that $M_{\phi_r} = M|_{I_r}$. Since $M|_{I_{r+1}}$ is a quasiarithmetic mean, we can choose a real interval L_{r+1} and a homeomorphism $\theta_{r+1} \colon I_{r+1} \to L_{r+1}$ such that $M_{\theta_{r+1}} = M|_{I_{r+1}}$. Then θ_{r+1} restricts to a homeomorphism of intervals $\theta'_{r+1} \colon I_r \to \theta_{r+1}I_r$, giving the top square of the commutative diagram in Figure 5.1.

To construct the bottom square, we need to define α'_{r+1}, J_{r+1}, and α_{r+1}. Put $\alpha'_{r+1} = \phi_r \circ \theta'^{-1}_{r+1}$, which is a homeomorphism. We have $M_{\theta_{r+1}} = M|_{I_{r+1}}$ by definition of θ_{r+1}, so

$$M_{\theta'_{r+1}} = M|_{I_r} = M_{\phi_r}.$$

Hence by Proposition 5.1.10, α'_{r+1} is affine. By Lemma 5.1.9, the affine injection α'_{r+1} on $\theta_{r+1}I_r$ extends uniquely to an affine injection defined on the larger interval L_{r+1}. Writing J_{r+1} for the image of this extended function (which is an interval), this gives an affine homeomorphism $\alpha_{r+1} \colon L_{r+1} \to J_{r+1}$ making the bottom square of Figure 5.1 commute.

Put $\phi_{r+1} = \alpha_{r+1} \circ \theta_{r+1}$. Then ϕ_{r+1} is a homeomorphism since α_{r+1} and θ_{r+1} are. Moreover, $M_{\phi_{r+1}} = M_{\theta_{r+1}}$ since α_{r+1} is affine. But $M_{\theta_{r+1}} = M|_{I_{r+1}}$, so $M_{\phi_{r+1}} = M|_{I_{r+1}}$, completing the inductive construction.

Finally, let J be the interval $\bigcup_{r=1}^{\infty} J_r$ and let $\phi \colon I \to J$ be the unique function extending all of the functions $\phi_r \colon I_r \to J_r$. Then ϕ is a homeomorphism since every ϕ_r is. Moreover, given $\mathbf{x} \in I^n$, we have $\mathbf{x} \in I_r^n$ for some $r \geq 1$, so

$$M_\phi(\mathbf{x}) = M_{\phi_r}(\mathbf{x}) = M|_{I_r}(\mathbf{x}) = M(\mathbf{x}),$$

where the middle equation is by construction of ϕ_r and the others are immediate from the definitions. Hence $M = M_\phi$ and M is a quasiarithmetic mean. □

Remark 5.1.14 Kolmogorov found an early characterization theorem for quasiarithmetic means on real intervals [192, 194]. He proved it for closed bounded intervals, and asserted that his argument could be extended to closed unbounded intervals with 'only a minor modification' ([194], p. 144). In fact, it can be extended to *all* intervals. Later authors used results similar to Lemma 5.1.13 to prove this and similar statements. For example, the argument above is an expansion of the argument on p. 291 of Aczél [2], and of part of the proof of Theorem 4.10 of Grabisch, Marichal, Mesiar and Pap [126].

5.2 Unweighted Means

In the next three sections, we focus exclusively on means that are unweighted, that is, weighted by the uniform distribution $\mathbf{u}_n$. Certainly this is a natural special case. But the real reason for this focus is that the results established will help us to prove theorems on *weighted* means (Section 5.5), which in turn will be used to prove unique characterizations of measures of value (Section 7.3) and measures of diversity (Section 7.4).

The pattern of argument in this chapter is broadly similar to that in the proof of Faddeev's theorem (Section 2.5). There, given a hypothetical entropy measure I satisfying some axioms, most of the work went into analysing the sequence $(I(\mathbf{u}_n))_{n\geq1}$, which then made it relatively easy to find $I(\mathbf{p})$ for distributions $\mathbf{p}$ with rational coordinates, and, in turn, for all $\mathbf{p}$. Here, we spend considerable time proving results on unweighted means $M(\mathbf{u}_n, -)$. This done, we will quickly be able to deduce results on weighted means $M(\mathbf{p}, -)$, first for rational $\mathbf{p}$ and then for all $\mathbf{p}$.

For simplicity, we adopt the abbreviated notation

$$M_t(\mathbf{x}) = M_t(\mathbf{u}_n, \mathbf{x})$$

($t \in [-\infty, \infty]$, $\mathbf{x} \in [0, \infty)^n$) for unweighted power means, as well as using the notation $M_\phi(\mathbf{x})$ as in equation (5.2).

Let $(M\colon I^n \to I)_{n\geq1}$ be a sequence of functions, where I is either $(0, \infty)$ or $[0, \infty)$. Over the next three sections, we answer the question:

> *What conditions on M guarantee that it is one of the unweighted power means M_t?*

The question can be interpreted in several ways, depending on whether I is $(0, \infty)$ or $[0, \infty)$, and also on whether we want to restrict the order t of the power mean to be positive, finite, etc.

We now list some of the conditions on M that might reasonably be imposed.

For many of them, we have already considered similar conditions for weighted means (Section 4.2). For Definitions 5.2.1–5.2.13, let I be a real interval and let $(M\colon I^n \to I)_{n\geq 1}$ be a sequence of functions.

Definition 5.2.1 M is **symmetric** if $M(\mathbf{x}) = M(\mathbf{x}\sigma)$ for all $n \geq 1$, $\mathbf{x} \in I^n$, and permutations σ of $\{1, \ldots, n\}$.

Examples 5.2.2 All quasiarithmetic means are symmetric. So too are all the power means M_t, including M_∞ and $M_{-\infty}$ (which are not quasiarithmetic).

Definition 5.2.3 M is **consistent** (or **idempotent**) if

$$M(\underbrace{x, \ldots, x}_{n}) = x$$

for all $n \geq 1$ and $x \in I$.

Example 5.2.4 All quasiarithmetic and power means are consistent.

Definition 5.2.5 M is **increasing** if for all $n \geq 1$ and $\mathbf{x}, \mathbf{y} \in I^n$,

$$\mathbf{x} \leq \mathbf{y} \implies M(\mathbf{x}) \leq M(\mathbf{y}).$$

It is **strictly increasing** if for all $n \geq 1$ and $\mathbf{x}, \mathbf{y} \in I^n$,

$$\mathbf{x} \leq \mathbf{y} \neq \mathbf{x} \implies M(\mathbf{x}) < M(\mathbf{y}).$$

Example 5.2.6 All quasiarithmetic means are strictly increasing.

Examples 5.2.7 Given a sequence of functions $(M\colon \Delta_n \times I^n \to I)_{n\geq 1}$, if M is increasing or strictly increasing in the sense of Definition 4.2.18 then $(M(\mathbf{u}_n, -) : I^n \to I)_{n\geq 1}$ is increasing or strictly increasing in the sense above. In particular, Lemma 4.2.19 implies that:

i. the unweighted power means M_t of all orders $t \in [-\infty, \infty]$ on $[0, \infty)$ are increasing;
ii. the unweighted power means M_t of *finite* orders $t \in (-\infty, \infty)$ on $(0, \infty)$ are strictly increasing;
iii. the unweighted power means M_t of *finite positive* orders $t \in (0, \infty)$ on $[0, \infty)$ are strictly increasing.

Examples 5.2.8 i. The power means $M_\infty = \max$ and $M_{-\infty} = \min$ are increasing but not strictly so, assuming that the interval I is nontrivial. (The counterexample of Remark 4.2.20 is easily adapted to I.) Hence $M_{\pm\infty}$ are not quasiarithmetic means.

ii. The power means M_t of order $t \in [-\infty, 0]$ are not strictly increasing on $[0, \infty)$ (again, as in Remark 4.2.20). So they are not quasiarithmetic on $[0, \infty)$, even though they are quasiarithmetic on $(0, \infty)$.

Definition 5.2.9 M is **decomposable** if for all $n, k_1, \ldots, k_n \geq 1$ and $x^i_j \in I$,

$$M(x^1_1, \ldots, x^1_{k_1}, \ \ldots, \ x^n_1, \ldots, x^n_{k_n}) = M(\underbrace{a_1, \ldots, a_1}_{k_1}, \ldots, \underbrace{a_n, \ldots, a_n}_{k_n}),$$

where $a_i = M(x^i_1, \ldots, x^i_{k_i})$ for $i \in \{1, \ldots, n\}$.

We adopt the shorthand

$$r * x = \underbrace{x, \ldots, x}_{r} \tag{5.3}$$

whenever $r \geq 1$ and $x \in \mathbb{R}$. Thus, the decomposability equation becomes

$$M(x^1_1, \ldots, x^1_{k_1}, \ \ldots, \ x^n_1, \ldots, x^n_{k_n}) = M(k_1 * a_1, \ldots, k_n * a_n).$$

Decomposability is an unweighted analogue of the chain rule for weighted means (Definition 4.2.23), as the following examples show.

Examples 5.2.10 i. For each $t \in [-\infty, \infty]$, the power mean M_t is decomposable. This can of course be shown by direct calculation, but instead we prove it using earlier results on weighted power means.

Take x^i_j and a_i as in Definition 5.2.9. Write

$$k = k_1 + \cdots + k_n, \quad \mathbf{p} = (k_1/k, \ldots, k_n/k) \in \Delta_n.$$

Then

$$\mathbf{p} \circ (\mathbf{u}_{k_1}, \ldots, \mathbf{u}_{k_n}) = \mathbf{u}_k,$$

so

$$\begin{aligned} &M_t(x^1_1, \ldots, x^1_{k_1}, \ \ldots, \ x^n_1, \ldots, x^n_{k_n}) \\ &\quad = M_t(\mathbf{p} \circ (\mathbf{u}_{k_1}, \ldots, \mathbf{u}_{k_n}), (x^1_1, \ldots, x^1_{k_1}) \oplus \cdots \oplus (x^n_1, \ldots, x^n_{k_n})) \\ &\quad = M_t(\mathbf{p}, (a_1, \ldots, a_n)), \end{aligned}$$

by the chain rule for power means (Proposition 4.2.24). On the other hand,

$$\begin{aligned} &M_t(k_1 * a_1, \ldots, k_n * a_n) \\ &\quad = M_t(\mathbf{p} \circ (\mathbf{u}_{k_1}, \ldots, \mathbf{u}_{k_n}), (k_1 * a_1) \oplus \cdots \oplus (k_n * a_n)) \\ &\quad = M_t\big(\mathbf{p}, (M_t(\mathbf{u}_{k_1}, k_1 * a_1), \ldots, M_t(\mathbf{u}_{k_n}, k_n * a_n))\big) \\ &\quad = M_t(\mathbf{p}, (a_1, \ldots, a_n)), \end{aligned}$$

by the chain rule again and consistency of M_t. Hence M_t is decomposable.

ii. In particular, M_1 is decomposable, from which it follows that all quasiarithmetic means are decomposable.

Remark 5.2.11 In the literature, decomposability is often stated in the asymmetric form

$$M(x_1, \ldots, x_k, y_1, \ldots, y_\ell) = M(k * a, y_1, \ldots, y_\ell)$$

($k, \ell \geq 1$, $x_i, y_j \in I$), where $a = M(x_1, \ldots, x_k)$. (This was the form used by both Kolmogorov [192, 194] and Nagumo [260], for instance.) Under the mild assumptions that M is symmetric and consistent, this is equivalent to the definition above, by a straightforward induction.

Definition 5.2.12 M is **modular** if for all

$$x_1^1, \ldots, x_{k_1}^1, \ \ldots, \ x_1^n, \ldots, x_{k_n}^n \in I, \qquad y_1^1, \ldots, y_{k_1}^1, \ \ldots, \ y_1^n, \ldots, y_{k_n}^n \in I$$

such that

$$M(x_1^i, \ldots, x_{k_i}^i) = M(y_1^i, \ldots, y_{k_i}^i)$$

for each i, we have

$$M(x_1^1, \ldots, x_{k_1}^1, \ \ldots, \ x_1^n, \ldots, x_{k_n}^n) = M(y_1^1, \ldots, y_{k_1}^1, \ \ldots, \ y_1^n, \ldots, y_{k_n}^n).$$

In other words, M is modular if

$$M(x_1^1, \ldots, x_{k_1}^1, \ \ldots, \ x_1^n, \ldots, x_{k_n}^n)$$

is determined by $k_1, \ldots, k_n$ and

$$M(x_1^1, \ldots, x_1^n), \ \ldots, \ M(x_1^n, \ldots, x_{k_n}^n).$$

Evidently this is true if M is decomposable.

Definition 5.2.13 Suppose that I is closed under multiplication. Then M is **homogeneous** if

$$M(c\mathbf{x}) = cM(\mathbf{x})$$

for all $n \geq 1$, $c \in I$, and $\mathbf{x} \in I^n$.

Examples 5.2.14 All the power means are homogeneous. But other quasiarithmetic means are not, as Theorem 5.1.12 shows.

It has already been mentioned that an important early result in the theory of means was proved by Kolmogorov and Nagumo, independently in 1930 [192, 260]. What they showed was that any continuous, symmetric, consistent, strictly increasing, decomposable sequence of functions $(M : I^n \to I)_{n\geq 1}$ on a real interval I is a quasiarithmetic mean.

One of the purposes of this book is to prove characterization theorems for diversity measures. The measures that we characterize are closely related to the power means M_t, where $t \in [-\infty, \infty]$. In particular, we want to include $M_{\pm\infty}$. Since Kolmogorov and Nagumo's theorem insists on a *strictly* increasing mean, it is inadequate for our purpose. So, we follow a different path.

There is another difference between the results below and those of Kolmogorov and Nagumo. Our focus on *power* means makes it natural to impose a homogeneity condition (in the light of Theorem 5.1.12). It turns out that when homogeneity is assumed, the continuity condition in the Kolmogorov–Nagumo theorem can be dropped.

In Section 5.3, we will characterize the power means of finite orders $t \in (-\infty, \infty)$. We will use this result in Section 5.4 to achieve our goal of characterizing the power means of all orders $t \in [-\infty, \infty]$. Our first steps are the same as the first steps of Kolmogorov's proof, and most of the lemmas in the remainder of this section can be found in his paper [192] (translated into English as [194]).

Our first lemma concerns repetition of terms.

Lemma 5.2.15 *Let I be an interval and let $(M\colon I^n \to I)_{n\geq 1}$ be a symmetric, consistent, decomposable sequence of functions. Then*

$$M(r * x_1, \dots, r * x_n) = M(x_1, \dots, x_n) \tag{5.4}$$

for all $r, n \geq 1$ and $x_1, \dots, x_n \in I$.

Proof Write $a = M(x_1, \dots, x_n)$. By symmetry, the left-hand side of equation (5.4) is equal to

$$M(x_1, \dots, x_n, \ \dots, \ x_1, \dots, x_n),$$

with rn terms in total. By decomposability, this is equal to $M(rn * a)$, which by consistency is equal to a. □

The next group of lemmas begins to answer the question: given a quasi-arithmetic mean M on an interval I, how can we construct from M a homeomorphism ϕ such that $M = M_\phi$? Proposition 5.1.10 tells us that there are many homeomorphisms with this property. But it also tells us that if I is of the form $[a, b]$ for some real $a < b$, then there is a unique homeomorphism $\phi\colon [a, b] \to [0, 1]$ such that $\phi(a) = 0$, $\phi(b) = 1$, and $M = M_\phi$. The function ψ constructed in the next lemma will turn out to be the inverse of ϕ, restricted to the rationals.

Lemma 5.2.16 *Let $a, b \in \mathbb{R}$ with $a < b$. Let $(M\colon [a, b]^n \to [a, b])_{n\geq 1}$ be a symmetric, consistent, decomposable sequence of functions.*

i. *There is a unique function*

$$\psi\colon [0,1] \cap \mathbb{Q} \to [a,b]$$

satisfying

$$\psi(r/s) = M((s-r) * a, r * b)$$

for all integers $0 \le r \le s$ *with* $s \ge 1$.

ii. $\psi(0) = a$ *and* $\psi(1) = b$.

iii. *For all* $n \ge 1$ *and* $q_1, \ldots, q_n \in [0,1] \cap \mathbb{Q}$,

$$M(\psi(q_1), \ldots, \psi(q_n)) = \psi\left(\frac{1}{n}\sum_{i=1}^{n} q_i\right).$$

iv. *If M is increasing then so is* ψ, *and if M is strictly increasing then so is* ψ.

Proof For (i), uniqueness is immediate. To prove existence, we must show that different representations r/s of the same rational number give the same value of $M((s-r) * a, r * b)$. Suppose that $r/s = r'/s'$. Then $s'r = sr'$, so using Lemma 5.2.15 twice,

$$\begin{aligned} M((s-r) * a, r * b) &= M(s'(s-r) * a, s'r * b) \\ &= M(s(s'-r') * a, sr' * b) \\ &= M((s'-r') * a, r' * b). \end{aligned}$$

This proves (i), and (ii) follows from the formula for ψ and consistency.

For (iii), express $q_1, \ldots, q_n$ as fractions over a common denominator, say $q_i = r_i/s$. Then

$$M(\psi(q_1), \ldots, \psi(q_n)) = M(s * \psi(q_1), \ldots, s * \psi(q_n)) \tag{5.5}$$

$$= M((s-r_1) * a, r_1 * b, \ldots, (s-r_n) * a, r_n * b) \tag{5.6}$$

$$= M((ns - r_1 - \cdots - r_n) * a, (r_1 + \cdots + r_n) * b) \tag{5.7}$$

$$= \psi\left(\frac{r_1 + \cdots + r_n}{ns}\right) = \psi\left(\frac{1}{n}\sum_{i=1}^{n} q_i\right),$$

where equation (5.5) uses Lemma 5.2.15, equation (5.6) follows from decomposability and the definition of ψ, and equation (5.7) is by symmetry.

For (iv), let $q, q' \in [0,1] \cap \mathbb{Q}$ with $q < q'$. We may write $q = r/s$ and $q' = r'/s$

for some integers $0 \leq r < r' \leq s$ with $s \geq 1$. Assuming that M is increasing,

$$\begin{aligned} M(q) &= M((s-r)*a, r*b) \\ &= M((s-r')*a, (r'-r)*a, r*b) \\ &\leq M((s-r')*a, (r'-r)*b, r*b) \qquad (5.8) \\ &= M((s-r')*a, r'*b) = M(q'), \end{aligned}$$

with strict inequality in (5.8) if M is strictly increasing. □

We will chiefly be working with decomposable, homogeneous means on $(0, \infty)$. Such a mean is automatically consistent:

Lemma 5.2.17 *Let $(M: (0, \infty)^n \to (0, \infty))_{n \geq 1}$ be a decomposable, homogeneous sequence of functions. Then M is consistent.*

Proof For $k \geq 1$, write $a_k = M(k*1)$. By decomposability, $a_k = M(k*a_k)$. (This follows from Definition 5.2.9 by taking $n = 1$, $k_1 = k$, and $x_1^1 = \cdots = x_k^1 = 1$.) Hence by homogeneity, $a_k = a_k M(k*1) = a_k^2$, giving $a_k \in \{0, 1\}$. But M takes values in $(0, \infty)$, so $a_k = 1$. Thus, for all $\mathbf{x} \in (0, \infty)^k$,

$$M(k*x) = xa_k = x$$

by homogeneity again. □

We will deduce that any symmetric such mean is multiplicative, in the following sense.

Definition 5.2.18 Let I be a real interval closed under multiplication. A sequence of functions $(M: I^n \to I)_{n \geq 1}$ is **multiplicative** if

$$M(\mathbf{x} \otimes \mathbf{y}) = M(\mathbf{x})M(\mathbf{y})$$

for all $n, m \geq 1$, $\mathbf{x} \in I^n$, and $\mathbf{y} \in I^m$.

For instance, if a *weighted* mean is multiplicative in the sense of Definition 4.2.27 then its unweighted part $(M(\mathbf{u}_n, -))_{n \geq 1}$ is multiplicative in the sense just defined.

Lemma 5.2.19 *Let $(M: (0, \infty)^n \to (0, \infty))_{n \geq 1}$ be a symmetric, decomposable, homogeneous sequence of functions. Then M is multiplicative.*

Proof By Lemma 5.2.17, M is consistent. Let $\mathbf{x} \in (0, \infty)^n$ and $\mathbf{y} \in (0, \infty)^m$. Writing

$$b_i = M(x_i y_1, \ldots, x_i y_m),$$

we have

$$M(\mathbf{x} \otimes \mathbf{y}) = M(m * b_1, \ldots, m * b_n) = M(b_1, \ldots, b_n), \tag{5.9}$$

by decomposability and Lemma 5.2.15 respectively. But by homogeneity, $b_i = x_i M(\mathbf{y})$. Substituting this into (5.9) and using homogeneity again gives the result. □

Lemmas 5.2.15–5.2.19 are largely taken from Kolmogorov [192, 194], who, assuming that M is continuous, went on to prove that the function ψ of Lemma 5.2.16 extends to a continuous function on $[0, 1]$. But this is where his path and ours diverge.

5.3 Strictly Increasing Homogeneous Means

Here we prove two theorems on strictly increasing, symmetric, decomposable, homogeneous, unweighted means (Table 5.1). First we show that on $(0, \infty)$, such means are exactly the power means of finite order. From this we deduce that on $[0, \infty)$, the means with these properties are exactly the power means of finite *positive* order.

To show that any sequence of functions $\left(M\colon (0, \infty)^n \to (0, \infty)\right)_{n \geq 1}$ with suitable properties is a power mean of finite order, the main challenge is to show that M is quasiarithmetic. We do this by showing that the restriction of M to each closed bounded subinterval $K \subset (0, \infty)$ is quasiarithmetic, then invoking Lemma 5.1.13. An important part of the proof that $M|_K$ is quasiarithmetic will be to take the map

$$\psi\colon [0, 1] \cap \mathbb{Q} \to K$$

provided by Lemma 5.2.16 and extend it to a map $[0, 1] \to K$. For this, we use a lemma of real analysis that has nothing intrinsically to do with means.

Lemma 5.3.1 *Let* $\psi\colon [0, 1] \cap \mathbb{Q} \to \mathbb{R}$ *be a strictly increasing function. Suppose that for all* $z \in [0, 1)$,

$$\sup\{\psi(p) : \text{rational } p \leq z\} = \inf\{\psi(q) : \text{rational } q > z\}, \tag{5.10}$$

and for all $z \in (0, 1]$,

$$\sup\{\psi(p) : \text{rational } p < z\} = \inf\{\psi(q) : \text{rational } q \geq z\}. \tag{5.11}$$

Then ψ *extends uniquely to a continuous function* $[0, 1] \to \mathbb{R}$, *and this extended function is strictly increasing.*

Equations (5.10) and (5.11) can be understood as follows. Taken over rational z, they together state that the function $\psi\colon [0,1]\cap\mathbb{Q}\to\mathbb{R}$ is continuous. When z is irrational, both (5.10) and (5.11) reduce to the equation

$$\sup\{\psi(p) : \text{rational } p < z\} = \inf\{\psi(q) : \text{rational } q > z\},$$

which states that ψ has no jump discontinuity at z. Thus, the result is that any continuous, strictly increasing function on $[0,1]\cap\mathbb{Q}$ extends to a function on $[0,1]$ with the same properties, as long as the original function has no jump discontinuities.

Proof Uniqueness is immediate. For existence, first note that

$$\sup\{\psi(p) : \text{rational } p \leq z\} = \inf\{\psi(q) : \text{rational } q \geq z\} \tag{5.12}$$

for all $z\in[0,1]$. Indeed, if $z\in\mathbb{Q}$ then both sides are equal to $\psi(z)$ (since ψ is increasing), and if $z\notin\mathbb{Q}$ then (5.12) is equivalent to both (5.10) and (5.11). Define a function $\overline{\psi}\colon [0,1]\to\mathbb{R}$ by taking $\overline{\psi}(z)$ to be either side of (5.12). Then $\overline{\psi}|_{\mathbb{Q}} = \psi$.

To see that $\overline{\psi}$ is strictly increasing, let $z,z'\in[0,1]$ with $z<z'$. We can choose rational q and p such that $z\leq q<p\leq z'$. Then by definition of $\overline{\psi}$ and the fact that ψ is strictly increasing,

$$\overline{\psi}(z)\leq\psi(q)<\psi(p)\leq\overline{\psi}(z'),$$

as required.

Finally, we show that $\overline{\psi}$ is continuous. Since $\overline{\psi}$ is increasing, it suffices to show that for all $z\in[0,1)$,

$$\overline{\psi}(z) = \inf\{\overline{\psi}(w) : w>z\},$$

and for all $z\in(0,1]$,

$$\overline{\psi}(z) = \sup\{\overline{\psi}(w) : w<z\}.$$

We prove just the first of these equations, the second being similar. Let $z\in[0,1)$. Then

$$\begin{aligned}
\inf\{\overline{\psi}(w) : w>z\} &= \inf\Big\{\inf\{\psi(q) : \text{rational } q\geq w\} : w>z\Big\} && (5.13)\\
&= \inf\bigcup_{w>z}\{\psi(q) : \text{rational } q\geq w\}\\
&= \inf\{\psi(q) : \text{rational } q>z\}\\
&= \inf\{\psi(q) : \text{rational } q\geq z\} && (5.14)\\
&= \overline{\psi}(z), && (5.15)
\end{aligned}$$

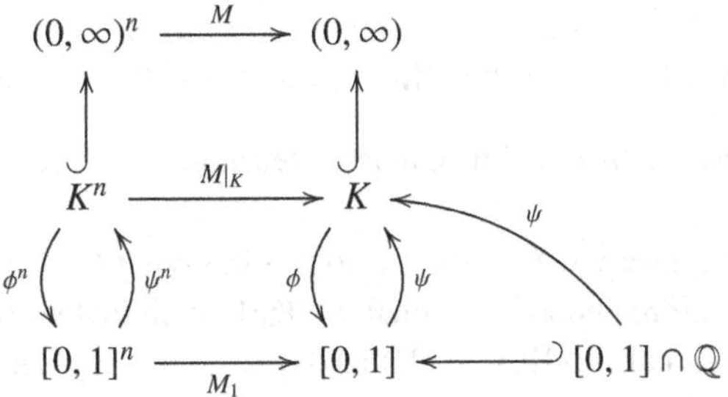

Figure 5.2 Maps involved in the proof of Theorem 5.3.2.

where (5.13) and (5.15) are by definition of $\overline{\psi}$, and (5.14) follows from (5.10) and (5.12). □

We now prove our characterization theorem for strictly increasing unweighted means on $(0,\infty)$ (Figure 5.2).

Theorem 5.3.2 *Let* $(M\colon (0,\infty)^n \to (0,\infty))_{n\geq 1}$ *be a sequence of functions. The following are equivalent:*

i. M is symmetric, strictly increasing, decomposable, and homogeneous;
ii. $M = M_t$ *for some* $t \in (-\infty,\infty)$.

Proof (ii) implies (i) by Examples 5.2.2, 5.2.7(ii), 5.2.10(i), and 5.2.14.

Now assume (i). The main part of the proof is to show that M restricts to a quasiarithmetic mean on each nontrivial closed bounded subinterval $K \subset (0,\infty)$. Let K be such an interval.

First note that for each $n \geq 1$, the function $M\colon (0,\infty)^n \to (0,\infty)$ restricts to a function $M|_K\colon K^n \to K$. Indeed, M is consistent by Lemma 5.2.17, and increasing, so for all $x_1,\ldots,x_n \in K$,

$$\min\{x_1,\ldots,x_n\} \leq M(x_1,\ldots,x_n) \leq \max\{x_1,\ldots,x_n\},$$

giving $M(x_1,\ldots,x_n) \in K$.

Next we show that the sequence of functions $(M|_K\colon K^n \to K)_{n\geq 1}$ is a quasiarithmetic mean. This sequence is symmetric, consistent, strictly increasing and decomposable, since M is. Let $\psi\colon [0,1]\cap\mathbb{Q} \to K$ be the function defined in Lemma 5.2.16. We will extend ψ to a continuous function $[0,1] \to K$ using Lemma 5.3.1. For this, we have to verify the hypotheses of that lemma: that ψ is strictly increasing (which is immediate from Lemma 5.2.16(iv)) and that ψ satisfies equations (5.10) and (5.11). We prove only (5.10), the proof of (5.11) being similar.

Let $z \in [0, 1)$. Put

$$u = \sup\{\psi(p) : \text{rational } p \leq z\}, \qquad v = \inf\{\psi(q) : \text{rational } q > z\}.$$

We have to show that $u = v$. Since ψ is increasing, $u \leq v$. It remains to show that $u \geq v$.

Let $C > 1$. We have $v \in K \subset (0, \infty)$, so $v > 0$, giving $Cv > v$. By definition of v, we can therefore choose a rational $q \in (z, 1]$ such that $\psi(q) \leq Cv$. We can then choose a rational $p \in [0, z]$ such that $\frac{1}{2}(p + q) > z$. By definition of u, we have $\psi(p) \leq u \leq Cu$. Now

$$CM(u, v) = M(Cu, Cv) \tag{5.16}$$
$$\geq M(\psi(p), \psi(q)) \tag{5.17}$$
$$= \psi(\tfrac{1}{2}(p + q)) \tag{5.18}$$
$$\geq v, \tag{5.19}$$

where (5.16) is by homogeneity, (5.17) is because M is increasing, (5.18) follows from Lemma 5.2.16(iii), and (5.19) is because $\frac{1}{2}(p+q) \in (z, 1] \cap \mathbb{Q}$. Hence $CM(u, v) \geq v$ for all $C > 1$, giving $M(u, v) \geq v$. But then

$$v = M(v, v) \geq M(u, v) \geq v$$

(using consistency), so $M(v, v) = M(u, v)$. Since M is *strictly* increasing, this forces $u = v$, proving equation (5.10) in Lemma 5.3.1.

We have now shown that the function $\psi\colon [0, 1] \cap \mathbb{Q} \to K$ satisfies the hypotheses of Lemma 5.3.1. By that lemma, ψ extends uniquely to a continuous, strictly increasing function $[0, 1] \to K$, which we also denote by ψ. The extended function ψ is endpoint-preserving by Lemma 5.2.16(ii), and is therefore a homeomorphism. Let $\phi\colon K \to [0, 1]$ be its inverse.

We will prove that $M|_K = M_\phi$, or equivalently that

$$M(\psi(z_1), \ldots, \psi(z_n)) = \psi(\tfrac{1}{n}(z_1 + \cdots + z_n)) \tag{5.20}$$

($z_i \in [0, 1]$). Indeed, for all $z_1, \ldots, z_n \in [0, 1]$,

$$M(\psi(z_1), \ldots, \psi(z_n)) \geq \sup\Big\{M(\psi(q_1), \ldots, \psi(q_n)) : \text{rational } q_i \leq z_i\Big\} \tag{5.21}$$
$$= \sup\Big\{\psi(\tfrac{1}{n}(q_1 + \cdots + q_n)) : \text{rational } q_i \leq z_i\Big\} \tag{5.22}$$
$$= \psi\Big(\sup\Big\{\tfrac{1}{n}(q_1 + \cdots + q_n) : \text{rational } q_i \leq z_i\Big\}\Big) \tag{5.23}$$
$$= \psi(\tfrac{1}{n}(z_1 + \cdots + z_n)), \tag{5.24}$$

where (5.21) holds because M and ψ are increasing, (5.22) follows from

Lemma 5.2.16(iii), equation (5.23) is a consequence of $\psi: [0,1] \to K$ being a strictly increasing bijection, and (5.24) is elementary. So

$$M(\psi(z_1), \ldots, \psi(z_n)) \geq \psi(\tfrac{1}{n}(z_1 + \cdots + z_n)).$$

The same argument with the inequalities reversed and the suprema changed to infima proves the opposite inequality, and equation (5.20) follows. So $M|_K = M_\phi$, as claimed.

We have now shown that M restricts to a quasiarithmetic mean on each nontrivial closed bounded subinterval of $(0, \infty)$. Hence by Lemma 5.1.13, M itself is a quasiarithmetic mean. But M is homogeneous, so Theorem 5.1.12 now implies that $M = M_t$ for some $t \in (-\infty, \infty)$. □

From this theorem about means on $(0, \infty)$, we deduce a theorem about means on $[0, \infty)$.

Theorem 5.3.3 *Let* $(M: [0, \infty)^n \to [0, \infty))_{n\geq 1}$ *be a sequence of functions. The following are equivalent:*

i. M is symmetric, strictly increasing, decomposable, and homogeneous;
ii. $M = M_t$ *for some* $t \in (0, \infty)$.

Proof Certainly (ii) implies (i), by Examples 5.2.2–5.2.14. Now assume (i). If $\mathbf{0} \neq \mathbf{x} \in [0, \infty)^n$ then $M(\mathbf{x}) > M(\mathbf{0}) \geq 0$, so $M(\mathbf{x}) > 0$. Hence M restricts to a sequence of functions

$$M|_{(0,\infty)}: (0, \infty)^n \to (0, \infty).$$

By Theorem 5.3.2, $M|_{(0,\infty)} = M_t$ for some $t \in (-\infty, \infty)$.

To show that $t > 0$, note that

$$0 < M(1, 0) \leq \inf_{\delta>0} M(1, \delta) = \inf_{\delta>0} M_t(1, \delta) = M_t(1, 0),$$

where in the last step we used the fact that M_t is continuous (Lemma 4.2.5) and increasing. Hence $M_t(1, 0) > 0$. But $M_t(1, 0) = 0$ for all $t \in [-\infty, 0]$ (Definition 4.2.1), so $t \in (0, \infty)$.

We now have to show that the equality $M(\mathbf{x}) = M_t(\mathbf{x})$, so far proved to hold for all $\mathbf{x} \in (0, \infty)^n$, holds for all $\mathbf{x} \in [0, \infty)^n$.

First I claim that

$$M(1, 0) = M_t(1, 0). \tag{5.25}$$

To prove this, we evaluate $M(2, 1, 0)$ in two ways. Put $a = M(1, 0) > 0$. Since

M is decomposable,

$$\begin{aligned} M(2,1,0) &= M(2,a,a) \\ &= M_t(2,a,a) \\ &= \left(\tfrac{1}{3}(2^t + 2a^t)\right)^{1/t}. \end{aligned}$$

On the other hand, $M(2,0) = 2a$ by homogeneity, so, using decomposability again,

$$\begin{aligned} M(2,1,0) &= M(1,2,0) \\ &= M(1,2a,2a) \\ &= M_t(1,2a,2a) \\ &= (\tfrac{1}{3}(1 + 2^{t+1}a^t))^{1/t}. \end{aligned}$$

Equating these two expressions for $M(2,1,0)$ gives $a = (1/2)^{1/t}$, or equivalently, $M(1,0) = M_t(1,0)$, as claimed.

Now we prove that $M(\mathbf{x}) = M_t(\mathbf{x})$ for all $n \geq 1$ and $\mathbf{x} \in [0,\infty)^n$. By symmetry, it suffices to prove this when

$$\mathbf{x} = (x_1, \ldots, x_m, k * 0)$$

for some $k, m \geq 0$ and $x_1, \ldots, x_m > 0$. The proof is by induction on k for all m simultaneously. We already have the result for $k = 0$. Suppose now that $k \geq 1$ and the result holds for $k - 1$. If $m = 0$ then the result is trivial (by homogeneity), so suppose that $m \geq 1$. We have

$$\begin{aligned} &M(x_1, \ldots, x_m, k * 0) \\ &\quad = M(x_1, \ldots, x_{m-1}, x_m, 0, (k-1) * 0) \\ &\quad = M(x_1, \ldots, x_{m-1}, x_m M(1,0), x_m M(1,0), (k-1) * 0) \qquad (5.26) \\ &\quad = M(x_1, \ldots, x_{m-1}, x_m M_t(1,0), x_m M_t(1,0), (k-1) * 0) \qquad (5.27) \\ &\quad = M_t(x_1, \ldots, x_{m-1}, x_m M_t(1,0), x_m M_t(1,0), (k-1) * 0) \qquad (5.28) \\ &\quad = M_t(x_1, \ldots, x_{m-1}, x_m, 0, (k-1) * 0) \qquad (5.29) \\ &\quad = M_t(x_1, \ldots, x_m, k * 0), \end{aligned}$$

where equations (5.26) and (5.29) are by decomposability and homogeneity of M and M_t, equation (5.27) follows from equation (5.25), and equation (5.28) is by inductive hypothesis. This completes the proof. □

5.4 Increasing Homogeneous Means

The extremal cases $M_{\pm\infty}$ of the power means are neither strictly increasing nor quasiarithmetic. Both of these factors put $M_{\pm\infty}$ outside the ambit of many characterizations of means. However, we will prove characterization theorems that include $M_{\pm\infty}$, mostly so as not to exclude the important Berger–Parker index D_∞ (Example 4.3.5(iv)) from a later characterization theorem for diversity measures.

We have already characterized the strictly increasing means, so our task now is to characterize the means M that are increasing but not strictly so. Assuming symmetry, we have

$$M(x_1, \ldots, x_m, u) = M(x_1, \ldots, x_m, v)$$

for some x_i, u, v with $u \neq v$. Our aim is to deduce from this equation, and the usual other hypotheses on M, that M is equal to either $M_\infty = \max$ or $M_{-\infty} = \min$.

Lemma 5.4.1 *Let I be a real interval. Let $(M\colon I^n \to I)_{n\geq 1}$ be a symmetric, decomposable sequence of functions. Let $m \geq 1$ and $x_1, \ldots, x_m, u, v \in I$ with*

$$M(x_1, \ldots, x_m, u) = M(x_1, \ldots, x_m, v).$$

Then

$$M(x_1, \ldots, x_m, n * u) = M(x_1, \ldots, x_m, n * v)$$

for all $n \geq 0$.

Proof This is trivial for $n = 0$. Suppose inductively that $n \geq 0$ and the result holds for n. Since M is decomposable, it is modular (Definition 5.2.12). Now

$$\begin{aligned} M(x_1, \ldots, x_m, (n+1) * u) &= M(x_1, \ldots, x_m, n * u, u) \\ &= M(x_1, \ldots, x_m, n * v, u) && (5.30) \\ &= M(x_1, \ldots, x_m, u, n * v) && (5.31) \\ &= M(x_1, \ldots, x_m, v, n * v) && (5.32) \\ &= M(x_1, \ldots, x_m, (n+1) * v), \end{aligned}$$

where (5.30) and (5.32) use modularity and (5.31) is by symmetry. □

We deduce the following result.

Lemma 5.4.2 *Let I be an interval and let $(M\colon I^n \to I)_{n\geq 1}$ be a sequence of functions that is symmetric, consistent, decomposable, and increasing but not strictly so. Then there exist $x, u, v \in I$ such that $u \neq v$ but $M(x, u) = M(x, v)$.*

Proof By symmetry, there exist $n \geq 0$ and $x_1, \ldots, x_n, u, v \in I$ such that $u \neq v$ and

$$M(x_1, \ldots, x_n, u) = M(x_1, \ldots, x_n, v).$$

By consistency, $n \geq 1$. Writing $x = M(x_1, \ldots, x_n)$, we have

$$M(n * x, u) = M(n * x, v)$$

by decomposability. Now

$$M(x, u) = M(n * x, n * u) = M(n * x, n * v) = M(x, v),$$

where the first and last equalities follow from Lemma 5.2.15 and the second from Lemma 5.4.1. □

Our next lemma contains the main substance of the argument that if M is increasing but not strictly so, then $M = M_{\pm\infty}$.

Lemma 5.4.3 *Let $(M\colon (0,\infty)^n \to (0,\infty))_{n\geq 1}$ be a symmetric, increasing, decomposable, homogeneous sequence of functions. If there exist $x \in (0,\infty)$ and distinct $u, v \geq x$ such that $M(x, u) = M(x, v)$, then there exist $a < b$ in $(0,\infty)$ such that $M(a, b) = a$.*

Proof Take x, u and v as described. We may assume without loss of generality that $u < v$ and (by homogeneity) that $x = 1$. We may therefore choose a real number $C > 1$ and an integer $N \geq 1$ such that $u \leq C^N < C^{N+1} \leq v$. We will prove that $M(1, C^N) = 1$.

Since M is increasing, the hypothesis that $M(1, u) = M(1, v)$ implies that

$$M(1, C^N) = M(1, C^{N+1}).$$

It follows from Lemma 5.4.1 that

$$M(1, r * C^N) = M(1, r * C^{N+1})$$

for all $r \geq 0$, then by homogeneity that

$$M(C^s, r * C^{s+N}) = M(C^s, r * C^{s+N+1}) \tag{5.33}$$

for all $r, s \geq 0$.

I claim that

$$M(1, r * C^k) \leq C^N \tag{5.34}$$

for all $k, r \geq 0$. To prove this, first note that M is consistent, by Lemma 5.2.17. When $k \leq N$, we have $C^k \leq C^N$, so (5.34) holds because M is consistent and increasing. Now let $k \geq N$ and suppose inductively that (5.34) holds for k, for all r. Then for all r,

$$\begin{aligned} M(1, r * C^{k+1}) &= M(1, 1, 2r * C^{k+1}) && (5.35)\\ &\leq M(1, C^{k-N}, 2r * C^{k+1}) && (5.36)\\ &= M(1, C^{k-N}, 2r * C^{k}) && (5.37)\\ &\leq M(1, (2r+1) * C^{k}) && (5.38)\\ &\leq C^N, && (5.39) \end{aligned}$$

where equation (5.35) follows from Lemma 5.2.15, inequality (5.36) from the fact that $C^{k-N} \geq 1$, equation (5.37) from (5.33) (with $s = k - N$) and decomposability, inequality (5.38) from the fact that $C^{k-N} \leq C^k$, and inequality (5.39) from the inductive hypothesis. This completes the induction and the proof of the claimed inequality (5.34).

It follows from (5.34) that

$$M(1, C^{k_1}, \dots, C^{k_r}) \leq C^N \tag{5.40}$$

for all $r, k_1, \dots, k_r \geq 0$. Indeed, since M is increasing, the left-hand side of (5.40) is at most $M(1, r * C^{\max k_i})$, which by (5.34) is at most C^N.

We finish by using the tensor power trick (Tao [325], Section 1.9). For all $r \geq 1$,

$$M(1, C^N) = M\big((1, C^N)^{\otimes r}\big)^{1/r}$$

by Lemma 5.2.19. Expanding the tensor power

$$(1, C^N)^{\otimes r} = (1, C^N) \otimes \cdots \otimes (1, C^N)$$

gives

$$\begin{aligned} M(1, C^N) &= M\left(1, \binom{r}{1} * C^N, \dots, \binom{r}{r-1} * C^{(r-1)N}, C^{rN}\right)^{1/r} \\ &\leq C^{N/r} \end{aligned}$$

for all $r \geq 1$, by symmetry of M and inequality (5.40). Letting $r \to \infty$, this proves that $M(1, C^N) \leq 1$. But also

$$M(1, C^N) \geq M(1, 1) = 1,$$

since M is increasing and consistent. Hence $M(1, C^N) = 1$ with $C^N > 1$, completing the proof. □

The lemma just proved states that under certain hypotheses, $M(a, b) = \min\{a, b\}$ for *some* distinct numbers a and b. The next lemma tells us that in that case, $M(a, b) = \min\{a, b\}$ for *all* a and b. Better still, $M(x_1, \ldots, x_n) = \min x_i$ for all $n \geq 1$ and all x_i.

Lemma 5.4.4 *Let* $(M: (0,\infty)^n \to (0,\infty))_{n\geq 1}$ *be a symmetric, increasing, decomposable, homogeneous sequence of functions. If* $M(a, b) = a$ *for some* $0 < a < b$ *then* $M = M_{-\infty}$.

Proof By homogeneity, we may assume that $a = 1$, so that $b > 1$ with $M(1, b) = 1$.

First I claim that $M(1, b^r) = 1$ for all $r \geq 0$. By Lemma 5.2.17, M is consistent, which gives the case $r = 0$. Now suppose inductively that $r \geq 1$ with $M(1, b^{r-1}) = 1$. By Lemma 5.2.15 and the fact that M is increasing,

$$M(1, b^r) = M(1, 1, b^r, b^r) \leq M(1, b, b^r, b^r). \tag{5.41}$$

By inductive hypothesis and homogeneity, $M(b, b^r) = b$. Hence, using decomposability twice,

$$M(1, b, b^r, b^r) = M(1, b, b, b^r) = M(1, b, b, b). \tag{5.42}$$

But $M(1, b) = 1 = M(1, 1)$ by hypothesis and consistency, so by Lemma 5.4.1,

$$M(1, b, b, b) = M(1, 1, 1, 1) = 1. \tag{5.43}$$

Putting together (5.41), (5.42) and (5.43) gives $M(1, b^r) \leq 1$. But also $M(1, b^r) \geq M(1, 1) = 1$, so $M(1, b^r) = 1$, completing the induction and proving the claim.

Next I claim that

$$M(x, y) = \min\{x, y\} \tag{5.44}$$

for all $x, y \in (0, \infty)$. By homogeneity, it is enough to prove this when $x = 1 \leq y$; then our task is to prove that $M(1, y) = 1$ for all $y \geq 1$. Certainly

$$M(1, y) \geq M(1, 1) = 1.$$

On the other hand, we can choose $r \geq 0$ such that $y \leq b^r$, and then

$$M(1, y) \leq M(1, b^r) = 1$$

by the claim above. Hence $M(1, y) = 1$, as claimed.

Finally, we prove that

$$M(x_1, \ldots, x_n) = \min\{x_1, \ldots, x_n\}$$

for all $n \geq 1$ and $\mathbf{x} \in (0, \infty)^n$. By symmetry, we may assume that $x_1 = \min_i x_i$.

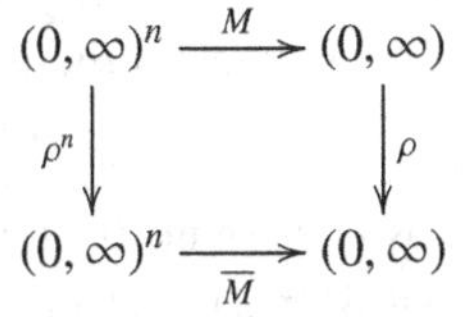

$$\begin{array}{ccc} (0,\infty)^n & \xrightarrow{M} & (0,\infty) \\ \downarrow{\scriptstyle \rho^n} & & \downarrow{\scriptstyle \rho} \\ (0,\infty)^n & \xrightarrow[\overline{M}]{} & (0,\infty) \end{array}$$

Figure 5.3 Commutative diagram showing the relationship between a mean M and its dual $\overline{M}$, where $\rho\colon x \mapsto 1/x$ is the reciprocal map.

Then $M(x_1, x_i) = x_1$ for all i, by equation (5.44). Hence

$$\begin{aligned} M(x_1, x_2, x_3, x_4, \ldots, x_n) &= M(x_1, x_1, x_3, x_4, \ldots, x_n) \\ &= M(x_1, x_1, x_1, x_4, \ldots, x_n) \\ &= \cdots \\ &= M(x_1, x_1, x_1, x_1, \ldots, x_1) \\ &= x_1, \end{aligned}$$

where the first equality follows from decomposability and the fact that $M(x_1, x_2) = x_1$, the second from decomposability and the fact that $M(x_1, x_3) = x_1$, and so on, while the last equality follows from the consistency of M. □

So far, we have focused on $M_{-\infty} = \min$ rather than $M_\infty = \max$. Of course, similar results hold for M_∞ by reversing all the inequalities, but the situation is handled most systematically by the following duality construction (Figure 5.3). Given a sequence of functions $(M\colon (0,\infty)^n \to (0,\infty))_{n\geq 1}$, define another such sequence $\overline{M}$ by

$$\overline{M}(x_1, \ldots, x_n) = \frac{1}{M(\frac{1}{x_1}, \ldots, \frac{1}{x_n})}$$

$(x_1, \ldots, x_n \in (0,\infty))$. For example, equation (4.11) (p. 105) implies that

$$\overline{M_t} = M_{-t}$$

for all $t \in [-\infty, \infty]$. Evidently $\overline{\overline{M}} = M$ for any M.

We will use without mention the following lemma, whose proof is trivial.

Lemma 5.4.5 *Let $(M\colon (0,\infty)^n \to (0,\infty))_{n\geq 1}$ be a sequence of functions. Then M is symmetric, consistent, increasing, strictly increasing, decomposable or homogeneous (respectively) if and only if $\overline{M}$ is.* □

The next result uses this duality.

Proposition 5.4.6 *Let* $(M : (0, \infty)^n \to (0, \infty))_{n \geq 1}$ *be a sequence of functions that is symmetric, decomposable, homogeneous, and increasing but not strictly so. Then* $M = M_{\pm\infty}$.

Proof By Lemma 5.2.17, M is consistent. By Lemma 5.4.2, we can choose $x, u, v \in (0, \infty)$ with $u \neq v$ but $M(x, u) = M(x, v)$. Without loss of generality, $u < v$. There are now three cases to consider, and we prove that in each, $M(a, b) \in \{a, b\}$ for some $a < b$ in $(0, \infty)$.

Case 1: $x \leq u < v$. By Lemma 5.4.3, $M(a, b) = a$ for some $a < b$ in $(0, \infty)$.

Case 2: $u < x < v$. We have

$$M(x, v) \geq M(x, x) = x,$$

but also

$$M(x, v) = M(x, u) \leq M(x, x) = x,$$

so $M(x, v) = x$. Putting $a = x$ and $b = v$ gives $M(a, b) = a$ with $a < b$.

Case 3: $u < v \leq x$. Then $1/x \leq 1/v < 1/u$ with $\overline{M}(1/x, 1/v) = \overline{M}(1/x, 1/u)$. Hence by Lemma 5.4.3 applied to $\overline{M}$, there exist $B < A$ in $(0, \infty)$ such that $\overline{M}(B, A) = B$. Putting $a = 1/A$ and $b = 1/B$ gives $M(b, a) = b$ with $a < b$.

So in all cases, we can choose $a < b$ in $(0, \infty)$ such that $M(a, b) \in \{a, b\}$. If $M(a, b) = a$ then $M = M_{-\infty}$ by Lemma 5.4.4. Otherwise, $M(a, b) = b$, so $\overline{M}(1/a, 1/b) = 1/b$ with $1/b < 1/a$. Applying Lemma 5.4.4 to $\overline{M}$ gives $\overline{M} = M_{-\infty}$, or equivalently, $M = M_{\infty}$. □

This brings us to our third characterization theorem for unweighted power means, this time including the extremal cases $M_{\pm\infty}$.

Theorem 5.4.7 *Let* $(M : (0, \infty)^n \to (0, \infty))_{n \geq 1}$ *be a sequence of functions. The following are equivalent:*

i. *M is symmetric, increasing, decomposable, and homogeneous;*
ii. *$M = M_t$ for some $t \in [-\infty, \infty]$.*

Proof (ii) implies (i) by Examples 5.2.2–5.2.14. Now assume (i). If M is strictly increasing then $M = M_t$ for some $t \in (-\infty, \infty)$, by Theorem 5.3.2. Otherwise, $M = M_{\pm\infty}$ by Proposition 5.4.6. □

Our fourth and final characterization theorem for unweighted power means captures all the power means M_t, including $M_{\pm\infty}$, on the larger interval $[0, \infty)$. It is an easy consequence of Theorem 5.4.7, but comes at the cost of a significant extra hypothesis: that for each $n \geq 1$, the function $M : [0, \infty)^n \to [0, \infty)$ is continuous.

We need one lemma in preparation.

Lemma 5.4.8 *Let $(M\colon [0,\infty)^n \to [0,\infty))_{n\geq 1}$ be a sequence of functions, none of which is identically zero. If M is increasing, decomposable, and homogeneous, then M is consistent.*

Proof Let $n \geq 1$. By the same argument as in Lemma 5.2.17, $M(n*1) \in \{0,1\}$. Suppose for a contradiction that $M(n*1) = 0$. Then by homogeneity, $M(n*x) = 0$ for all $x \in [0,\infty)$. For all $\mathbf{x} \in [0,\infty)^n$,

$$M(\mathbf{x}) \leq M(n * \max_i x_i) = 0$$

since M is increasing. Hence $M\colon [0,\infty)^n \to [0,\infty)$ is identically zero, contrary to our hypothesis. Thus, $M(n * 1) = 1$. It follows by homogeneity that M is consistent. □

Theorem 5.4.9 *Let $(M\colon [0,\infty)^n \to [0,\infty))_{n\geq 1}$ be a sequence of functions. The following are equivalent:*

i. M is symmetric, increasing, decomposable, homogeneous and continuous, and none of the functions $M\colon [0,\infty)^n \to [0,\infty)$ is identically zero;
ii. $M = M_t$ for some $t \in [-\infty,\infty]$.

Proof It is straightforward that (ii) implies (i), with the continuity coming from Lemma 4.2.5. Now assume (i).

By Lemma 5.4.8, M is consistent. Since M is also increasing, $M(\mathbf{x}) \geq \min_i x_i > 0$ for all $\mathbf{x} \in (0,\infty)^n$. Hence M restricts to a sequence of functions

$$(M|_{(0,\infty)}\colon (0,\infty)^n \to (0,\infty))_{n\geq 1}.$$

The functions $M|_{(0,\infty)}$ are symmetric, increasing, decomposable and homogeneous, so by Theorem 5.4.7, there exists $t \in [-\infty,\infty]$ such that $M = M_t$ on $(0,\infty)$. For each $n \geq 1$, the functions

$$M, M_t\colon [0,\infty)^n \to [0,\infty)$$

are continuous and are equal on the dense subset $(0,\infty)^n$, so they are equal everywhere. □

Remarks 5.4.10 i. The continuity condition in Theorem 5.4.9 cannot be dropped. Indeed, take any $t \in (0,\infty]$ and define a function $M\colon [0,\infty)^n \to [0,\infty)$ for each $n \geq 1$ by

$$M(\mathbf{x}) = \begin{cases} M_t(\mathbf{x}) & \text{if } \mathbf{x} \in (0,\infty)^n, \\ 0 & \text{otherwise.} \end{cases}$$

Then M satisfies all the conditions of Theorem 5.4.9(i) apart from continuity, and is not a power mean.

	Strictly increasing	Increasing
$(0, \infty)$	$t \in (-\infty, \infty)$ Theorem 5.5.8	$t \in [-\infty, \infty]$ Theorem 5.5.10
$[0, \infty)$	$t \in (0, \infty)$ Theorem 5.5.9	$t \in [-\infty, \infty]$ Theorem 5.5.11 (also assume continuous in second argument)

Table 5.2 Summary of characterization theorems for symmetric, absence-invariant, consistent, modular, homogeneous, weighted means. For instance, the top-left entry indicates that the strictly increasing such means on $(0, \infty)$ are exactly the weighted power means M_t of order $t \in (-\infty, \infty)$. Table 5.1 (p. 134) gives the corresponding results on unweighted means.

ii. The hypothesis that none of the functions $M\colon [0, \infty)^n \to [0, \infty)$ is identically zero cannot be dropped either. Indeed, take any $t \in [-\infty, \infty]$ and any integer $k \geq 1$, and for $\mathbf{x} \in \mathbb{R}^n$, define

$$M(\mathbf{x}) = \begin{cases} M_t(\mathbf{x}) & \text{if } n \leq k, \\ 0 & \text{if } n > k. \end{cases}$$

Then M satisfies all the other conditions of Theorem 5.4.9(i), and is not a power mean.

5.5 Weighted Means

So far, this chapter has been directed towards characterization theorems for unweighted means (Theorems 5.3.2, 5.3.3, 5.4.7 and 5.4.9, summarized in Table 5.1). But we can now deduce characterization theorems for weighted means with comparatively little work.

We do this in three steps. First, we record some elementary implications between properties that a notion of weighted mean may or may not satisfy, and between conditions on weighted and unweighted means. Second, we create a method for converting characterization theorems for unweighted means into characterization theorems for weighted means. Third, we apply that method to the theorems just mentioned. This produces four theorems for weighted means, summarized in Table 5.2.

The elementary implications (Lemmas 5.5.1–5.5.4) are shown in Figure 5.4. In these lemmas, I denotes a real interval and M denotes a sequence of func-

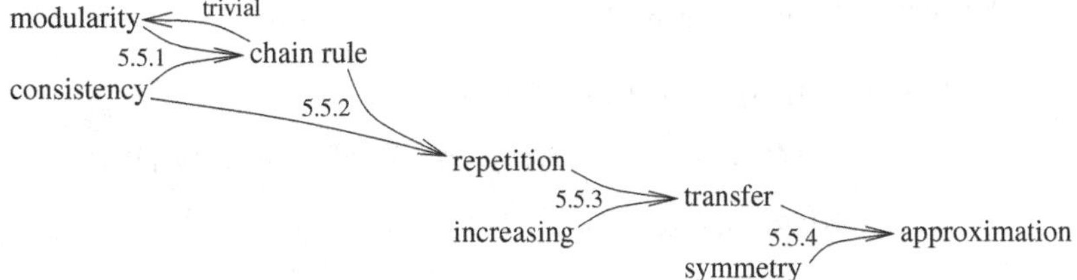

Figure 5.4 Implications between properties of weighted means (Lemmas 5.5.1–5.5.4). The labels on the arrows indicate lemma numbers.

tions $(M \colon \Delta_n \times I^n \to I)_{n\geq 1}$. The properties of means mentioned there were all defined in Section 4.2 (apart from transfer and approximation, defined below) and are summarized in Appendix B.

Plainly the chain rule implies modularity. There is also a kind of converse.

Lemma 5.5.1 *If M is consistent and modular then M satisfies the chain rule.*

Proof Let $\mathbf{w} \in \Delta_n$, let $\mathbf{p}^1 \in \Delta_{k_1}, \ldots, \mathbf{p}^n \in \Delta_{k_n}$, and let $\mathbf{x}^1 \in I^{k_1}, \ldots, \mathbf{x}^n \in I^{k_n}$, where $n, k_i \geq 1$ are integers. Write $a_i = M(\mathbf{p}^i, \mathbf{x}^i)$. By consistency,

$$M(\mathbf{p}^i, \mathbf{x}^i) = a_i = M(\mathbf{u}_1, (a_i))$$

for each i. Hence by modularity,

$$M(\mathbf{w} \circ (\mathbf{p}^1, \ldots, \mathbf{p}^n), \mathbf{x}^1 \oplus \cdots \oplus \mathbf{x}^n) = M(\mathbf{w} \circ (\mathbf{u}_1, \ldots, \mathbf{u}_1), (a^1) \oplus \cdots \oplus (a^n)).$$

But the right-hand side is $M(\mathbf{w}, (a_1, \ldots, a_n))$, so the result is proved. □

Lemma 5.5.2 *If M is consistent and satisfies the chain rule then M has the repetition property.*

Proof Let $\mathbf{p} \in \Delta_n$ and $\mathbf{x} \in I^n$; suppose that $x_i = x_{i+1}$ for some $i < n$. We must prove that

$$\begin{aligned} M(\mathbf{p}, \mathbf{x}) = M((&p_1, \ldots, p_{i-1}, p_i + p_{i+1}, p_{i+2}, \ldots, p_n), \\ &(x_1, \ldots, x_{i-1}, x_i, x_{i+2}, \ldots, x_n)). \end{aligned}$$

For ease of notation, let us assume that $i = n - 1$. (The general case is similar.) By Lemma 2.1.9,

$$\mathbf{p} = (p_1, \ldots, p_{n-2}, p_{n-1} + p_n) \circ (\mathbf{u}_1, \ldots, \mathbf{u}_1, \mathbf{r})$$

for some $\mathbf{r} \in \Delta_2$. Then

$$\begin{aligned} M(\mathbf{p}, \mathbf{x}) = M((&p_1, \ldots, p_{n-2}, p_{n-1} + p_n) \circ (\mathbf{u}_1, \ldots, \mathbf{u}_1, \mathbf{r}), \\ &(x_1) \oplus \cdots \oplus (x_{n-2}) \oplus (x_{n-1}, x_{n-1})), \end{aligned}$$

so by the chain rule and consistency,

$$\begin{aligned}M(\mathbf{p},\mathbf{x}) &= M\Big((p_1,\ldots,p_{n-2},p_{n-1}+p_n),\\ &\qquad (M(\mathbf{u}_1,(x_1)),\ldots,M(\mathbf{u}_1,(x_{n-2})),M(\mathbf{r},(x_{n-1},x_{n-1})))\Big)\\ &= M((p_1,\ldots,p_{n-2},p_{n-1}+p_n),(x_1,\ldots,x_{n-2},x_{n-1})),\end{aligned}$$

as required. □

Lemma 5.5.3 *If M has the repetition property and is increasing, then M also has the* **transfer** *property:*

$$M(\mathbf{p},\mathbf{x}) \le M((p_1,\ldots,p_{i-1},p_i-\delta,p_{i+1}+\delta,p_{i+2},\ldots,p_n),\mathbf{x})$$

whenever $1 \le i < n$, $\mathbf{p} \in \Delta_n$, $\mathbf{x} \in I^n$ with $x_i \le x_{i+1}$, and $0 \le \delta \le p_i$.

The transfer property states that a weighted mean increases when weight is transferred from a smaller argument to a larger one.

Proof As in the last proof, we may harmlessly assume that $i = n - 1$. We have

$$\begin{aligned}M(\mathbf{p},\mathbf{x}) &= M((p_1,\ldots,p_{n-2},p_{n-1}-\delta,\delta,p_n),(x_1,\ldots,x_{n-2},x_{n-1},x_{n-1},x_n)) &(5.45)\\ &\le M((p_1,\ldots,p_{n-2},p_{n-1}-\delta,\delta,p_n),(x_1,\ldots,x_{n-2},x_{n-1},x_n,x_n)) &(5.46)\\ &= M((p_1,\ldots,p_{n-2},p_{n-1}-\delta,p_n+\delta),\mathbf{x}), &(5.47)\end{aligned}$$

where (5.45) and (5.47) hold by repetition and (5.46) holds because M is increasing. □

Lemma 5.5.4 *Suppose that M is symmetric and has the transfer property. Then M also has the following* **approximation** *property: for all $\mathbf{p} \in \Delta_n$, $\mathbf{x} \in I^n$ and $\delta > 0$, there exist $\mathbf{p}^-, \mathbf{p}^+ \in \Delta_n$ such that all the coordinates of $\mathbf{p}^-$ and $\mathbf{p}^+$ are rational,*

$$\max_i |p_i^- - p_i| < \delta, \quad \max_i |p_i^+ - p_i| < \delta,$$

and

$$M(\mathbf{p}^-,\mathbf{x}) \le M(\mathbf{p},\mathbf{x}) \le M(\mathbf{p}^+,\mathbf{x}).$$

Proof We just prove the existence of such a $\mathbf{p}^+$, the argument for $\mathbf{p}^-$ being similar. By symmetry, we may assume that $x_1 \le \cdots \le x_n$.

Choose $\delta_1 \in [0,\delta)$ with $0 \le p_1 - \delta \in \mathbb{Q}$. By the transfer property,

$$M(\mathbf{p},\mathbf{x}) \le M((p_1-\delta_1,p_2+\delta_1,p_3,\ldots,p_n),\mathbf{x}).$$

Next, choose $\delta_2 \in [0, \delta)$ such that $0 \leq p_2+\delta_1-\delta_2 \in \mathbb{Q}$. By the transfer property,

$$M((p_1 - \delta_1, p_2 + \delta_1, p_3, \ldots, p_n), \mathbf{x}) \\ \leq M((p_1 - \delta_1, p_2 + \delta_1 - \delta_2, p_3 + \delta_2, p_4, \ldots, p_n), \mathbf{x}).$$

Continuing in this way, we obtain $n - 1$ inequalities that together imply that

$$M(\mathbf{p}, \mathbf{x}) \leq M((p_1 - \delta_1, p_2 + \delta_1 - \delta_2, \ldots, p_{n-1} + \delta_{n-2} - \delta_{n-1}, p_n + \delta_{n-1}), \mathbf{x}).$$

The result follows by taking $\mathbf{p}^+$ to be the distribution on the right-hand side. □

Many properties of weighted means $M(-, -)$ imply the corresponding property of their unweighted counterparts $M(\mathbf{u}_n, -)$.

Lemma 5.5.5 *Let I be an interval and let $(M\colon \Delta_n \times I^n \to I)_{n\geq 1}$ be a sequence of functions. If M is symmetric, consistent, increasing, or strictly increasing (respectively), then so is the sequence of functions $(M(\mathbf{u}_n, -)\colon I^n \to I)_{n\geq 1}$. Moreover, if I is closed under multiplication and M is homogeneous then so is $(M(\mathbf{u}_n, -))_{n\geq 1}$.*

Proof Trivial. □

It was stated on p. 144 that decomposability is an unweighted analogue of the chain rule. The following lemma supports that claim.

Lemma 5.5.6 *Let I be a real interval and let $(M\colon \Delta_n \times I^n \to I)_{n\geq 1}$ be a sequence of functions that is consistent and satisfies the chain rule. Then $(M(\mathbf{u}_n, -)\colon I^n \to I)_{n\geq 1}$ is decomposable.*

Proof Let $n, k_1, \ldots, k_n \geq 1$ and $\mathbf{x}^1 \in I^{k_1}, \ldots, \mathbf{x}^n \in I^{k_n}$. Write $a_i = M(\mathbf{u}_{k_i}, \mathbf{x}^i)$ and $k = \sum k_i$. We must show that

$$M(\mathbf{u}_k, \mathbf{x}^1 \oplus \cdots \oplus \mathbf{x}^n) = M(\mathbf{u}_k, (k_1 * a_1, \ldots, k_n * a_n)).$$

We have

$$\mathbf{u}_k = (k_1/k, \ldots, k_n/k) \circ (\mathbf{u}_{k_1}, \ldots, \mathbf{u}_{k_n}),$$

so by the chain rule,

$$M(\mathbf{u}_k, \mathbf{x}^1 \oplus \cdots \oplus \mathbf{x}^n) = M((k_1/k, \ldots, k_n/k), (a_1, \ldots, a_n)). \tag{5.48}$$

But by Lemma 5.5.2, M has the repetition property, which implies by induction that the right-hand side of (5.48) is equal to

$$M(\mathbf{u}_k, (k_1 * a_1, \ldots, k_n * a_n)).$$

This completes the proof. □

We now make a tool for converting theorems on unweighted means into theorems on weighted means.

Proposition 5.5.7 *Let I be a real interval and let*

$$(M, M' : \Delta_n \times I^n \to I)_{n \geq 1}$$

be two sequences of functions. Suppose the following:

i. *both M and M' have the absence-invariance and repetition properties;*
ii. *M is symmetric and increasing;*
iii. *for each $\mathbf{x} \in I^n$, the function $M'(-, \mathbf{x})$ is continuous on the open simplex Δ_n°.*

Suppose also that

$$M(\mathbf{u}_n, -) = M'(\mathbf{u}_n, -) : I^n \to I$$

for all $n \geq 1$. Then $M = M'$.

Proof First we prove that $M(\mathbf{p}, -) = M'(\mathbf{p}, -)$ when the coordinates of $\mathbf{p}$ are rational and nonzero. Write

$$\mathbf{p} = (k_1/k, \dots, k_n/k),$$

where $k_1, \dots, k_n$ are positive integers and $k = \sum k_i$. Let $\mathbf{x} \in I^n$. Then by the repetition property of M and induction,

$$M(\mathbf{p}, \mathbf{x}) = M(\mathbf{u}_k, (k_1 * x_1, \dots, k_n * x_n)). \tag{5.49}$$

The same argument applied to M' gives

$$M'(\mathbf{p}, \mathbf{x}) = M'(\mathbf{u}_k, (k_1 * x_1, \dots, k_n * x_n)). \tag{5.50}$$

But the right-hand sides of (5.49) and (5.50) are equal by hypothesis, so $M(\mathbf{p}, \mathbf{x}) = M'(\mathbf{p}, \mathbf{x})$.

Now we show by induction on $n \geq 1$ that $M(\mathbf{p}, \mathbf{x}) = M'(\mathbf{p}, \mathbf{x})$ for all $\mathbf{p} \in \Delta_n$ and $\mathbf{x} \in I^n$.

For $n = 1$, we must have $\mathbf{p} = \mathbf{u}_1$, hence $M(\mathbf{p}, \mathbf{x}) = M'(\mathbf{p}, \mathbf{x})$ by hypothesis.

Let $n \geq 2$ and assume the result for $n - 1$. If $p_i = 0$ for some i then $M(\mathbf{p}, \mathbf{x}) = M'(\mathbf{p}, \mathbf{x})$ by inductive hypothesis and absence-invariance of M and M'. Suppose, then, that $\mathbf{p} \in \Delta_n^\circ$.

Let $\varepsilon > 0$. Since $M'(-, \mathbf{x})$ is continuous at $\mathbf{p}$, we can choose $\delta \in (0, \min_i p_i)$ such that for $\mathbf{r} \in \Delta_n^\circ$,

$$\max_i |p_i - r_i| < \delta \implies |M'(\mathbf{p}, \mathbf{x}) - M'(\mathbf{r}, \mathbf{x})| < \varepsilon.$$

By Lemma 5.5.3, M has the transfer property, so by Lemma 5.5.4, M also has the approximation property. Choose $\mathbf{p}^+$ as in Lemma 5.5.4; then

$$|M'(\mathbf{p}, \mathbf{x}) - M'(\mathbf{p}^+, \mathbf{x})| < \varepsilon. \tag{5.51}$$

Also, since $\mathbf{p} \in \Delta_n^\circ$ and

$$\max_i |p_i - p_i^+| < \delta < \min_i p_i,$$

we have $\mathbf{p}^+ \in \Delta_n^\circ$ too. Now

$$\begin{aligned} M(\mathbf{p}, \mathbf{x}) &\leq M(\mathbf{p}^+, \mathbf{x}) && (5.52)\\ &= M'(\mathbf{p}^+, \mathbf{x}) && (5.53)\\ &< M'(\mathbf{p}, \mathbf{x}) + \varepsilon, && (5.54) \end{aligned}$$

where inequality (5.52) is one of the defining properties of $\mathbf{p}^+$, equation (5.53) holds because the coordinates of $\mathbf{p}^+$ are rational and nonzero (using the first step of the proof), and inequality (5.54) follows from (5.51). But this holds for all $\varepsilon > 0$, so

$$M(\mathbf{p}, \mathbf{x}) \leq M'(\mathbf{p}, \mathbf{x}).$$

A very similar argument, using the distribution $\mathbf{p}^-$ of Lemma 5.5.4, proves the opposite inequality. Hence $M(\mathbf{p}, \mathbf{x}) = M'(\mathbf{p}, \mathbf{x})$, completing the proof. □

We can now simply read off four characterization theorems for weighted power means. They are summarized in Table 5.2, and are derived from the four theorems on unweighted means shown in Table 5.1.

Theorem 5.5.8 *Let* $(M\colon \Delta_n \times (0, \infty)^n \to (0, \infty))_{n \geq 1}$ *be a sequence of functions. The following are equivalent:*

i. M is symmetric, absence-invariant, consistent, strictly increasing, modular, and homogeneous;
ii. $M = M_t$ *for some* $t \in (-\infty, \infty)$.

Proof Part (ii) implies part (i) by the results in Section 4.2. Now assume (i). The unweighted mean

$$(M(\mathbf{u}_n, -)\colon (0, \infty)^n \to (0, \infty))_{n \geq 1}$$

is symmetric, strictly increasing, decomposable, and homogeneous (using Lemmas 5.5.1 and 5.5.6 for decomposability). Hence by Theorem 5.3.2, there is some $t \in (-\infty, \infty)$ such that

$$M(\mathbf{u}_n, -) = M_t(\mathbf{u}_n, -)$$

for all $n \geq 1$. By Lemmas 5.5.1 and 5.5.2, M has the repetition property. Hence by the previously established properties of M_t, we can apply Proposition 5.5.7 with $M' = M_t$, giving $M = M_t$. □

Theorem 5.5.8 is essentially due to Hardy, Littlewood and Pólya [137]. Some minor details aside, it is the conjunction of their Theorems 84 and 215, translated out of the language of Stieltjes integrals and into elementary terms. Section 6.21 of [137] gives details.

Theorem 5.5.9 *Let* $(M\colon \Delta_n \times [0,\infty)^n \to [0,\infty))_{n\geq 1}$ *be a sequence of functions. The following are equivalent:*

i. *M is symmetric, absence-invariant, consistent, strictly increasing, modular, and homogeneous;*
ii. *$M = M_t$ for some $t \in (0,\infty)$.*

Proof This follows by exactly the same argument as for the last theorem, but using Theorem 5.3.3 instead of Theorem 5.3.2. □

Theorem 5.5.10 *Let* $(M\colon \Delta_n \times (0,\infty)^n \to (0,\infty))_{n\geq 1}$ *be a sequence of functions. The following are equivalent:*

i. *M is symmetric, absence-invariant, consistent, increasing, modular, and homogeneous;*
ii. *$M = M_t$ for some $t \in [-\infty,\infty]$.*

Proof This follows from Theorem 5.4.7 by the same argument. □

Theorem 5.5.11 *Let* $(M\colon \Delta_n \times [0,\infty)^n \to [0,\infty))_{n\geq 1}$ *be a sequence of functions. The following are equivalent:*

i. *M is symmetric, absence-invariant, consistent, increasing, modular, homogeneous, and continuous in its second argument;*
ii. *$M = M_t$ for some $t \in [-\infty,\infty]$.*

Proof This follows from Theorem 5.4.9 by the same argument again, this time also noting that by consistency, none of the functions $M(\mathbf{u}_n, -)$ is identically zero. □

We will use Theorem 5.5.10 to prove an axiomatic characterization of measures of the value of a community (Section 7.3) and, building on this, to characterize the Hill numbers (Section 7.4).

6

Species Similarity and Magnitude

Alfred Russel Wallace, who in parallel with Charles Darwin discovered what we now call the theory of evolution, spent much of the 1850s travelling in tropical south-east Asia and South America. On his return, he wrote widely on what he had experienced, including the following description of the diversity of a tropical forest ([346], p. 65):

> If the traveller notices a particular species and wishes to find more like it, he may often turn his eyes in vain in every direction. Trees of varied forms, dimensions, and colours are around him, but he rarely sees any one of them repeated. Time after time he goes towards a tree which looks like the one he seeks, but a closer examination proves it to be distinct. He may at length, perhaps, meet with a second specimen half a mile off, or may fail altogether, till on another occasion he stumbles on one by accident.

One of Wallace's observations was that besides there being a large number of species, mostly rare, there was also a great deal of similarity between different species. Clearly, any comprehensive account of the variety or diversity of life has to incorporate the varying degrees of similarity between species. All else being equal, a community of species that are closely related to one another should be judged as less diverse than if they were highly dissimilar.

This is not an abstract concern. The Organization for Economic Co-operation and Development's guide to biodiversity for policy makers recognizes this same point, stating that

> associated with the idea of diversity is the concept of *distance*, i.e., some measure of the dissimilarity of the resources in question

([268], p. 25). With global biodiversity now being lost at historically unprecedented rates, it is crucial that politicians and scientists speak the same language. However, most conventional measures of diversity, and all the ones discussed in this text so far, fail to take the different dissimilarities between species into account.

Here we solve this problem, defining a system of measures that depend not only on the relative abundances of the species, but also on the varying similarity between them (Sections 6.1 and 6.2). It was first introduced in a 2012 article of Leinster and Cobbold [220]. We make no assumption about *how* similarity is measured: it could be genetic, phylogenetic, functional, etc., leading to measures of genetic diversity, phylogenetic diversity, functional diversity, etc. As such, the system is adaptable to a wide variety of scientific needs.

More specifically, we will encode the similarities between species as a real matrix Z, continuing to represent the relative abundances of the species as a probability distribution $\mathbf{p}$. With this model of a community, we will define for each $q \in [0, \infty]$ a measure $D_q^Z(\mathbf{p})$ of the diversity of the community. As for the Hill numbers, the parameter q controls the extent to which the measure emphasizes the common species at the expense of the rare ones. Under the extreme hypothesis that different species never have anything in common, Z is the identity matrix I and the diversity $D_q^I(\mathbf{p})$ reduces to the Hill number $D_q(\mathbf{p})$. In that sense, these similarity-sensitive diversity measures generalize the Hill numbers.

Let $\mathbf{p}$ be a probability distribution on a finite set. We saw in Section 4.3 that the Hill numbers $D_q(\mathbf{p})$, the Rényi entropies $H_q(\mathbf{p})$ and the q-logarithmic entropies $S_q(\mathbf{p})$ are all simple increasing transformations of one another. The same is true in the more general context here. Thus, accompanying the similarity-sensitive diversity measures $D_q^Z(\mathbf{p})$ are similarity-sensitive Rényi entropies $H_q^Z(\mathbf{p})$ and q-logarithmic entropies $S_q^Z(\mathbf{p})$. Any metric on our finite set gives rise naturally to a similarity matrix Z, as we shall see. So, we obtain definitions of the Rényi and q-logarithmic entropies of a probability distribution on a finite metric space, extending the classical definitions on a finite set.

How is diversity maximized? For a fixed similarity matrix Z (and in particular, for a finite metric space), we can seek the probability distribution $\mathbf{p}$ that maximizes the diversity or entropy of a given order q. As we saw in the special case of the Hill numbers, different values of q can lead to different judgements on which of two communities is the more diverse. So in principle, both the maximizing distribution and the value of the maximum diversity depend on q. However, it is a theorem that neither does. Every similarity matrix has an unambiguous maximum diversity, independent of q, and a distribution that maximizes the diversity of all orders q simultaneously. This is the subject of Section 6.3.

The maximum diversity of a matrix Z is closely related to another quantity, the magnitude of a matrix. The general concept of magnitude, expressed in the formalism of enriched categories, brings together a wide range of size-like invariants in mathematics, including cardinality, Euler characteristic, volume, surface area, dimension, and other geometric measures. Sections 6.4 and 6.5 are a broad-brush survey of magnitude, and demonstrate that maximum diversity – far from being tethered to ecology – has profound connections with fundamental invariants of geometry.

6.1 The Importance of Species Similarity

Here we introduce a family of measures of the diversity of an ecological community that take into account the varying similarities between species, following work of Leinster and Cobbold [220].

These diversity measures will be almost completely neutral as to what 'similarity' means or how it is quantified, just as the diversity measures discussed earlier were neutral as to the meaning of abundance (Example 2.1.1). The following examples illustrate some of the ways in which similarity can be quantified. In these examples, the similarity z between two species is measured on a scale of 0 to 1, with 0 representing complete dissimilarity and 1 representing identical species.

Examples 6.1.1 i. The similarity z between two species can be interpreted as percentage genetic similarity (in any of several senses; typically one would restrict to a particular part of the genome). With the rapid fall in the cost of DNA sequencing, this way of quantifying similarity is increasingly common. It can be used even when the taxonomic classification of the organisms concerned is unclear or incomplete, as is often the case for microbial communities (a problem discussed by Johnson [161] and Watve and Gangal [349], for instance).

ii. Functional similarity can also be quantified. For instance, suppose that we have a list of k functional traits satisfied by some species but not others. We can then define the similarity z between two species as j/k, where j is the number of traits possessed by either both species or neither. (For an overview of functional diversity, see Petchey and Gaston [280].)

iii. Similarity can also be measured phylogenetically, that is, in terms of an evolutionary tree. For instance, z can be defined as the proportion of evolutionary time before the two species diverged, relative to some fixed start time.

iv. In the absence of better data, we can measure similarity crudely using taxonomy. For instance, we could define the similarity z between two species by

$$z = \begin{cases} 1 & \text{if the species are the same,} \\ 0.8 & \text{if the species are different but of the same genus,} \\ 0.5 & \text{if the species are of different genera but the same family,} \\ 0 & \text{otherwise,} \end{cases}$$

or similarly for any other choice of constants and number of taxonomic levels.

v. More crudely still, we can define the similarity z between two species by

$$z = \begin{cases} 1 & \text{if the species are the same,} \\ 0 & \text{if the species are different.} \end{cases}$$

This definition embodies the assumption that different species never have anything in common. Unrealistic as this is, we will see that it is implicit in all of the measures of diversity defined in this book so far, and most of the diversity measures common in the ecological literature.

Now consider a list of species, numbered as $1, \ldots, n$, and suppose that we have fixed a way of quantifying the similarity between them. We obtain an $n \times n$ matrix

$$Z = (Z_{ij})_{1 \leq i,j \leq n},$$

where Z_{ij} is the similarity between species i and j.

Formally, a real square matrix Z is a **similarity matrix** if $Z_{ij} \geq 0$ for all i, j and $Z_{ii} > 0$ for all i. The examples above suggest additional hypotheses: that $Z_{ij} \leq 1$ for all i, j, that $Z_{ii} = 1$ for all i, and that Z is symmetric. (Indeed, in the paper [220] on which this section is based, the term 'similarity matrix' included the first two of these additional hypotheses.) But in most of what follows, we will not need these extra assumptions, so we do not make them.

Examples 6.1.2 i. The genetic, functional, phylogenetic and taxonomic similarity measures of Examples 6.1.1 give genetic, functional, phylogenetic and taxonomic similarity matrices Z, taking Z_{ij} to be any of the quantities z described there.

ii. The very crude similarities z of Example 6.1.1(v), where distinct species are taken to be completely dissimilar, give the identity similarity matrix $Z = I$. We will call this the **naive model** of a community.

Example 6.1.3 Given any finite metric space, with distance d and points labelled as $1, \ldots, n$, we obtain an $n \times n$ similarity matrix Z by setting

$$Z_{ij} = e^{-d(i,j)}.$$

Thus, large distances correspond to small similarities. In the extreme, the metric defined by $d(i, j) = \infty$ for all $i \neq j$ corresponds to the naive model. (We allow ∞ as a distance in our metric spaces.) Any taxonomic similarity matrix of the general type indicated in Example 6.1.1(iv) corresponds to an **ultrametric space**, that is, a metric space satisfying the stronger form

$$d(i,k) \leq \max\{d(i, j), d(j,k)\}$$

of the triangle inequality.

From a purely mathematical viewpoint, this matrix $Z = (e^{-d(i,j)})$ associated with a finite metric space is highly significant, as we will discover when we come to the theory of magnitude (Sections 6.4 and 6.5). From a biological viewpoint, we may find ourselves starting with a measure of inter-species difference δ on a scale of 0 to ∞ (as in Warwick and Clarke [348], for instance), in which case the transformation $z = e^{-\delta}$ converts it into a similarity z on a scale of 0 to 1. From both viewpoints, the choice of constant e is arbitrary, and one should consider replacing it by any other constant, or equivalently, scaling the distance by a linear factor. Again, this is a fundamental point in the theory of magnitude, as demonstrated by the theorems in Section 6.5.

Example 6.1.4 A symmetric similarity matrix whose entries are all 0 or 1 corresponds to a finite reflexive graph with no multiple edges. Here, **reflexive** means that there is an edge from each vertex to itself (a **loop**). The correspondence works as follows: labelling the vertices of the graph as $1, \ldots, n$, we put $Z_{ij} = 1$ whenever there is an edge between i and j, and $Z_{ij} = 0$ otherwise. One says that Z is the **adjacency matrix** of the graph. The reflexivity means that $Z_{ii} = 1$ for all i.

No ecological relevance is claimed for this family of examples, but mathematically it is a natural special case, and it sheds light on computational aspects of calculating maximum diversity (Remark 6.3.24).

Our earlier discussions of diversity modelled an ecological community crudely as a finite probability distribution $\mathbf{p} = (p_1, \ldots, p_n)$. Our new and less crude model of a community has two components: a relative abundance distribution $\mathbf{p} \in \Delta_n$ and an $n \times n$ similarity matrix Z. We now build up to the definition of the diversity of a community modelled in this way.

Treating $\mathbf{p}$ as a column vector, we can form the matrix product $Z\mathbf{p}$, which has entries

$$(Z\mathbf{p})_i = \sum_{j=1}^{n} Z_{ij} p_j \tag{6.1}$$

$(1 \leq i \leq n)$. The quantity (6.1) is the expected similarity between an individual of species i and an individual chosen at random. It can therefore be understood as the ordinariness of species i. If the diagonal entries of Z are all 1 (as in every example above) then

$$(\mathbf{Zp})_i = \sum_j Z_{ij}p_j \geq Z_{ii}p_i = p_i. \tag{6.2}$$

This inequality states that a species appears more ordinary when the similarities between species are recognized than when they are ignored.

By inequality (6.2), any species that is highly abundant is also highly ordinary: large p_i implies large $(\mathbf{Zp})_i$. But even if species i is rare, its ordinariness $(\mathbf{Zp})_i$ will be high if there is some common species very similar to it. The ordinariness of species i will even be high if it is similar to several species that are each individually rare, but whose total abundance is large. (For example, in Wallace's tropical forest, many tree species have much higher ordinariness $(\mathbf{Zp})_i$ than relative abundance p_i.) This makes intuitive sense: the more thorny bushes a region contains, the more ordinary any thorny bush will seem, even if its particular species is rare.

Judgements about what is 'ordinary' depend on one's perception of similarity. If one wishes to make a strong distinction between different species, one should use a similarity matrix Z whose off-diagonal entries are small, and this will have the effect of lowering the ordinariness of every species.

Since $(\mathbf{Zp})_i$ measures how ordinary the ith species is, $1/(\mathbf{Zp})_i$ measures how special it is. In the case $Z = I$ (the naive model of Example 6.1.2(ii)), this reduces to $1/p_i$, which in Sections 2.4 and 4.3 we called the specialness or rarity of species i. We have now extended that concept to our more refined model.

When we modelled a community as a simple probability distribution, we defined the diversity of a community to be the average specialness of an individual within it. We do the same again now in our new model.

Definition 6.1.5 Let $\mathbf{p} \in \Delta_n$, let Z be an $n \times n$ similarity matrix, and let $q \in [0, \infty]$. The **diversity of p of order** q, with respect to Z, is

$$D_q^Z(\mathbf{p}) = M_{1-q}(\mathbf{p}, 1/\mathbf{Zp}).$$

Here, the vector $1/\mathbf{Zp}$ is defined as

$$(1/(\mathbf{Zp})_1, \ldots, 1/(\mathbf{Zp})_n).$$

Although there may be some values of i for which $(\mathbf{Zp})_i = 0$, this can only occur when $p_i = 0$: for $Z_{ii} > 0$ by definition of similarity matrix, so if $p_i > 0$

then

$$(\mathbf{Zp})_i = \sum_j Z_{ij} p_j \geq Z_{ii} p_i > 0.$$

So by the convention in Remark 4.2.15, $M_{1-q}(\mathbf{p}, 1/\mathbf{Zp})$ is well defined. Explicitly,

$$D_q^Z(\mathbf{p}) = \left(\sum_{i \in \text{supp}(\mathbf{p})} p_i (\mathbf{Zp})_i^{q-1} \right)^{1/(1-q)}$$

for $q \neq 1, \infty$, and

$$D_1^Z(\mathbf{p}) = \prod_{i \in \text{supp}(\mathbf{p})} (\mathbf{Zp})_i^{-p_i} = \frac{1}{(\mathbf{Zp})_1^{p_1} \cdots (\mathbf{Zp})_n^{p_n}},$$

$$D_\infty^Z(\mathbf{p}) = \frac{1}{\max_{i \in \text{supp}(\mathbf{p})} (\mathbf{Zp})_i}.$$

We could extend Definition 6.1.5 to negative q, but it would be misleading to call $D_q^Z(\mathbf{p})$ 'diversity' when q is negative, for the reasons given in Remark 4.4.4(ii). We therefore restrict to $q \in [0, \infty]$.

Examples 6.1.6 Here we consider some special values of Z and q, and in doing so recover various earlier measures of diversity.

i. In the naive model $Z = I$, where distinct species are taken to be completely dissimilar, $\mathbf{Zp} = \mathbf{p}$ and so $D_q^Z(\mathbf{p})$ is just the Hill number $D_q(\mathbf{p})$. In this sense, the Hill numbers implicitly use the naive model of a community.
ii. For a general similarity matrix, the diversity of order 0 is

$$D_0^Z(\mathbf{p}) = \sum_{i \in \text{supp}(\mathbf{p})} \frac{p_i}{(\mathbf{Zp})_i}.$$

This is a sum of contributions from all species present. The contribution made by the ith species, $p_i/(\mathbf{Zp})_i$, is between 0 and 1, by inequality (6.2) (assuming that $Z_{ii} = 1$). It is large when, relative to the size of the ith species, there are not many individuals of other similar species – that is, when the ith species is unusual. We discuss the quantity $p_i/(\mathbf{Zp})_i$ in greater depth in Example 7.1.7.
iii. In the naive model, the diversity of order ∞ is the Berger–Parker index

$$D_\infty^I(\mathbf{p}) = D_\infty(\mathbf{p}) = 1\Big/\max_i p_i$$

(Example 4.3.5(iv)). It measures the dominance of the most common species, the idea being that in a diverse community, no species should be too dominant. For a general similarity matrix, the diversity

$$D_\infty^Z(\mathbf{p}) = 1\Big/\max_{i\in\operatorname{supp}(\mathbf{p})} (Z\mathbf{p})_i$$

of order ∞ can be interpreted in the same way, but now with sensitivity to species similarity: $D_\infty^Z(\mathbf{p})$ is low not only if there is a single highly abundant species, but also if there is some highly abundant *cluster* of species.

iv. The diversity of order 2 is

$$D_2^Z(\mathbf{p}) = 1\Big/\sum_{i,j=1}^{n} p_i Z_{ij} p_j = 1/\mathbf{p}^{\mathrm{T}} Z\mathbf{p}.$$

(We continue to regard $\mathbf{p}$ as a column vector, so that its transpose $\mathbf{p}^{\mathrm{T}}$ is a row vector.) The number $\mathbf{p}^{\mathrm{T}} Z\mathbf{p}$ is the expected similarity between a pair of individuals chosen at random. This is a measure of a community's *lack* of diversity, and its reciprocal $D_2^Z(\mathbf{p})$ therefore measures diversity itself.

For instance, take a probability distribution $\mathbf{p}$ on the vertices of a graph, and let Z be the adjacency matrix (as in Example 6.1.4). Then $D_2^Z(\mathbf{p})$ is the reciprocal of the probability that two vertices chosen at random are **adjacent** (joined by an edge). Equivalently, if pairs of vertices are repeatedly chosen at random, $D_2^Z(\mathbf{p})$ is the expected number of trials needed in order to find an adjacent pair.

Example 6.1.7 By Example 6.1.6(iv), one can estimate the diversity of order 2 of a community by sampling pairs of individuals at random, recording the similarity between them, calculating the mean of these similarities, then taking the reciprocal. More generally, for any integer $q \geq 2$, one can estimate $D_q^Z(\mathbf{p})$ as follows. Sample q individuals at random from the community (with replacement). Supposing that they are of species $i_1, \ldots, i_q$, let us temporarily refer to the product

$$Z_{i_1 i_2} Z_{i_1 i_3} \cdots Z_{i_1 i_q}$$

as their 'group similarity'. Let μ_q be the expected group similarity of q individuals from the community. Then

$$D_q^Z(\mathbf{p}) = \mu_q^{1/(1-q)}.$$

This was first proved as Proposition A3 of the appendix to [220], and the proof is also given here as Appendix A.5.

For instance, in the naive model, μ_q is the probability that q random individuals are all of the same species, which is $\sum_i p_i^q$. In this case, it is immediate that $D_q(\mathbf{p}) = \mu_q^{1/(1-q)}$.

This procedure for estimating the diversity of orders $2, 3, \ldots$ has the advantage that it does not require the organisms to be classified into species. All we require is a measure of similarity between any pair of individuals. This is potentially very useful in studies of microbial systems, where there is often no complete taxonomic classification; all we have is a way of measuring the similarity between two samples. We can estimate μ_q, hence $D_q^Z(\mathbf{p})$, by repeatedly drawing q samples from the community, recording their group similarity, and then taking the mean.

Both relative abundance and similarity can be quantified in whatever way is appropriate to the scientific problem at hand. This makes the diversity measures $D_q^Z(\mathbf{p})$ highly versatile. For example, if the similarity coefficients Z_{ij} are defined genetically then D_q^Z measures genetic diversity, and in the same way, a phylogenetic, functional or taxonomic similarity matrix will produce a measure of phylogenetic, functional or taxonomic diversity.

The different diversity measures arising from different choices of similarity matrix may produce opposing results. This is a feature, not a bug. For instance, if over a period of time, a community undergoes an increase in genetic diversity but a decrease in morphological diversity, the opposite trends are a point of scientific interest.

When selecting a similarity matrix, a useful observation is that if

$$Z = \begin{pmatrix} 1 & z \\ z & 1 \end{pmatrix}$$

then

$$D_q^Z\left(\tfrac{1}{2}, \tfrac{1}{2}\right) = \frac{2}{1+z},$$

or equivalently

$$z = \frac{2}{D_q^Z\left(\tfrac{1}{2}, \tfrac{1}{2}\right)} - 1,$$

for all $q \in [0, \infty]$. So, deciding on the similarity Z_{ij} between species i and j is equivalent to deciding on the diversity d of a community consisting of species i and j in equal proportions:

$$Z_{ij} = \frac{2}{d} - 1.$$

Taking $d = 1$ embodies the viewpoint that this two-species community consists of effectively only one species, giving a similarity coefficient $Z_{ij} = 1$: the species are deemed to be identical. At the opposite extreme, if one decides that

Species	Abundance in canopy	Abundance in understorey
Prepona laertes	15	0
Archaeoprepona demophon	14	37
Zaretis itys	25	11
Memphis arachne	89	23
Memphis offa	21	3
Memphis xenocles	32	8

Table 6.1 Counts of butterflies of species in the subfamily Charaxinae, in the canopy and understorey of an Ecuadorian rain forest site (Example 6.1.8; data from Table 5 of DeVries et al. [83]).

such a community should have diversity 2 ('effectively 2 species') for all i and j, this produces the naive matrix $Z = I$.

The flexibility afforded by the choice of similarity matrix may make it tempting to reject the measures D_q^Z in favour of the simpler Hill numbers D_q, where no such choice is necessary. However, it is a mathematical fact that doing so amounts to choosing the naive model $Z = I$ (Example 6.1.6(i)), which represents the extreme position that distinct species have nothing whatsoever in common. This always leads to an overestimate of diversity (Lemma 6.2.3). The framework of similarity matrices forces us to be transparent: using the naive similarity matrix I is a *choice*, embodying ecological assumptions, just as much as for any other similarity matrix.

The next example, adapted from [220] (Example 3), demonstrates how ecological judgements can be altered by taking species similarity into account. Extending the terminology of Chapter 4, we refer to the graph of $D_q^Z(\mathbf{p})$ against q as a **diversity profile**.

Example 6.1.8 DeVries et al. ([83], Table 5) counted butterflies in the canopy and understorey at a certain site in the Ecuadorian rain forest. In the subfamily Charaxinae, the abundances were as shown in Table 6.1. We will compare the diversity profiles of the canopy and the understorey in two ways, once using the naive similarity matrix and once using a non-naive matrix.

With the naive similarity matrix I, the diversity profiles are as shown in Figure 6.1(a). The profile of the canopy lies above that of the understorey until about $q = 5$, after which the two profiles are near-identical. So, whatever emphasis we may place on rare or common species, the canopy is at least as diverse as the understorey.

Now let us compare the communities using a taxonomic similarity matrix.

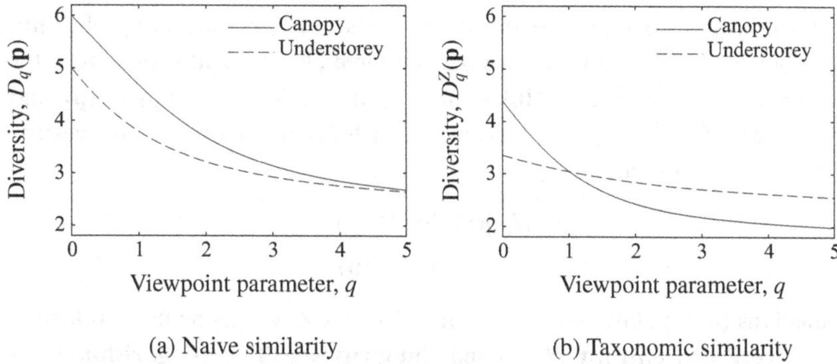

Figure 6.1 Diversity profiles of butterflies in the canopy and understorey of a rain forest site, using (a) the naive similarity matrix I; (b) a taxonomic similarity matrix. Graphs adapted from Figure 3 of Leinster and Cobbold [220].

Put

$$Z_{ij} = \begin{cases} 1 & \text{if } i = j, \\ 0.5 & \text{if species } i \text{ and } j \text{ are different but of the same genus,} \\ 0 & \text{otherwise.} \end{cases}$$

The resulting diversity profiles, shown in Figure 6.1(b), tell a different story. For most values of q, it is the understorey that is more diverse. This can be explained as follows. Most of the canopy population belongs to the three species in the *Memphis* genus, so when we build into the model the principle that species of the same genus tend to be somewhat similar, the canopy looks less diverse than it did before. On the other hand, the understorey population does not contain large numbers of individuals of different species but the same genus, so factoring in taxonomic similarity does not cause its diversity to decrease so much.

The measures D_q^Z, as well as unifying into one family many older diversity measures, have also found application in a variety of ecological systems at many scales, from microbes (Bakker et al. [26]), fungi (Veresoglou et al. [344]) and crustacean zooplankton (Jeziorski et al. [160]) to alpine plants (Chalmandrier et al. [66]) and large arctic predators (Bromaghin et al. [50]). As one would expect, incorporating similarity has been found to improve inferences about the diversity of natural systems [344]. The measures have also been applied in non biological contexts such as computer network security (Wang et al. [347]).

We now turn from diversity to entropy. In the simpler context of probability distributions $\mathbf{p}$ on a finite set, we defined three closely related quantities for each parameter value q: the Hill number $D_q(\mathbf{p})$, the Rényi entropy $H_q(\mathbf{p})$, and the q-logarithmic entropy $S_q(\mathbf{p})$. They are related to one another by increasing, invertible transformations:

$$H_q(\mathbf{p}) = \log D_q(\mathbf{p}),$$
$$S_q(\mathbf{p}) = \ln_q D_q(\mathbf{p})$$

(equations (4.20)). Given also a similarity matrix Z, we define the **similarity-sensitive Rényi entropy** $H_q^Z(\mathbf{p})$ and **similarity-sensitive q-logarithmic entropy** $S_q^Z(\mathbf{p})$ by the same transformations:

$$H_q^Z(\mathbf{p}) = \log D_q^Z(\mathbf{p}), \tag{6.3}$$
$$S_q^Z(\mathbf{p}) = \ln_q D_q^Z(\mathbf{p}). \tag{6.4}$$

In the first definition, $q \in [0, \infty]$, and in the second, $q \in [0, \infty)$.

Let us be explicit. For $q \neq 1, \infty$, the similarity-sensitive Rényi entropy is

$$H_q^Z(\mathbf{p}) = \frac{1}{1-q} \log \sum_{i \in \mathrm{supp}(\mathbf{p})} p_i (Z\mathbf{p})_i^{q-1},$$

and in the exceptional cases,

$$H_1^Z(\mathbf{p}) = - \sum_{i \in \mathrm{supp}(\mathbf{p})} p_i \log(Z\mathbf{p})_i$$

(generalizing the Shannon entropy) and

$$H_\infty^Z(\mathbf{p}) = -\log \max_{i \in \mathrm{supp}(\mathbf{p})} (Z\mathbf{p})_i.$$

We now derive an explicit expression for $S_q^Z(\mathbf{p})$. By Lemma 4.2.29,

$$S_q^Z(\mathbf{p}) = \ln_q M_{1-q}(\mathbf{p}, 1/Z\mathbf{p}) = \sum_{i \in \mathrm{supp}(\mathbf{p})} p_i \ln_q \frac{1}{(Z\mathbf{p})_i}.$$

Then applying the definition of $\ln_q$ gives

$$S_q^Z(\mathbf{p}) = \frac{1}{1-q} \left(\sum_{i \in \mathrm{supp}(\mathbf{p})} p_i (Z\mathbf{p})_i^{q-1} - 1 \right)$$

when $q \neq 1$, and

$$S_1^Z(\mathbf{p}) = H_1^Z(\mathbf{p}) = - \sum_{i \in \mathrm{supp}(\mathbf{p})} p_i \log(Z\mathbf{p})_i.$$

Figure 4.1, which schematically depicts the two families of deformed entropies in the special case $Z = I$, applies equally to an arbitrary similarity matrix Z.

Example 6.1.9 The definitions above specialize to definitions of Rényi and q-logarithmic entropies for a probability distribution on a finite metric space. Indeed, let $A = \{1, \ldots, n\}$ be a finite metric space and write $Z = (e^{-d(i,j)})$, as in Example 6.1.3. For any probability distribution $\mathbf{p}$ on A, and for any parameter value q, we have an associated Rényi entropy $H_q^Z(\mathbf{p})$ and q-logarithmic entropy $S_q^Z(\mathbf{p})$. Naturally, these quantities depend on the metric. In the extreme case where $d(i, j) = \infty$ for all $i \neq j$, we recover the standard definitions of the Rényi and q-logarithmic entropies of a probability distribution on a finite set.

(One can speculate about extending the results of classical information theory to the metric context. As usual, the elements of the set $A = \{1, \ldots, n\}$ represent the source symbols and the distribution $\mathbf{p}$ specifies their frequencies, but now we also have a metric d on the source symbols. It could be defined in such a way that $d(i, j)$ is small when the ith and jth symbols are easily mistaken for one another, or alternatively when one is an acceptable substitute for the other, for applications such as the encoding of colour images.)

Example 6.1.10 For any similarity matrix Z, we can define a **dissimilarity matrix** Δ by $\Delta_{ij} = 1 - Z_{ij}$. (Let us assume here that $Z_{ij} \leq 1$ for all i and j.) In these terms, the 2-logarithmic entropy is

$$S_2^Z(\mathbf{p}) = 1 - \sum_{i,j} p_i Z_{ij} p_j = \sum_{i,j} p_i \Delta_{ij} p_j = \mathbf{p}^{\mathrm{T}} \Delta \mathbf{p}.$$

Thus, $S_2^Z(\mathbf{p})$ is the dissimilarity between a pair of individuals chosen at random. This quantity, studied by the statistician C. R. Rao [286, 287], is known as **Rao's quadratic entropy**.

Of course, anything that can be expressed in terms of Z can also be expressed in terms of Δ, and vice versa. An important early step towards similarity-sensitive diversity measures was taken by Ricotta and Szeidl [296], who gave a version of the entropy $S_q^Z(\mathbf{p})$ expressed in terms of a dissimilarity matrix Δ.

Example 6.1.11 Let Z be the adjacency matrix of a finite reflexive graph G with vertex-set $\{1, \ldots, n\}$, as in Example 6.1.4. Write $i \sim j$ to mean that vertices i and j are joined by an edge, and $i \nsim j$ otherwise. Then the dissimilarity matrix Δ of the last example has entries 1 for non-adjacent pairs and 0 for adjacent pairs. Hence the 2-logarithmic entropy of a probability distribution $\mathbf{p}$ on G is given by

$$S_2^Z(\mathbf{p}) = \sum_{i,j:\, i \nsim j} p_i p_j.$$

This is the probability that two vertices chosen at random according to $\mathbf{p}$ are *not* joined by an edge. Thus, entropy is high when vertices of high probability tend not to be adjacent. We will make more precise statements of this type in Section 6.3, where we will solve the problem of maximizing entropy on a set with similarities – and, in particular, maximizing entropy on a graph.

6.2 Properties of the Similarity-Sensitive Diversity Measures

Here we establish the algebraic and analytic properties of the similarity-sensitive diversity measures $D_q^Z(\mathbf{p})$, extending the results already proved in Section 4.4 for the Hill numbers (the case $Z = I$). Mathematically speaking, most of the properties of the diversity measures are easy consequences of the properties of means. However, they are given new significance by the ecological interpretation.

Each of the listed properties of $D_q^Z(\mathbf{p})$ is a piece of evidence that these measures behave logically, in the way that should be required of any diversity measure. Contrast, for instance, the behaviour of Shannon entropy in the oil company argument of Example 2.4.11. Nearly all of the properties were first established in the 2012 paper of Leinster and Cobbold [220].

In the naive model, diversity profiles are either strictly decreasing or constant (Proposition 4.4.1), and we will show that this is also true in the general case. But whereas in the naive model, the condition for the profile to be constant is that all species present have equal *abundance* p_i, in the general case, the condition is that all species present have equal *ordinariness* $(Z\mathbf{p})_i$.

Proposition 6.2.1 *Let Z be an $n \times n$ similarity matrix and let $\mathbf{p} \in \Delta_n$. Then $D_q^Z(\mathbf{p})$ is a decreasing function of $q \in [0, \infty]$. It is constant if $(Z\mathbf{p})_i = (Z\mathbf{p})_j$ for all $i, j \in \mathrm{supp}(\mathbf{p})$, and strictly decreasing otherwise.*

Proof Since $D_q^Z(\mathbf{p}) = M_{1-q}(\mathbf{p}, 1/Z\mathbf{p})$, this follows from Theorem 4.2.8. □

The more similar the species in a population are perceived to be, the less the perceived diversity. Our diversity measures conform to this intuition:

Lemma 6.2.2 *Let Z' and Z be $n \times n$ similarity matrices with $Z'_{ij} \leq Z_{ij}$ for all i, j. Then $D_q^{Z'}(\mathbf{p}) \geq D_q^Z(\mathbf{p})$ for all $\mathbf{p} \in \Delta_n$ and $q \in [0, \infty]$.*

Proof Since $D_q^Z(\mathbf{p}) = M_{1-q}(\mathbf{p}, 1/Z\mathbf{p})$, this follows from the fact that the power means are increasing (Lemma 4.2.19). □

All of our examples of similarity matrices (Examples 6.1.2–6.1.4) have the additional properties that all similarities are at most 1 and the similarity of each species to itself is 1. Assuming those properties, we can bound the range of possible diversities.

Lemma 6.2.3 (Range) *Let Z be an $n \times n$ similarity matrix such that $Z_{ij} \leq 1$ for all i, j and $Z_{ii} = 1$ for all i. Then*

$$1 \leq D_q^Z(\mathbf{p}) \leq D_q(\mathbf{p}) \leq n$$

for all $\mathbf{p} \in \Delta_n$ and $q \in [0, \infty]$.

Proof Taking $Z' = I$ in Lemma 6.2.2 and using the hypotheses on Z gives

$$D_q^Z(\mathbf{p}) \leq D_q^I(\mathbf{p}) = D_q(\mathbf{p}),$$

and we already showed in Lemma 4.4.3(ii) that $D_q(\mathbf{p}) \leq n$. It remains to prove that $D_q^Z(\mathbf{p}) \geq 1$. For each $i \in \{1, \ldots, n\}$, we have

$$(Z\mathbf{p})_i = \sum_{j=1}^{n} Z_{ij} p_j.$$

This is a mean, weighted by $\mathbf{p}$, of numbers $Z_{ij} \in [0, 1]$; hence $(Z\mathbf{p})_i \in [0, 1]$ and so $1/(Z\mathbf{p})_i \geq 1$. It follows that

$$D_q^Z(\mathbf{p}) = M_{1-q}(\mathbf{p}, 1/Z\mathbf{p}) \geq 1.$$

□

Fix a matrix Z satisfying the hypotheses of Lemma 6.2.3. The minimum diversity $D_q^Z(\mathbf{p}) = 1$ is attained by any distribution in which only one species is present:

$$\mathbf{p} = (0, \ldots, 0, 1, 0, \ldots, 0).$$

Much more difficult is to *maximize* $D_q^Z(\mathbf{p})$ for fixed Z and variable $\mathbf{p}$. We do this in Section 6.3.

Since all of our examples of similarity matrices satisfy the hypotheses of Lemma 6.2.3, the corresponding diversities always lie in the range $[1, n]$. The maximum value of n is attained just when $Z = I$ and $\mathbf{p} = \mathbf{u}_n$, by Lemmas 4.4.3 and 6.2.3.

In the case $Z = I$, we interpreted $D_q(\mathbf{p})$ as the effective number of species in the community (Section 4.3). The bounds in the previous paragraph encourage us to interpret $D_q^Z(\mathbf{p})$ as the effective number of species for a general matrix Z (at least if it satisfies the hypotheses of Lemma 6.2.3). More precisely, $D_q^Z(\mathbf{p})$ is the effective number of *completely dissimilar* species, since a community of n equally abundant, completely dissimilar species has diversity n.

It is nearly true that the diversity $D_q^Z(\mathbf{p})$ is continuous in each of q, Z and $\mathbf{p}$. The precise statement is as follows.

Lemma 6.2.4 *i. Let Z be an $n \times n$ similarity matrix and $\mathbf{p} \in \Delta_n$. Then $D_q^Z(\mathbf{p})$ is continuous in $q \in [0, \infty]$.*

ii. Let $q \in [0, \infty]$ and $\mathbf{p} \in \Delta_n$. Then $D_q^Z(\mathbf{p})$ is continuous in $n \times n$ similarity matrices Z.

iii. Let $q \in (0, \infty)$ and let Z be an $n \times n$ similarity matrix. Then $D_q^Z(\mathbf{p})$ is continuous in $\mathbf{p} \in \Delta_n$.

Part (i) states that diversity profiles are continuous, just as in the naive model.

The first two parts follow immediately from results on power means, but the third does not. The subtlety is that the sum

$$D_q^Z(\mathbf{p}) = \left(\sum_{i \in \text{supp}(\mathbf{p})} p_i (Z\mathbf{p})_i^{q-1} \right)^{1/(1-q)} \tag{6.5}$$

in the definition of diversity is taken only over $\text{supp}(\mathbf{p})$. Thus, if $p_i = 0$ then the contribution of the ith species to the sum is 0. However, if p_i is nonzero but small then $(Z\mathbf{p})_i$ may be small, which if $q < 1$ means that $(Z\mathbf{p})_i^{q-1}$ is large; we need to show that, nevertheless, $p_i(Z\mathbf{p})_i^{q-1}$ is close to 0.

Proof Part (i) follows from Lemma 4.2.7, and part (ii) from Lemma 4.2.5.

For part (iii), we split into three cases: $q \in (1, \infty)$, $q \in (0, 1)$, and $q = 1$.

If $q \in (1, \infty)$ then the sum in equation (6.5) can equivalently be taken over $i \in \{1, \ldots, n\}$ (and the summands are still well defined), so the result is clear.

Now let $q \in (0, 1)$. Define functions $\phi_1, \ldots, \phi_n \colon \Delta_n \to \mathbb{R}$ by

$$\phi_i(\mathbf{p}) = \begin{cases} p_i (Z\mathbf{p})_i^{q-1} & \text{if } p_i > 0, \\ 0 & \text{otherwise.} \end{cases}$$

Then $D_q^Z(\mathbf{p}) = \left(\sum_{i=1}^n \phi_i(\mathbf{p})\right)^{1/(1-q)}$, so it suffices to show that each ϕ_i is continuous.

Fix $i \in \{1, \ldots, n\}$. Write

$$\Delta_n^{(i)} = \{\mathbf{p} \in \Delta_n : p_i > 0\}.$$

Then ϕ_i is continuous on $\Delta_n^{(i)}$ and zero on its complement, so all we have to prove is that if $\mathbf{p} \in \Delta_n$ with $p_i = 0$ then $\phi_i(\mathbf{r}) \to 0$ as $\mathbf{r} \to \mathbf{p}$, and we may as well constrain $\mathbf{r}$ to lie in $\Delta_n^{(i)}$. We have $(Z\mathbf{r})_i \geq Z_{ii} r_i$, so

$$0 \leq \phi_i(\mathbf{r}) \leq r_i (Z_{ii} r_i)^{q-1} = Z_{ii}^{q-1} r_i^q \tag{6.6}$$

(since $q < 1$). Note that $Z_{ii} > 0$ by definition of similarity matrix, so Z_{ii}^{q-1} is finite. As $\mathbf{r} \to \mathbf{p}$, we have $r_i^q \to p_i^q = 0$ (since $q > 0$). Hence the bounds (6.6) give $\phi_i(\mathbf{r}) \to 0$, as required.

Finally, consider $q = 1$. Define functions $\psi_1, \ldots, \psi_n \colon \Delta_n \to \mathbb{R}$ by

$$\psi_i(\mathbf{p}) = \begin{cases} (Z\mathbf{p})_i^{-p_i} & \text{if } p_i > 0, \\ 1 & \text{otherwise.} \end{cases}$$

Then $D_q^Z(\mathbf{p}) = \prod_{i=1}^n \psi_i(\mathbf{p})$, so it suffices to show that each ψ_i is continuous.

Fix $i \in \{1, \ldots, n\}$. As in the previous case, it suffices to show that if $\mathbf{p} \in \Delta_n$ with $p_i = 0$, then $\psi_i(\mathbf{r}) \to 1$ as $\mathbf{r} \to \mathbf{p}$ with $\mathbf{r} \in \Delta_n^{(i)}$. Writing $K = \max_j Z_{ij}$, we have

$$Z_{ii} r_i \leq (Z\mathbf{r})_i = \sum_{j=1}^n Z_{ij} r_j \leq \sum_{j=1}^n K r_j = K,$$

so

$$K^{-r_i} \leq \psi_i(\mathbf{r}) \leq Z_{ii}^{-r_i} r_i^{-r_i}. \tag{6.7}$$

Now $K \geq Z_{ii} > 0$, so $K^{-r_i} \to 1$ and $Z_{ii}^{-r_i} \to 1$ as $\mathbf{r} \to \mathbf{p}$. Also, $\lim_{x \to 0+} x^x = 1$, so $r_i^{-r_i} \to 1$ as $\mathbf{r} \to \mathbf{p}$. Hence the bounds (6.7) give $\psi_i(\mathbf{r}) \to 1$ as $\mathbf{r} \to \mathbf{p}$. □

Remark 6.2.5 The cases $q = 0$ and $q = \infty$ were excluded from the statement of Lemma 6.2.4(iii) because $D_q^Z(\mathbf{p})$ is *not* continuous in $\mathbf{p}$ when q is 0 or ∞. We have already seen that D_0^Z is discontinuous even in the naive case $Z = I$, where $D_0^Z(\mathbf{p}) = D_0(\mathbf{p})$ is the species richness $|\mathrm{supp}(\mathbf{p})|$. The diversity

$$D_\infty^Z(\mathbf{p}) = 1 \Big/ \max_{i \in \mathrm{supp}(\mathbf{p})} (Z\mathbf{p})_i$$

of order ∞ is continuous when $Z = I$, but not in general. For example, let

$$Z = \begin{pmatrix} 1 & 1 & 0 \\ 1 & 1 & 1 \\ 0 & 1 & 1 \end{pmatrix}$$

(a similarity matrix that we will meet again in Example 6.3.20). For $0 \leq t < 1/2$, put

$$\mathbf{p} = \begin{pmatrix} \frac{1}{2} - t \\ 2t \\ \frac{1}{2} - t \end{pmatrix}.$$

Then

$$Z\mathbf{p} = \begin{pmatrix} \frac{1}{2} + t \\ 1 \\ \frac{1}{2} + t \end{pmatrix},$$

so

$$D_\infty^Z(\mathbf{p}) = \begin{cases} 1 & \text{if } t > 0, \\ 2 & \text{if } t = 0. \end{cases}$$

Hence D_∞^Z is discontinuous.

The idea behind this counterexample is that the second species is so closely related to the other two that it appears more ordinary than them ($(Z\mathbf{p})_2 = \max_i(Z\mathbf{p})_i$) even if it is very rare itself (t is small). However, if the second species disappears entirely ($t = 0$) then its ordinariness $(Z\mathbf{p})_2$ is excluded from the maximum that defines $D_\infty^Z(\mathbf{p})$, causing the discontinuity.

Next we establish three properties of the measures that are logically fundamental. We will deduce all of them from a naturality property (in the categorical sense of natural transformations), following a strategy similar to the one we used for power means (Section 4.2). Let

$$\theta\colon \{1, \ldots, m\} \to \{1, \ldots, n\} \tag{6.8}$$

be a map of sets ($m, n \geq 1$), let $\mathbf{p} \in \Delta_m$, and let Z be an $n \times n$ similarity matrix. Then we obtain a pushforward distribution $\theta\mathbf{p} \in \Delta_m$ (Definition 2.1.10) and an $m \times m$ similarity matrix $Z\theta$ defined by

$$(Z\theta)_{ii'} = Z_{\theta(i),\theta(i')}$$

($i, i' \in \{1, \ldots, m\}$).

Lemma 6.2.6 (Naturality) *With θ, $\mathbf{p}$ and Z as above,*

$$D_q^{Z\theta}(\mathbf{p}) = D_q^Z(\theta\mathbf{p})$$

for all $q \in [0, \infty]$.

Proof We use the naturality property of the power means (Lemma 4.2.14), which implies that

$$M_{1-q}(\theta\mathbf{p}, \mathbf{x}) = M_{1-q}(\mathbf{p}, \mathbf{x}\theta) \tag{6.9}$$

for all $\mathbf{x} \in [0, \infty)^n$. Let us adopt the convention that unless indicated otherwise, the indices i and i' range over $\{1, \ldots, m\}$ and the indices j and j' range over $\{1, \ldots, n\}$. Then

$$((Z\theta)\mathbf{p})_i = \sum_{i'} (Z\theta)_{ii'} p_{i'} = \sum_{i'} Z_{\theta(i),\theta(i')} p_{i'},$$

$$(Z(\theta\mathbf{p}))_j = \sum_{j'} Z_{jj'} (\theta\mathbf{p})_{j'} = \sum_{j'} \sum_{i' \in \theta^{-1}(j')} Z_{jj'} p_{i'} = \sum_{i'} Z_{j,\theta(i')} p_{i'}.$$

Hence

$$((Z\theta)\mathbf{p})_i = Z(\theta\mathbf{p})_{\theta(i)}$$

for all i, or equivalently,

$$(Z\theta)\mathbf{p} = (Z(\theta\mathbf{p}))\theta. \tag{6.10}$$

Now

$$\begin{aligned} D_q^Z(\theta\mathbf{p}) &= M_{1-q}\left(\theta\mathbf{p}, \frac{1}{Z(\theta\mathbf{p})}\right) \\ &= M_{1-q}\left(\mathbf{p}, \frac{1}{Z(\theta\mathbf{p})}\theta\right) \\ &= M_{1-q}\left(\mathbf{p}, \frac{1}{(Z(\theta\mathbf{p}))\theta}\right), \end{aligned}$$

where the second equality follows from equation (6.9) and the others are immediate. Equation (6.10) now gives

$$D_q^Z(\theta\mathbf{p}) = M_{1-q}\left(\mathbf{p}, \frac{1}{(Z\theta)\mathbf{p}}\right) = D_q^{Z\theta}(\mathbf{p}),$$

as required. □

From naturality, we deduce three elementary properties of the diversity measures. (In the special case of the Hill numbers, $Z = I$, the first two already appeared as Lemma 4.4.8.) First, diversity is independent of the order in which the species are listed.

Lemma 6.2.7 (Symmetry) *Let Z be an $n \times n$ similarity matrix, let $\mathbf{p} \in \Delta_n$, and let σ be a permutation of $\{1, \ldots, n\}$. Define Z' and $\mathbf{p}'$ by $Z'_{ij} = Z_{\sigma(i),\sigma(j)}$ and $p'_i = p_{\sigma(i)}$. Then $D_q^{Z'}(\mathbf{p}') = D_q^Z(\mathbf{p})$ for all $q \in [0, \infty]$.*

Proof By definition, $Z' = Z\sigma$ and $\mathbf{p} = \sigma\mathbf{p}'$, so the result follows from Lemma 6.2.6. □

Diversity is also unchanged by ignoring any species with abundance 0.

Lemma 6.2.8 (Absence-invariance) *Let Z be an $n \times n$ similarity matrix, and let $\mathbf{p} \in \Delta_n$ with $p_n = 0$. Write Z' for the restriction of Z to the first $n-1$ species, and write $\mathbf{p}' = (p_1, \ldots, p_{n-1}) \in \Delta_{n-1}$. Then $D_q^{Z'}(\mathbf{p}') = D_q^Z(\mathbf{p})$ for all $q \in [0, \infty]$.*

Proof Let θ be the inclusion $\{1, \ldots, n-1\} \hookrightarrow \{1, \ldots, n\}$. Then $Z' = Z\theta$ and $\mathbf{p} = \theta\mathbf{p}'$, so the result follows from Lemma 6.2.6. □

Third and finally, if two species are identical, then merging them into one leaves the diversity unchanged.

Lemma 6.2.9 (Identical species) *Let Z be an* $n \times n$ *similarity matrix such that*

$$Z_{in} = Z_{i,n-1}, \qquad Z_{ni} = Z_{n-1,i}$$

for all $i \in \{1, \ldots, n\}$. *Let* $\mathbf{p} \in \Delta_n$. *Write* Z' *for the restriction of Z to the first* $n-1$ *species, and define* $\mathbf{p}' \in \Delta_{n-1}$ *by*

$$p'_j = \begin{cases} p_j & \text{if } j < n-1, \\ p_{n-1} + p_n & \text{if } j = n-1. \end{cases}$$

Then $D_q^{Z'}(\mathbf{p}') = D_q^Z(\mathbf{p})$ *for all* $q \in [0, \infty]$.

Proof Define a function $\theta\colon \{1, \ldots, n\} \to \{1, \ldots, n-1\}$ by

$$\theta(i) = \begin{cases} i & \text{if } i < n, \\ n-1 & \text{if } i = n. \end{cases}$$

Then $Z = Z'\theta$ and $\mathbf{p}' = \theta\mathbf{p}$, so the result follows from Lemma 6.2.6. □

The identical species property means that 'a community of 100 species that are identical in every way is no different from a community of only one species' (Ives [151], p. 102).

The boundaries between species can be changeable and somewhat arbitrary, not only for microscopic life but even for well-studied large mammals. (For example, the classification of the lemurs of Madagascar has changed frequently; see Mittermeier et al. [254].) The challenge this poses for the quantification of diversity has long been recognized. Good wrote in 1982 of the need to measure diversity in a way that resolves 'the difficult "species problem" ' and avoids 'the Platonic all-or-none approach to the definition of species' ([123], p. 562).

Incorporating species similarity into diversity measurement, as we have done, allows these challenges to be met. In particular, our measures behave reasonably when species are reclassified, as the following example shows.

Example 6.2.10 This hypothetical example is from [220]. Consider a system of three totally dissimilar species with relative abundances $\mathbf{p} = (0.1, 0.3, 0.6)$. Suppose that on the basis of new genetic evidence, the last species is reclassified into two separate species of equal abundance, so that the relative abundances become $(0.1, 0.3, 0.3, 0.3)$.

If the two new species are assumed to be totally dissimilar to one another, the diversity profile changes dramatically (Figure 6.2). For example, the diversity of order ∞ jumps by 100%, from $1.66\ldots$ to $3.33\ldots$ Of course, it is wholly unrealistic to assume that the new species are totally dissimilar, given that until recently they were thought to be identical. But if, more realistically, the two new species are assigned a high similarity, the diversity profile changes only

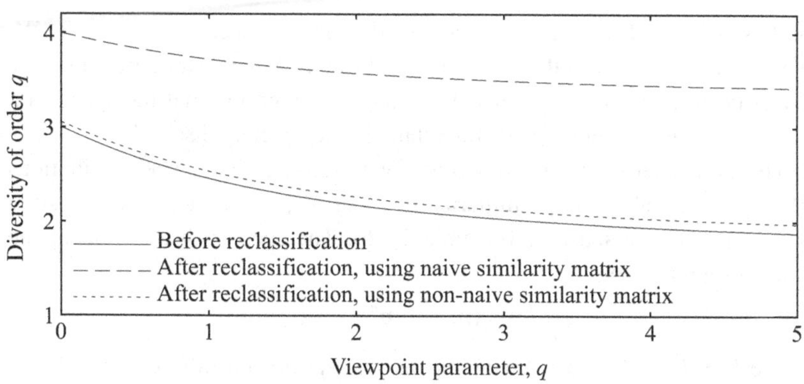

Figure 6.2 Diversity profiles of a hypothetical community before and after a species is reclassified (Example 6.2.10). (Figure adapted from [220], Figure 1.)

slightly. Figure 6.2 shows the profile based on similarities $Z_{34} = Z_{43} = 0.9$ between the two new species (and $Z_{ij} = 0$ for $i \neq j$ otherwise).

This sensible behaviour is guaranteed by two features of the diversity measures: the identical species property and continuity in Z. Indeed, if the two new species were deemed to be identical then the profile would be unchanged. So by continuity, if the new species are deemed to be nearly identical then the profile is nearly unchanged.

For similar reasons, the diversity measures D_q^Z behave reasonably under changes of the level of resolution in the data. For example, suppose that an initial, crude, survey of a community gathers population abundance data at the genus level, a second survey records abundance at the species level, and a third records abundance at the subspecies level. Provided that similarity is measured coherently, the resulting three diversities will be comparable, in the sense of being measured on the same scale. The more fine-grained the data is, the more variation becomes visible, so the diversity will be greater for the later surveys. But for the same reasons as in Example 6.2.10, it will not jump disproportionately from one survey to the next. There will only be a large difference between the diversities calculated from the first and second surveys if there is a large amount of variation within genera. Similarly, the difference between the diversities obtained from the second and third surveys faithfully reflects the amount of intraspecific variation.

In Propositions 4.4.10 and 4.4.12, we proved two forms of the chain rule for the Hill numbers, interpreting them as formulas for the diversity of a community spread across several islands in terms of the diversities and relative sizes of those islands. The islands were assumed to have no species in common. We now derive two forms of the chain rule for the more general similarity-sensitive diversity measures $D_q^Z(\mathbf{p})$, under the stronger assumption that the species on different islands are not only distinct, but also completely dissimilar.

Thus, consider n island communities with relative abundance distributions $\mathbf{p}^1 \in \Delta_{k_1}, \ldots, \mathbf{p}^n \in \Delta_{k_n}$, similarity matrices $Z^1, \ldots, Z^n$, and relative sizes $w_1, \ldots, w_n$ (in the sense of Example 2.1.6). The species distribution of the whole group is, then,

$$\mathbf{w} \circ (\mathbf{p}^1, \ldots, \mathbf{p}^n) \in \Delta_k,$$

where $k = k_1 + \cdots + k_n$. Assuming that the species on different islands are completely dissimilar, the $k \times k$ similarity matrix Z for the whole group is the block sum

$$Z = Z^1 \oplus \cdots \oplus Z^n = \begin{pmatrix} Z^1 & 0 & \cdots & 0 \\ 0 & Z^2 & \ddots & \vdots \\ \vdots & \ddots & \ddots & 0 \\ 0 & \cdots & 0 & Z^n \end{pmatrix}.$$

So, the diversity of the whole is

$$D_q^Z(\mathbf{w} \circ (\mathbf{p}^1, \ldots, \mathbf{p}^n)),$$

and our task is to express this in terms of the islands' diversities $D_q^{Z^i}(\mathbf{p}^i)$ and their relative sizes w_i.

Proposition 6.2.11 (Chain rule) *Let $q \in [0, \infty]$ and $n, k_1, \ldots, k_n \geq 1$. For each $i \in \{1, \ldots, n\}$, let Z^i be a $k_i \times k_i$ similarity matrix and let $\mathbf{p}^i \in \Delta_{k_i}$; also, let $\mathbf{w} \in \Delta_n$. Write $Z = Z^1 \oplus \cdots \oplus Z^n$ and $d_i = D_q^{Z^i}(\mathbf{p}^i)$.*

i. We have

$$D_q^Z(\mathbf{w} \circ (\mathbf{p}^1, \ldots, \mathbf{p}^n)) = M_{1-q}(\mathbf{w}, \mathbf{d}/\mathbf{w}) = \begin{cases} \left(\sum w_i^q d_i^{1-q}\right)^{1/(1-q)} & \text{if } q \neq 1, \infty, \\ \prod (d_i/w_i)^{w_i} & \text{if } q = 1, \\ \min d_i/w_i & \text{if } q = \infty, \end{cases}$$

where $\mathbf{d}/\mathbf{w} = (d_1/w_1, \ldots, d_n/w_n)$ and the sum, product, and minimum are over all $i \in \operatorname{supp}(\mathbf{w})$.

ii. For $q < \infty$,

$$D_q^Z(\mathbf{w} \circ (\mathbf{p}^1, \ldots, \mathbf{p}^n)) = D_q(\mathbf{w}) \cdot M_{1-q}(\mathbf{w}^{(q)}, \mathbf{d}),$$

where $\mathbf{w}^{(q)}$ is the escort distribution defined after Proposition 4.4.10.

Proof For (i), an elementary calculation shows that

$$Z(\mathbf{w} \circ (\mathbf{p}^1, \ldots, \mathbf{p}^n)) = w_1(Z^1\mathbf{p}^1) \oplus \cdots \oplus w_n(Z^n\mathbf{p}^n).$$

Hence, using the chain rule for the power means and then the homogeneity of the power means,

$$\begin{aligned}
&D_q^Z(\mathbf{w} \circ (\mathbf{p}^1, \ldots, \mathbf{p}^n)) \\
&\quad = M_{1-q}\left(\mathbf{w} \circ (\mathbf{p}^1, \ldots, \mathbf{p}^n), \frac{1}{w_1(Z^1\mathbf{p}^1)} \oplus \cdots \oplus \frac{1}{w_n(Z^n\mathbf{p}^n)}\right) \\
&\quad = M_{1-q}\left(\mathbf{w}, \left(M_{1-q}\left(\mathbf{p}^1, \frac{1}{w_1(Z^1\mathbf{p}^1)}\right), \ldots, M_{1-q}\left(\mathbf{p}^n, \frac{1}{w_n(Z^n\mathbf{p}^n)}\right)\right)\right) \\
&\quad = M_{1-q}\left(\mathbf{w}, \left(\frac{d_1}{w_1}, \ldots, \frac{d_n}{w_n}\right)\right).
\end{aligned}$$

This proves the first equality in (i), and the second follows from the explicit formulas for the power means. Lemma 4.4.11 then gives (ii). □

In particular, the diversity of the overall community depends only on the sizes and diversities of the islands:

Corollary 6.2.12 (Modularity) *In the situation of Proposition 6.2.11, the total diversity $D_q^Z(\mathbf{w} \circ (\mathbf{p}^1, \ldots, \mathbf{p}^n))$ depends only on q, $\mathbf{w}$, and $D_q^Z(\mathbf{p}^1)$, …, $D_q^Z(\mathbf{p}^n)$.* □

A further consequence of the chain rule is also important. Suppose that the islands all have the same size and the same diversity, d. (For example, the islands will have the same diversity if they all have the same species distributions, but on disjoint sets of species; or formally, if $k_1 = \cdots = k_n$ and $\mathbf{p}^1 = \cdots = \mathbf{p}^n$.) Then in the notation of Proposition 6.2.11,

$$\mathbf{d}/\mathbf{w} = (d/(1/n), \ldots, d/(1/n)) = (nd, \ldots, nd),$$

so

$$D_q^Z(\mathbf{w} \circ (\mathbf{p}^1, \ldots, \mathbf{p}^n)) = nd.$$

In other words, the diversity of a group of n islands, each having diversity d, is nd. This is the **replication principle** for the similarity-sensitive measures D_q^Z. It generalizes the replication principle for the Hill numbers, noted at the end of Section 4.4. The fact that our diversity measures satisfy it means that they do not suffer from the problems described in the oil company example (Example 2.4.11).

Since the diversity D_q^Z, Rényi entropies H_q^Z, and q-logarithmic entropies S_q^Z are all related to one another by invertible transformations, the chain rule for

D_q^Z can be translated into chain rules for H_q^Z and S_q^Z. In the case of S_q^Z, it takes a simple form, generalizing the chain rule for q-logarithmic entropy (equation (4.2)).

Proposition 6.2.13 (Chain rule) *Let* $q \in [0, \infty)$*. For* $\mathbf{w}$*,* $\mathbf{p}^i$*,* Z^i *and* Z *as in Proposition 6.2.11,*

$$S_q^Z(\mathbf{w} \circ (\mathbf{p}^1, \ldots, \mathbf{p}^n)) = S_q(\mathbf{w}) + \sum_{i \in \operatorname{supp}(\mathbf{w})} w_i^q \cdot S_q^{Z^i}(\mathbf{p}^i).$$

Proof Proposition 6.2.11(i) gives

$$\ln_q(D_q^Z(\mathbf{w} \circ (\mathbf{p}^1, \ldots, \mathbf{p}^n))) = \ln_q(M_{1-q}(\mathbf{w}, \mathbf{d}/\mathbf{w})).$$

By definition of S_q^Z (equation (6.4)) and Lemma 4.2.29, an equivalent statement is that

$$S_q^Z(\mathbf{w} \circ (\mathbf{p}^1, \ldots, \mathbf{p}^n)) = \sum_{i \in \operatorname{supp}(\mathbf{w})} w_i \ln_q \frac{d_i}{w_i}.$$

But writing $d_i/w_i = (1/w_i)d_i$ and applying the formula (1.18) for the q-logarithm of a product, the right-hand side is

$$\begin{aligned}\sum_{i \in \operatorname{supp}(\mathbf{w})} w_i \left(\ln_q \frac{1}{w_i} + \left(\frac{1}{w_i} \right)^{1-q} \ln_q d_i \right) &= \sum_{i \in \operatorname{supp}(\mathbf{w})} w_i \ln_q \frac{1}{w_i} + \sum_{i \in \operatorname{supp}(\mathbf{w})} w_i^q \ln_q d_i \\ &= S_q(\mathbf{w}) + \sum_{i \in \operatorname{supp}(\mathbf{w})} w_i^q \cdot S_q^Z(\mathbf{p}^i). \qquad \square\end{aligned}$$

6.3 Maximizing Diversity

Consider a community made up of organisms drawn from a fixed list of species, whose similarities to one another are known. Suppose that we can control the abundances of the species within the community. How should we choose those abundances in order to maximize the diversity, and what is the maximum diversity achievable?

In mathematical terms, fix an $n \times n$ similarity matrix Z. The fundamental questions are these.

- Which distributions $\mathbf{p}$ maximize the diversity $D_q^Z(\mathbf{p})$ of order q?
- What is the value of the maximum diversity, $\sup_{\mathbf{p} \in \Delta_n} D_q^Z(\mathbf{p})$?

In principle, the answers to both questions depend on q. After all, we have seen that when comparing two abundance distributions, different values of q

may produce different judgements on which of the distributions is more diverse (as in Examples 4.3.9 and 6.1.8). For instance, there seems no reason to suppose that a distribution maximizing diversity of order 1 will also maximize diversity of order 2. Similarly, we have seen nothing to suggest that the maximum diversity $\sup_{\mathbf{p}} D_1^Z(\mathbf{p})$ of order 1 should be equal to the maximum diversity of order 2.

However, it is a theorem that as long as Z is symmetric, the answers to both questions are indeed independent of q. That is, every symmetric similarity matrix has an unambiguous maximum diversity, and there is a distribution $\mathbf{p}$ that maximizes $D_q^Z(\mathbf{p})$ for all q simultaneously.

This result was first stated and proved by Leinster [211]. An improved proof and further results were given in a paper of Leinster and Meckes [221], from which much of this section is adapted. We omit most proofs, referring to [221].

Before stating the theorem, let us explore the maximum diversity problem informally.

Example 6.3.1 If there is only one species ($n = 1$) then the problem is trivial. If there are two then, assuming that Z is symmetric, their roles are interchangeable, so the distribution that maximizes diversity will clearly be $(1/2, 1/2)$.

Now consider a three-species pond community consisting of two highly similar species of frog and one species of newt. If we ignore the similarity between the species of frog and give the three species equal status, then the maximizing distribution should be uniform: $(1/3, 1/3, 1/3)$. But intuitively, this is not the distribution that maximizes diversity, since it is $2/3$ frog and $1/3$ newt. At the other extreme, if we treat the two frog species as identical, then diversity is maximized when there are equal quantities of frogs and newts (as in the two-species example); so, the distribution $(1/4, 1/4, 1/2)$ should maximize diversity. In reality, with a reasonable measure of similarities between species, the distribution that maximizes diversity should be somewhere between these two extremes. We will see in Example 6.3.16 that this is indeed the case.

For the rest of this section, fix an integer $n \geq 1$ and an $n \times n$ symmetric similarity matrix Z. The symmetry hypothesis matters, as we will see in Example 6.3.17.

Theorem 6.3.2 (Maximum diversity) *i. There exists a probability distribution on* $\{1, \ldots, n\}$ *that maximizes* D_q^Z *for all* $q \in [0, \infty]$ *simultaneously.*
ii. The maximum diversity $\sup_{\mathbf{p} \in \Delta_n} D_q^Z(\mathbf{p})$ *is independent of* $q \in [0, \infty]$.

Proof This is Theorem 1 of [221]. □

Let us say that a probability distribution $\mathbf{p} \in \Delta_n$ is **maximizing** (with respect

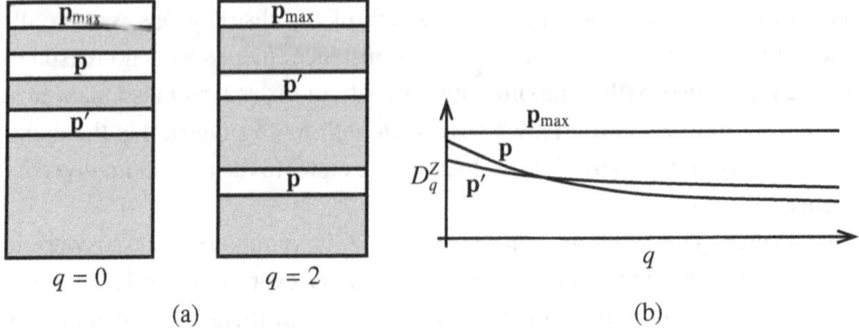

Figure 6.3 Visualizations of Theorem 6.3.2: (a) in terms of how different values of q rank the set of distributions, and (b) in terms of diversity profiles.

to Z) if $D_q^Z(\mathbf{p})$ maximizes D_q^Z for each $q \in [0, \infty]$. Theorem 6.3.2(ii) immediately implies that the diversity profile of a maximizing distribution is flat:

Corollary 6.3.3 *Let* $\mathbf{p}$ *be a maximizing distribution. Then* $D_q^Z(\mathbf{p}) = D_{q'}^Z(\mathbf{p})$ *for all* $q, q' \in [0, \infty]$. □

Theorem 6.3.2 can be understood as follows (Figure 6.3(a)). Each particular value of the viewpoint parameter q ranks the set of all distributions $\mathbf{p}$ in order of diversity, with $\mathbf{p}$ placed above $\mathbf{p}'$ when $D_q^Z(\mathbf{p}) > D_q^Z(\mathbf{p}')$. Different values of q rank the set of distributions differently. Nevertheless, there is a distribution $\mathbf{p}_{\max}$ that is at the top of every ranking. This is the content of Theorem 6.3.2(i).

Alternatively, we can visualize the theorem in terms of diversity profiles (Figure 6.3(b)). Diversity profiles may cross, reflecting the different priorities embodied by different values of q. But there is at least one distribution $\mathbf{p}_{\max}$ whose profile is above every other profile; moreover, its profile is constant. If diversity is seen as a positive quality, then $\mathbf{p}_{\max}$ is the best of all possible worlds.

Associated with the matrix Z is a real number: the constant value of any maximizing distribution.

Definition 6.3.4 The **maximum diversity** of the matrix Z is $D_{\max}(Z) = \sup_{\mathbf{p} \in \Delta_n} D_q^Z(\mathbf{p})$, for any $q \in [0, \infty]$.

By Theorem 6.3.2(ii), $D_{\max}(Z)$ is independent of q.

Later, we will see how to compute the maximizing distributions and maximum diversity of a matrix. For now, we just note a trivial example.

Example 6.3.5 Let Z be the $n \times n$ identity matrix I. We have already seen that $D_q^I(\mathbf{p}) = D_q(\mathbf{p})$ is maximized when $\mathbf{p}$ is the uniform distribution $\mathbf{u}_n$, and that the maximum value is n (Lemma 4.4.3(ii)). It is a special case of Theorem 6.3.2(ii) that this maximum value, n, is independent of q. In the notation just introduced, $D_{\max}(I) = n$.

If a distribution $\mathbf{p}$ maximizes diversity of order 2, must it also maximize diversity of orders 1 and ∞, for instance? The answer turns out to be yes:

Corollary 6.3.6 *Let* $\mathbf{p} \in \Delta_n$. *If* $\mathbf{p}$ *maximizes* D_q^Z *for some* $q \in (0, \infty]$ *then* $\mathbf{p}$ *maximizes* D_q^Z *for all* $q \in [0, \infty]$.

Proof This is Corollary 2 of [221]. □

The significance of this result is that if we wish to find a distribution that maximizes diversity of all orders q, all we need to do is to find one that maximizes diversity of whichever nonzero order is most convenient.

The hypothesis that $q > 0$ cannot be dropped from Corollary 6.3.6. Indeed, take $Z = I$. Then $D_0^I(\mathbf{p})$ is species richness (the cardinality of $\mathrm{supp}(\mathbf{p})$), which is maximized by any distribution $\mathbf{p}$ of full support. On the other hand, when $q > 0$, the diversity $D_q^I(\mathbf{p}) = D_q(\mathbf{p})$ is maximized only when $\mathbf{p}$ is uniform (Lemma 4.4.3(ii)).

Remark 6.3.7 Since the similarity-sensitive Rényi entropy H_q^Z and similarity-sensitive q-logarithmic entropy S_q^Z are increasing transformations of D_q^Z, the same distributions that maximize D_q^Z for all q also maximize H_q^Z and S_q^Z for all q. And since $H_q^Z = \log D_q^Z$, the maximum similarity-sensitive Rényi entropy, $\sup_{\mathbf{p}} H_q^Z(\mathbf{p})$, is also independent of q: it is simply $\log D_{\max}(Z)$.

In contrast, the maximum similarity-sensitive q-logarithmic entropy, $\sup_{\mathbf{p}} S_q^Z(\mathbf{p})$, is *not* independent of q. It is $\ln_q D_{\max}(Z)$, which varies with q. This is one advantage of the Rényi entropy (and its exponential) over the q-logarithmic entropy.

Theorem 6.3.2 guarantees the existence of a maximizing distribution $\mathbf{p}_{\max}$, but does not tell us how to find one. It also states that $D_q^Z(\mathbf{p}_{\max})$ is independent of q, but does not tell us its value. Our next theorem repairs both omissions. To state it, we need some definitions.

Definition 6.3.8 A **weighting** on a matrix M is a column vector $\mathbf{w}$ such that

$$M\mathbf{w} = \begin{pmatrix} 1 \\ \vdots \\ 1 \end{pmatrix}.$$

Lemma 6.3.9 *Let* M *be a matrix. Suppose that* M *and its transpose* M^{T} *each have at least one weighting. Then* $\sum_i w_i$ *is independent of the choice of weighting* $\mathbf{w}$ *on* M.

Proof Let $\mathbf{w}$ and $\mathbf{w}'$ be weightings on M. Choose a weighting $\mathbf{v}$ on M^{T}. Then

$$\sum_i w_i = (1 \ \cdots \ 1)\mathbf{w} = (M^{\mathrm{T}}\mathbf{v})^{\mathrm{T}}\mathbf{w} = \mathbf{v}^{\mathrm{T}}M\mathbf{w} = \mathbf{v}^{\mathrm{T}}\begin{pmatrix}1\\ \vdots \\ 1\end{pmatrix} = \sum_j v_j.$$

Similarly, $\sum_i w'_i = \sum_j v_j$. Hence $\sum_i w_i = \sum_i w'_i$. □

Definition 6.3.10 Let M be a matrix such that both M and M^{T} have at least one weighting. Its **magnitude** $|M|$ is $\sum_i w_i$, where $\mathbf{w}$ is any weighting on M.

By Lemma 6.3.9, the magnitude is independent of the choice of weighting.

Remarks 6.3.11 i. When M is symmetric (the case of interest here), $|M|$ is defined as long as M has at least one weighting.
ii. When M is invertible, M has exactly one weighting. Its entries are the row-sums of M^{-1}. Thus, $|M|$ is the sum of all the entries of M^{-1}.

Definition 6.3.12 A vector $\mathbf{v} = (v_i)$ over $\mathbb{R}$ is **nonnegative** if $v_i \geq 0$ for all i, and **positive** if $v_i > 0$ for all i.

For a nonempty subset $B \subseteq \{1, \ldots, n\}$, let Z_B denote the submatrix $(Z_{ij})_{i,j \in B}$ of Z. This is also a symmetric similarity matrix. Suppose that we have a nonnegative weighting $\mathbf{w}$ on Z_B. Then $\mathbf{w} \neq \mathbf{0}$, so $\sum_{j \in B} w_j \neq 0$. We can therefore define a probability distribution $\hat{\mathbf{w}} \in \Delta_n$ by normalizing and extending by 0:

$$\hat{w}_i = \begin{cases} w_i/|Z_B| & \text{if } i \in B, \\ 0 & \text{otherwise} \end{cases}$$

($i \in \{1, \ldots, n\}$).

Theorem 6.3.13 (Computation of maximum diversity) *i. The maximum diversity of Z is given by*

$$D_{\max}(Z) = \max_B |Z_B|, \tag{6.11}$$

where the maximum is over all nonempty $B \subseteq \{1, \ldots, n\}$ such that Z_B admits a nonnegative weighting.
ii. The set of maximizing distributions is

$$\bigcup_B \{\hat{\mathbf{w}} : \textit{nonnegative weightings } \mathbf{w} \textit{ on } B\},$$

where the union is over all B attaining the maximum in (6.11).

Proof This is Theorem 2 of [221]. □

Remark 6.3.14 Let $B \subseteq \{1, \ldots, n\}$ be a subset attaining the maximum in (6.11), and let $\mathbf{w}$ be a nonnegative weighting on Z_B, so that $\hat{\mathbf{w}} \in \Delta_n$ is a maximizing distribution. A short calculation shows that

$$(Z\hat{\mathbf{w}})_i = \frac{1}{|Z_B|}$$

for all $i \in B$. In particular, $(Z\hat{\mathbf{w}})_i$ is constant over $i \in B$. This can be understood as follows. For the Hill numbers (the case $Z = I$), the maximizing distribution takes the relative abundances p_i to be the same for all species i. This is no longer true when inter-species similarities are taken into account. Instead, the maximizing distributions have the property that the *ordinariness* $(Z\mathbf{p})_i$ is the same for all species i that are present.

Determining which species are present in a maximizing distribution is not straightforward. In particular, maximizing distributions do not always have full support, a phenomenon discussed at the end of this section.

Theorem 6.3.13 provides a finite-time algorithm for computing the maximizing diversity of Z, as well as all its maximizing distributions, as follows.

For each of the $2^n - 1$ nonempty subsets B of $\{1, \ldots, n\}$, perform some simple linear algebra to find the space of nonnegative weightings on Z_B. If this space is nonempty, call B **feasible** and record the magnitude $|Z_B|$. Then $D_{\max}(Z)$ is the maximum of all the recorded magnitudes. For each feasible B such that $|Z_B| = D_{\max}(Z)$, and each nonnegative weighting $\mathbf{w}$ on Z_B, the distribution $\hat{\mathbf{w}}$ is maximizing. This generates all of the maximizing distributions.

This algorithm takes exponentially many steps in n, and Remark 6.3.24 provides strong evidence that no algorithm can compute maximum diversity in polynomial time. But the situation is not as hopeless as it might appear, for two reasons.

First, each step of the algorithm is fast, consisting as it does of solving a system of linear equations. For instance, using a standard laptop and a standard computer algebra package, with no attempt at optimization, the maximizing distributions of 25×25 matrices were computed in a few seconds. Second, for certain classes of matrices Z, the computing time can be reduced dramatically, as we will see.

We first consider some examples, starting with the most simple cases.

Example 6.3.15 Take a 2×2 similarity matrix

$$Z = \begin{pmatrix} 1 & z \\ z & 1 \end{pmatrix},$$

where $0 \leq z < 1$. Let us run the algorithm just described.

$$Z = \begin{pmatrix} 1 & 0.9 & 0.4 \\ 0.9 & 1 & 0.4 \\ 0.4 & 0.4 & 1 \end{pmatrix}$$

Figure 6.4 Hypothetical three-species system. Distances between species indicate degrees of dissimilarity between them (not to scale).

- First we determine for which nonempty $B \subseteq \{1, \ldots, n\}$ the submatrix Z_B has a nonnegative weighting, and record the magnitudes of those that do.

  When $B = \{1\}$, the submatrix Z_B is (1); this has a unique nonnegative weighting $\mathbf{w} = (1)$, so $|Z_B| = 1$. The same is true for $B = \{2\}$. When $B = \{1, 2\}$, we have $Z_B = Z$, which has a unique nonnegative weighting

$$\mathbf{w} = \frac{1}{1+z}\begin{pmatrix} 1 \\ 1 \end{pmatrix} \tag{6.12}$$

  and magnitude $|Z_B| = 2/(1+z)$.
- The maximum diversity of Z is given by

$$D_{\max}(Z) = \max\left\{1, 1, \frac{2}{1+z}\right\},$$

  and $2/(1+z) > 1$, so $D_{\max}(Z) = 2/(1+z)$. The unique maximizing distribution is the normalization of the weighting (6.12), which is the uniform distribution $\mathbf{u}_2$.

That the maximizing distribution is uniform conforms to the intuitive expectation of Example 6.3.1. The computed value of $D_{\max}(Z)$ also conforms to the expectation that the maximum diversity should be a decreasing function of the similarity between the species.

Example 6.3.16 Now consider the three-species pond community of Example 6.3.1, with similarities as shown in Figure 6.4. Implementing the algorithm or using Proposition 6.3.25 below reveals that the unique maximizing distribution is $(0.478, 0.261, 0.261)$ (to 3 decimal places). This confirms the intuitive guess of Example 6.3.1.

One of our standing hypotheses on Z is symmetry. Without it, the main theorem fails in every respect:

Example 6.3.17 Let $Z = \left(\begin{smallmatrix} 1 & 1/2 \\ 0 & 1 \end{smallmatrix}\right)$, which is a similarity matrix but not symmetric. Consider a distribution $\mathbf{p} = (p_1, p_2) \in \Delta_2$. If $\mathbf{p}$ is $(1, 0)$ or $(0, 1)$ then

$D_q^Z(\mathbf{p}) = 1$ for all q. Otherwise,

$$D_0^Z(\mathbf{p}) = 3 - \frac{2}{1 + p_1}, \tag{6.13}$$

$$D_2^Z(\mathbf{p}) = \frac{2}{3(p_1 - 1/2)^2 + 5/4}, \tag{6.14}$$

$$D_\infty^Z(\mathbf{p}) = \begin{cases} \frac{1}{1-p_1} & \text{if } p_1 \leq 1/3, \\ \frac{2}{1+p_1} & \text{if } p_1 \geq 1/3. \end{cases} \tag{6.15}$$

It follows that $\sup_{\mathbf{p}\in\Delta_2} D_0^Z(\mathbf{p}) = 2$. However, no distribution maximizes D_0^Z; we have $D_0^Z(\mathbf{p}) \to 2$ as $\mathbf{p} \to (1, 0)$, but $D_0^Z(1, 0) = 1$. Equations (6.14) and (6.15) imply that

$$\sup_{\mathbf{p}\in\Delta_2} D_2^Z(\mathbf{p}) = 1.6, \qquad \sup_{\mathbf{p}\in\Delta_2} D_\infty^Z(\mathbf{p}) = 1.5,$$

with unique maximizing distributions $(1/2, 1/2)$ and $(1/3, 2/3)$ respectively.

Thus, when Z is not symmetric, the main theorem fails comprehensively: the supremum $\sup_{\mathbf{p}\in\Delta_n} D_0^Z(\mathbf{p})$ may not be attained; there may be no distribution maximizing $\sup_{\mathbf{p}\in\Delta_n} D_q^Z(\mathbf{p})$ for all q simultaneously; and that supremum may vary with q.

Perhaps surprisingly, nonsymmetric similarity matrices Z do have practical uses. For example, it is shown in Proposition A7 of the appendix to Leinster and Cobbold [220] that the mean phylogenetic diversity measures of Chao, Chiu and Jost [67] are a special case of the measures $D_q^Z(\mathbf{p})$, obtained by taking a particular Z constructed from the phylogenetic tree concerned. This Z is usually nonsymmetric, reflecting the asymmetry of evolutionary time. More generally, the case for dropping the symmetry axiom for metric spaces was made by Lawvere (pp. 138–139 of [204]), and Gromov has argued that symmetry 'unpleasantly limits many applications' (p. xv of [131]). So, the fact that the maximum diversity theorem fails for nonsymmetric Z is an important restriction.

Now consider finite, undirected graphs with no multiple edges (henceforth, **graphs** for short). As in Example 6.1.4, any such graph corresponds to a symmetric similarity matrix. What, then, is the maximum diversity of the adjacency matrix of a graph?

The answer requires some terminology. Recall that vertices x and y of a graph are said to be adjacent, written as $x \sim y$, if there is an edge between them. (In particular, every vertex of a reflexive graph is adjacent to itself.) A set of vertices is **independent** if no two distinct vertices are adjacent. The

independence number $\alpha(G)$ of a graph G is the number of vertices in an independent set of greatest cardinality.

Proposition 6.3.18 *Let G be a reflexive graph with adjacency matrix Z. Then the maximum diversity $D_{\max}(Z)$ is equal to the independence number $\alpha(G)$.*

Proof We will maximize the diversity of order ∞. For any probability distribution $\mathbf{p}$ on the vertex-set $\{1, \ldots, n\}$,

$$D_\infty^Z(\mathbf{p}) = 1 \Big/ \max_{i \in \operatorname{supp}(\mathbf{p})} \sum_{j:\, i \sim j} p_j. \tag{6.16}$$

First we show that $D_{\max}(Z) \geq \alpha(G)$. Choose an independent set B of cardinality $\alpha(G)$, and define $\mathbf{p} \in \Delta_n$ by

$$p_i = \begin{cases} 1/\alpha(G) & \text{if } i \in B, \\ 0 & \text{otherwise.} \end{cases}$$

For each $i \in \operatorname{supp}(\mathbf{p}) = B$, the sum on the right-hand side of equation (6.16) is $1/\alpha(G)$. Hence $D_\infty^Z(\mathbf{p}) = \alpha(G)$, giving $D_{\max}(Z) \geq \alpha(G)$.

Now we show that $D_{\max}(Z) \leq \alpha(G)$. By equation (6.16), an equivalent statement is that for each $\mathbf{p} \in \Delta_n$, there is some $i \in \operatorname{supp}(\mathbf{p})$ such that

$$\sum_{j:\, i \sim j} p_j \geq \frac{1}{\alpha(G)}. \tag{6.17}$$

Let $\mathbf{p} \in \Delta_n$. Choose an independent set $B \subseteq \operatorname{supp}(\mathbf{p})$ with maximal cardinality among all independent subsets of $\operatorname{supp}(\mathbf{p})$. Then every vertex in $\operatorname{supp}(\mathbf{p})$ is adjacent to at least one vertex in B, otherwise we could adjoin it to B to make a larger independent subset. This gives the inequality

$$\sum_{i \in B} \sum_{j:\, i \sim j} p_j = \sum_{i \in B} \sum_{j \in \operatorname{supp}(\mathbf{p}):\, i \sim j} p_j \geq \sum_{j \in \operatorname{supp}(\mathbf{p})} p_j = 1.$$

So we can choose some $i \in B$ such that $\sum_{j:\, i \sim j} p_j \geq 1/\#B$, where # denotes cardinality. But $\#B \leq \alpha(G)$ since B is independent, so the desired inequality (6.17) follows. □

Remark 6.3.19 The first part of the proof (together with Corollary 6.3.6) shows that a maximizing distribution on a reflexive graph can be constructed by taking the uniform distribution on some independent set of greatest cardinality, then extending by zero to the whole vertex-set. Except in the trivial case of a graph with no edges between distinct vertices, this maximizing distribution never has full support.

Example 6.3.20 The reflexive graph $G =$ •–•–• (loops not shown) has adjacency matrix $Z = \begin{pmatrix} 1 & 1 & 0 \\ 1 & 1 & 1 \\ 0 & 1 & 1 \end{pmatrix}$. The independence number of G is 2; this, then, is the maximum diversity of Z. There is a unique independent set of cardinality 2, and a unique maximizing distribution, $(1/2, 0, 1/2)$.

Example 6.3.21 The reflexive graph •–•–•–• also has independence number 2. There are three independent sets of maximal cardinality, so by Remark 6.3.19, there are at least three maximizing distributions,

$$(\tfrac{1}{2}, 0, \tfrac{1}{2}, 0), \qquad (\tfrac{1}{2}, 0, 0, \tfrac{1}{2}), \qquad (0, \tfrac{1}{2}, 0, \tfrac{1}{2}),$$

all with different supports. (The possibility of multiple maximizing distributions was also observed in the case $q = 2$ by Pavoine and Bonsall [275].) In fact, there are further maximizing distributions not constructed in the proof of Proposition 6.3.18, namely, $(\frac{1}{2}, 0, t, \frac{1}{2} - t)$ and $(\frac{1}{2} - t, t, 0, \frac{1}{2})$ for all $t \in (0, \frac{1}{2})$.

Example 6.3.22 Kolmogorov's notion of the ε-entropy of a metric space [193] is approximately an instance of maximum diversity, assuming that one is interested in its behaviour as $\varepsilon \to 0$ rather than for individual values of ε.

Let A be a finite metric space. Given $\varepsilon > 0$, the **ε-covering number** $N_\varepsilon(A)$ is the smallest number of closed ε-balls needed to cover A. But also associated with ε is the graph $G_\varepsilon(A)$ whose vertices are the points of A and with an edge between a and b whenever $d(a, b) \leq \varepsilon$. Write $Z_\varepsilon(A)$ for the adjacency matrix of $G_\varepsilon(A)$. From Proposition 6.3.18, it is not hard to deduce that

$$N_\varepsilon(A) \leq D_{\max}(Z_\varepsilon(A)) \leq N_{\varepsilon/2}(A)$$

(Example 11 of [221]).

We have repeatedly seen that quantities called entropy tend to be the logarithms of quantities called diversity. Kolmogorov's **ε-entropy** of A is $\log N_\varepsilon(A)$, and, by the inequalities above, is closely related to the logarithm of maximum diversity.

The moral of the proof of Proposition 6.3.18 is that by performing the simple task of maximizing diversity of order ∞, we automatically maximize diversity of all other orders. Here is an example of how this observation can be exploited.

Every graph G has a **complement** $\overline{G}$, with the same vertex-set as G; two vertices are adjacent in $\overline{G}$ if and only if they are not adjacent in G. Thus, the complement of a reflexive graph is **irreflexive** (has no loops), and vice versa. A set B of vertices in an irreflexive graph X is a **clique** if all pairs of distinct elements of B are adjacent in X. The **clique number** $\omega(X)$ of X is the maximal cardinality of a clique in X. Thus, $\omega(X) = \alpha(\overline{X})$.

We now recover a result first proved by Motzkin and Straus (Theorem 1 of [256]).

Corollary 6.3.23 (Motzkin and Straus) *Let X be an irreflexive graph. Then we have*

$$\sup_{\mathbf{p}} \sum_{(i,j):\, i\sim j} p_i p_j = 1 - \frac{1}{\omega(X)},$$

where the supremum is over probability distributions $\mathbf{p}$ *on the vertex-set of* X *and the sum is over ordered pairs of adjacent vertices of* X.

Proof Write $\{1,\ldots,n\}$ for the vertex-set of X, and Z for the adjacency matrix of the reflexive graph $\overline{X}$. Then for all $\mathbf{p} \in \Delta_n$,

$$\begin{aligned}\sum_{(i,j):\, i\sim j \text{ in } X} p_i p_j &= \sum_{i,j=1}^{n} p_i p_j - \sum_{(i,j):\, i\sim j \text{ in } \overline{X}} p_i p_j \\ &= 1 - \sum_{i,j=1}^{n} p_i Z_{ij} p_j \\ &= 1 - \frac{1}{D_2^Z(\mathbf{p})}.\end{aligned}$$

Hence by Proposition 6.3.18,

$$\sup_{\mathbf{p}\in\Delta_n} \sum_{(i,j):\, i\sim j \text{ in } X} p_i p_j = 1 - \frac{1}{D_{\max}(\mathbf{p})} = 1 - \frac{1}{\alpha(\overline{X})} = 1 - \frac{1}{\omega(X)}.$$

□

It follows from this proof and Remark 6.3.19 that $\sum_{(i,j):\, i\sim j} p_i p_j$ can be maximized as follows: take the uniform distribution on some clique in X of maximal cardinality, then extend by zero to the whole vertex-set. This distribution maximizes the probability that two vertices chosen at random are adjacent, as in Example 6.1.6(iv).

Remark 6.3.24 Proposition 6.3.18 implies that computationally, finding the maximum diversity of an arbitrary symmetric $n \times n$ similarity matrix is at least as hard as finding the independence number of a reflexive graph with n vertices. This is a very well-studied problem, usually presented in its dual form (find the clique number of an irreflexive graph) and called the **maximum clique problem** [181]. It is **NP**-hard. Hence, assuming that $\mathbf{P} \neq \mathbf{NP}$, there is no polynomial-time algorithm for computing maximum diversity, nor even for computing the support of a maximizing distribution.

We now return to general symmetric similarity matrices, addressing two remaining questions: when are maximizing distributions unique, and when do they have full support?

Recall that a real symmetric matrix Z is **positive definite** if $\mathbf{x}^{\mathrm{T}} Z \mathbf{x} > 0$ for all $\mathbf{0} \neq \mathbf{x} \in \mathbb{R}^n$, and **positive semidefinite** if $\mathbf{x}^{\mathrm{T}} Z \mathbf{x} \geq 0$ for all $\mathbf{x} \in \mathbb{R}^n$. Equivalently, Z is positive definite if all its eigenvalues are positive, and positive semidefinite if they are all nonnegative. A positive definite matrix is invertible and therefore has a unique weighting.

Proposition 6.3.25 *i. If Z is positive semidefinite and has a nonnegative weighting* $\mathbf{w}$*, then $D_{\max}(Z) = |Z|$ and $\mathbf{w}/|Z|$ is a maximizing distribution.*

ii. If Z is positive definite and its unique weighting $\mathbf{w}$ *is positive then $\mathbf{w}/|Z|$ is the unique maximizing distribution.*

Proof This is Proposition 3 of [221]. □

In particular, if $|Z|$ is positive semidefinite and has a nonnegative weighting, then computing its maximum diversity is trivial.

When Z is positive definite and its unique weighting is positive, its unique maximizing distribution eliminates no species. Here are two classes of such matrices Z.

Example 6.3.26 Call Z **ultrametric** if $Z_{ik} \geq \min\{Z_{ij}, Z_{jk}\}$ for all i, j, k and $Z_{ii} > Z_{jk}$ for all i, j, k with $j \neq k$. For instance, the matrix $Z = (e^{-d(i,j)})$ of any ultrametric space is ultrametric; see Example 6.1.3. If Z is ultrametric then Z is positive definite with positive weighting, by Proposition 2.4.18 of Leinster [216].

(The positive definiteness of ultrametric matrices was also proved, earlier, by Varga and Nabben [343], and a different proof still was given in Theorem 3.6 of Meckes [250]. An earlier, indirect proof of the positivity of the weighting can be found in Pavoine, Ollier and Pontier [276].)

Such matrices arise in practice. For instance, Z is ultrametric if it is defined from a phylogenetic or taxonomic tree as in Examples 6.1.1(iii) and (iv).

Example 6.3.27 The identity matrix $Z = I$ is certainly positive definite with positive weighting. By topological arguments, there is a neighbourhood U of I in the space of symmetric matrices such that every matrix in U also has these properties. (See the proofs of Propositions 2.2.6 and 2.4.6 of Leinster [216].)

Quantitative versions of this result are also available. For instance, suppose that $Z_{ii} = 1$ for all i, j and that Z is **strictly diagonally dominant**, that is, $Z_{ii} > \sum_{j \neq i} Z_{ij}$ for all i. Then Z is positive definite with positive weighting (Proposition 4 of Leinster and Meckes [221]).

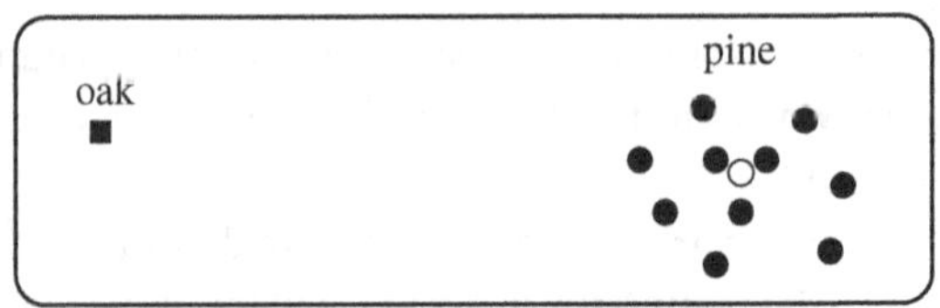

Figure 6.5 Hypothetical community consisting of one species of oak (▪) and ten species of pine (•), to which one further species of pine is then added (∘). Distances between species indicate degrees of dissimilarity (not to scale).

In summary, if our similarity matrix Z is ultrametric, or if it is close to the matrix I that encodes the naive model, then it enjoys many special properties: the maximum diversity is equal to the magnitude, there is a unique maximizing distribution, the maximizing distribution has full support, and both the maximizing distribution and the maximum diversity can be computed in polynomial time.

We saw in Examples 6.3.20 and 6.3.21 that for some similarity matrices Z, no maximizing distribution has full support. Mathematically, this simply means that maximizing distributions sometimes lie on the boundary of Δ_n. But ecologically, it may sound shocking: is it reasonable that diversity can be increased by *eliminating* some species?

We argue that it is. For example, consider a forest consisting of one species of oak and ten species of pine, with all species equally abundant. Suppose that an eleventh species of pine is added, with the same abundance as all the existing species (Figure 6.5). This makes the forest even more heavily dominated by pine than it was before, so it is intuitively reasonable that the diversity should decrease. But now running time backwards, the conclusion is that if we start with a forest containing the oak and all eleven pine species, then eliminating the eleventh should *increase* the diversity.

To clarify further, recall that diversity is defined in terms of the *relative* abundances only. Thus, eliminating the ith species causes not only a decrease in p_i, but also an increase in the other relative abundances p_j. If the ith species is particularly ordinary within the community (like the eleventh species of pine), then eliminating it increases the relative abundances of less ordinary species, resulting in a community that is more diverse.

The instinct that maximizing diversity should not eliminate any species is based on the assumption that the distinction between species is of high value. (After all, if two species were very nearly identical – or in the extreme, actually identical – then losing one would be of little importance.) If one wishes to make that assumption, one must build it into the model. This is done by choosing a similarity matrix Z with a low similarity coefficient Z_{ij} for each $i \neq j$. Thus, Z is close to the identity matrix I (assuming that similarity is measured on a scale of 0 to 1). Example 6.3.27 guarantees that in this case, there is a unique maximizing distribution and it does not, in fact, eliminate any species.

The fact that maximizing distributions can eliminate some species has previously been discussed in the ecological literature in the case of Rao's quadratic entropy ($q = 2$): see Izsák and Szeidl [152], Pavoine and Bonsall [275], and references therein. The same phenomenon was observed and explored by Shimatani in genetics [312], again in the case $q = 2$.

We finish by stating necessary and sufficient conditions for a symmetric similarity matrix Z to admit at least one maximizing distribution of full support, so that diversity can be maximized without eliminating any species. We also state necessary and sufficient conditions for *every* maximizing distribution to have full support. The latter conditions are genuinely more restrictive. For instance, if $Z = \left(\begin{smallmatrix} 1 & 1 \\ 1 & 1 \end{smallmatrix}\right)$ then every distribution is maximizing, so some but not all maximizing distributions have full support.

Proposition 6.3.28 *The following are equivalent:*

i. *there exists a maximizing distribution for Z of full support;*
ii. *Z is positive semidefinite and has at least one positive weighting.*

Proof This is Proposition 5 of [221]. □

Proposition 6.3.29 *The following are equivalent:*

i. *every maximizing distribution for Z has full support;*
ii. *Z has exactly one maximizing distribution, which has full support;*
iii. *Z is positive definite with positive weighting;*
iv. *$D_{\max}(Z_B) < D_{\max}(Z)$ for every nonempty proper subset B of $\{1, \ldots, n\}$.*

Proof This is Proposition 6 of [221]. □

Let us put our results on maximum diversity into context. First, they belong to the huge body of work on maximum entropy. For example, among all probability distributions on $\mathbb{R}$ with a given mean and variance, the one with the maximum entropy is the normal distribution [232, 29]. Given the fundamental nature of the normal distribution, this fact alone would be motivation enough to seek maximum entropy distributions in other settings (such as the one at hand), quite apart from the importance of maximum entropy in thermodynamics, machine learning, and so on.

Second, the maximum diversity theorem (Theorem 6.3.2) is stated for probability distributions on *finite* sets equipped with a similarity matrix, but it can be generalized to compact Hausdorff spaces A equipped with a suitable function $Z\colon A\times A \to [0,\infty)$ measuring similarity between points. This more general theorem was recently proved by Leinster and Roff [223], along with a version of the computation theorem (Theorem 6.3.13). It encompasses both the finite case and the case of a compact metric space with similarities $Z(a,b) = e^{-d(a,b)}$.

Third, maximum diversity is closely related to the emerging invariant known as magnitude, as now described.

6.4 Introduction to Magnitude

In the solution to the maximum diversity problem, a supporting role was played by the notion of the magnitude of a matrix (Definition 6.3.10). Theorem 6.3.13 implies that the maximum diversity of a symmetric similarity matrix Z is always equal to the magnitude of one of its principal submatrices Z_B, and Examples 6.3.26 and 6.3.27 describe classes of matrix for which the maximum diversity is actually equal to the magnitude.

The definition of magnitude was introduced without motivation, and may appear to be nothing but a technicality. But in fact, magnitude is an answer to the following very broad conceptual challenge.

For many types of objects in mathematics, there is a canonical notion of size. For example:

- Every set A (finite, say) has a cardinality $|A|$, which satisfies the inclusion-exclusion formula

  $$|A \cup B| = |A| + |B| - |A \cap B|$$

  (for subsets A and B of some larger set) and the multiplicativity formula

  $$|A \times B| = |A| \cdot |B|.$$

- Every measurable subset A of Euclidean space has a volume $\mathrm{Vol}(A)$, which satisfies similar formulas:

  $$\mathrm{Vol}(A \cup B) = \mathrm{Vol}(A) + \mathrm{Vol}(B) - \mathrm{Vol}(A \cap B),$$
  $$\mathrm{Vol}(A \times B) = \mathrm{Vol}(A) \cdot \mathrm{Vol}(B).$$

- Every sufficiently well-behaved topological space A has an Euler characteristic $\chi(A)$, which again satisfies

  $$\chi(A \cup B) = \chi(A) + \chi(B) - \chi(A \cap B),$$
  $$\chi(A \times B) = \chi(A) \cdot \chi(B).$$

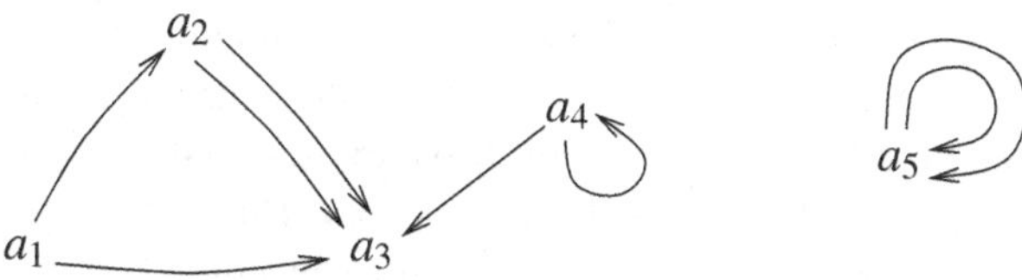

Figure 6.6 The objects and maps of a finite category (identity maps not shown).

(Here, inclusion-exclusion holds for subspaces A and B of some larger space, under suitable hypotheses. Technically, it is best to work in the setting of either cohomology with compact supports, as in Section 3.3 of Hatcher [140], or tame topology, as in Chapter 4 of van den Dries [341] or Chapter 3 of Ghrist [116].) The insight that Euler characteristic is the topological analogue of cardinality is principally due to Schanuel, who compared Euler's investigation of spaces of negative 'cardinality' (Euler characteristic) with Cantor's investigation of sets of infinite cardinality:

> Euler's analysis, which demonstrated that in counting suitably 'finite' spaces one can get well-defined negative integers, was a revolutionary advance in the idea of cardinal number – perhaps even more important than Cantor's extension to infinite sets, if we judge by the number of areas in mathematics where the impact is pervasive.

(Schanuel [304], Section 3).

The close resemblance between these invariants suggests a challenge: find a general notion of the size of a mathematical object, encompassing these three invariants and others. And this challenge has a solution: the magnitude of an enriched category.

Enriched categories are very general structures, and the theory of the magnitude of an enriched category sweeps across many parts of mathematics, most of them very distant from diversity measurement. This section and the next paint a broad-brush picture, omitting all details. General references for this material are Leinster and Meckes [222] and Leinster [216].

We begin with ordinary categories. A finite category $\mathbf{A}$ consists of, first of all, a finite directed multigraph, that is, a finite collection of objects $a_1, \ldots, a_n$ together with a finite set $\mathrm{Hom}(a_i, a_j)$ for each i and j, whose elements are to be thought of as maps or arrows from a_i to a_j (Figure 6.6). It is also equipped with an associative operation of composition of maps and an identity map on each object. (See Mac Lane [236], for instance.)

Any finite category $\mathbf{A}$ gives rise to an $n \times n$ matrix $Z_{\mathbf{A}}$ whose (i, j)-entry is

$|\mathrm{Hom}(a_i, a_j)|$, the number of maps from a_i to a_j. The **magnitude** $|\mathbf{A}| \in \mathbb{Q}$ of the category $\mathbf{A}$ is defined to be the magnitude $|Z_{\mathbf{A}}|$ of the matrix $Z_{\mathbf{A}}$, if it exists.

Here we have used the notation $|\cdot|$ for two purposes: first for the cardinality of a finite set, then for the magnitude of a category. This is deliberate. In both cases, $|\cdot|$ is a measure of the size of the structure concerned.

For example, if $\mathbf{A}$ has no maps except for identities then $Z_{\mathbf{A}}$ is the $n \times n$ identity matrix, so the magnitude $|\mathbf{A}|$ is the cardinality n of its set of objects. Less trivially, any (small) category $\mathbf{A}$ has a classifying space $B\mathbf{A}$ (also called its nerve or geometric realization), which is a topological space constructed from $\mathbf{A}$ by starting with one 0-simplex for each object of $\mathbf{A}$, then pasting in one 1-simplex for each map in $\mathbf{A}$, one 2-simplex for each commutative triangle in $\mathbf{A}$, and so on (Segal [308]). It is a theorem that

$$|\mathbf{A}| = \chi(B\mathbf{A}), \tag{6.18}$$

under finiteness conditions to ensure that the Euler characteristic of $B\mathbf{A}$ is well defined (Proposition 2.11 of Leinster [210]). So, the situation is similar to group homology: the homology of a group G can be defined either through a direct algebraic formula (as for $|\mathbf{A}|$) or as the homology of its classifying space (as for $\chi(B\mathbf{A})$), and it is a theorem that the two are equal.

Example 6.4.1 Let $\mathbf{A} = (\bullet \rightrightarrows \bullet)$ (identity maps not shown). Then

$$Z_{\mathbf{A}} = \begin{pmatrix} 1 & 2 \\ 0 & 1 \end{pmatrix},$$

giving

$$Z_{\mathbf{A}}^{-1} = \begin{pmatrix} 1 & -2 \\ 0 & 1 \end{pmatrix}$$

and so

$$|\mathbf{A}| = |Z_{\mathbf{A}}| = 1 + (-2) + 0 + 1 = 0.$$

On the other hand, $B\mathbf{A} = S^1$, so $\chi(B\mathbf{A}) = 0$, confirming equation (6.18).

Equation (6.18) shows how, under hypotheses, magnitude for categories can be derived from Euler characteristic for topological spaces. In the other direction, we can derive topological Euler characteristic from categorical magnitude. Let M be a finitely triangulated manifold. Then associated with the triangulation, there is a category $\mathbf{A}_M$ whose objects are the simplices $s_1, \ldots, s_n$ of the triangulation, with one map $s_i \to s_j$ whenever $s_i \subseteq s_j$, and with no maps $s_i \to s_j$ otherwise. Then

$$\chi(M) = |\mathbf{A}_M|$$

(Section 3.8 of Stanley [320] and Section 2 of Leinster [210]).

The moral of the last two results is that topological Euler characteristic and categorical magnitude each determine the other (under suitable hypotheses). Indeed, the magnitude of a category is often called its Euler characteristic; see, for instance, Leinster [210], Berger and Leinster [36], Fiore, Lück and Sauer [105, 106], Noguchi [264, 265, 266], and Tanaka [323].

Further theorems connect the magnitude of a category to the Euler characteristic of an orbifold (Proposition 2.12 of [210]) and to the Baez–Dolan cardinality of a groupoid (Example 2.7 of [210] and Section 3 of Baez and Dolan [23]), both of which are rational numbers, not usually integers. The notion of magnitude can also be seen as an extension of the theory of Möbius inversion for posets (most commonly associated with the name of Rota [301]), which itself generalizes the classical Möbius function of number theory; see [210, 215] for explanation.

The definition of the magnitude of a category involved $|\mathrm{Hom}(a_i, a_j)|$, the cardinality of the set of maps from a_i to a_j. Thus, we used the notion of the cardinality of a finite set to define the magnitude of a finite category. We can envisage that if $\mathrm{Hom}(a_i, a_j)$ were some other kind of structure with a preexisting notion of size, a similar definition could be made. And indeed, this idea can be implemented in the language of enriched categories, as follows.

A **monoidal category** is, loosely speaking, a category $\mathscr{V}$ equipped with a product operation satisfying reasonable conditions. Section VII.1 of Mac Lane [236] gives the full definition, but the following examples will be all that we need here.

Example 6.4.2 Typical examples of monoidal categories are the category **Set** of sets with the cartesian product $\times$ and the category **Vect** of vector spaces with the tensor product $\otimes$.

A less obvious example is the category whose objects are the elements of the interval $[0, \infty]$, with one map $x \to y$ whenever $x \geq y$, and with no maps $x \to y$ otherwise. Here we take $+$ as the 'product' operation. (We could also take ordinary multiplication as the product, but it is $+$ that will be of interest here.)

Now fix a monoidal category $\mathscr{V}$, with product denoted by $\otimes$. Loosely, a **category enriched in** $\mathscr{V}$, or $\mathscr{V}$**-category**, **A**, consists of:

- a set $a, b, \ldots$ of **objects** of **A**;
- for each pair (a, b) of objects of **A**, an object $\mathrm{Hom}(a, b)$ of $\mathscr{V}$;
- for each triple (a, b, c) of objects of **A**, a map

$$\mathrm{Hom}(a, b) \otimes \mathrm{Hom}(b, c) \to \mathrm{Hom}(a, c) \tag{6.19}$$

in $\mathscr{V}$ (called **composition** in **A**),

subject to conditions. (For the full definition, see Section 1.2 of Kelly [184] or Section 6.2 of Borceux [44].)

Examples 6.4.3 The following examples of enriched categories are depicted in Figure 6.7.

i. A category enriched in (**Set**, $\times$) is just an ordinary category. So, an enriched category is not a category with special properties; it is something *more general* than a category.
ii. A category enriched in (**Vect**, $\otimes$) is a **linear category**: a category equipped with a vector space structure on each of the sets $\text{Hom}(a, b)$, in such a way that composition is bilinear.
iii. As first observed by Lawvere [204], any metric space A can be viewed as a category $\mathbf{A}$ enriched in $([0, \infty], +)$: the objects of $\mathbf{A}$ are the points of A, while $\text{Hom}(a, b) \in [0, \infty]$ is the distance $d(a, b)$, and the composition (6.19) is the triangle inequality

$$d(a, b) + d(b, c) \geq d(a, c).$$

Thus, categories, linear categories and metric spaces are all instances of a single general concept: enriched category. This enables constructions and insights to be passed backwards and forwards between them, a strategy that proves to have great power.

In particular, it is straightforward to generalize the definition of the magnitude of a finite category to finite enriched categories. Let $\mathscr{V}$ be a monoidal category equipped with a function $|\cdot|$ that assigns to each object X of $\mathscr{V}$ an element $|X|$ of some ring. This function $|\cdot|$ is to play the role of the cardinality of a finite set, and we therefore impose the requirements that it is isomorphism-invariant and multiplicative:

$$X \cong Y \implies |X| = |Y|, \qquad |X \otimes Y| = |X| \cdot |Y|.$$

(Section 1.3 of Leinster [216] gives details.) Then any $\mathscr{V}$-category $\mathbf{A}$ with finitely many objects, $a_1, \ldots, a_n$, gives rise to a matrix

$$Z_{\mathbf{A}} = (|\text{Hom}(a_i, a_j)|)_{i,j}.$$

The **magnitude** $|\mathbf{A}|$ of $\mathbf{A}$ is defined to be the magnitude of $Z_{\mathbf{A}}$, if it exists.

Example 6.4.4 If we begin with the monoidal category $\mathscr{V}$ of finite sets equipped with the cartesian product, with the cardinality function $|\cdot|$ on finite sets, then we recover the notion of the magnitude of a finite category.

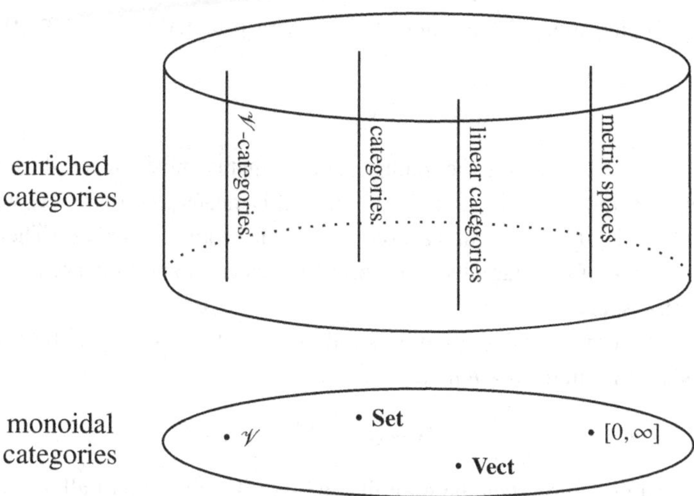

Figure 6.7 Schematic diagram of monoidal and enriched categories.

Example 6.4.5 Let $\mathscr{V}$ be the category of finite-dimensional vector spaces over some field, with the tensor product, and put $|X| = \dim X$ for finite-dimensional vector spaces X. Then we obtain a notion of the magnitude $|\mathbf{A}|$ of a linear category $\mathbf{A}$ with finitely many objects and finite-dimensional hom-spaces $\mathrm{Hom}(a, b)$. By definition, $|\mathbf{A}|$ is the magnitude of the matrix

$$(\dim \mathrm{Hom}(a, b))_{a,b\in\mathbf{A}}.$$

For instance, let E be an associative algebra over an algebraically closed field. In the representation theory of algebras, an important linear category associated with E is $\mathbf{IP}(E)$, the category of indecomposable projective E-modules. Under finiteness hypotheses on E, it is a theorem that $\mathbf{IP}(E)$ has magnitude

$$|\mathbf{IP}(E)| = \sum_{n=0}^{\infty}(-1)^n \dim \mathrm{Ext}^n_E(S, S), \tag{6.20}$$

where S is the direct sum of the simple E-modules (Theorem 1.1 of Chuang, King and Leinster [69]). The matrix $Z_{\mathbf{IP}(E)}$ is better known as the **Cartan matrix** of E. The right-hand side of equation (6.20) is called the **Euler form** $\chi(S, S)$ of the pair (S, S), and is another manifestation of the concept of Euler characteristic.

The examples so far of the magnitude of an enriched category have been closely related to other, older invariants. But when we apply the definition to metric spaces, we obtain something new.

Let $\mathscr{V}$ be the monoidal category $[0, \infty]$, with product $+$. For $x \in [0, \infty]$, define

$$|x| = e^{-x} \in \mathbb{R}.$$

(Recall that $|\cdot|$ is required to be 'multiplicative', that is, must convert the tensor product on $\mathscr{V}$ into multiplication. In this case, this means $|x+y| = |x| \cdot |y|$, which by Corollary 1.1.11(i) essentially forces $|x| = c^x$ for some constant c.) Then we obtain a notion of the magnitude of a finite $\mathscr{V}$-category, and in particular, of a finite metric space.

In explicit terms, the definition is as follows. Let $A = \{a_1, \ldots, a_n\}$ be a finite metric space. Form the $n \times n$ matrix

$$Z_A = (e^{-d(a_i, a_j)}).$$

Invert Z_A (if possible); then the **magnitude** $|A|$ of A is the sum of all n^2 entries of Z_A^{-1}.

Here we have used Remark 6.3.11(ii) on the magnitude of a matrix in terms of its inverse. Since Z_A is a square matrix of real numbers, it is usually invertible, and in fact it is *always* invertible when A is a subspace of Euclidean space (Theorem 2.5.3 of Leinster [216] or Section 4 of Meckes [250]).

Examples 6.4.6 i. The magnitude of the zero-point space is 0, and the magnitude of the one-point space is 1.

ii. Consider the metric space A consisting of two points distance ℓ apart:

$$\bullet \xleftrightarrow{\ \ell\ } \bullet$$

Then

$$|A| = \text{sum of entries of } \begin{pmatrix} e^{-0} & e^{-\ell} \\ e^{-\ell} & e^{-0} \end{pmatrix}^{-1} = \frac{2}{1 + e^{-\ell}},$$

as illustrated in Figure 6.8.

This example can be understood as follows. When ℓ is small, the two points are barely distinguishable, and may appear to be only one point (at poor resolution, for instance). As ℓ increases, the two points acquire increasingly separate identities, and correspondingly, the magnitude increases towards 2. In the extreme, when $\ell = \infty$, the two points are entirely separated and the magnitude is exactly 2. This example and others suggest that we can usefully think of the magnitude of a finite metric space as the 'effective number of points', or, more fully, the effective number of completely separate points.

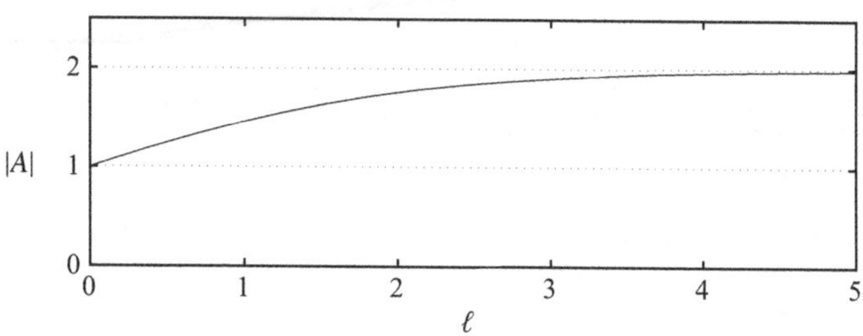

Figure 6.8 The magnitude of a two-point space A, with points distance ℓ apart (Example 6.4.6(ii)).

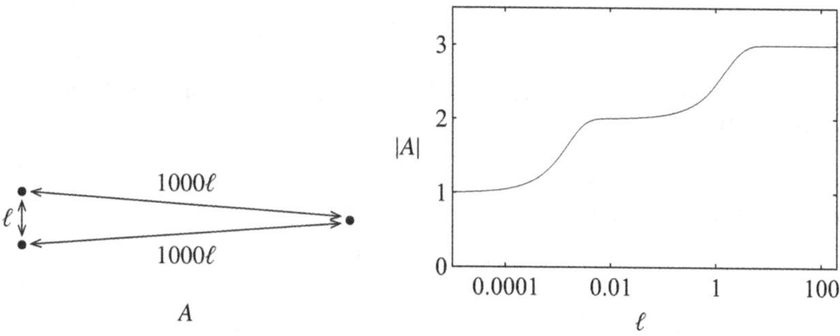

Figure 6.9 The magnitude of a certain three-point metric space A (Example 6.4.6(iv)). Note the logarithmic scale.

iii. Let A be a finite metric space in which all nonzero distances are ∞. Then $Z_A = I$ and $|A|$ is just the cardinality of A. This also fits with the interpretation of magnitude as the effective number of points.

iv. This example is adapted from Willerton ([357], Figure 1). Let A be a three-point space with the points arranged in a long thin triangle, as in Figure 6.9. When ℓ is small, the space appears to be just a single point, and the magnitude is close to 1. When ℓ is moderate, the space appears to have two points, and the magnitude is about 2. When ℓ is large, the distinction between all three points is clearly visible, and the magnitude is close to 3.

Empirical data such as this suggests a connection between magnitude and persistent homology. Indeed, results of Otter [269] have begun to establish such a connection. We return to this topic at the end of the section.

Every metric space A belongs to a one-parameter family $(tA)_{t>0}$ of spaces, where tA denotes A scaled up by a factor of t. So, magnitude assigns to each finite metric space A not just a *number* $|A|$, but a (partially defined) *function*:

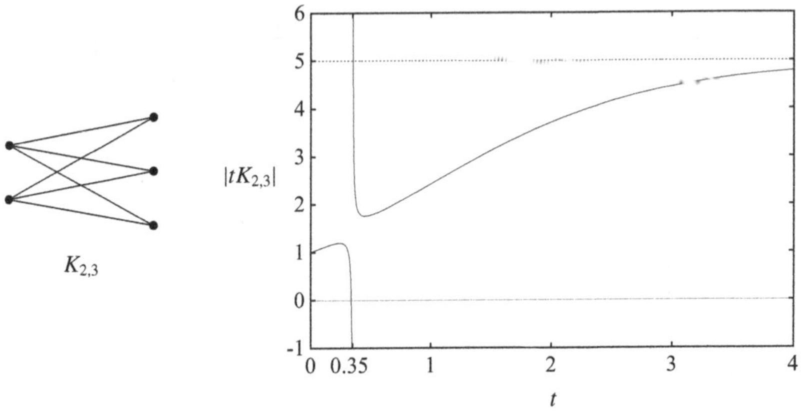

Figure 6.10 The complete bipartite graph $K_{2,3}$ and its magnitude function (Example 6.4.7). The singularity is at $\log \sqrt{2} \approx 0.35$.

its **magnitude function**

$$\begin{array}{ccc} (0, \infty) & \to & \mathbb{R} \\ t & \mapsto & |tA|. \end{array}$$

For instance, Figures 6.8 and 6.9 show the magnitude functions of a certain two-point space and a certain three-point space.

Example 6.4.7 Magnitude functions can behave wildly. Consider the complete bipartite graph $K_{2,3}$ (Figure 6.10), regarded as a metric space as follows: the points of the space are the vertices of the graph, and the distance between two vertices is the number of edges in a shortest path between them.

The magnitude function of $K_{2,3}$ has several striking features: it is sometimes negative, sometimes greater than the number of points, sometimes undefined, and sometimes *decreasing* in the scale factor t. Example 2.2.7 of Leinster [216] gives details.

However, the magnitude function of a finite metric space never behaves *too* badly. It can be shown that the magnitude function has only finitely many singularities (none for subspaces of Euclidean space), that it is increasing for $t \gg 0$, and that $|tA|$ converges to the cardinality of A as $t \to \infty$ (Proposition 2.2.6 of Leinster [216]). In particular, this last statement implies that the magnitude function of a space knows its cardinality.

In Example 6.4.7, we started from a graph, constructed the metric space whose points are the vertices and whose distances are shortest path-lengths, and considered the magnitude of that space. This is a construction of general interest, investigated in Leinster [217]. In this context, we replace the real number e^{-1} in the definition of magnitude by a formal variable x. The magnitude of a graph can then be expressed as either a rational function or a power series in x with integer coefficients (Section 2 of [217]). For example, the graphs

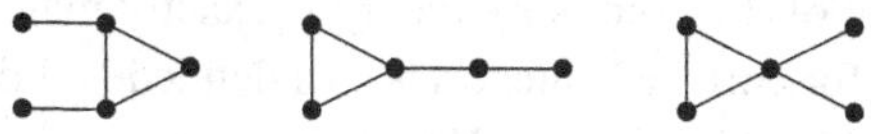

all have magnitude

$$\frac{5+5x-4x^2}{(1+x)(1+2x)} = 5 - 10x + 16x^2 - 28x^3 + 52x^4 - 100x^5 + \cdots$$

(Example 4.11 of [217]). The magnitude of a graph shares some invariance properties with one of the most important graph invariants of all, the Tutte polynomial. For instance, it is invariant under Whitney twists when the points of identification are adjacent. But it is not a specialization of the Tutte polynomial: it carries information that the Tutte polynomial does not.

Graph magnitude satisfies multiplicativity and inclusion-exclusion principles:

$$|G \times H| = |G| \cdot |H|,$$
$$|G \cup H| = |G| + |H| - |G \cap H|$$

(where the latter is under quite strict hypotheses), shown as Lemma 3.6 and Theorem 4.9 of Leinster [217]. As such, it has a reasonable claim to being the graph-theoretic analogue of cardinality.

As additional evidence for this claim, Hepworth and Willerton [144] constructed a graded homology theory of graphs whose Euler characteristic is magnitude. In more detail: since their homology theory is graded, the Euler characteristic of a graph is not a single number but a sequence of numbers, which when construed as a power series is exactly the graph's magnitude. Thus, their homology theory is a categorification of magnitude in the same sense that the Khovanov homology of knots and links [189] is a categorification of the Jones polynomial. It is a finer invariant than magnitude, in that there are graphs with the same magnitude but different homology groups (Gu [133], Appendix A; see also Summers [321]).

Not only the definition of magnitude for graphs, but also some theorems about it, can be categorified. For instance, Hepworth and Willerton proved that the multiplicativity and inclusion-exclusion theorems for magnitude lift to Künneth and Mayer–Vietoris theorems in homology. In this sense, known properties of the magnitude of graphs are shadows of functorial results in homology.

Hepworth and Willerton's idea even works in the full generality of enriched categories. That is, the magnitude of an enriched category (a numerical invariant) can be categorified to a graded homology theory for enriched categories (an algebraic invariant). As in the case of graphs, 'categorified' means that the Euler characteristic of the homology theory is exactly magnitude. This **magnitude homology** for enriched categories was defined and developed in work led by Shulman [224]. It is a kind of Hochschild homology.

Since metric spaces are a special kind of enriched category, this construction provides a new homology theory for metric spaces. It is genuinely metric rather than topological. For example, the first magnitude homology of a closed subset X of $\mathbb{R}^n$ is trivial if and only if X is convex ([224], Section 4). Indeed, all of the magnitude homology groups of a convex subset of $\mathbb{R}^n$ are trivial, a metric analogue of the topological fact that the homology of a contractible space is trivial. This was proved independently by Kaneta and Yoshinaga (Corollary 5.3 of [176]) and Jubin (Theorem 7.2 of [174]). Gomi [119] states a slogan:

The more geodesics are unique, the more magnitude homology is trivial.

Methods for computing the magnitude homology of metric spaces have recently been developed and applied to calculate specific homology groups. Gomi developed spectral sequence techniques and used them to prove results on the magnitude homology groups of circles ([120], Section 4). Kaneta and Yoshinaga [176] showed that while ordinary topological homology detects the existence of holes, magnitude homology detects the *diameter* of holes, in a sense made precise in their Theorem 5.7. Asao proved that if a space contains a closed geodesic then its second magnitude homology group is nontrivial (Theorem 5.3 of [19]), while Gomi [119] proved general results on the second and third magnitude homology groups of metric spaces.

Magnitude homology is not the first homology theory of metric spaces: there is also persistent homology, fundamental in the field of topological data analysis. (For expository accounts of persistent homology, see Ghrist [115] or Carlsson [59].) Otter [269] has proved results relating the two homology theories, introducing for this purpose a notion of 'blurred magnitude homology'; see also Govc and Hepworth [125] and Cho [68].

Finally, Hepworth [143] has introduced a theory of magnitude *cohomology* for enriched categories. It carries a product that formally resembles the ordinary cup product, but is noncommutative. For finite metric spaces, magnitude cohomology is a complete invariant: the cohomology ring of such a space determines it uniquely up to isometry.

6.5 Magnitude in Geometry and Analysis

Most metric spaces of geometric interest are not finite. The general enriched-categorical concept of magnitude provides no definition of the magnitude of an infinite metric space. On the other hand, there are several plausible strategies for extending the definition of magnitude from finite to compact metric spaces. Meckes [250, 251] showed that as long as the space satisfies a certain classical condition, they all give the same outcome.

The condition is that the space must be of **negative type**. We do not need the original definition here, but Meckes refined old results of Schoenberg [307] to show that A is of negative type if and only if the matrix Z_{tB} is positive definite for every finite $B \subseteq A$ and real $t > 0$ (Theorem 3.3 of [250]). A great many spaces are of negative type, including all subspaces of $\mathbb{R}^n$ with the Euclidean or ℓ^1 (taxicab) metric, all ultrametric spaces, real and complex hyperbolic space, and spheres with the geodesic metric. A list can be found in Theorem 3.6 of [250].

The most direct way to state the extended definition of magnitude is as follows.

Definition 6.5.1 Let A be a compact metric space of negative type. The **magnitude** of A is

$$|A| = \sup\{|B| : \text{finite subsets } B \subseteq A\} \in [0, \infty].$$

Equivalently, one can choose a sequence $B_1 \subseteq B_2 \subseteq \cdots \subseteq A$ of finite subspaces B_n of A such that $\bigcup B_n$ is dense in A, then put $|A| = \lim_{n\to\infty} |B_n|$. Equivalently again, one can define the magnitude of A by the variational formula

$$|A| = \sup_{\mu} \frac{\mu(A)^2}{\int_A \int_A e^{-d(a,b)}\, d\mu(a)\, d\mu(b)}, \tag{6.21}$$

where the supremum is over the finite signed Borel measures μ on A for which the denominator is nonzero.

This last characterization is related to yet another formulation. A **weight measure** on A is a finite signed Borel measure μ such that

$$\int_A e^{-d(a,b)}\, d\mu(b) = 1$$

for all $a \in A$. This definition was introduced by Willerton ([355], Section 1.1), and is the continuous analogue of the notion of weighting (Definition 6.3.8). If μ is a weight measure on A then $|A| = \mu(A)$, by Theorem 2.3 of Meckes [250] or Proposition 5.3.6 of Leinster and Meckes [222]. However, not every compact metric space of negative type admits a weight measure. Most weightings are distributions of a more general kind, defined in Meckes [251].

The equivalence of these and other definitions of magnitude was established by Meckes [250, 251], using techniques of harmonic and functional analysis.

We now give some examples of compact spaces A and their magnitude functions $t \mapsto |tA|$.

Example 6.5.2 The magnitude function of a straight line $[0, \ell]$ of length ℓ is given by

$$|t \cdot [0, \ell]| = 1 + \tfrac{1}{2}\ell \cdot t.$$

Several proofs are known, as in Theorem 7 of Leinster and Willerton [225], Theorem 2 of Willerton [355], and Proposition 3.2.1 of Leinster [216]. The easiest proof uses weight measures. Let δ_0 and δ_ℓ denote the point measures at 0 and ℓ, let $\lambda_{[0,\ell]}$ denote Lebesgue measure on $[0, \ell]$, and put

$$\mu = \tfrac{1}{2}(\delta_0 + \lambda_{[0,\ell]} + \delta_\ell).$$

It is easily verified that μ is a weight measure on $[0, \ell]$. Hence

$$|[0, \ell]| = 1 + \tfrac{1}{2}\ell,$$

and so the magnitude function of $[0, \ell]$ is given by

$$|t \cdot [0, \ell]| = 1 + \tfrac{1}{2}\ell \cdot t.$$

Example 6.5.3 Magnitude is multiplicative with respect to the ℓ^1 product of metric spaces, that is, the product space with the metric given by adding the distances in the two factors (Proposition 3.1.4 of Leinster [216]). This has the following consequence. Equip $\mathbb{R}^n$ with the ℓ^1 metric:

$$d(\mathbf{x}, \mathbf{y}) = \sum_{i=1}^{n} |x_i - y_i|$$

($\mathbf{x}, \mathbf{y} \in \mathbb{R}^n$). Then by the previous example, the magnitude function of a rectangle

$$A = [0, \ell] \times [0, m] \subseteq \mathbb{R}^2$$

is given by

$$\begin{aligned} |tA| &= (1 + \tfrac{1}{2}\ell \cdot t)(1 + \tfrac{1}{2}m \cdot t) \\ &= 1 + \tfrac{1}{2}(\ell + m) \cdot t + \tfrac{1}{4}\ell m \cdot t^2. \end{aligned}$$

Up to a constant factor, the coefficient of t^2 is the area of A, the coefficient of t is the perimeter of A, and the constant term is the Euler characteristic of

A. Similar statements apply to higher-dimensional boxes (Corollary 3.4.3 of Leinster [216]).

For rectangles, and for nonempty convex sets in general, the Euler characteristic is always 1. As such, it may seem pretentious to call the constant term the 'Euler characteristic'. This usage will be justified shortly.

To begin to explain the geometric content of magnitude, we need to recall the concept of intrinsic volumes (Klain and Rota [190] or Section 4.1 of Schneider [306]), which with different normalizations are also known as quermassintegrals or Minkowski functionals.

Consider all reasonable ways of measuring the size of compact convex subsets of $\mathbb{R}^n$ (which in the present discussion will just be called **convex** sets). In the plane $\mathbb{R}^2$, there are at least three ways to measure a set: take its area, its perimeter, or its Euler characteristic. These are 2-, 1-, and 0-dimensional measures, respectively. The general fact is that there are $n + 1$ canonical ways of measuring convex subsets of $\mathbb{R}^n$, which define functions

$$V_0, \ldots, V_n \colon \{\text{convex subsets of } \mathbb{R}^n\} \to \mathbb{R}.$$

Here V_i is i-dimensional, in the sense that $V_i(tA) = t^i V_i(A)$, and $V_i(A)$ is called the ith **intrinsic volume** of A.

The ith intrinsic volume of a convex set $A \subseteq \mathbb{R}^n$ can be defined as follows. Choose at random an i-dimensional linear subspace L of $\mathbb{R}^n$, take the orthogonal projection $\pi_L(A)$ of A onto L, then take its i-dimensional volume $\mathrm{Vol}(\pi_L(A))$. Up to a constant factor, $V_i(A)$ is the expected value of $\mathrm{Vol}(\pi_L(A))$.

Example 6.5.4 Let A be a convex subset of $\mathbb{R}^3$. Then $V_0(A)$ is 0 if A is empty, and 1 otherwise. (In both cases, $V_0(A)$ is the Euler characteristic of A.) The first intrinsic volume $V_1(A)$ is proportional to the expected length of the projection of A onto a random line, and is called the **mean width** of A. The second intrinsic volume $V_2(A)$ is proportional to the expected area of the projection of A onto a random plane, and it is a theorem of Cauchy that this is proportional to the surface area of A (Klain and Rota [190], Theorem 5.5.2). Finally, $V_3(A)$ is just the volume of A.

Each of the intrinsic volumes V_i on convex sets is isometry-invariant, continuous with respect to the Hausdorff metric, and a **valuation**: $V_i(\varnothing) = 0$ and

$$V_i(A \cup B) = V_i(A) + V_i(B) - V_i(A \cap B)$$

whenever A, B and $A \cup B$ are convex. The same is, therefore, true of any linear combination of the intrinsic volumes. A celebrated theorem of Hadwiger [134] states that such linear combinations are the *only* isometry-invariant continuous valuations on convex sets.

The intrinsic volumes can be adapted to more general classes of space and to different geometries. For instance, we can speak of the volume or surface area of a sufficiently smooth subset of $\mathbb{R}^n$, and in that context, the intrinsic volumes are closely related to curvature measures. (See Section 2.1.1 of Alesker and Fu [8] for a concise review of the relationship, Morvan [255] or Gray [127] for full accounts of curvature measures, and Alesker [6] for a survey of some more recent developments.) The intrinsic volumes can also be defined on any finite union of convex sets (as in Klain and Rota [190]). At these levels of generality, V_0 is no longer trivial; it is the Euler characteristic. This justifies 'Euler characteristic' as the right name for V_0 even in the case of convex sets.

The next example uses a notion of intrinsic volume adapted to $\mathbb{R}^n$ with the ℓ^1 metric.

Example 6.5.5 Generalizing Example 6.5.3, let $A \subseteq \mathbb{R}^n$ be a **convex body**, that is, a convex set with nonempty interior. Give A the ℓ^1 metric. Then the magnitude function of A is the polynomial

$$|tA| = \sum_{i=0}^{n} \frac{1}{2^i} V_i'(A) \cdot t^i$$

(Theorem 5.4.6(2) of Leinster and Meckes [222]). Here $V_i'(A)$ is the ℓ^1 analogue of the ith intrinsic volume of A, discussed in [222] and in Section 5 of Leinster [213]. Explicitly, it is the sum of the i-dimensional volumes of the projections of A onto the i-dimensional coordinate subspaces of $\mathbb{R}^n$. The significance of 2^i is that the volume of the unit ball in the 1-norm on $\mathbb{R}^n$ is $2^i/i!$.

So, for convex bodies in $\mathbb{R}^n$ equipped with the ℓ^1 metric, the magnitude function is a polynomial whose degree is the dimension and whose ith coefficient is an i-dimensional geometric measure.

For the Euclidean rather than ℓ^1 metric on $\mathbb{R}^n$, results on magnitude are harder. Until 2015, the only convex subset of $\mathbb{R}^n$ whose magnitude was known was the line segment. But a significant advance was made by Barceló and Carbery [28], who used PDE methods to prove the following.

Theorem 6.5.6 (Barceló and Carbery) *Let $n \geq 1$ be odd. Then:*

i. *the magnitude function $t \mapsto |tB^n|$ of the n-dimensional unit Euclidean ball B^n is a rational function over $\mathbb{Z}$ of the radius t;*

ii. the magnitude functions of B^1, B^3 and B^5 are given by

$$|tB^1| = 1 + t,$$

$$|tB^3| = \frac{1}{3!}(6 + 12t + 6t^2 + t^3),$$

$$|tB^5| = \frac{1}{5!}\frac{360 + 1080t + 525t^2 + 135t^4 + 18t^5 + t^6}{3 + t}.$$

Proof Part (i) is Theorem 4 of [28]. For part (ii), the formulas for $|tB^3|$ and $|tB^5|$ are Theorems 2 and 3 of [28], and the formula for $|tB^1|$ is Example 6.5.2 (not due to Barceló and Carbery, but included in the statement for completeness).□

In the ℓ^1 metric on $\mathbb{R}^n$, the magnitude of a ball is a polynomial in its radius, by Example 6.5.5. In the Euclidean metric, it is no longer a polynomial, but it is the next best thing: a rational function. Subsequent work of Willerton [359, 358] identified exactly which rational function $|tB^n|$ is, in terms of Bessel polynomials and Hankel determinants.

Theorem 6.5.6 is stated under the hypothesis that n is odd, a condition imposed in order to put the proof into the realm of differential rather than pseudodifferential equations. The magnitude of even-dimensional balls remains unknown. Even the 2-dimensional disc B^2 has unknown magnitude, although numerical experiments suggest that it is a certain quadratic polynomial in the radius (Willerton [353], Section 3.2).

Barceló and Carbery also proved the following result on general compact sets (Theorem 1 of [28]).

Theorem 6.5.7 (Barceló and Carbery) *For all $n \geq 1$ and compact $A \subseteq \mathbb{R}^n$,*

$$\mathrm{Vol}(A) = c_n \lim_{t\to\infty} \frac{|tA|}{t^n},$$

where the constant c_n is $n!\,\mathrm{Vol}(B^n)$. □

The volume of the Euclidean unit ball B^n is given by a standard classical formula, as in Propositions 6.2.1 and 6.2.2 of Klain and Rota [190], for instance.

By Theorem 6.5.7, we can extract the volume of a set from its magnitude function. This substantiates the earlier claim that the general notion of the magnitude of an enriched category encompasses the notion of volume.

Better still, using methods of global analysis, Gimperlein and Goffeng proved the following result (Theorem 2(d) of [117]).

Theorem 6.5.8 (Gimperlein and Goffeng) *Let $n \geq 1$ be odd, and let $A \subseteq \mathbb{R}^n$ be a bounded set with smooth boundary such that A is the closure of its interior.*

Then the magnitude function of A has an asymptotic expansion

$$|tA| \sim \sum_{i=0}^{\infty} m_i(A)t^{n-i} \text{ as } t \to \infty,$$

and up to a known constant factor (depending on n and i but not A), the coefficient $m_i(A)$ is equal to the intrinsic volume $V_{n-i}(A)$ for $i = 0, 1, 2$. □

Recent work of Gimperlein, Goffeng and Louca, so far unpublished, removes the restriction that n is odd.

For instance, $m_0(A) = \mathrm{Vol}(A)/n!\,\mathrm{Vol}(B^n)$ (as in Theorem 6.5.7) and $m_1(A)$ is proportional to the $(n-1)$-dimensional surface area of A. In the statement of Theorem 6.5.8, the term 'intrinsic volume' has been extended beyond its usual context of convex sets. A more precise statement for $i = 2$ is that $m_2(A)$ is proportional to the integral over ∂A of the mean curvature of ∂A (which when A is convex is itself proportional to $V_{n-2}(A)$).

The magnitude of a metric space does not satisfy the inclusion-exclusion principle in the strongest conceivable sense, since otherwise, every n-point space would have magnitude n. But Gimperlein and Goffeng showed that magnitude does satisfy inclusion-exclusion in an asymptotic sense, using techniques related to the heat equation proof of the Atiyah–Singer index theorem and making essential use of *complex* scale factors t. Indeed, for subsets A, B and $A \cap B$ of $\mathbb{R}^n$ satisfying the regularity conditions of Theorem 6.5.8,

$$|t(A \cup B)| + |t(A \cap B)| - |tA| - |tB| \to 0 \text{ as } t \to \infty$$

(Remark 3 of [117]). This is further evidence for the claim that magnitude should be regarded as a measure of size.

Finally, we return to diversity. Meckes defined the **maximum diversity** of a compact space A of negative type as

$$D_{\max}(A) = \sup_{\mu} \frac{1}{\int_A \int_A e^{-d(a,b)}\, d\mu(a)\, d\mu(b)},$$

which is similar to the formula (6.21) for magnitude, except that now the supremum runs over only the Borel *probability* measures μ, as opposed to all signed measures. (In principle, the formula is for the maximum diversity of order 2, but Theorem 7.1 of Leinster and Roff [223] implies that the maximum diversity of every order is the same.) Evidently $D_{\max}(A) \leq |A|$.

When A is a subset of Euclidean space, $D_{\max}(A)$ is equal to a classical quantity, the Bessel capacity $C_{(n+1)/2}(A)$. As Meckes showed, a deep result from the theory of capacities provides an upper bound on $|A|/D_{\max}(A)$ in terms of n alone (Corollary 6.2 of [251]). Thus, magnitude is never very different from this Bessel capacity.

Meckes [251] exploited the connection between magnitude and maximum diversity to extract information about the dimension of a compact set $A \subseteq \mathbb{R}^n$ from its magnitude function. We have already met some families of spaces where the magnitude function is a polynomial whose degree is the dimension (Example 6.5.5). But here we allow non-integer dimensions too.

One of the most important notions of fractional dimension is the **Minkowski** or **box-counting dimension** (Section 3.1 of Falconer [98]). The Minkowski dimension of a subset of $\mathbb{R}^n$ is always greater than or equal to the Hausdorff dimension, and equality often holds. (See p. 43 of [98] for a summary of how the two dimensions are related.) For instance, both the Minkowski and the Hausdorff dimensions of the middle-thirds Cantor set are $\log 2/\log 3$. Write $\dim_M A$ for the Minkowski dimension of a compact set $A \subseteq \mathbb{R}^n$, if defined.

Roughly speaking, the following result states that $|tA|$ grows like $t^{\dim_M A}$ when t is large. Thus, we can can recover the Minkowski dimension of a space from its magnitude function. This result is due to Meckes (Corollary 7.4 of [251]).

Theorem 6.5.9 (Meckes) *Let A be a compact subset of* $\mathbb{R}^n$. *Then*

$$\dim_M A = \lim_{t\to\infty} \frac{\log|tA|}{\log t},$$

with one side of the equation defined if and only if the other is. □

For instance, if A is a subset of $\mathbb{R}^n$ with nonzero volume, then $|tA|$ grows like t^n when t is large, and by the volume theorem of Barceló and Carbery, the ratio $|tA|/t^n$ converges to a known constant times the volume of A. When A is the middle-thirds Cantor set, $|tA|$ grows like $t^{\log 2/\log 3}$. (In fact, the magnitude function of the Cantor set also has a kind of hidden periodicity, as shown in Section 3 of Leinster and Willerton [225].) For convex subsets of $\mathbb{R}^n$, more precise statements can be made; Meckes bounds the magnitude function of a convex set by a polynomial whose coefficients are proportional to its intrinsic volumes ([252], Theorem 1).

Theorem 6.5.9 demonstrates the usefulness of the concept of maximum diversity for pure-mathematical purposes in geometry and analysis, independently of any biological application.

7

Value

> Quite apart from many theoretical and practical problems that continue to affect the species concept and its application, is it appropriate for conservation purposes to regard all species as equal in this manner? To a conservationist, regardless of relative abundance, is *Welwitschia* equal to a species of *Taraxacum*? Is the panda equivalent to one species of rat?
>
> – Vane-Wright et al. ([342], p. 237)

Putting aside entropy and diversity, let us consider a very general question:

What is the value of the whole in terms of its parts?

Although the question in this form is far too vague to admit a mathematical treatment, we will see that once posed precisely, it has a complete answer. From that answer, the concept of diversity arises automatically. The answer also leads to a unique characterization of the Hill numbers (or equivalently, the Rényi entropies), more powerful than the characterization theorem in Section 4.5.

We will consider a 'whole' divided into n 'parts' of relative sizes $p_1, \ldots, p_n$, which are assigned values $v_1, \ldots, v_n$ respectively (Figure 7.1). The question is how to aggregate those values into a single value $\sigma(\mathbf{p}, \mathbf{v})$ for the whole, mea-

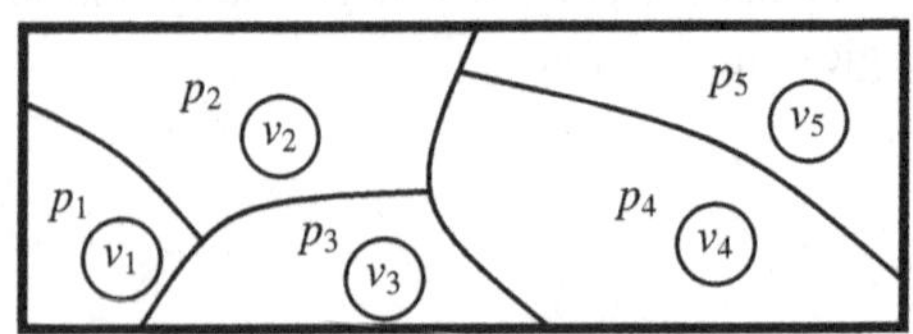

Figure 7.1 A whole divided into 5 parts, with relative sizes $(p_1, \ldots, p_5) \in \Delta_5$ and values $(v_1, \ldots, v_5)$.

sured in the same units as the values v_i of the parts. This aggregation method should have sensible properties. For instance, if we put together two parts of equal size and equal value, v, the result should have value $2v$.

One simple method is to ignore the sizes of the parts and just sum their values, so that

$$\sigma(\mathbf{p}, \mathbf{v}) = v_1 + \cdots + v_n$$

(or better, $\sigma(\mathbf{p}, \mathbf{v}) = \sum_{i \in \operatorname{supp}(\mathbf{p})} v_i$). But there are many other possibilities. In fact, we will define a one-parameter family (σ_q) of value measures. They include as special cases the Hill numbers D_q, the more general similarity-sensitive diversity measures D_q^Z of Chapter 6, certain phylogenetic diversity measures (due to Chao, Chiu and Jost [67]) and, essentially, the ℓ^p norms. For example, when a community is divided into n species in proportions $p_1, \ldots, p_n$, and each species is assigned the same value, 1, the value of the whole according to σ_q is the Hill number of order q:

$$\sigma_q(\mathbf{p}, (1, \ldots, 1)) = D_q(\mathbf{p}).$$

In most of the cases just listed, the whole is taken to be an ecological community and the parts are its species. But there is an important complementary situation, in which the whole is still a community but the parts are taken to be subcommunities. For instance, the community might be divided geographically into regions, and we might attempt to evaluate the community as a whole based on the sizes and values of those regions. In the case where value is interpreted as diversity, that is exactly what we did when we derived the chain rule for diversity (Propositions 4.4.10 and 6.2.11). Indeed, the function σ_q can be seen as an embodiment of the chain rules for D_q and D_q^Z, in a sense explained in Example 7.1.8.

We begin by defining the value measures σ_q and analysing some special cases (Section 7.1), with important examples from both ecology and the analysis of social welfare. We then introduce the Rényi relative entropies, which are very closely related to the value measures σ_q. (The q-logarithmic relative entropies were already covered in Section 4.1.) As a bonus, we use the Rényi and q-logarithmic relative entropies to provide further evidence for the canonical nature of the Fisher metric on probability distributions (Remark 7.2.3(i)).

Using our earlier results on means, we then prove that the only value measures with reasonable properties are those belonging to the family (σ_q) (Section 7.3). From this we deduce that for communities modelled as their relative abundance distributions, the only reasonable measures of diversity are the Hill numbers (Section 7.4).

We have already proved a characterization theorem for the Hill numbers D_q

in Section 4.5, showing that for a *fixed* q, if a diversity measure D has certain properties *depending on* q, then it must be equal to D_q. But in the theorem proved in this chapter, there is no 'q' mentioned in the hypotheses, and the conclusion is that D must be equal to D_q *for some* q. In short, the earlier theorem characterized the Hill numbers individually, but this theorem characterizes them as a family.

7.1 Introduction to Value

Here we consider sequences of functions

$$(\sigma\colon \Delta_n \times [0,\infty)^n \to [0,\infty))_{n\geq 1},$$

which will be referred to as **value measures**. We regard a pair $(\mathbf{p},\mathbf{v}) \in \Delta_n \times [0,\infty)^n$ as representing a whole made up of n disjoint parts with relative sizes $p_1,\ldots,p_n$ and values $v_1,\ldots,v_n$, and we regard $\sigma(\mathbf{p},\mathbf{v})$ as the value that σ assigns to the whole.

A special role is played by the family

$$(\sigma_q)_{q\in[-\infty,\infty]}$$

of value measures, defined by

$$\sigma_q(\mathbf{p},\mathbf{v}) = M_{1-q}(\mathbf{p},\mathbf{v}/\mathbf{p})$$

($n \geq 1, \mathbf{p} \in \Delta_n, \mathbf{v} \in [0,\infty)^n$). The convention adopted in Remark 4.2.15 ensures that $\sigma_q(\mathbf{p},\mathbf{v})$ is always well defined. We call σ_q the **value measure of order** q. Explicitly, when $q \neq 1, \pm\infty$,

$$\sigma_q(\mathbf{p},\mathbf{v}) = \left(\sum_{i\in\mathrm{supp}(\mathbf{p})} p_i^q v_i^{1-q}\right)^{1/(1-q)},$$

unless $q > 1$ and $v_i = 0$ for some $i \in \mathrm{supp}(\mathbf{p})$, in which case $\sigma_q(\mathbf{p},\mathbf{v}) = 0$. For $q \in \{1, \pm\infty\}$,

$$\sigma_{-\infty}(\mathbf{p},\mathbf{v}) = \max_{i\in\mathrm{supp}(\mathbf{p})} \frac{v_i}{p_i},$$

$$\sigma_1(\mathbf{p},\mathbf{v}) = \prod_{i\in\mathrm{supp}(\mathbf{p})} \left(\frac{v_i}{p_i}\right)^{p_i},$$

$$\sigma_\infty(\mathbf{p},\mathbf{v}) = \min_{i\in\mathrm{supp}(\mathbf{p})} \frac{v_i}{p_i}.$$

Examples 7.1.1 i. Consider a set of k individuals, divided into n equivalence classes ('parts'), with the ith part consisting of k_i individuals. Let $p_i = k_i/k$

be the proportion of individuals in the ith part. Let $v_1, \ldots, v_n \in [0, \infty)$ be any values assigned to the parts. Then

$$\sigma_q(\mathbf{p}, \mathbf{v}) = M_{1-q}\left(\mathbf{p}, \left(\frac{kv_1}{k_1}, \ldots, \frac{kv_n}{k_n}\right)\right),$$

or equivalently,

$$\sigma_q(\mathbf{p}, \mathbf{v}) = k \cdot M_{1-q}\left(\mathbf{p}, \left(\frac{v_1}{k_1}, \ldots, \frac{v_n}{k_n}\right)\right). \tag{7.1}$$

This can be understood as follows. If the value v_i of the ith part is shared out evenly among its k_i members, then the value per individual in the ith part is v_i/k_i. Hence the mean value per individual in the whole is

$$M_{1-q}\left(\mathbf{p}, \left(\frac{v_1}{k_1}, \ldots, \frac{v_n}{k_n}\right)\right).$$

So, equation (7.1) states that

value of whole = number of individuals × mean value per individual.

This is the basic conceptual relationship between value measures and means.

ii. If in (i) we interpret 'mean' as arithmetic mean, then we are in the case $q = 0$, and σ_0 is simply given by

$$\sigma_0(\mathbf{p}, \mathbf{v}) = \sum_{i \in \text{supp}(\mathbf{p})} v_i$$

(as in the introduction to this chapter). But we have seen repeatedly in this book that the *arithmetic* mean is not the only useful kind. The other power means should always be considered alongside it, and in this case, they give the whole family (σ_q).

Remark 7.1.2 The value measures σ_q and the power means M_t are sequences of functions of the same type:

$$\left(\sigma_q, M_t \colon \Delta_n \times [0, \infty)^n \to [0, \infty)\right)_{n \geq 1}.$$

However, Example 7.1.1(i) makes clear that there should be no overlap between the classes of value measures and means. Indeed, a reasonable value measure σ should satisfy

$$\sigma(\mathbf{u}_n, (v, \ldots, v)) = nv,$$

whereas a minimal requirement of a mean M is the consistency condition

$$M(\mathbf{u}_n, (x, \ldots, x)) = x.$$

So, no reasonable mean is a reasonable value measure. We return to the relationship between means and value measures in Section 7.3.

For positive parameters q, the value of the whole is never more than the sum of the values of its parts:

Lemma 7.1.3 *For all $q \geq 0$, $\mathbf{p} \in \Delta_n$, and $\mathbf{v} \in [0,\infty)^n$,*

$$\sigma_q(\mathbf{p},\mathbf{v}) \leq \sum_{i=1}^{n} v_i.$$

For $q > 0$, equality holds if and only if $\mathbf{v}$ is a scalar multiple of $\mathbf{p}$.

So for fixed $\sum v_i$, the value of the whole is maximized when value is spread evenly across the constituent parts, in proportion to their sizes.

Proof For all $q \geq 0$,

$$\sigma_q(\mathbf{p},\mathbf{v}) = M_{1-q}(\mathbf{p},\mathbf{v}/\mathbf{p}) \leq M_1(\mathbf{p},\mathbf{v}/\mathbf{p}) = \sum_{i\in\mathrm{supp}(\mathbf{p})} v_i \leq \sum_{i=1}^{n} v_i.$$

Assuming now that $q > 0$, equality holds in the first inequality if and only if v_i/p_i is constant over $i \in \mathrm{supp}(\mathbf{p})$ (by Theorem 4.2.8), and in the second if and only if $v_i = 0$ for all $i \notin \mathrm{supp}(\mathbf{p})$. The result follows. □

The next two examples illuminate the meaning of the parameter q. They concern the case where the parts are of equal size ($\mathbf{p} = \mathbf{u}_n$), so that the value measures σ_q are given by

$$\sigma_q(\mathbf{u}_n,\mathbf{v}) = n \cdot M_{1-q}(\mathbf{u}_n,\mathbf{v})$$

($q \in [-\infty,\infty]$, $\mathbf{v} \in [0,\infty)^n$).

Example 7.1.4 A classical question in welfare economics is how to take a group of agents, each of which has an assigned utility, and aggregate their individual utilities into a measure of the utility of the group as a whole. For instance, the agents might be the citizens of a society, and the utility of a citizen might be their individual level of welfare, wealth or well-being. The challenge, then, is to combine them into a single number representing the collective welfare of the society. (As a general reference for all of this example, we refer to Section 1.2 and Chapter 3 of Moulin [257].)

Specifically, fix n, and take a group of n individuals with respective utilities $v_1,\ldots,v_n \geq 0$. A **collective utility function** assigns a real number $f(\mathbf{v})$ to each such tuple $\mathbf{v} = (v_1,\ldots,v_n)$. For example,

$$\sigma_q(\mathbf{u}_n,-)\colon [0,\infty)^n \to \mathbb{R}$$

is a collective utility function for each $q \in [-\infty, \infty]$.

More important than the collective utility function f itself is its associated **social welfare ordering**, which is the relation $\preceq$ on $[0, \infty)^n$ defined by

$$\mathbf{v} \preceq \mathbf{v}' \iff f(\mathbf{v}) \leq f(\mathbf{v}').$$

In the case of the welfare of the citizens of a society, $\mathbf{v} \preceq \mathbf{v}'$ is interpreted as the judgement that when the welfare levels of the citizens are $v_1, \ldots, v_n$, society is in a poorer state than when they are $v'_1, \ldots, v'_n$.

Of course, such judgements depend on a choice of collective utility function f. When $f = \sigma_q(\mathbf{u}_n, -)$, different values of q correspond to different viewpoints, some of which are associated with particular schools of political philosophy. The case $q = 0$ is

$$\sigma_0(\mathbf{u}_n, -): \mathbf{v} \mapsto \sum v_i,$$

so that the collective welfare is simply the sum of the individual welfares. This function is associated with classical utilitarianism, with its roots in the philosophy of Jeremy Bentham and in John Stuart Mill's 'sum total of happiness'. When $q = \infty$, the collective utility function is

$$\sigma_\infty(\mathbf{u}_n, -): \mathbf{v} \mapsto n \min v_i,$$

so that

$$\mathbf{v} \preceq \mathbf{v}' \iff \min v_i \leq \min v'_i.$$

This viewpoint on collective welfare is associated with the philosophy of John Rawls: a society should be judged by the welfare of its most miserable citizen. An intermediate position is $q = 1$, where

$$\sigma_1(\mathbf{u}_n, -): \mathbf{v} \mapsto n \cdot \left(\prod v_i\right)^{1/n}$$

and so

$$\mathbf{v} \preceq \mathbf{v}' \iff \prod v_i \leq \prod v'_i.$$

In this context, the product operation $\mathbf{v} \mapsto \prod v_i$ is known as the **Nash collective utility function**, and has special properties not shared by any other collective utility function (unsurprisingly, given the special role played by the case $q = 1$ in the context of entropy).

An important property of collective utility functions is the Pigou–Dalton principle. In the language of wealth, this states that transferring a small amount of wealth from a richer citizen to a poorer one is beneficial to the overall welfare of society. Formally, let $\mathbf{v} \in [0, \infty)^n$ and $i, j \in \{1, \ldots, n\}$ with $v_i < v_j$, and let $0 \leq \delta \leq (v_j - v_i)/2$; define $\mathbf{v}' \in [0,\infty)^n$ by

$$v'_k = \begin{cases} v_i + \delta & \text{if } k = i, \\ v_j - \delta & \text{if } k = j, \\ v_k & \text{otherwise.} \end{cases}$$

The **Pigou–Dalton principle** is that $\mathbf{v} \leq \mathbf{v}'$ for all such $\mathbf{v}$, i, j, and δ.

When $q \in [0,\infty]$, an elementary calculation shows that $\sigma_q(\mathbf{u}_n, -)$ satisfies the Pigou–Dalton principle. Thus, redistribution is regarded positively. On the other hand, the Pigou–Dalton principle fails for all $q \in [-\infty, 0)$. In fact, for $q \in (-\infty, 0)$, redistribution from richer to poorer always strictly *decreases* overall welfare. In the extreme case $q = -\infty$, the collective utility function is

$$\sigma_{-\infty}(\mathbf{u}_n, -) \colon \mathbf{v} \mapsto n \max_i v_i,$$

so that the welfare of a society is proportional to the welfare of its most privileged citizen. (Recall that n is fixed.) Thus, from the viewpoint of $q = -\infty$, collective welfare is optimized when all the wealth is transferred to a single individual. In the welfare economics literature, negative values of q are often excluded.

The family $(\sigma_q(\mathbf{u}_n, -))$ of collective utility functions that we have used is different from the family used in economics texts such as Moulin [257], but only superficially. In the literature, it is conventional to use the functions

$$\begin{array}{lll} \mathbf{v} \mapsto \sum v_i^t & & (t \in (0,\infty)), \\ \mathbf{v} \mapsto \sum \log v_i, & & \\ \mathbf{v} \mapsto -\sum v_i^t & & (t \in (-\infty, 0)), \end{array}$$

whereas we have been using

$$\mathbf{v} \mapsto \sigma_q(\mathbf{u}_n, \mathbf{v}) = \begin{cases} n^{q/(q-1)} \left(\sum v_i^{1-q} \right)^{1/(1-q)} & \text{if } 1 \neq q \in (-\infty,\infty), \\ n \prod v_i^{1/n} & \text{if } q = 1. \end{cases} \tag{7.2}$$

But reparametrizing with $q = 1 - t$, the induced social welfare orderings are identical.

Example 7.1.5 In contexts such as collective welfare and diversity, it is natural to restrict the parameter q to be positive. But for negative parameters q, the value measures σ_q also define something important, at least when the parts are of equal size: the ℓ^p norms. Indeed, for $-\infty < q \leq 0$, equation (7.2) gives

$$\sigma_q(\mathbf{u}_n, \mathbf{v}) = n^{q/(q-1)} \|\mathbf{v}\|_{1-q},$$

where the norm $\|\cdot\|_{1-q}$ is as defined in Example 9.3.2.

We now show that all of the diversity measures discussed in previous chapters are encompassed by the value measures σ_q.

Example 7.1.6 Consider an ecological community made up of species with relative abundances $p_1, \ldots, p_n$. In the absence of other information, it is natural to give all the species the same value, 1. We have

$$\sigma_q(\mathbf{p}, (1, \ldots, 1)) = M_{1-q}(\mathbf{p}, 1/\mathbf{p}) = D_q(\mathbf{p}),$$

so the value assigned to the community by σ_q is the Hill number $D_q(\mathbf{p})$.

Example 7.1.7 Now let us enrich our model of the community with an $n \times n$ similarity matrix Z. Assume that the diagonal entries of Z are all 1 (as discussed on p. 172). Based on this model, what value v_i can we reasonably assign to each species?

In Section 6.1, we considered the quantity

$$(Z\mathbf{p})_i = \sum_{j=1}^{n} Z_{ij} p_j$$

associated with the ith species. This is the expected similarity between an individual of species i and an individual chosen from the community at random. We called $(Z\mathbf{p})_i$ the ordinariness of species i, and $1/(Z\mathbf{p})_i$ its specialness.

This might seem to suggest using $1/(Z\mathbf{p})_i$ as the value of the ith species. However, $1/(Z\mathbf{p})_i$ is a measure of the specialness of an *individual* of the ith species, whereas v_i is supposed to measure the value of the ith part (species) *as a whole*. We therefore define v_i to be the specialness per individual in the species multiplied by the size of the species:

$$v_i = \frac{p_i}{(Z\mathbf{p})_i}.$$

When Z is the naive similarity matrix I, this formula reduces to $v_i = 1$, as in Example 7.1.6. More generally, if species i is completely dissimilar to all other species ($Z_{ij} = 0$ for all $i \neq j$) then $v_i = 1$. In any case, $v_i \leq 1$, since $(Z\mathbf{p})_i \geq p_i$ (inequality (6.2), p. 174). Lower values v_i indicate that in comparison to the size of the ith species, there are many individuals belonging to species similar to it. This agrees with the intuition that such a species contributes little to the diversity of the whole.

With this definition of $\mathbf{v}$ as $\mathbf{p}/Z\mathbf{p}$, we recover the similarity-sensitive diversity measures D_q^Z of Chapter 6:

$$\sigma_q(\mathbf{p}, \mathbf{p}/Z\mathbf{p}) = M_{1-q}(\mathbf{p}, 1/Z\mathbf{p}) = D_q^Z(\mathbf{p}),$$

by definition of D_q^Z.

Example 7.1.8 Now take a community of individuals that are not only classified into a number of species (with similarities encoded in a matrix Z), but also divided into n disjoint subcommunities. Thus, each individual belongs to exactly one species and exactly one subcommunity. We will assume that the different subcommunities share no species, and that species in different subcommunities are completely dissimilar (as in Example 2.4.9 and Propositions 4.4.10 and 6.2.11, where the subcommunities were called 'islands').

Write w_i for the population size of the ith subcommunity relative to the whole community, so that $\sum w_i = 1$. Also write d_i for the diversity of order q of the ith subcommunity. Then by the chain rule for the similarity-sensitive diversities (Proposition 6.2.11), the diversity of order q of the whole community is

$$\sigma_q(\mathbf{w}, \mathbf{d}).$$

This is the fundamental relationship between value and diversity. If value is taken to mean diversity of order q, then σ_q correctly aggregates the values of the parts of a community to give the value of the whole.

Example 7.1.9 In the ecological settings discussed, we have only ever considered the *relative* abundances of species. But absolute abundances sometimes matter. What happens if we measure the value of a species within a community as its absolute abundance?

Consider a community of individuals divided into n species, with absolute abundances $A_1, \ldots, A_n$. Writing $A = \sum A_i$, the relative abundances are $p_i = A_i/A$. For all $q \in [-\infty, \infty]$,

$$\begin{aligned}
\sigma_q(\mathbf{p}, (A_1, \ldots, A_n)) &= M_{1-q}\left(\mathbf{p}, \left(\frac{A_1}{p_1}, \ldots, \frac{A_n}{p_n}\right)\right) \\
&= M_{1-q}(\mathbf{p}, (A, \ldots, A)) \\
&= A.
\end{aligned}$$

So, the value of the whole is simply the total abundance.

In this example, the value measures σ_q give us no interesting new quantity. The answer to the question posed is trivial. But it is also reasonable: if the value of each part of a community is taken to be just the number of individuals it contains, it is natural that the value of the whole community is measured in that way too.

We conclude this introduction to value with a more substantial example.

Example 7.1.10 Here we describe the phylogenetic diversity measures of

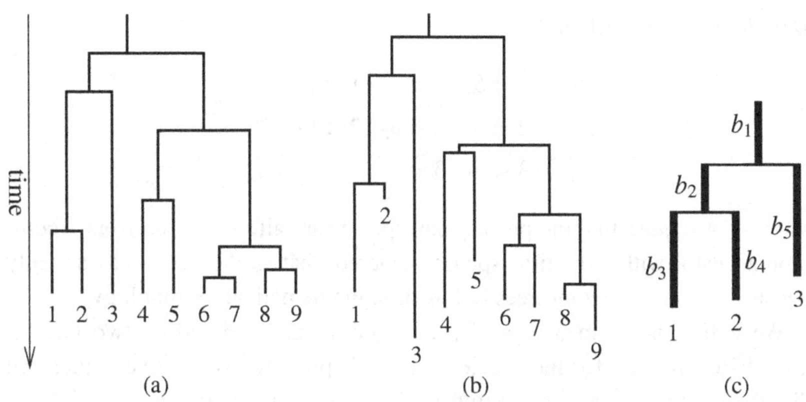

Figure 7.2 Simple examples of phylogenetic trees, with present-day species labelled as $1, 2, \ldots$ Tree (a) is ultrametric, but trees (b) and (c) are not. Tree (c) has five branches, shown as thick lines and labelled as $b_1, \ldots, b_5$.

Chao, Chiu and Jost [67] and show that they too are a special case of the value measures σ_q.

A **phylogenetic tree** is a depiction of the evolutionary history of a group of species, as in Figure 7.2. (For a general guide to the subject, see Lemey et al. [226].) The vertical axis indicates time, or some proxy for time; the horizontal distances in the trees mean nothing. Figure 7.2(a) shows nine species descended from a single species. In that example, the tree is **ultrametric**, meaning that the tips of the tree (the present-day species) are all at the same height.

Evolutionary history is often inferred from genetic data, with the number of genetic mutations used as a means of estimating time. Because the rate of genetic mutation is not constant (and for other reasons), the trees produced in this way are generally not ultrametric. Figure 7.2(b) shows an example.

From a phylogenetic tree, we can extract the following information:

- the set of present-day species, which we label as $1, \ldots, S$;
- the set B of branches;
- the binary relation $\lhd$, where for a present-day species r and a branch b, we write $r \lhd b$ if r is descended from b;
- the length $L(b) \geq 0$ of each branch b.

These four pieces of information are the only aspects of a tree that we will need for the present purposes. For instance, the tree of Figure 7.2(c) has $S = 3$,

$B = \{b_1, b_2, b_3, b_4, b_5\}$, and

$$1 \lhd b_1,\ 1 \lhd b_2,\ 1 \lhd b_3,$$
$$2 \lhd b_1,\ 2 \lhd b_2,\ 2 \lhd b_4,$$
$$3 \lhd b_1,\ 3 \lhd b_5.$$

We do not require that the present-day species are all descended from a common ancestor within the time span considered; that is, the 'tree' may actually consist of several disjoint trees (a forest, in mathematical terminology).

We will consider measures of community diversity based on two factors: a phylogenetic tree for the species, and their present-day relative abundance distribution $(\pi_1, \ldots, \pi_S)$. To do this, we introduce some notation.

For each branch b, write

$$\pi(b) = \sum_{r:\, r \lhd b} \pi_r, \tag{7.3}$$

which is the total relative abundance of present-day species descended from branch b. So if the tree is ultrametric then whenever we draw a horizontal line across the tree (representing a particular point t in evolutionary time), the sum of $\pi(b)$ over all branches b intersecting that line is 1. For any given point in evolutionary time, we therefore have a probability distribution on the set of species present then, although as Chao et al. warn, 'These abundances [$\pi(b)$] are not estimates of the actual abundances of these ancestral species at time t, but rather measures of their importance for the present-day assemblage' ([67], Section 4(a)).

For each present-day species $r \in \{1, \ldots, S\}$, write

$$L_r = \sum_{b:\, r \lhd b} L(b).$$

This is the length of the lineage of species r within the tree. For the tree to be ultrametric means that $L_1 = \cdots = L_S$. Whether or not it is ultrametric, we can define the average lineage length $\overline{L}$ by any of three equivalent formulas:

$$\overline{L} = \sum_r \pi_r L_r = \sum_{r,b:\, r \lhd b} \pi_r L(b) = \sum_b \pi(b) L(b).$$

Hence $\overline{L}$ is the expected lineage length of an individual chosen at random from the present-day community.

Chao, Chiu and Jost defined a phylogenetic diversity measure as follows. For each time point t in the period under consideration, they took the abundance distribution described below equation (7.3). They then took its Hill number of order q, and formed the average of these Hill numbers over all times t. After some simplification, the result is the diversity measure

$$\mathrm{CCJ}_q = \left(\sum_b \frac{L(b)}{\overline{L}} \pi(b)^q \right)^{1/(1-q)}$$

for $1 \neq q \in [0, \infty)$, and

$$\mathrm{CCJ}_1 = \prod_b \pi(b)^{-\frac{L(b)}{\overline{L}}\pi(b)}.$$

(Their derivation is on its surest footing when the tree is ultrametric. Discussion of what can go wrong otherwise is in the supplement to Chao et al. [67] and in Example A20 of the appendix to Leinster and Cobbold [220].) For example, the case $q = 0$ is simply

$$\mathrm{CCJ}_0 = \frac{1}{\overline{L}} \sum_b L(b).$$

Up to a factor of $\overline{L}$, this is the total length of all the branches in the tree, which is known as **Faith's phylogenetic diversity** [97].

We now show that Chao, Chiu and Jost's measure CCJ_q is a simple instance of the value measure σ_q. For this, we consider the phylogenetic tree as our whole and the branches as its parts. The value of a branch b is defined as

$$v(b) = \frac{L(b)}{\overline{L}},$$

the proportion of evolutionary time over which the branch extends. It is purely a measure of the branch's historical duration, and is independent of the abundances of present-day species. We define the relative size $p(b)$ of the branch to be

$$p(b) = \frac{\pi(b)L(b)}{\overline{L}}.$$

In other words, $p(b)$ is the product of $\pi(b)$, the proportion of present-day individuals descended from branch b, and $L(b)/\overline{L}$, the relative length of the branch. Then $\sum_b p(b) = 1$.

With these definitions, the value $\sigma_q(\mathbf{p}, \mathbf{v})$ of the community is

$$\begin{aligned}\sigma_q(\mathbf{p}, \mathbf{v}) &= \left(\sum_b p(b)^q v(b)^{1-q}\right)^{1/(1-q)} \\ &= \left(\sum_b \frac{\pi(b)^q L(b)^q}{\overline{L}^q} \frac{L(b)^{1-q}}{\overline{L}^{1-q}}\right)^{1/(1-q)} \\ &= \left(\sum_b \frac{L(b)}{\overline{L}} \pi(b)^q\right)^{1/(1-q)} \\ &= \mathrm{CCJ}_q\end{aligned}$$

$(q \neq 1, \infty)$. Similarly, $\sigma_1(\mathbf{p}, \mathbf{v}) = \mathrm{CCJ}_1$.

The community value $\sigma_q(\mathbf{p}, \mathbf{v}) = \mathrm{CCJ}_q$ is unitless, since the individual branch values $v(b) = L(b)/\overline{L}$ are unitless. We could alternatively put $v(b) = L(b)$, which might be measured in years or number of mutations. Then $\sigma_q(\mathbf{p}, \mathbf{v})$ would be measured in the same units, and $\sigma_0(\mathbf{p}, \mathbf{v})$ would be exactly Faith's phylogenetic diversity, without the factor of $1/\overline{L}$.

In summary, the value measures σ_q unify not only the Hill numbers D_q, the similarity-sensitive diversity measures D_q^Z, and the diversity of a community divided into completely dissimilar subcommunities (Example 7.1.8), but also some known measures of phylogenetic diversity.

One could also assign a value to each species in a more literal, utilitarian sense (perhaps monetary). Solow and Polasky noted that 'one justification for species conservation is that some species may provide a future medical benefit' ([319], p. 98), and analysed diversity from that viewpoint. This line of enquiry is worthwhile not only for the evident scientific reasons, but also because it is how Solow and Polasky arrived at the mathematically profound invariant now called magnitude (as related on p. 9). But we will not pursue it, instead making a connection between value measures and established quantities in information theory.

7.2 Value and Relative Entropy

The value measure σ_q is a simple transformation of a classical object of study, the Rényi relative entropy or Rényi divergence (Rényi [294], Section 3). In this short section, we describe the relationship between value, relative entropy, and some of the other quantities that we have considered. This provides useful context, although nothing here is logically necessary for anything that follows.

For $q \in [-\infty, \infty]$ and probability distributions $\mathbf{p}, \mathbf{r} \in \Delta_n$, the **Rényi entropy of order q of p relative to r** is defined as

$$H_q(\mathbf{p} \,\|\, \mathbf{r}) = \frac{1}{q-1} \log \sum_{i \in \mathrm{supp}(\mathbf{p})} p_i^q r_i^{1-q}$$

when $q \neq 1, \pm\infty$, and in the exceptional cases by

$$H_{-\infty}(\mathbf{p} \,\|\, \mathbf{r}) = \log \min_{i \in \mathrm{supp}(\mathbf{p})} \frac{p_i}{r_i},$$
$$H_1(\mathbf{p} \,\|\, \mathbf{r}) = \sum_{i \in \mathrm{supp}(\mathbf{p})} p_i \log \frac{p_i}{r_i} = H(\mathbf{p} \,\|\, \mathbf{r}),$$
$$H_{\infty}(\mathbf{p} \,\|\, \mathbf{r}) = \log \max_{i \in \mathrm{supp}(\mathbf{p})} \frac{p_i}{r_i}.$$

In all cases,

$$H_q(\mathbf{p} \,\|\, \mathbf{r}) = \log M_{q-1}(\mathbf{p}, \mathbf{p}/\mathbf{r}) = -\log M_{1-q}(\mathbf{p}, \mathbf{r}/\mathbf{p})$$

(by the duality equation (4.11), p. 105), giving

$$H_q(\mathbf{p} \,\|\, \mathbf{r}) = -\log \sigma_q(\mathbf{p}, \mathbf{r}). \tag{7.4}$$

Rényi relative entropy can take the value ∞. But as for classical relative entropy ($q = 1$, p. 65), it is convenient to restrict to pairs $(\mathbf{p}, \mathbf{r})$ such that $p_i = 0$ whenever $r_i = 0$; then $H_q(\mathbf{p} \,\|\, \mathbf{r}) < \infty$ for all q.

The Rényi relative entropies share with the classical version the basic property that $H_q(\mathbf{p} \,\|\, \mathbf{p}) = 0$ for all distributions $\mathbf{p}$. When $q > 0$, they also share its positive definiteness property, stated in the classical case as Lemma 3.1.4:

Lemma 7.2.1 *For all $q > 0$ and $\mathbf{p}, \mathbf{r} \in \Delta_n$,*

$$H_q(\mathbf{p} \,\|\, \mathbf{r}) \geq 0,$$

with equality if and only if $\mathbf{p} = \mathbf{r}$.

Proof This follows from equation (7.4) and Lemma 7.1.3, since $\sum_{i=1}^n r_i = 1$.□

In the definition above of Rényi relative entropy, both arguments were required to be probability distributions, whereas the second argument $\mathbf{v}$ of the value measure σ_q can be any vector of nonnegative reals. In fact, when Rényi introduced his relative entropies, he allowed $\mathbf{p}$ and $\mathbf{r}$ to be 'generalized probability distributions' (vectors of nonnegative reals summing to *at most* 1), and he inserted a normalizing factor of $\sum p_i$ accordingly (Section 3 of [294]). But we will consider relative entropy only for pairs of genuine probability distributions.

Just as Rényi relative entropy of order q is closely related to the value measure σ_q, so too is q-logarithmic relative entropy

$$S_q(\mathbf{p} \,\|\, \mathbf{r}) = -\sum_{i \in \text{supp}(\mathbf{p})} p_i \ln_q \frac{r_i}{p_i}$$

(Definition 4.1.7). The formula for q-logarithmic relative entropy in terms of value is the same as the formula (7.4) for Rényi relative entropy in terms of value, but with the logarithm replaced by the q-logarithm:

$$S_q(\mathbf{p} \,\|\, \mathbf{r}) = -\ln_q \sigma_q(\mathbf{p}, \mathbf{r})$$

($-\infty < q < \infty$). To prove this, we use Lemma 4.2.29:

$$\begin{aligned} S_q(\mathbf{p} \,\|\, \mathbf{r}) &= -M_1(\mathbf{p}, \ln_q(\mathbf{r}/\mathbf{p})) \\ &= -\ln_q M_{1-q}(\mathbf{p}, \mathbf{r}/\mathbf{p}) \\ &= -\ln_q \sigma_q(\mathbf{p}, \mathbf{r}). \end{aligned}$$

Hence $\sigma_q(-,-)$, $H_q(- \,\|\, -)$ and $S_q(- \,\|\, -)$ are all simple transformations of one another.

Rényi relative entropy shares with ordinary relative entropy the property that

$$H_q(\mathbf{p} \,\|\, \mathbf{u}_n) = H_q(\mathbf{u}_n) - H_q(\mathbf{p})$$

($q \in [-\infty, \infty]$, $\mathbf{p} \in \Delta_n$). In this respect, Rényi relative entropy has slightly more convenient algebraic properties than q-logarithmic relative entropy: compare the formula for $S_q(\mathbf{p} \,\|\, \mathbf{u}_n)$ in Remark 4.1.8.

Remark 7.2.2 In Remark 4.3.3, we observed that for any differentiable function $\lambda\colon (0, \infty) \to \mathbb{R}$ satisfying $\lambda(1) = 0$ and $\lambda'(1) = 1$, the formula

$$\frac{1}{1-q} \lambda\Bigg(\sum_{i \in \text{supp}(\mathbf{p})} p_i^q \Bigg)$$

defines a one-parameter family of deformations of Shannon entropy, in the sense that it converges to $H(\mathbf{p})$ as $q \to 1$. A similar statement holds for relative entropy: for any such function λ, the generalized relative entropy

$$\frac{1}{q-1} \lambda\Bigg(\sum_{i \in \text{supp}(\mathbf{p})} p_i^q r_i^{1-q} \Bigg)$$

converges to the ordinary relative entropy $H(\mathbf{p} \,\|\, \mathbf{r})$ as $q \to 1$. Taking $\lambda = \log$ gives Rényi relative entropy, and taking $\lambda(x) = x - 1$ gives q-logarithmic relative entropy.

Remarks 7.2.3 Here we relate the deformed relative entropies to the Fisher metric on probability distributions.

i. In Section 3.4, we showed that although the square root of ordinary relative entropy is not a distance function on the open simplex Δ_n° (that is, not a metric in the sense of metric spaces), it is an *infinitesimal* metric in the Riemannian sense. As we saw, it is proportional to the Fisher metric, which itself is proportional to the standard Riemannian metric on the positive orthant of the unit sphere, transferred to Δ_n° via the bijection $\mathbf{p} \leftrightarrow \sqrt{\mathbf{p}}$.

   It is natural to ask what happens if we apply the same procedure to the Rényi relative entropy of order q, or the q-logarithmic relative entropy, for some $q \neq 1$. Do we obtain some new, deformed, Fisher-like metric on Δ_n°?

   The answer turns out to be no. Using $H_q(- \| -)$ or $S_q(- \| -)$ instead of the ordinary relative entropy $H(- \| -)$ simply multiplies the induced metric on Δ_n° by a constant factor of q. More generally, the same is true of any family of deformations of relative entropy of the type constructed in Remark 7.2.2. (We omit the proof, but it is similar to the argument for ordinary relative entropy; compare also Section 2.7 of Ay, Jost, Lê and Schwachhöfer [22] and Chapter 3 of Amari [12].) It follows that the q-analogues of Fisher distance and Fisher information (defined as in equation (3.17)) are proportional to the classical Fisher distance and information, and that the q-analogue of the Jeffreys prior is exactly equal to the classical notion.

   The moral is that the Fisher metric on probability distributions is a very stable, canonical concept. However we may choose to deform relative entropy, the induced metric is always essentially the same.

ii. The parameter value $q = 1/2$ plays a special role. The Rényi and q-logarithmic relative entropies of order $1/2$ are

$$H_{1/2}(\mathbf{p} \,\|\, \mathbf{r}) = -2 \log \sum \sqrt{p_i r_i}, \qquad S_{1/2}(\mathbf{p} \,\|\, \mathbf{r}) = 2\left(1 - \sum \sqrt{p_i r_i}\right).$$

   Both are symmetric in $\mathbf{p}$ and $\mathbf{r}$ (and $q = 1/2$ is the only parameter value with this property). In fact, both are increasing, invertible transformations of the Fisher distance

$$d_{\mathrm{F}}(\mathbf{p}, \mathbf{r}) = 2 \cos^{-1}\left(\sum \sqrt{p_i r_i}\right).$$

   Thus, the Rényi relative entropy of order $1/2$ of a pair of distributions determines the Fisher distance between them. Similarly, knowing either the $(1/2)$-logarithmic entropy of $(\mathbf{p}, \mathbf{r})$ or the value of order $1/2$,

$$\sigma_{1/2}(\mathbf{p}, \mathbf{r}) = \left(\sum \sqrt{p_i r_i}\right)^2,$$

   determines the Fisher distance between $\mathbf{p}$ and $\mathbf{r}$.

7.3 Characterization of Value

Here we show that the only value measures with reasonable properties are those of the form σ_q for some $q \in [-\infty, \infty]$.

We defined the value measure σ_q on the nonnegative half-line $[0, \infty)$, but it restricts to a sequence of functions

$$(\sigma_q\colon \Delta_n \times (0, \infty)^n \to (0, \infty))_{n\geq 1}$$

on the strictly positive reals. It is this family $(\sigma_q)_{q\in[-\infty,\infty]}$ that we will characterize. A similar theorem on $[0, \infty)$ can be proved, at the cost of an extra hypothesis (Remark 7.3.5), but we will focus on strictly positive values. Thus, we will identify a list of conditions on a sequence of functions

$$(\sigma\colon \Delta_n \times (0, \infty)^n \to (0, \infty))_{n\geq 1} \tag{7.5}$$

that are satisfied by σ_q for each $q \in [-\infty, \infty]$, but not by any other σ.

We begin by describing those conditions.

Recall that a weighted mean M on $(0, \infty)$ is a sequence of functions of the same type as a value measure on $(0, \infty)$:

$$(M\colon \Delta_n \times (0, \infty)^n \to (0, \infty))_{n\geq 1}.$$

Although the classes of reasonable means and reasonable value measures are intended to be disjoint (Remark 7.1.2), some of the properties that one expects of a mean can also be expected of a value measure. We therefore reuse some of the terminology defined previously for weighted means, and summarized in Appendix B.

In what follows, let σ denote a sequence of functions as in (7.5). Then σ may or may not have the following properties, all defined previously in the context of weighted means.

Symmetry For σ to be symmetric (Definition 4.2.10(i)) means that the value of the whole is independent of the order in which the parts are listed.

Absence-invariance For σ to be absence-invariant (Definition 4.2.10(ii)) means that a part that is absent ($p_i = 0$) makes no contribution to the value of the whole, and might as well be ignored.

Increasing For σ to be increasing (Definition 4.2.18) means that the parts make a positive (or at least, nonnegative) contribution to the whole: if the value of one part increases and the rest stay the same, this does not cause the value of the whole to become smaller.

Homogeneity Homogeneity of σ (Definition 4.2.21) means that the value of the whole and the values of the parts are measured in the same units. For instance, if the value of each part is measured in kilograms then

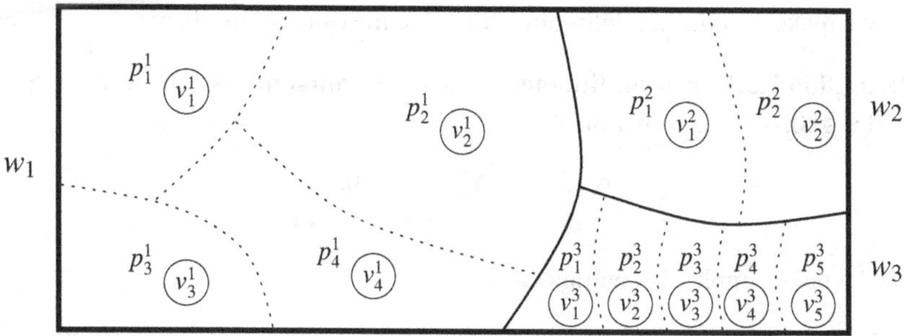

Figure 7.3 The chain rule for value measures, as in equation (7.6). Here, the whole is divided into $n = 3$ parts, the first part is divided into $k_1 = 4$ subparts, the second into $k_2 = 2$ subparts, and the third into $k_3 = 5$ subparts.

so is the value of the whole. Converting to grams multiplies both by 1000.

Chain rule The chain rule for σ (Definition 4.2.23) is the most complicated of the properties that we will need, but it is logically fundamental. It states that

$$\sigma(\mathbf{w} \circ (\mathbf{p}^1, \dots, \mathbf{p}^n), \mathbf{v}^1 \oplus \cdots \oplus \mathbf{v}^n) \\ = \sigma\big(\mathbf{w}, (\sigma(\mathbf{p}^1, \mathbf{v}^1), \dots, \sigma(\mathbf{p}^n, \mathbf{v}^n))\big) \tag{7.6}$$

for all $n, k_1, \dots, k_n \geq 1$, $\mathbf{w} \in \Delta_n$, $\mathbf{p}^i \in \Delta_{k_i}$, and $\mathbf{v}^i \in (0, \infty)^{k_i}$.

This is a recursivity property (Figure 7.3). It means that our method σ of aggregating value behaves consistently when the whole is divided into parts which are further divided into subparts.

Suppose, for example, that we are performing some evaluation of the whole planetary landmass, and that we have already assigned a value to each country. We could first use σ to compute the value of each continent, then use σ again on those continental values to compute the global value. This is the right-hand side of equation (7.6), if v_j^i denotes the value of the jth country on the ith continent, $\mathbf{p}^i$ is the relative size distribution of the countries on the ith continent, and $\mathbf{w}$ is the relative size distribution of the continents. Alternatively, we could ignore the intermediate level of continents and use σ to compute the global value directly from the country values. This gives the left-hand side of equation (7.6). The two methods for computing the global value should give the same result, and the chain rule states that they do.

We make two further definitions for value measures σ on $(0, \infty)$.

Definition 7.3.1 σ is **continuous in positive probabilities** if for each $n \geq 1$ and $\mathbf{v} \in (0, \infty)^n$, the function

$$\begin{array}{rccc} \sigma(-, \mathbf{v}): & \Delta_n^\circ & \to & (0, \infty) \\ & \mathbf{p} & \mapsto & \sigma(\mathbf{p}, \mathbf{v}) \end{array}$$

on the open simplex is continuous.

This condition only involves continuity in the *sizes* of the parts present, not their values. The restriction to the interior of the simplex means that we do not forbid value measures that make a sharp distinction between presence and absence.

Definition 7.3.2 σ is an **effective number** if

$$\sigma(\mathbf{u}_n, (1, \dots, 1)) = n$$

for all $n \geq 1$.

Assuming homogeneity, the effective number property is equivalent to

$$\sigma(\mathbf{u}_n, (v, \dots, v)) = nv \tag{7.7}$$

for all $n \geq 1$ and $v \in (0, \infty)$. That is, if we put together n parts of equal size and equal value, v, the result has value nv.

Remark 7.3.3 Let σ be an absence-invariant value measure. Then $\sigma(\mathbf{p}, \mathbf{v})$ is independent of v_i for $i \notin \text{supp}(\mathbf{p})$. Indeed, writing $\text{supp}(\mathbf{p}) = \{i_1, \dots, i_k\}$ with $i_1 < \cdots < i_k$, absence-invariance implies that

$$\sigma(\mathbf{p}, \mathbf{v}) = \sigma((p_{i_1}, \dots, p_{i_k}), (v_{i_1}, \dots, v_{i_k})). \tag{7.8}$$

So we can consistently extend the definition of $\sigma(\mathbf{p}, \mathbf{v})$ to pairs $(\mathbf{p}, \mathbf{v})$ where v_i need not be within the permissible range $(0, \infty)$, or even defined at all, when $i \notin \text{supp}(\mathbf{p})$. In that case, we define $\sigma(\mathbf{p}, \mathbf{v})$ to be the right-hand side of equation (7.8). This convention is exactly analogous to the convention for means introduced in Remark 4.2.15, and to the usual convention for integrals of functions undefined on a set of measure zero.

We now prove that the properties listed above uniquely characterize the family of value measures (σ_q).

Theorem 7.3.4 *Let* $(\sigma\colon \Delta_n \times (0, \infty)^n \to (0, \infty))_{n \geq 1}$ *be a sequence of functions. The following are equivalent:*

i. σ is symmetric, absence-invariant, increasing, homogeneous, continuous in positive probabilities and an effective number, and satisfies the chain rule;
ii. $\sigma = \sigma_q$ for some $q \in [-\infty, \infty]$.

Proof To prove that (ii) implies (i), let $q \in [-\infty, \infty]$. That σ_q is symmetric, absence-invariant, increasing, homogeneous, and continuous in positive probabilities follows from the definition

$$\sigma_q(\mathbf{p}, \mathbf{v}) = M_{1-q}(\mathbf{p}, \mathbf{v}/\mathbf{p})$$

of σ_q and the corresponding properties of M_{1-q} (Lemmas 4.2.11, 4.2.19, 4.2.22 and 4.2.6(i)). That σ_q is an effective number follows from the consistency of M_{1-q}, and the chain rule for σ_q follows from the chain rule for M_{1-q} (Proposition 4.2.24).

Conversely, assume that σ satisfies the conditions in (i). Define a sequence of functions

$$\left(M\colon \Delta_n \times (0,\infty)^n \to (0,\infty)\right)_{n\geq 1}$$

by

$$M(\mathbf{p}, \mathbf{x}) = \sigma(\mathbf{p}, \mathbf{px})$$

($\mathbf{p} \in \Delta_n$, $\mathbf{x} \in (0,\infty)^n$). Although it may be that $(\mathbf{px})_i = 0$ for some i, in which case $\sigma(\mathbf{p}, \mathbf{px})$ is strictly speaking undefined, this can only happen when $p_i = 0$; hence $\sigma(\mathbf{p}, \mathbf{px})$ can be interpreted according to the convention of Remark 7.3.3.

We will prove that M is a power mean. We do this by showing that M satisfies the hypotheses of Theorem 5.5.10: M is symmetric, absence-invariant, increasing, homogeneous, modular, and consistent. The first four follow from the corresponding properties of σ. It remains to prove that M is modular and consistent.

For modularity, let $\mathbf{w} \in \Delta_n$, $\mathbf{p}^i \in \Delta_{k_i}$, and $\mathbf{x}^i \in (0,\infty)^{k_i}$. Using the chain rule and homogeneity properties of σ, we find that

$$\begin{aligned}
&M(\mathbf{w} \circ (\mathbf{p}^1, \dots, \mathbf{p}^n), \mathbf{x}^1 \oplus \cdots \oplus \mathbf{x}^n) \\
&= \sigma(\mathbf{w} \circ (\mathbf{p}^1, \dots, \mathbf{p}^n), w_1\mathbf{p}^1\mathbf{x}^1 \oplus \cdots \oplus w_n\mathbf{p}^n\mathbf{x}^n) \\
&= \sigma\big(\mathbf{w}, (\sigma(\mathbf{p}^1, w_1\mathbf{p}^1\mathbf{x}^1), \dots, \sigma(\mathbf{p}^n, w_n\mathbf{p}^n\mathbf{x}^n))\big) \\
&= \sigma\big(\mathbf{w}, (w_1 M(\mathbf{p}^1, \mathbf{x}^1), \dots, w_n M(\mathbf{p}^n, \mathbf{x}^n))\big) \\
&= M\big(\mathbf{w}, (M(\mathbf{p}^1, \mathbf{x}^1), \dots, M(\mathbf{p}^n, \mathbf{x}^n))\big).
\end{aligned}$$

Hence M satisfies the chain rule, and is therefore modular.

Proving that M is consistent is equivalent, by homogeneity, to showing that

$$\sigma(\mathbf{p}, \mathbf{p}) = 1$$

for all $n \geq 1$ and $\mathbf{p} \in \Delta_n$. We do this in three steps.

First suppose that the coordinates of $\mathbf{p}$ are positive and rational, so that $\mathbf{p} = (k_1/k, \ldots, k_n/k)$ for some positive integers k_i summing to k. Then

$$\mathbf{u}_k = \mathbf{p} \circ (\mathbf{u}_{k_1}, \ldots, \mathbf{u}_{k_n}),$$

so by the chain rule for σ,

$$\sigma(\mathbf{u}_k, k * 1) = \sigma\Big(\mathbf{p}, (\sigma(\mathbf{u}_{k_1}, k_1 * 1), \ldots, \sigma(\mathbf{u}_{k_n}, k_n * 1))\Big).$$

By the effective number property of σ, this means that

$$k = \sigma(\mathbf{p}, (k_1, \ldots, k_n)).$$

Dividing through by k and using the homogeneity of σ gives $1 = \sigma(\mathbf{p}, \mathbf{p})$, as required.

For the second step, let $\mathbf{p}$ be any point in Δ_n°. Let $\varepsilon > 0$. Since σ is continuous in positive probabilities, there is some $\delta > 0$ such that for $\mathbf{r} \in \Delta_n^\circ$,

$$\|\mathbf{p} - \mathbf{r}\| < \delta \implies |\sigma(\mathbf{p}, \mathbf{p}) - \sigma(\mathbf{r}, \mathbf{p})| < \varepsilon/2, \tag{7.9}$$

where $\|\cdot\|$ denotes Euclidean length. We can choose $\mathbf{r} \in \Delta_n^\circ$ with rational coordinates such that

$$\|\mathbf{p} - \mathbf{r}\| < \delta, \quad \max_i \frac{p_i}{r_i} \leq 1 + \frac{\varepsilon}{2}, \quad \min_i \frac{p_i}{r_i} \geq 1 - \frac{\varepsilon}{2}.$$

Since σ is increasing and homogeneous,

$$\sigma(\mathbf{r}, \mathbf{p}) \leq \sigma\left(\mathbf{r}, \left(\max_i \frac{p_i}{r_i}\right)\mathbf{r}\right) = \left(\max_i \frac{p_i}{r_i}\right)\sigma(\mathbf{r}, \mathbf{r}),$$

which by the first step gives

$$\sigma(\mathbf{r}, \mathbf{p}) \leq \max_i \frac{p_i}{r_i} \leq 1 + \frac{\varepsilon}{2}.$$

Similarly, $\sigma(\mathbf{r}, \mathbf{p}) \geq 1 - \varepsilon/2$, so

$$|\sigma(\mathbf{r}, \mathbf{p}) - 1| \leq \varepsilon/2.$$

This, together with (7.9) and the triangle inequality, implies that $|\sigma(\mathbf{p}, \mathbf{p}) - 1| < \varepsilon$. But ε was arbitrary, so $\sigma(\mathbf{p}, \mathbf{p}) = 1$.

Third and finally, take any $\mathbf{p} \in \Delta_n$. Write $\operatorname{supp}(\mathbf{p}) = \{i_1, \ldots, i_k\}$ with $i_1 < \cdots < i_k$, and write $\mathbf{r} = (p_{i_1}, \ldots, p_{i_k}) \in \Delta_k^\circ$. Then

$$\sigma(\mathbf{p}, \mathbf{p}) = \sigma(\mathbf{r}, \mathbf{r}) = 1,$$

where the first equality holds for the reasons given in Remark 7.3.3, and the second follows from the second step above.

This completes the proof that M is consistent. We have now shown that M

satisfies the hypotheses of Theorem 5.5.10. By that theorem, $M = M_{1-q}$ for some $q \in [-\infty, \infty]$. It follows that

$$\sigma(\mathbf{p}, \mathbf{v}) = M_{1-q}(\mathbf{p}, \mathbf{v}/\mathbf{p}) = \sigma_q(\mathbf{p}, \mathbf{v})$$

for all $n \geq 1$, $\mathbf{p} \in \Delta_n$, and $\mathbf{v} \in (0, \infty)^n$. □

Remark 7.3.5 A similar characterization theorem can be proved for values in $[0, \infty)$ instead of $(0, \infty)$, using Theorem 5.5.11 on means on $[0, \infty)$. In this case, we have to strengthen the continuity requirement, also asking that $\sigma(\mathbf{p}, \mathbf{v})$ is continuous in $\mathbf{v}$ for each fixed $\mathbf{p}$.

Theorem 7.3.4 can be translated into a characterization theorem for either the Rényi relative entropies or the q-logarithmic relative entropies, using the observations in Section 7.2. This translation exercise is left to the reader.

7.4 Total Characterization of the Hill Numbers

The axiomatic approach to diversity measurement is to specify mathematical properties that we want the concept of diversity to possess, then to prove a theorem classifying all the diversity measures with the specified properties.

Here we do this for the simple but very commonplace model of a community as its relative abundance distribution $\mathbf{p} = (p_1, \ldots, p_n)$. We prove that any measure $\mathbf{p} \mapsto D(\mathbf{p})$ satisfying a handful of intuitive properties must be one of the Hill numbers D_q. To do this, we use the characterization theorem for value measures (Theorem 7.3.4). The strategy is to construct from our hypothetical diversity measure D a value measure σ, apply Theorem 7.3.4 to show that $\sigma = \sigma_q$ for some q, and deduce from this that $D = D_q$.

This is the second characterization theorem for the Hill numbers that we have proved, and it is more powerful than the first (Theorem 4.5.1), in the sense that the hypotheses are simpler and have more direct ecological explanations. Another difference is that the previous theorem fixed a parameter value q, whereas the one below characterizes D_q for all q simultaneously. Further discussion of the differences can be found at the end of the introduction to this chapter.

Consider, then, a sequence of functions

$$\left(D \colon \Delta_n \to (0, \infty)\right)_{n \geq 1},$$

intended to measure the diversity $D(\mathbf{p})$ of any community of n species with relative abundances $\mathbf{p} = (p_1, \ldots, p_n)$. What properties would we expect D to possess?

We already discussed some desirable properties in Section 4.4, arguing that any reasonable diversity measure D ought to be symmetric, absence-invariant, and continuous in positive probabilities, and that it should obey the replication principle. To fix the scale on which we are working, we also ask that a community consisting of only one species has diversity 1. Formally, D is **normalized** if $D(\mathbf{u}_1) = 1$.

We impose one further condition on our hypothetical diversity measure. Consider a pair of islands, perhaps with different population sizes, with no species in common. Replace the population of the first island by a population of the same abundance but greater or equal diversity, still sharing no species with the second island. Then the diversity of the two-island community should be greater than or equal to what it was originally.

More generally, consider a group of several islands, perhaps with different population sizes, with no species shared between islands. Replace the population of each island by a population of the same abundance but greater or equal diversity, and still with no shared species between islands. Then the diversity of the whole island group should be greater than or equal to what it was originally. Although this condition is superficially stronger than the special case described in the previous paragraph, it is equivalent by induction. We formalize it as follows.

Definition 7.4.1 A sequence of functions $(D\colon \Delta_n \to (0,\infty))_{n\geq 1}$ is **modular-monotone** if

$$\begin{aligned} &D(\mathbf{p}^i) \leq D(\widetilde{\mathbf{p}}^i) \text{ for all } i \in \{1,\ldots,n\} \\ \implies &D(\mathbf{w}\circ(\mathbf{p}^1,\ldots,\mathbf{p}^n)) \leq D(\mathbf{w}\circ(\widetilde{\mathbf{p}}^1,\ldots,\widetilde{\mathbf{p}}^n)) \end{aligned}$$

for all $n, k_i, \widetilde{k}_i \geq 1$ and $\mathbf{w} \in \Delta_n$, $\mathbf{p}^i \in \Delta_{k_i}$ and $\widetilde{\mathbf{p}}^i \in \Delta_{\widetilde{k}_i}$.

For comparison, recall that by definition, D is modular if and only if

$$\begin{aligned} &D(\mathbf{p}^i) = D(\widetilde{\mathbf{p}}^i) \text{ for all } i \in \{1,\ldots,n\} \\ \implies &D(\mathbf{w}\circ(\mathbf{p}^1,\ldots,\mathbf{p}^n)) = D(\mathbf{w}\circ(\widetilde{\mathbf{p}}^1,\ldots,\widetilde{\mathbf{p}}^n)) \end{aligned}$$

(Definition 4.4.14). Modular-monotonicity implies modularity (Lemma 7.4.4), and like modularity, it is a basic requirement for a diversity measure.

Example 7.4.2 Let $q \in [-\infty,\infty]$. The Hill number D_q is modular-monotone, since by the chain rule for D_q (Proposition 4.4.10),

$$D_q(\mathbf{w}\circ(\mathbf{p}^1,\ldots,\mathbf{p}^n)) = M_{1-q}\big(\mathbf{w}, (D_q(\mathbf{p}^1)/w_1,\ldots,D_q(\mathbf{p}^n)/w_n)\big),$$

and the power mean M_{1-q} is increasing.

We will prove:

Theorem 7.4.3 *Let* $(D\colon \Delta_n \to (0,\infty))_{n\geq 1}$ *be a sequence of functions. The following are equivalent:*

i. D is symmetric, absence-invariant, continuous in positive probabilities, normalized and modular-monotone, and satisfies the replication principle;
ii. $D = D_q$ *for some* $q \in [-\infty, \infty]$.

The rest of this section is devoted to the proof, and to a refinement of the theorem that excludes negative values of q. We have already shown that (ii) implies (i), so it remains to prove the converse.

For the rest of this section, let

$$(D\colon \Delta_n \to (0,\infty))_{n\geq 1}$$

be a sequence of functions satisfying the six conditions in Theorem 7.4.3(i).

We begin our proof by proving that the assumed properties of D imply some of the other desirable properties discussed in Section 4.4.

Lemma 7.4.4 *D is an effective number and modular.*

Proof For the effective number property, we have

$$D(\mathbf{u}_n) = D(\mathbf{u}_n \otimes \mathbf{u}_1) = nD(\mathbf{u}_1) = n$$

for each $n \geq 1$, by replication and normalization.

Modularity follows from modular-monotonicity, since in the notation of Definition 4.4.14, if $D(\mathbf{p}^i) = D(\widetilde{\mathbf{p}}^i)$ then $D(\mathbf{p}^i) \leq D(\widetilde{\mathbf{p}}^i) \leq D(\mathbf{p}^i)$. □

The next few results establish that D is multiplicative. This is harder. First we prove the weaker statement that $D(\mathbf{p} \otimes \mathbf{r})$ depends only on $D(\mathbf{p})$ and $D(\mathbf{r})$.

Lemma 7.4.5 *Let* $\mathbf{p} \in \Delta_m$, $\mathbf{p}' \in \Delta_{m'}$, $\mathbf{r} \in \Delta_n$, *and* $\mathbf{r}' \in \Delta_{n'}$. *Then*

$$D(\mathbf{p}) = D(\mathbf{p}'),\ D(\mathbf{r}) = D(\mathbf{r}') \implies D(\mathbf{p} \otimes \mathbf{r}) = D(\mathbf{p}' \otimes \mathbf{r}').$$

Proof Suppose that $D(\mathbf{p}) = D(\mathbf{p}')$ and $D(\mathbf{r}) = D(\mathbf{r}')$. By definition of $\otimes$ and modularity,

$$\begin{aligned} D(\mathbf{p} \otimes \mathbf{r}) &= D(\mathbf{p} \circ (\mathbf{r}, \ldots, \mathbf{r})) \\ &= D(\mathbf{p} \circ (\mathbf{r}', \ldots, \mathbf{r}')) \\ &= D(\mathbf{p} \otimes \mathbf{r}'). \end{aligned}$$

By symmetry of D, the order of the factors in the tensor product is irrelevant, so $D(\mathbf{p} \otimes \mathbf{r}') = D(\mathbf{p}' \otimes \mathbf{r}')$ by the same argument. The result follows. □

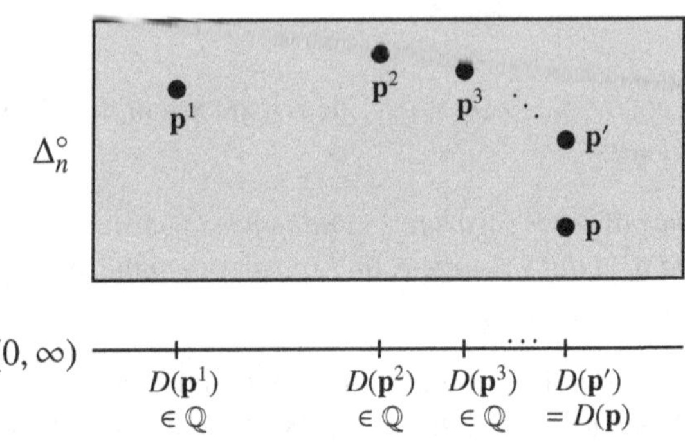

Figure 7.4 Schematic illustration of Lemma 7.4.6.

As the next step in showing that D is multiplicative, we prove a technical lemma (Figure 7.4).

Lemma 7.4.6 *Let $n \geq 1$ and $\mathbf{p} \in \Delta_n^\circ$. Then there exists a sequence $(\mathbf{p}^j)_{j=1}^\infty$ in Δ_n° converging to a point $\mathbf{p}' \in \Delta_n^\circ$, such that $D(\mathbf{p}^j)$ is rational for all j and $D(\mathbf{p}') = D(\mathbf{p})$.*

Proof We can choose a continuous map $\gamma\colon [0, 1] \to \Delta_n^\circ$ such that $\gamma(0) = \mathbf{u}_n$ and $\gamma(1) = \mathbf{p}$. (For example, take $\gamma(t) = (1 - t)\mathbf{u}_n + t\mathbf{p}$.) By continuity in positive probabilities, $D\gamma[0, 1]$ is connected and is therefore a subinterval of $(0, \infty)$. It contains $D(\gamma(0))$, which by the effective number property is n, and also contains $D(\gamma(1)) = D(\mathbf{p})$. Hence $D\gamma[0, 1]$ contains all real numbers between n and $D(\mathbf{p})$. Either $D(\mathbf{p}) = n$ or $D(\mathbf{p}) \neq n$, and in either case, there is some sequence $(d_j)_{j=1}^\infty$ of rational numbers in $D\gamma[0, 1]$ that converges to $D(\mathbf{p})$ and is either increasing or decreasing. (In the case $D(\mathbf{p}) = n$, we can simply take $d_j = n$ for all j.)

Since $d_1 \in D\gamma[0, 1]$, we can choose $t_1 \in [0, 1]$ such that $D(\gamma(t_1)) = d_1$. Then by continuity in positive probabilities, $D\gamma[t_1, 1]$ is an interval containing d_1 and $D(\gamma(1)) = D(\mathbf{p})$. But (d_j) is an increasing or decreasing sequence converging to $D(\mathbf{p})$, so the interval $D\gamma[t_1, 1]$ also contains d_2. Hence we can choose $t_2 \in [t_1, 1]$ such that $D(\gamma(t_2)) = d_2$. Continuing in this way, we obtain an increasing sequence $(t_j)_{j=1}^\infty$ in $[0, 1]$ with $D(\gamma(t_j)) = d_j$ for all $j \geq 1$.

Put $\mathbf{p}^j = \gamma(t_j) \in \Delta_n^\circ$ for each $j \geq 1$. Then $D(\mathbf{p}^j) = d_j \in \mathbb{Q}$ for all j. Also put $t = \sup_j t_j \in [0, 1]$ and $\mathbf{p}' = \gamma(t) \in \Delta_n^\circ$. Then $t_j \to t$ as $j \to \infty$, so

$$\mathbf{p}^j = \gamma(t_j) \to \gamma(t) = \mathbf{p}'$$

as $j \to \infty$. Since D is continuous in positive probabilities, this implies that

$D(\mathbf{p}^j) \to D(\mathbf{p}')$ as $j \to \infty$. But also $D(\mathbf{p}^j) = d_j \to D(\mathbf{p})$ as $j \to \infty$, by definition of the sequence (d_j). Hence $D(\mathbf{p}') = D(\mathbf{p})$, as required. □

Lemma 7.4.7 *D is multiplicative.*

Proof Let $\mathbf{p} \in \Delta_m$ and $\mathbf{r} \in \Delta_n$. We have to show that $D(\mathbf{p} \otimes \mathbf{r}) = D(\mathbf{p})D(\mathbf{r})$.

First suppose that $D(\mathbf{p})$ is rational, say $D(\mathbf{p}) = a/b$ for positive integers a and b. Since D is an effective number, $bD(\mathbf{p}) = D(\mathbf{u}_a)$. Hence by replication,

$$D(\mathbf{u}_b \otimes \mathbf{p}) = D(\mathbf{u}_a). \tag{7.10}$$

Now

$$bD(\mathbf{p} \otimes \mathbf{r}) = D(\mathbf{u}_b \otimes \mathbf{p} \otimes \mathbf{r}) \tag{7.11}$$
$$= D(\mathbf{u}_a \otimes \mathbf{r}) \tag{7.12}$$
$$= aD(\mathbf{r}), \tag{7.13}$$

where (7.11) and (7.13) follow from the replication principle for D, and (7.12) follows from (7.10) and Lemma 7.4.5. Hence

$$D(\mathbf{p} \otimes \mathbf{r}) = (a/b)D(\mathbf{r}) = D(\mathbf{p})D(\mathbf{r}),$$

as required.

Next we prove that $D(\mathbf{p} \otimes \mathbf{r}) = D(\mathbf{p})D(\mathbf{r})$ in the case that $\mathbf{p} \in \Delta_m^\circ$ and $\mathbf{r} \in \Delta_n^\circ$. Choose a sequence $(\mathbf{p}^j)$ in Δ_m° converging to $\mathbf{p}' \in \Delta_m^\circ$ as in Lemma 7.4.6. By the previous paragraph,

$$D(\mathbf{p}^j \otimes \mathbf{r}) = D(\mathbf{p}^j)D(\mathbf{r}) \tag{7.14}$$

for all $j \geq 1$. Now $\mathbf{p}^j \otimes \mathbf{r} \in \Delta_{mn}^\circ$ for all j, and $\mathbf{p}^j \otimes \mathbf{r} \to \mathbf{p}' \otimes \mathbf{r}$ as $j \to \infty$. Hence, taking the limit as $j \to \infty$ in equation (7.14) and using continuity in positive probabilities,

$$D(\mathbf{p}' \otimes \mathbf{r}) = D(\mathbf{p}')D(\mathbf{r}).$$

But $D(\mathbf{p}') = D(\mathbf{p})$, so by Lemma 7.4.5,

$$D(\mathbf{p} \otimes \mathbf{r}) = D(\mathbf{p})D(\mathbf{r}),$$

as required.

Finally, we prove multiplicativity for an arbitrary $\mathbf{p} \in \Delta_m$ and $\mathbf{r} \in \Delta_n$. By symmetry, we may suppose that $\mathbf{p} = (p_1, \ldots, p_{m'}, 0, \ldots, 0)$ with $p_1, \ldots, p_{m'} > 0$. Write $\mathbf{p}' = (p_1, \ldots, p_{m'}) \in \Delta_{m'}$, and similarly $\mathbf{r}' \in \Delta_{n'}$. By the previous paragraph, $D(\mathbf{p}' \otimes \mathbf{r}') = D(\mathbf{p}')D(\mathbf{r}')$. On the other hand, by absence-invariance, $D(\mathbf{p}') = D(\mathbf{p})$ and $D(\mathbf{r}') = D(\mathbf{r})$. Hence by Lemma 7.4.5, $D(\mathbf{p} \otimes \mathbf{r}) = D(\mathbf{p})D(\mathbf{r})$, completing the proof. □

The plan for the rest of the proof of Theorem 7.4.3 is as follows. We wish to show that $D = D_q$ for some q. We know that the Hill number D_q satisfies the chain rule

$$D_q(\mathbf{w} \circ (\mathbf{p}^1, \ldots, \mathbf{p}^n)) = \sigma_q\big(\mathbf{w}, (D_q(\mathbf{p}^1), \ldots, D_q(\mathbf{p}^n))\big)$$

(Example 7.1.8). Our diversity measure D is modular, which means that $D(\mathbf{w} \circ (\mathbf{p}^1, \ldots, \mathbf{p}^n))$ is some function of $\mathbf{w}$ and $D(\mathbf{p}^1), \ldots, D(\mathbf{p}^n)$. We will therefore be able to define a function σ by

$$D(\mathbf{w} \circ (\mathbf{p}^1, \ldots, \mathbf{p}^n)) = \sigma\big(\mathbf{w}, (D(\mathbf{p}^1), \ldots, D(\mathbf{p}^n))\big). \tag{7.15}$$

Roughly speaking, we then show that the assumed good properties of the diversity measure D imply good properties of σ, deduce from our earlier characterization of value measures that $\sigma = \sigma_q$ for some q, and conclude that $D = D_q$.

There is a subtlety. In order to use the characterization of value measures (Theorem 7.3.4), we need σ to be defined on all pairs $\sigma(\mathbf{p}, \mathbf{v})$ with $\mathbf{p} \in \Delta_n$ and $\mathbf{v} \in (0, \infty)^n$, whereas equation (7.15) only defines $\sigma(\mathbf{p}, \mathbf{v})$ on vectors $\mathbf{v}$ whose coordinates v_i can be expressed as values of the diversity measure D. And it may happen that some elements of $(0, \infty)$ do not arise as values of D. Indeed, if $D = D_q$ then $D_q(\mathbf{r}) \geq 1$ for all distributions $\mathbf{r}$.

For this reason, we now analyse the set of real numbers that arise as diversities $D(\mathbf{p})$. Write

$$\operatorname{im} D = \bigcup_{n=1}^{\infty} D\Delta_n \subseteq (0, \infty).$$

The case of the Hill numbers shows that the situation is not entirely simple:

Example 7.4.8 For $q \in [-\infty, \infty]$, the Hill number D_q has image

$$\operatorname{im} D_q = \begin{cases} [1, \infty) & \text{if } q > 0, \\ \{1, 2, 3, \ldots\} & \text{if } q = 0, \\ \{1\} \cup [2, \infty) & \text{if } q < 0. \end{cases} \tag{7.16}$$

The statement for $q > 0$ follows from the facts that $D_q(\mathbf{p}) \geq 1$ for all $\mathbf{p}$ (Lemma 4.4.3(i)), D_q is an effective number (equation (4.25)), and D_q is continuous (Lemma 4.4.6(ii)). For $q = 0$, the result is immediate, since $D_0(\mathbf{p}) = |\operatorname{supp}(\mathbf{p})|$.

Now let $q < 0$. Since diversity profiles are decreasing (Proposition 4.4.1),

$$D_q(\mathbf{p}) \geq |\operatorname{supp}(\mathbf{p})|$$

for all $\mathbf{p}$. If $|\operatorname{supp}(\mathbf{p})| = 1$ then $\mathbf{p} = (0, \ldots, 0, 1, 0, \ldots, 0)$ and so $D_q(\mathbf{p}) = 1$. Otherwise, $|\operatorname{supp}(\mathbf{p})| \geq 2$, so $D_q(\mathbf{p}) \in [2, \infty)$. Hence $\operatorname{im} D_q \subseteq \{1\} \cup [2, \infty)$. To prove the opposite inclusion, first note that both $1 = D_q(\mathbf{u}_1)$ and $2 = D_q(\mathbf{u}_2)$ belong to $\operatorname{im} D_q$. An elementary calculation shows that

$$D_q(t, 1-t) \to \infty \text{ as } t \to 0+.$$

Since $D_q \colon \Delta_2^\circ \to (0, \infty)$ is continuous (Lemma 4.4.6(i)), $D_q\Delta_2^\circ$ is an interval that contains 2 and is unbounded above. Hence $D_q\Delta_2^\circ \supseteq [2, \infty)$, completing the proof of the last clause of equation (7.16).

Lemma 7.4.9 $\operatorname{im} D$ *is closed under multiplication.*

Proof This follows from the multiplicativity of D (Lemma 7.4.7). □

Lemma 7.4.10 *Suppose that $D \neq D_0$. Then* $\operatorname{im} D \supseteq [L, \infty)$ *for some $L > 0$.*

Proof If $D\Delta_n^\circ$ is a one-element set for each $n \geq 1$ then by the effective number property, $D\Delta_n^\circ = \{n\}$ for each n. Hence by absence-invariance, $D = D_0$, a contradiction.

We can therefore choose $n \geq 1$ such that $D\Delta_n^\circ$ has more than one element, which by continuity in positive probabilities implies that $D\Delta_n^\circ$ is a nontrivial interval. Since D is an effective number, this interval contains n. Now $n \neq 1$ (since $D\Delta_1^\circ$ is trivial), so $n \geq 2$, so $\operatorname{im} D \cap [1, \infty)$ contains a nontrivial interval. Since both $\operatorname{im} D$ and $[1, \infty)$ are closed under multiplication, so is $\operatorname{im} D \cap [1, \infty)$.

It is now enough to prove that any subset B of $[1, \infty)$ that is closed under multiplication and contains a nontrivial interval must contain $[L, \infty)$ for some $L \geq 1$. Indeed, since B contains a nontrivial interval, $B \supseteq [b, b^{1+1/r}]$ for some real $b > 1$ and positive integer r. Since B is closed under multiplication, it is closed under positive integer powers, so for every integer $m \geq r$,

$$B \supseteq [b^m, b^{m+m/r}] \supseteq [b^m, b^{m+1}].$$

Hence

$$B \supseteq \bigcup_{m \geq r} [b^m, b^{m+1}] = [b^r, \infty),$$

using $b > 1$ in the last step. □

We now construct from D a value measure σ. The construction proceeds in two steps. First, since D is modular, we can consistently define a sequence of functions

$$(\rho \colon \Delta_n \times (\operatorname{im} D)^n \to \operatorname{im} D)_{n \geq 1}$$

by

$$\rho\big(\mathbf{w}, (D(\mathbf{p}^1), \ldots, D(\mathbf{p}^n))\big) = D(\mathbf{w} \circ (\mathbf{p}^1, \ldots, \mathbf{p}^n))$$

for all $n, k_1, \ldots, k_n \geq 1$, $\mathbf{w} \in \Delta_n$ and $\mathbf{p}^i \in \Delta_{k_i}$. Second, we extend ρ to a sequence of functions defined on not just $\Delta_n \times (\operatorname{im} D)^n$, but the whole of $\Delta_n \times (0, \infty)^n$:

Lemma 7.4.11 *Suppose that $D \neq D_0$. Then there is a unique homogeneous sequence of functions*

$$\left(\sigma\colon \Delta_n \times (0,\infty)^n \to (0,\infty)\right)_{n\geq 1}$$

such that

$$\sigma\big(\mathbf{w}, (D(\mathbf{p}^1), \ldots, D(\mathbf{p}^n))\big) = D(\mathbf{w} \circ (\mathbf{p}^1, \ldots, \mathbf{p}^n))$$

for all $n, k_1, \ldots, k_n \geq 1$, $\mathbf{w} \in \Delta_n$, and $\mathbf{p}^i \in \Delta_{k_i}$.

In brief, there is a unique homogeneous extension of ρ from $\operatorname{im} D$ to $(0, \infty)$.

Proof We begin by establishing a homogeneity property of ρ:

$$\rho(\mathbf{w}, c\mathbf{x}) = c\rho(\mathbf{w}, \mathbf{x}) \tag{7.17}$$

for all $\mathbf{w} \in \Delta_n$, $\mathbf{x} \in (\operatorname{im} D)^n$, and $c \in \operatorname{im} D$. (The left-hand side is well defined since $\operatorname{im} D$ is closed under multiplication, by Lemma 7.4.9.) To prove this, for each $i \in \{1, \ldots, n\}$, choose $\mathbf{p}^i \in \Delta_{k_i}$ such that $D(\mathbf{p}^i) = x_i$, and choose $\mathbf{r} \in \Delta_m$ such that $D(\mathbf{r}) = c$. Then

$$\begin{aligned}
\rho(\mathbf{w}, c\mathbf{x}) &= \rho\big(\mathbf{w}, (D(\mathbf{p}^1)D(\mathbf{r}), \ldots, D(\mathbf{p}^n)D(\mathbf{r}))\big) && (7.18)\\
&= \rho\big(\mathbf{w}, (D(\mathbf{p}^1 \otimes \mathbf{r}), \ldots, D(\mathbf{p}^n \otimes \mathbf{r}))\big) && (7.19)\\
&= D(\mathbf{w} \circ (\mathbf{p}^1 \otimes \mathbf{r}, \ldots, \mathbf{p}^n \otimes \mathbf{r})) && (7.20)\\
&= D\big((\mathbf{w} \circ (\mathbf{p}^1, \ldots, \mathbf{p}^n)) \otimes \mathbf{r}\big) && (7.21)\\
&= D(\mathbf{w} \circ (\mathbf{p}^1, \ldots, \mathbf{p}^n))D(\mathbf{r}) && (7.22)\\
&= c\rho(\mathbf{w}, \mathbf{x}), && (7.23)
\end{aligned}$$

where equation (7.18) is by definition of $\mathbf{p}^i$ and $\mathbf{r}$, equations (7.19) and (7.22) are by multiplicativity of D (Lemma 7.4.7), equations (7.20) and (7.23) are by definition of ρ, and (7.21) is by associativity of composition of distributions (Remark 2.1.8). This proves the claimed homogeneity equation (7.17).

We now prove the uniqueness and existence stated in the lemma.

Uniqueness Let $\mathbf{p} \in \Delta_n$ and $\mathbf{v} \in (0, \infty)^n$. By Lemma 7.4.10, $\operatorname{im} D$ contains all sufficiently large real numbers, so we can choose $c \in (0, \infty)$ such that $c\mathbf{v} \in (\operatorname{im} D)^n$. Then $\rho(\mathbf{p}, c\mathbf{v})$ is defined, and any sequence of homogeneous functions σ extending ρ satisfies

$$\sigma(\mathbf{p}, \mathbf{v}) = \tfrac{1}{c}\rho(\mathbf{p}, c\mathbf{v}).$$

This proves uniqueness.

Existence First I claim that for all $\mathbf{p} \in \Delta_n$, $\mathbf{v} \in (0,\infty)^n$, and $c, d \in (0,\infty)$ such that $c\mathbf{v}, d\mathbf{v} \in (\operatorname{im} D)^n$,

$$\tfrac{1}{c}\rho(\mathbf{p}, c\mathbf{v}) = \tfrac{1}{d}\rho(\mathbf{p}, d\mathbf{v}). \tag{7.24}$$

Indeed, since $\operatorname{im} D$ contains all sufficiently large real numbers, we can choose $a > 0$ such that $ac, ad \in \operatorname{im} D$. Then

$$ad \cdot \rho(\mathbf{p}, c\mathbf{v}) = \rho(\mathbf{p}, acd\mathbf{v})$$

by the homogeneity property (7.17) of ρ. Similarly,

$$ac \cdot \rho(\mathbf{p}, d\mathbf{v}) = \rho(\mathbf{p}, acd\mathbf{v}).$$

Combining the last two equations gives equation (7.24), as claimed.

It follows that there is a unique sequence of functions

$$\big(\sigma\colon \Delta_n \times (0,\infty)^n \to (0,\infty)\big)_{n\geq 1}$$

satisfying

$$\sigma(\mathbf{p}, \mathbf{v}) = \tfrac{1}{c}\rho(\mathbf{p}, c\mathbf{v}) \tag{7.25}$$

whenever $\mathbf{p} \in \Delta_n$, $\mathbf{v} \in (0,\infty)^n$ and $c \in (0,\infty)$ with $c\mathbf{v} \in (\operatorname{im} D)^n$.

It remains to prove that σ is homogeneous. Let $\mathbf{p} \in \Delta_n$, $\mathbf{v} \in (0,\infty)^n$, and $a \in (0,\infty)$; we must show that

$$\sigma(\mathbf{p}, a\mathbf{v}) = a\sigma(\mathbf{p}, \mathbf{v}). \tag{7.26}$$

Choose $d \in (0,\infty)$ such that $ad\mathbf{v}, d\mathbf{v} \in (\operatorname{im} D)^n$. By the claim just proved,

$$\tfrac{1}{ad}\rho(\mathbf{p}, ad\mathbf{v}) = \tfrac{1}{d}\rho(\mathbf{p}, d\mathbf{v}),$$

or equivalently,

$$\tfrac{1}{d}\rho(\mathbf{p}, d \cdot a\mathbf{v}) = a \cdot \tfrac{1}{d}\rho(\mathbf{p}, d\mathbf{v}).$$

But by the defining property (7.25) of σ, this is exactly the desired equation (7.26). □

Example 7.4.12 Consider the case $D = D_q$. We have

$$\sigma_q\big(\mathbf{w}, (D_q(\mathbf{p}^1), \dots, D_q(\mathbf{p}^n))\big) = D_q(\mathbf{w} \circ (\mathbf{p}^1, \dots, \mathbf{p}^n))$$

for all $\mathbf{w}$ and $\mathbf{p}^i$, by Example 7.1.8. Moreover, σ_q is homogeneous. Hence by the uniqueness part of Lemma 7.4.11, $\sigma = \sigma_q$.

We have now constructed from our diversity measure D a value measure σ. From our standing assumption that D has certain good properties, it follows that σ has good properties too:

Lemma 7.4.13 *Suppose that $D \neq D_0$. Then σ is symmetric, absence-invariant, increasing, homogeneous, continuous in positive probabilities and an effective number, and satisfies the chain rule.*

Proof The symmetry, absence-invariance and effective number properties of D imply the corresponding properties of σ. The modular-monotonicity of D implies that ρ, hence σ, is increasing. Homogeneity is one of the defining properties of σ (Lemma 7.4.11). It remains to prove continuity in positive probabilities and the chain rule.

To prove that σ is continuous in positive probabilities, let $\mathbf{v} \in (0,\infty)^n$; we wish to prove that

$$\sigma(-,\mathbf{v})\colon \Delta_n^\circ \to (0,\infty)$$

is continuous. Choose $c \in (0,\infty)$ such that $c\mathbf{v} \in (\operatorname{im} D)^n$. Then $\sigma(-,\mathbf{v}) = \frac{1}{c}\rho(-,c\mathbf{v})$. It therefore suffices to prove that

$$\rho(-,\mathbf{x})\colon \Delta_n^\circ \to (0,\infty)$$

is continuous for every $\mathbf{x} \in (\operatorname{im} D)^n$. For each $i \in \{1,\ldots,n\}$, choose $\mathbf{p}^i \in \Delta_{k_i}$ such that $x_i = D(\mathbf{p}^i)$. By absence-invariance, we may assume that each $\mathbf{p}^i$ has full support. For all $\mathbf{w} \in \Delta_n$, we have

$$\rho(\mathbf{w},\mathbf{x}) = D(\mathbf{w} \circ (\mathbf{p}^1,\ldots,\mathbf{p}^n)),$$

and if $\mathbf{w}$ has full support then so does $\mathbf{w} \circ (\mathbf{p}^1,\ldots,\mathbf{p}^n)$. Thus, the restriction of $\rho(-,\mathbf{x})$ to Δ_n° is the composite of the continuous maps

$$\begin{array}{ccccc} \Delta_n^\circ & \longrightarrow & \Delta_{k_1+\cdots+k_n}^\circ & \xrightarrow{D} & (0,\infty) \\ \mathbf{w} & \longmapsto & \mathbf{w} \circ (\mathbf{p}^1,\ldots,\mathbf{p}^n). & & \end{array}$$

It is therefore continuous, as claimed.

To prove that σ satisfies the chain rule, we first prove a chain rule for ρ:

$$\rho(\mathbf{w} \circ (\mathbf{p}^1,\ldots,\mathbf{p}^n), \mathbf{x}^1 \oplus \cdots \oplus \mathbf{x}^n) = \rho\Big(\mathbf{w}, (\rho(\mathbf{p}^1,\mathbf{x}^1),\ldots,\rho(\mathbf{p}^n,\mathbf{x}^n))\Big) \tag{7.27}$$

for all $\mathbf{w} \in \Delta_n$, $\mathbf{p}^i \in \Delta_{k_i}$, and $\mathbf{x}^i \in (\operatorname{im} D)^{k_i}$. To see this, begin by choosing for each $i \in \{1,\ldots,n\}$ and $j \in \{1,\ldots,k_i\}$ a probability distribution $\mathbf{r}^i_j$ such that $D(\mathbf{r}^i_j) = x^i_j$. Then by definition of ρ, the left-hand side of equation (7.27) is equal to

$$D\Big((\mathbf{w} \circ (\mathbf{p}^1,\ldots,\mathbf{p}^n)) \circ (\mathbf{r}^1_1,\ldots,\mathbf{r}^1_{k_1}, \ \ldots, \ \mathbf{r}^n_1,\ldots,\mathbf{r}^n_{k_n})\Big).$$

By associativity of composition of distributions (Remark 2.1.8), this is equal to

$$D\Big(\mathbf{w} \circ (\mathbf{p}^1 \circ (\mathbf{r}^1_1,\ldots,\mathbf{r}^1_{k_1}), \ \ldots, \ \mathbf{p}^n \circ (\mathbf{r}^n_1,\ldots,\mathbf{r}^n_{k_n}))\Big).$$

By definition of ρ, this in turn is equal to

$$\rho\Big(\mathbf{w}, \big(D(\mathbf{p}^1 \circ (\mathbf{r}^1_1, \dots, \mathbf{r}^1_{k_1})), \ \dots, \ D(\mathbf{p}^n \circ (\mathbf{r}^n_1, \dots, \mathbf{r}^n_{k_n}))\big)\Big),$$

which by definition of ρ again is equal to the right-hand side of (7.27). This proves the claimed chain rule (7.27) for ρ.

We now want to prove the chain rule for σ:

$$\sigma(\mathbf{w} \circ (\mathbf{p}^1, \dots, \mathbf{p}^n), \mathbf{v}^1 \oplus \cdots \oplus \mathbf{v}^n) = \sigma\Big(\mathbf{w}, (\sigma(\mathbf{p}^1, \mathbf{v}^1), \dots, \sigma(\mathbf{p}^n, \mathbf{v}^n))\Big)$$

for all $\mathbf{w} \in \Delta_n$, $\mathbf{p}^i \in \Delta_{k_i}$, and $\mathbf{v}^i \in (0, \infty)^{k_i}$. We may choose $c \in (0, \infty)$ such that $cv^i_j \in \operatorname{im} D$ for all i, j. Then by definition of σ and the chain rule (7.27) for ρ,

$$\begin{aligned}
\sigma(\mathbf{w} \circ (\mathbf{p}^1, \dots, \mathbf{p}^n), \mathbf{v}^1 \oplus \cdots \oplus \mathbf{v}^n) &= \tfrac{1}{c}\rho(\mathbf{w} \circ (\mathbf{p}^1, \dots, \mathbf{p}^n), c\mathbf{v}^1 \oplus \cdots \oplus c\mathbf{v}^n) \\
&= \tfrac{1}{c}\rho\Big(\mathbf{w}, (\rho(\mathbf{p}^1, c\mathbf{v}^1), \dots, \rho(\mathbf{p}^n, c\mathbf{v}^n))\Big) \\
&= \tfrac{1}{c}\rho\Big(\mathbf{w}, (c\sigma(\mathbf{p}^1, \mathbf{v}^1), \dots, c\sigma(\mathbf{p}^n, \mathbf{v}^n))\Big) \\
&= \sigma\Big(\mathbf{w}, (\sigma(\mathbf{p}^1, \mathbf{v}^1), \dots, \sigma(\mathbf{p}^n, \mathbf{v}^n))\Big),
\end{aligned}$$

as required. □

We are now ready to prove that when a community is modelled as a probability distribution, the Hill numbers are the only sensible measures of diversity.

Proof of Theorem 7.4.3 We have to show that $D = D_q$ for some $q \in [-\infty, \infty]$. If $D = D_0$, this is immediate. Otherwise, by Lemma 7.4.13, σ is a value measure satisfying the hypotheses of Theorem 7.3.4. By that theorem, $\sigma = \sigma_q$ for some $q \in [-\infty, \infty]$. Let $\mathbf{p} \in \Delta_n$. Then

$$\begin{aligned}
D(\mathbf{p}) &= D(\mathbf{p} \circ \underbrace{(\mathbf{u}_1, \dots, \mathbf{u}_1)}_{n}) \\
&= \rho\Big(\mathbf{p}, (D(\mathbf{u}_1), \dots, D(\mathbf{u}_1))\Big) && (7.28)\\
&= \rho(\mathbf{p}, (1, \dots, 1)) && (7.29)\\
&= \sigma_q(\mathbf{p}, (1, \dots, 1)) && (7.30)\\
&= D_q(\mathbf{p}), && (7.31)
\end{aligned}$$

where equation (7.28) is by definition of ρ, equation (7.29) holds because D is normalized, equation (7.30) holds because σ extends ρ and $\sigma = \sigma_q$, and equation (7.31) is from Example 7.1.6. Hence $D = D_q$. □

The theorem axiomatically characterizes the whole family $(\sigma_q)_{q \in [-\infty, \infty]}$ of Hill numbers. But as argued in Remark 4.4.4(ii), D_q probably does not deserve to be called a measure of diversity when q is negative. We may therefore wish to characterize the Hill numbers D_q for which $q \geq 0$, and the following result achieves this.

Lemma 7.4.14 *Let* $q \in [-\infty, \infty]$. *The following are equivalent:*

i. $D_q(\mathbf{p}) \leq D_q(\mathbf{u}_n)$ *for all* $n \geq 1$ *and* $\mathbf{p} \in \Delta_n$*;*
ii. $D_q(\mathbf{p}) \leq 2$ *for all* $\mathbf{p} \in \Delta_2$*;*
iii. $q \in [0, \infty]$.

Proof (i) implies (ii) trivially, (ii) implies (iii) by Remark 4.4.4(ii), and (iii) implies (i) by Lemma 4.4.3(ii). □

Remark 7.4.15 Our characterization theorem for the Hill numbers can easily be translated into a characterization theorem for the Rényi or q-logarithmic entropies, using the transformations of Section 7.2. However, the hypotheses of Theorem 7.4.3 are particularly natural in the context of diversity.

When translated into terms of q-logarithmic entropy, Theorem 7.4.3 is of the same general type as a result of Forte and Ng [109] (also stated as Theorem 6.3.12 of Aczél and Daróczy [3]). Apart from some differences in hypotheses, Forte and Ng's characterization excludes the case $q = 0$, which from the point of view of diversity measurement is a serious drawback: the Hill number D_0 is species richness, the most common diversity measure of all.

8

Mutual Information and Metacommunities

From the viewpoint of information theory, there is a conspicuous omission from this text so far. Given a random variable X taking values in a finite set, we have a measure of the information associated with X: its entropy $H(X)$. But suppose that we are also given another random variable Y, not necessarily independent of X, taking values in another finite set. If we know the value of X, how much information does that give us about the value of Y?

For instance, Y might be a function of X, in which case knowing the value of X gives complete information about Y. Or, at the other extreme, X and Y might be independent, in which case knowing the value of X tells us nothing about Y. We would like to quantify the dependence between the two variables. The covariance and correlation coefficients will not do, since they are usually only defined for random variables taking values in $\mathbb{R}^n$; and while they can be defined in greater generality, there is no definition for an arbitrary pair of finite sets.

From the viewpoint of diversity measurement, there is also something missing. We know how to quantify the diversity of a single community. But when we have several associated or adjacent communities – for instance, the gut microbiomes of healthy and unhealthy adults, or the aquatic life in areas of different salinity near the mouth of a river – some natural questions present themselves. How much variation is there between the communities? Which contribute most to the overall diversity? Which are most or least typical in the context of the system as a whole? The diversity measures discussed so far give no answers to such questions.

We will see that these two problems, one information-theoretic and one ecological, have the same solution.

Our starting point is the classical information-theoretic concept of mutual information (a measure of the dependence between two random variables) and the closely related concepts of conditional and joint entropy. These are introduced in Section 8.1. Then we take exponentials of all these quantities, which produces a suite of meaningful measures of an ecological metacommunity (large community) divided into smaller subcommunities. The two random variables in play here correspond to a choice of species and a choice of subcommunity. Some of the measures reflect features of individual subcommunities (Section 8.2), while others encapsulate information about the entire metacommunity (Section 8.3). We establish the many good logical properties of these measures in Section 8.4.

All of the entropies and diversities in this chapter can be reduced to relative entropy (Section 8.5). In the diversity case, they are also usefully expressed in terms of value, in the sense of Chapter 7. Reducing the various metacommunity and subcommunity measures to one single concept provides new insights into their ecological meaning.

The diversity measures treated in this chapter are a very special case of those introduced in work of Reeve et al. [293]. In the terminology of Chapter 6, it is the case $q = 1$ (no deformation) and $Z = I$ (no inter-species similarity). The framework of Reeve et al. allows a general q (variable emphasis on rare or common species) and a general Z (to model the varying similarities between species). Section 8.6 is a sketch of the development for a general q, the details of which lie outside the scope of this book.

8.1 Joint Entropy, Conditional Entropy and Mutual Information

Shannon entropy H assigns a real number $H(\mathbf{p})$ to each probability distribution $\mathbf{p}$, but information theory also associates several quantities with any *pair* of probability distributions. To organize them, it is helpful to distinguish between two types of quantity: those defined for a pair of distributions on the *same* set, and those defined for a pair of distributions on potentially *different* sets.

We have already met two quantities of the first type: the relative entropy $H(\mathbf{p} \parallel \mathbf{r})$ and cross entropy $H^{\times}(\mathbf{p} \parallel \mathbf{r})$ of two distributions $\mathbf{p}$ and $\mathbf{r}$ on the same finite set (Chapter 3).

We now introduce the standard information-theoretic quantities of the second type. The material in this section is all classical, and can be found in texts such as Cover and Thomas ([71], Chapter 2) and MacKay ([238], Chapter 8). As usual, we only consider probability distributions on *finite* sets. But it is convenient to switch from the language of probability distributions to that of random variables.

So, ***for the rest of this section***, we consider a random variable X taking values in a finite set $\mathcal{X}$, and another random variable Y taking values in a finite set $\mathcal{Y}$. Assuming that X and Y have the same sample space, we also have the random variable (X, Y), which takes values in $\mathcal{X} \times \mathcal{Y}$.

Given $x \in \mathcal{X}$ and $y \in \mathcal{Y}$, we write $\mathbb{P}((X, Y) = (x, y))$ as $\mathbb{P}(X = x, Y = y)$. We usually abbreviate $\mathbb{P}(X = x)$ as $\mathbb{P}(x)$, etc. Thus, by definition, X and Y are independent if and only if

$$\mathbb{P}(x, y) = \mathbb{P}(x)\mathbb{P}(y)$$

for all $x \in \mathcal{X}$ and $y \in \mathcal{Y}$. The conditional probability of x given y is

$$\mathbb{P}(x \mid y) = \frac{\mathbb{P}(x, y)}{\mathbb{P}(y)},$$

and is defined as long as $\mathbb{P}(y) > 0$.

The **Shannon entropy** of the random variable X is the Shannon entropy of its distribution:

$$H(X) = \sum_{x:\, \mathbb{P}(x)>0} \mathbb{P}(x) \log \frac{1}{\mathbb{P}(x)}.$$

Here and below, the variable x in summations is assumed to run over the set $\mathcal{X}$ unless indicated otherwise, and similarly for y in $\mathcal{Y}$.

Joint Entropy

The general definition of the entropy of a random variable can be applied to the random variable (X, Y), giving

$$H(X, Y) = \sum_{x,y:\, \mathbb{P}(x,y)>0} \mathbb{P}(x, y) \log \frac{1}{\mathbb{P}(x, y)},$$

the **joint entropy** of X and Y.

Examples 8.1.1 i. Suppose that X and Y are independent. If X has distribution $\mathbf{p}$ and Y has distribution $\mathbf{r}$ then (X, Y) has distribution $\mathbf{p} \otimes \mathbf{r}$, so

$$H(X, Y) = H(X) + H(Y)$$

by Corollary 2.2.10.

ii. Suppose that $\mathcal{Y}$ is a one-element set. Then the distribution of Y is uniquely determined, $H(Y) = 0$, and $H(X, Y) = H(X)$.

iii. Suppose that $\mathcal{X} = \mathcal{Y}$ and $X = Y$. Then $H(X, Y) = H(X) = H(Y)$.

iv. Generalizing the last two examples, let us say that Y is **determined by** X if for all $x \in \mathcal{X}$ such that $\mathbb{P}(x) > 0$, there is a unique $y \in \mathcal{Y}$ such that

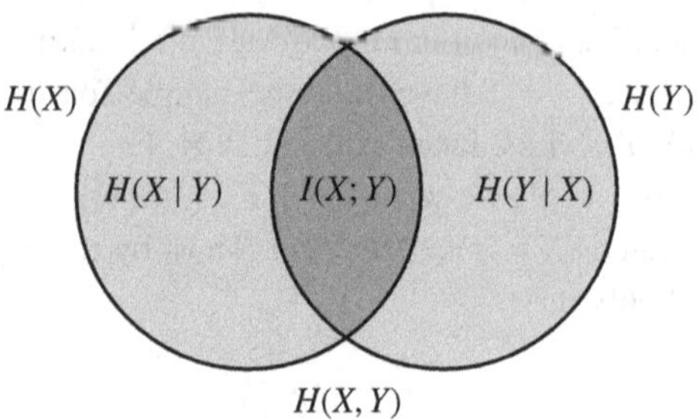

Figure 8.1 Venn diagram showing entropic quantities associated with a pair of random variables taking values in different sets: the Shannon entropies $H(X)$ and $H(Y)$, the joint entropy $H(X, Y)$, the conditional entropies $H(X \mid Y)$ and $H(Y \mid X)$, and the mutual information $I(X; Y)$.

$\mathbb{P}(x, y) > 0$. Writing this element y as $f(x)$, we then have $\mathbb{P}(x, f(x)) = \mathbb{P}(x)$, or equivalently, $\mathbb{P}(f(x) \mid x) = 1$. The joint entropy is given by

$$\begin{aligned} H(X, Y) &= \sum_{x,y:\, \mathbb{P}(x,y)>0} \mathbb{P}(x, y) \log \frac{1}{\mathbb{P}(x, y)} \\ &= \sum_{x:\, \mathbb{P}(x)>0} \mathbb{P}(x) \log \frac{1}{\mathbb{P}(x)} \\ &= H(X). \end{aligned}$$

Conditional Entropy

The definitions of the conditional entropies $H(X|Y)$ and $H(Y|X)$ and the mutual information $I(X; Y)$ are suggested by the schematic diagram of Figure 8.1. The diagram depicts the joint entropy $H(X, Y)$ as the union of the two discs and $H(X \mid Y)$ as the complement of the second disc in the union. This suggests the following definition.

Definition 8.1.2 The **conditional entropy** of X given Y is

$$H(X \mid Y) = H(X, Y) - H(Y).$$

We now explore this definition. For each $y \in \mathcal{Y}$ such that $\mathbb{P}(y) > 0$, there is a random variable $X \mid y$ taking values in $\mathcal{X}$, with distribution

$$\mathbb{P}((X \mid y) = x) = \mathbb{P}(x \mid y)$$

($x \in \mathcal{X}$). Like all random variables, it has an entropy, $H(X \mid y)$. The name 'conditional entropy' is explained by part (ii) of the following result.

Lemma 8.1.3 *i.* $H(X \mid Y) = \sum_{x,y:\, \mathbb{P}(x,y)>0} \mathbb{P}(x,y) \log \frac{1}{\mathbb{P}(x \mid y)}$.

ii. $H(X \mid Y) = \sum_{y:\, \mathbb{P}(y)>0} \mathbb{P}(y) H(X \mid y)$.

Proof For (i), first note that $\mathbb{P}(y) = \sum_x \mathbb{P}(x,y)$ for each $y \in \mathcal{Y}$, and in particular, $\mathbb{P}(y) > 0$ if there exists an x such that $\mathbb{P}(x,y) > 0$. Hence

$$H(Y) = \sum_{x,y:\, \mathbb{P}(x,y)>0} \mathbb{P}(x,y) \log \frac{1}{\mathbb{P}(y)}. \tag{8.1}$$

It follows that

$$\begin{aligned}
H(X \mid Y) &= \sum_{x,y:\, \mathbb{P}(x,y)>0} \mathbb{P}(x,y) \log \frac{1}{\mathbb{P}(x,y)} - \sum_{x,y:\, \mathbb{P}(x,y)>0} \mathbb{P}(x,y) \log \frac{1}{\mathbb{P}(y)} \\
&= \sum_{x,y:\, \mathbb{P}(x,y)>0} \mathbb{P}(x,y) \log \frac{\mathbb{P}(y)}{\mathbb{P}(x,y)} \\
&= \sum_{x,y:\, \mathbb{P}(x,y)>0} \mathbb{P}(x,y) \log \frac{1}{\mathbb{P}(x \mid y)},
\end{aligned}$$

proving (i). This in turn is equal to

$$\sum_{y:\, \mathbb{P}(y)>0} \mathbb{P}(y) \sum_{x:\, \mathbb{P}(x|y)>0} \mathbb{P}(x \mid y) \log \frac{1}{\mathbb{P}(x \mid y)} = \sum_{y:\, \mathbb{P}(y)>0} \mathbb{P}(y) H(X \mid y),$$

proving (ii). □

The conditional entropy $H(X \mid Y)$ is, therefore, the expected entropy of the conditional random variable $X \mid y$ when y is chosen at random. It follows that $H(X \mid Y) \geq 0$, or equivalently, that $H(X, Y) \geq H(Y)$.

Examples 8.1.4 Consider again the four scenarios of Examples 8.1.1.

i. Suppose that X and Y are independent. Then by Example 8.1.1(i),

$$H(X \mid Y) = H(X), \qquad H(Y \mid X) = H(Y).$$

Knowing the value of Y gives no information on the value of X, nor vice versa. This is the situation shown in Figure 8.2(a).

ii. Suppose that $\mathcal{Y}$ is a one-element set. Then by Example 8.1.1(ii),

$$\begin{aligned}
H(X \mid Y) &= H(X, Y) - H(Y) = H(X), \\
H(Y \mid X) &= H(X, Y) - H(X) = 0.
\end{aligned}$$

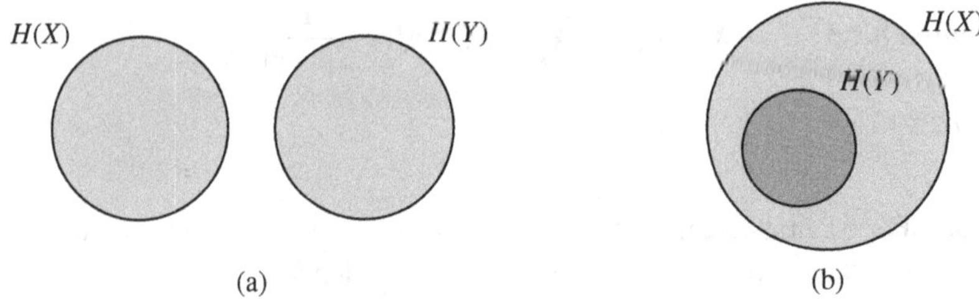

Figure 8.2 Venn diagrams for the cases in which (a) the random variables X and Y are independent; (b) Y is determined by X.

iii. Suppose that $\mathcal{X} = \mathcal{Y}$ and $X = Y$. By Example 8.1.1(iii), $H(X \mid Y) = 0$. This is intuitively plausible: once we know the value of Y, we know the value of X with certainty, so its probability distribution is concentrated on a single element and therefore has entropy 0. Similarly, $H(Y \mid X) = 0$.

iv. More generally, suppose that Y is determined by X (Figure 8.2(b)). Then by Example 8.1.1(iv),

$$H(X \mid Y) = H(X) - H(Y), \tag{8.2}$$
$$H(Y \mid X) = 0.$$

Since $H(X \mid Y) \geq 0$, we have

$$H(Y) \leq H(X)$$

whenever Y is determined by X.

Remark 8.1.5 In most texts, Lemma 8.1.3(ii) is taken as the *definition* of conditional entropy, and the equation $H(X \mid Y) = H(X, Y) - H(Y)$ that we took as our definition is proved as a theorem. This theorem is called the **chain rule**. In other words, the chain rule states that

$$H(X, Y) = H(Y) + \sum_{y:\, \mathbb{P}(y)>0} \mathbb{P}(y) H(X \mid y). \tag{8.3}$$

It is essentially the same as what we have been calling the chain rule throughout this text (beginning in Proposition 2.2.8). This can be seen as follows.

Write $\mathcal{X} = \{1, \ldots, k\}$ and $\mathcal{Y} = \{1, \ldots, n\}$. Write $\mathbf{w} = (w_1, \ldots, w_n) \in \Delta_n$ for the distribution of Y; thus, $w_i = \mathbb{P}(Y = i)$ for each $i \in \mathcal{Y}$. Also, for each $i \in \mathcal{Y}$ and $j \in \mathcal{X}$, define p^i_j by

$$w_i p^i_j = \mathbb{P}(X = j, Y = i),$$

so that $\mathbf{p}^i = (p^i_1, \ldots, p^i_k) \in \Delta_k$ is the distribution of the random variable $X \mid i$.

(Here we have assumed that $\mathbb{P}(i) > 0$; otherwise, choose $\mathbf{p}^i \in \Delta_k$ arbitrarily.) Then $\mathbf{w} \circ (\mathbf{p}^1, \ldots, \mathbf{p}^n) \in \Delta_{nk}$ is the joint distribution of X and Y. In this notation, equation (8.3) states that

$$H(\mathbf{w} \circ (\mathbf{p}^1, \ldots, \mathbf{p}^n)) = H(\mathbf{w}) + \sum_{i:\, w_i > 0} w_i H(\mathbf{p}^i).$$

This is exactly the chain rule in our usual sense.

Mutual Information

In Figure 8.1, the intersection of the two discs is labelled as $I(X;Y)$, and the inclusion-exclusion principle suggests the formula

$$H(X,Y) = H(X) + H(Y) - I(X;Y).$$

We define $I(X;Y)$ to make this true.

Definition 8.1.6 The **mutual information** of X and Y is

$$I(X;Y) = H(X) + H(Y) - H(X,Y).$$

Evidently I is symmetric:

$$I(X;Y) = I(Y;X). \tag{8.4}$$

Alternative expressions for I, in terms of conditional rather than joint entropy, follow immediately from the definitions and are also suggested by the Venn diagram:

$$I(X;Y) = H(X) - H(X \mid Y) = H(Y) - H(Y \mid X). \tag{8.5}$$

Mutual information can be expressed in two further ways still.

Lemma 8.1.7 *i.* $I(X;Y) = \displaystyle\sum_{x,y:\, \mathbb{P}(x,y)>0} \mathbb{P}(x,y) \log \frac{\mathbb{P}(x,y)}{\mathbb{P}(x)\mathbb{P}(y)}$.

ii. $I(X;Y) = \displaystyle\sum_{y:\, \mathbb{P}(y)>0} \mathbb{P}(y) H((X \mid y) \,\|\, X)$.

The right-hand side of (ii) refers to the random variables $X \mid y$ and X taking values in $\mathcal{X}$, and the relative entropy of the first with respect to the second.

Proof For (i), we have

$$\begin{aligned} I(X;Y) &= H(X) - H(X \mid Y) \\ &= \sum_{x,y:\, \mathbb{P}(x,y)>0} \mathbb{P}(x,y) \log \frac{1}{\mathbb{P}(x)} - \sum_{x,y:\, \mathbb{P}(x,y)>0} \mathbb{P}(x,y) \log \frac{\mathbb{P}(y)}{\mathbb{P}(x,y)}, \end{aligned}$$

by equation (8.1) and Lemma 8.1.3(i). Collecting terms, the result follows.

To prove (ii), we use (i) and the equation $\mathbb{P}(x, y) = \mathbb{P}(y)\mathbb{P}(x \mid y)$:

$$\begin{aligned} I(X;Y) &= \sum_{x,y:\, \mathbb{P}(y)>0,\, \mathbb{P}(x|y)>0} \mathbb{P}(y)\mathbb{P}(x \mid y) \log \frac{\mathbb{P}(x \mid y)}{\mathbb{P}(x)} \\ &= \sum_{y:\, \mathbb{P}(y)>0} \mathbb{P}(y) H((X \mid y) \,\|\, X), \end{aligned}$$

as required. □

The formula in (ii) can be interpreted as follows. For probability distributions $\mathbf{p}$ and $\mathbf{r}$ on the same finite set, $H(\mathbf{p} \,\|\, \mathbf{r})$ can be understood as the information gained when learning that the distribution of a random variable is $\mathbf{p}$, when one had previously believed that it was $\mathbf{r}$. Thus, $H((X \mid y) \,\|\, X)$ is the information gained about X by learning that $Y = y$. Consequently,

$$\sum_{y:\, \mathbb{P}(y)>0} \mathbb{P}(y) H((X \mid y) \,\|\, X)$$

is the expected information about X gained by learning the value of Y. This is the mutual information $I(X; Y)$. Briefly put, it is the information that Y gives about X.

For instance, if X and Y are independent, then knowing the value of Y gives us no clue as to the value of X, so one would expect that $I(X; Y) = 0$. And indeed, $X \mid y$ has the same distribution as X (for each y), so $H((X \mid y) \,\|\, X) = 0$, giving $I(X; Y) = 0$. We examine the extremal cases more systematically in Proposition 8.1.12.

Of course, Lemma 8.1.7(ii) has a counterpart with X and Y interchanged, and the symmetry property $I(X; Y) = I(Y; X)$ of mutual information (equation (8.4)) implies that

$$\sum_{y:\, \mathbb{P}(y)>0} \mathbb{P}(y) H((X \mid y) \,\|\, X) = \sum_{x:\, \mathbb{P}(x)>0} \mathbb{P}(x) H((Y \mid x) \,\|\, Y).$$

That is, the information that Y gives about X is equal to the information that X gives about Y. This explains the word 'mutual'.

Examples 8.1.8 We consider again the four cases of Examples 8.1.1 and 8.1.4, using the results derived there.

i. If X and Y are independent then $I(X; Y) = 0$: neither variable gives any information about the other.
ii. If $\mathcal{Y}$ is a one-element set then $I(X; Y) = 0$. From one viewpoint, knowing the value of X gives no information about the value of Y, since the value of

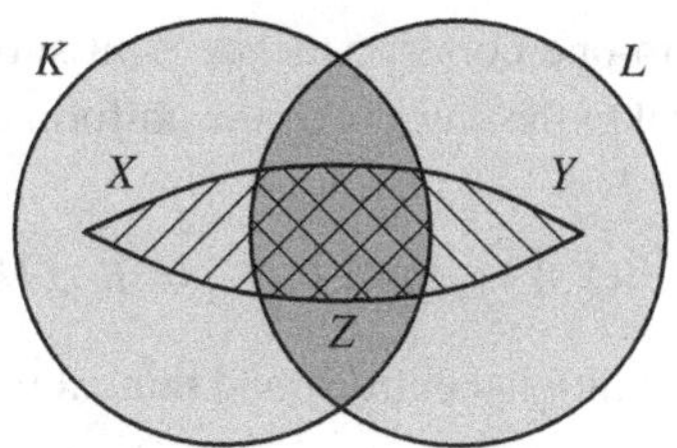

Figure 8.3 Random subsets (Example 8.1.9). The subset Z is the whole lozenge, $X = Z \cap K$ is shaded as ◪, and $Y = Z \cap L$ is shaded as ◩.

Y is predetermined anyway. From the other, knowing the value of Y gives no information about the value of X (or indeed, about anything).

iii. If $\mathcal{X} = \mathcal{Y}$ and $X = Y$ then $I(X; Y) = H(X) = H(Y)$. This is the maximal value that $I(X; Y)$ can take (by Proposition 8.1.12(iii) below), which is intuitively plausible: knowing X gives complete information about Y.

iv. Generally, if Y is determined by X, then $I(X; Y) = H(Y)$. As in (iii), this tells us that knowledge of X gives certain knowledge of Y (even though knowledge of Y does not, in this case, give certain knowledge of X).

The Venn diagram of Figure 8.1 is not merely a metaphor or an analogy. It depicts the following specific example.

Example 8.1.9 For this example, first note that joint entropy, conditional entropy and mutual information can be defined using logarithms to any base. Just as we write $H^{(2)}(X) = H(X)/\log 2$ (Remark 2.2.1), let us write the base 2 version of joint entropy as $H^{(2)}(X, Y) = H(X, Y)/\log 2$, and similarly for $H^{(2)}(X|Y)$ and $I^{(2)}(X; Y)$.

Fix finite subsets K and L of some set. Let Z denote a subset of $K \cup L$ chosen uniformly at random, and put

$$X = Z \cap K, \qquad Y = Z \cap L$$

(Figure 8.3). Then X and Y are uniformly distributed random variables taking values in the power sets $\mathscr{P}(K)$ and $\mathscr{P}(L)$ respectively, so

$$H^{(2)}(X) = \log_2(2^{|K|}) = |K|,$$
$$H^{(2)}(Y) = \log_2(2^{|L|}) = |L|.$$

The random variable (X, Y), which takes values in $\mathscr{P}(K) \times \mathscr{P}(L)$, is uniformly distributed on the set of pairs

$$\{(A, B) \in \mathscr{P}(K) \times \mathscr{P}(L) : A = C \cap K \text{ and } B = C \cap L \text{ for some } C \subseteq K \cup L\}.$$

Such pairs are in one-to-one correspondence with subsets of $K \cup L$, so the entropy of (X, Y) is equal to the entropy of the uniform distribution on $\mathscr{P}(K \cup L)$. Hence

$$H^{(2)}(X, Y) = \log_2(2^{|K \cup L|}) = |K \cup L|.$$

Then by definition of conditional entropy and mutual information,

$$\begin{aligned} H^{(2)}(X \mid Y) &= |K \cup L| - |L| = |K \setminus L|, \\ H^{(2)}(Y \mid X) &= |K \cup L| - |K| = |L \setminus K|, \\ I^{(2)}(X; Y) &= |K| + |L| - |K \cup L| = |K \cap L|. \end{aligned}$$

So, this example realizes the various entropies shown in the Venn diagram of Figure 8.1 as actual cardinalities.

Extremal Cases

We finish this introduction to joint entropy, conditional entropy and mutual information by finding their maximal and minimal values in terms of ordinary entropy. Here is the central fact.

Lemma 8.1.10 *$I(X; Y) \geq 0$, with equality if and only if X and Y are independent.*

Proof Lemma 8.1.7(ii) states that

$$I(X; Y) = \sum_{y:\, \mathbb{P}(y)>0} \mathbb{P}(y) H((X \mid y) \,\|\, X).$$

Given $y \in \mathcal{Y}$ such that $\mathbb{P}(y) > 0$, Lemma 3.1.4 implies that $H((X \mid y) \,\|\, X) \geq 0$, with equality if and only if $\mathbb{P}(x \mid y) = \mathbb{P}(x)$ for all $x \in \mathcal{X}$. Thus, $I(X; Y) \geq 0$, with equality if and only if $\mathbb{P}(x \mid y) = \mathbb{P}(x)$ for all x, y such that $\mathbb{P}(y) > 0$. But this condition is equivalent to X and Y being independent. □

Remark 8.1.11 Given three random variables X, Y and Z with the same sample space, one can define a threefold mutual information $I(X; Y; Z)$ by the same inclusion-exclusion principle that has guided us so far:

$$I(X; Y; Z) = (H(X) + H(Y) + H(Z)) - (H(X, Y) + H(X, Z) + H(Y, Z)) + H(X, Y, Z).$$

But in contrast to the binary case, $I(X; Y; Z)$ is sometimes negative. For example, this is the case when all three random variables take values in $\{0, 1\}$ and (X, Y, Z) is uniformly distributed on the four triples

$$(0, 0, 0),\ (0, 1, 1),\ (1, 0, 1),\ (1, 1, 0),$$

with probability zero on the other four. For discussion of this and other multivariate information measures, see Timme et al. [326], especially Section 4.2.

The following proposition gathers together the various maximal and minimal values and the conditions under which they are attained. All the results are as one would guess from the Venn diagrams of Figures 8.1 and 8.2.

Proposition 8.1.12 *i. Joint entropy is bounded as follows:*

a. $\max\{H(X), H(Y)\} \le H(X, Y) \le H(X) + H(Y)$*;*
b. $H(X, Y) = \max\{H(X), H(Y)\}$ *if and only if X is determined by Y or Y is determined by X;*
c. $H(X, Y) = H(X) + H(Y)$ *if and only if X and Y are independent.*

ii. Conditional entropy is bounded as follows:

a. $0 \le H(X \mid Y) \le H(X)$*;*
b. $H(X \mid Y) = 0$ *if and only if X is determined by Y;*
c. $H(X \mid Y) = H(X)$ *if and only if X and Y are independent.*

iii. Mutual information is bounded as follows:

a. $0 \le I(X; Y) \le \min\{H(X), H(Y)\}$*;*
b. $I(X; Y) = 0$ *if and only if X and Y are independent;*
c. $I(X; Y) = \min\{H(X), H(Y)\}$ *if and only if X is determined by Y or Y is determined by X.*

Proof We begin with (ii), using Lemma 8.1.3(ii):

$$H(X \mid Y) = \sum_{y\,:\,\mathbb{P}(y)>0} \mathbb{P}(y) H(X \mid y).$$

For each y such that $\mathbb{P}(y) > 0$, Lemma 2.2.4(i) implies that $H(X \mid y) \ge 0$, with equality if and only if there is some x such that $\mathbb{P}(x \mid y) = 1$. So, $H(X \mid Y) \ge 0$, with equality if and only if X is determined by Y. For the upper bound, Lemma 8.1.10 gives

$$H(X) - H(X \mid Y) = I(X; Y) \ge 0,$$

with equality if and only if X and Y are independent.

For (i), we have

$$H(X, Y) - H(X) = H(Y \mid X) \ge 0,$$

with equality if and only if Y is determined by X (by (ii)). Hence $H(X, Y) \ge \max\{H(X), H(Y)\}$, and if equality holds then Y is determined by X or vice versa. Conversely, suppose without loss of generality that Y is determined by

X. We have $H(X, Y) = H(X)$ by Example 8.1.1(iv) and $H(Y) \leq H(X)$ by Example 8.1.4(iv), so $H(X, Y) = \max\{H(X), H(Y)\}$, as required. For the upper bound on $H(X, Y)$, we have

$$H(X) + H(Y) - H(X, Y) = I(X; Y) \geq 0$$

with equality if and only if X and Y are independent, by Lemma 8.1.10.

For (iii), the lower bound and its equality condition were proved as Lemma 8.1.10. The upper bound follows from the lower bound in (i) by subtracting from $H(X) + H(Y)$:

$$\begin{aligned} & \max\{H(X), H(Y)\} \leq H(X, Y) \\ \iff & H(X) + H(Y) - \max\{H(X), H(Y)\} \geq H(X) + H(Y) - H(X, Y) \\ \iff & \min\{H(X), H(Y)\} \geq I(X; Y), \end{aligned}$$

with the same condition for equality as in (i). □

Remark 8.1.13 Given random variables X and Y taking values in finite sets $\mathcal{X}$ and $\mathcal{Y}$ respectively, there is a random variable $X \otimes Y$ taking values in $\mathcal{X} \times \mathcal{Y}$, the **independent coupling** of X and Y, with distribution

$$\mathbb{P}(X \otimes Y = (x, y)) = \mathbb{P}(X = x)\mathbb{P}(Y = y)$$

($x \in \mathcal{X}$, $y \in \mathcal{Y}$). That is, if X has distribution $\mathbf{p}$ and Y has distribution $\mathbf{r}$ then $X \otimes Y$ has distribution $\mathbf{p} \otimes \mathbf{r}$. Then

$$H(X \otimes Y) = H(X) + H(Y)$$

by Corollary 2.2.10, so the upper bound in Proposition 8.1.12(i) is equivalent to

$$H(X, Y) \leq H(X \otimes Y).$$

Another way to state this is as follows. Take probability distributions $\mathbf{p}$ on $\mathcal{X}$ and $\mathbf{r}$ on $\mathcal{Y}$. Then among all probability distributions on $\mathcal{X} \times \mathcal{Y}$ with marginals $\mathbf{p}$ and $\mathbf{r}$, none has greater entropy than $\mathbf{p} \otimes \mathbf{r}$.

This is a special property of *Shannon* entropy. It does not hold for any of the other Rényi entropies H_q or q-logarithmic entropies S_q except, trivially, when $q = 0$. Counterexamples are given in Appendix A.6. There is a substantial literature on the entropy of couplings; see, for instance, Sason [303], Kovačević, Stanojević and Šenk [197], and references therein.

8.2 Diversity Measures for Subcommunities

In the next two sections, we introduce quantities measuring features of a large community of organisms (a **metacommunity**) divided into smaller communities (**subcommunities**), to answer the ecological questions posed in the introduction to this chapter. As before, we use terminology inspired by ecology, even though the mathematics applies far more generally to any types of object.

We have already discussed several times a special type of metacommunity, namely, a group of islands (Examples 2.1.6, 2.4.9, 2.4.11, etc.). There, the subcommunities are the islands, the metacommunity is the union of all of them, and a very strong assumption is made: that no species are shared between islands. Although this is a useful hypothetical extreme case, it is not realistic. In the metacommunities that we are about to consider, each species may be present in one, many, or all of the subcommunities, in any proportions.

In ecology, there is established terminology for measures of metacommunity diversity:

- the **alpha-diversity** is the average diversity of the subcommunities (in some sense of 'average');
- the **beta-diversity** is the variation between the subcommunities;
- the **gamma-diversity** is the diversity of the whole metacommunity (the global diversity), ignoring its division into subcommunities.

These terms were introduced by the ecologist Robert Whittaker in an influential paper of 1960 ([351], p. 320). As Tuomisto observed in a survey paper on beta-diversity, 'Obviously, Whittaker (1960) did not have an exact definition of beta diversity in mind' ([334], p. 2). However, a large number of specific proposals have been made for defining these three quantities mathematically. Some early work on the subject may have been inspired by analysis of variance (ANOVA) in statistics, where one seeks to quantify within-group and between-group variation. But the broad concepts of alpha-, beta- and gamma-diversity acquired their own independent standing long ago.

This section and the next are largely based on a paper of Reeve et al. [293] that sets out a comprehensive and non-traditional suite of diversity measures for metacommunities and their subcommunities. The system of measures is highly flexible, incorporating both the parameter q (to allow different emphasis on rare and common species) and the similarity matrix Z (to encode the different similarities between species). Here, we confine ourselves to the very special case where $q = 1$ and $Z = I$ (thus, ignoring inter-species similarity). Even so, we will be able to see some of the power and subtlety of the system.

We begin by fixing our notation (Figure 8.4). The metacommunity consists

subcommunities

species

$$P = \begin{pmatrix} P_{11} & \cdots & P_{1N} \\ \vdots & & \vdots \\ P_{S1} & \cdots & P_{SN} \end{pmatrix} \qquad \mathbf{p} = \begin{pmatrix} p_1 \\ \vdots \\ p_S \end{pmatrix}$$

$$\mathbf{w} = \begin{pmatrix} w_1 & \cdots & w_N \end{pmatrix}$$

Figure 8.4 Notation for the relative abundances in a metacommunity.

of a collection of individuals, each of which belongs to exactly one of S species (numbered as $1, \ldots, S$) and exactly one of N subcommunities (numbered as $1, \ldots, N$). We write P_{ij} for the proportion or relative abundance of individuals belonging to the ith species and the jth subcommunity. Thus, $\sum_{i,j} P_{ij} = 1$. We adopt the convention that the index i ranges over the set $\{1, \ldots, S\}$ of species and the index j ranges over the set $\{1, \ldots, N\}$ of subcommunities.

For each species i, write

$$p_i = \sum_j P_{ij},$$

which is the relative abundance of species i in the whole metacommunity. Then $\sum_i p_i = 1$. For each subcommunity j, write

$$w_j = \sum_i P_{ij},$$

which is the relative size of subcommunity j in the metacommunity. Then $\sum_j w_j = 1$.

In purely mathematical terms, the matrix P defines a probability distribution on the set $\{1, \ldots, S\} \times \{1, \ldots, N\}$, with marginal distributions

$$\mathbf{p} = (p_1, \ldots, p_S), \qquad \mathbf{w} = (w_1, \ldots, w_N).$$

To translate into the language of random variables, we will consider a random variable (X, Y) taking values in $\{1, \ldots, S\} \times \{1, \ldots, N\}$, with distribution P. Then X is a random variable with values in $\{1, \ldots, S\}$ and distribution $\mathbf{p}$, and Y is a random variable with values in $\{1, \ldots, N\}$ and distribution $\mathbf{w}$. Thus, X is a random species and Y is a random subcommunity.

What are the ecological meanings of the joint entropy, conditional entropy and mutual information of the random variables X and Y? And what are the roles of relative and cross entropy? We have seen that when measuring diversity, it is more appropriate to use the *exponential* of entropy than entropy itself

(Section 2.4, especially Example 2.4.7). So it is better to ask: what are the ecological meanings of the exponentials of relative entropy, mutual information, and so on?

We now proceed to answer these questions, following throughout the notation and terminology of Reeve et al. [293].

First, consider the two entropies defined for a pair of distributions on the *same* set: relative and cross entropy. The *j*th subcommunity has species distribution

$$P_{\bullet j}/w_j = (P_{1j}/w_j, \ldots, P_{Sj}/w_j),$$

which is the normalization of the *j*th column $P_{\bullet j}$ of the matrix P. (Assume that the *j*th subcommunity is nonempty: $w_j > 0$.) We write

$$\overline{\alpha}_j(P) = D(P_{\bullet j}/w_j) = \exp H(P_{\bullet j}/w_j)$$

for the diversity of order 1 of the *j*th subcommunity, and call it the **subcommunity alpha-diversity**. Thus, $\overline{\alpha}_j(P)$ depends on the *j*th subcommunity only, and is unaffected by the rest of the metacommunity. Here D denotes the Hill number D_1 of order 1 (as in Section 2.4); no other value of the parameter q is under consideration.

As well as considering the *j*th subcommunity in isolation, we can compare its species distribution $P_{\bullet j}/w_j$ with the species distribution $\mathbf{p}$ of the whole metacommunity, using the relative entropy $H(P_{\bullet j}/w_j \parallel \mathbf{p})$. Better, we can use the exponential of relative entropy, which is a relative diversity in the sense of Section 3.3. Thus, we define the **subcommunity beta-diversity** $\overline{\beta}_j(P)$ by

$$\overline{\beta}_j(P) = D(P_{\bullet j}/w_j \parallel \mathbf{p}) = \prod_i \left(\frac{P_{ij}}{p_i w_j}\right)^{P_{ij}/w_j}. \tag{8.6}$$

This is the diversity of the species distribution of the *j*th subcommunity relative to that of the metacommunity. As established in Section 3.3, it reflects the unusualness or atypicality of the subcommunity in the context of the metacommunity as a whole. For example, if the subcommunity is exactly representative of the whole metacommunity then $\overline{\beta}_j(P)$ takes its minimum possible value, 1.

(Reeve et al. also defined quantities called α_j and β_j, not discussed here. The bars are used in that work to indicate normalization by subcommunity size.)

Alternatively, we can compare a subcommunity with the metacommunity using cross entropy rather than relative entropy. The exponential $D^\times(P_{\bullet j}/w_j \| \mathbf{p})$ of the cross entropy is a cross diversity (again in the sense of Section 3.3), and is called the **subcommunity gamma-diversity**,

$$\gamma_j(P) = D^{\times}(P_{\bullet j}/w_j \parallel \mathbf{p}) = \prod_i \left(\frac{1}{p_i}\right)^{P_{ij}/w_j}. \tag{8.7}$$

Thus, $\gamma_j(P)$ is the cross diversity of the species distribution of the jth subcommunity with respect to that of the metacommunity. It is the average rarity of species in the subcommunity, measuring rarity by the standards of the metacommunity (and using the geometric mean as our notion of average). For example, if the subcommunity is exactly representative of the metacommunity then $\gamma_j(P)$ is just the diversity of the metacommunity.

Other examples of relative diversity and cross diversity were given in Section 3.3, illustrating the ecological meanings of high or low values of $\overline{\beta}_j(P)$ or $\gamma_j(P)$.

Equation (3.6) (p. 67) implies that

$$\overline{\alpha}_j(P) \cdot \overline{\beta}_j(P) = \gamma_j(P). \tag{8.8}$$

This identity can be understood as follows:

- $\overline{\alpha}_j(P)$ measures how unusual the average individual is within the subcommunity;
- $\overline{\beta}_j(P)$ measures how unusual the subcommunity is within the metacommunity;
- $\gamma_j(P)$ measures how unusual the average individual in the subcommunity is within the metacommunity.

Thus, equation (8.8) partitions the global diversity measure $\gamma_j(P)$ into components measuring diversity at different levels of resolution.

In the next section, we explain the connection between, on the one hand, the *subcommunity* alpha-, beta- and gamma-diversities just defined, and, on the other, what ecologists usually call alpha-, beta- and gamma-diversity, which are quantities associated with the *metacommunity*. We will use the language of random variables. In that language, the subcommunity measures that we have just defined are given by

$$\begin{aligned}
\overline{\alpha}_j(P) &= \exp(H(X \mid j)),\\
\overline{\beta}_j(P) &= \exp\big(H((X \mid j) \parallel X)\big),\\
\gamma_j(P) &= \exp\big(H^{\times}((X \mid j) \parallel X)\big),
\end{aligned}$$

since the distribution of the conditional random variable $X \mid j$ is the species distribution $P_{\bullet j}/w_j$ in the jth subcommunity.

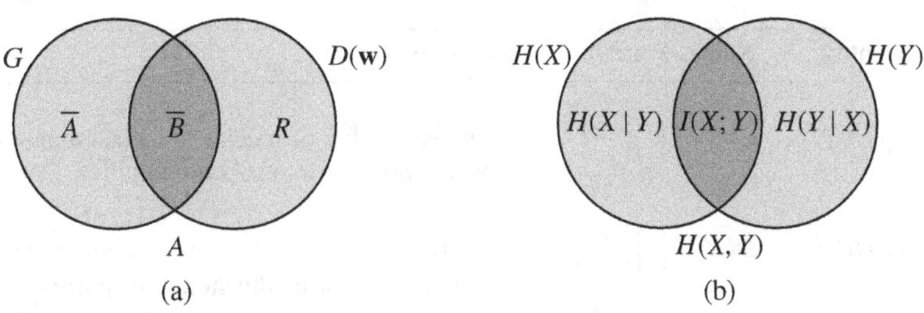

Figure 8.5 The metacommunity gamma-diversities, shown in (a), are the exponentials of the entropies shown in (b). For instance, $\overline{A}(P) = \exp H(X \mid Y)$.

8.3 Diversity Measures for Metacommunities

In the last section, the alpha-, beta- and gamma-diversities of the jth subcommunity were defined by comparing two random variables taking values in the set of species: $X \mid j$, which is the species of an individual chosen at random from the jth subcommunity, and X, which is the species of an individual chosen at random from the whole metacommunity.

In this section, we derive measures of the metacommunity by comparing two random variables taking values in different sets: the species X and the subcommunity Y of an individual chosen at random from the metacommunity. Specifically, we consider the exponentials of their joint entropy, conditional entropies, and mutual information.

Figure 8.5(a) summarizes the situation, with the previous Venn diagram for entropies (Figure 8.1) reproduced in (b) for reference. The notation in (a) is again taken from Reeve et al. [293], and each term is now explained in turn. Tables 8.1 and 8.2 give summaries.

Metacommunity Gamma-Diversity

First consider the random variable X for species. The exponential of its Shannon entropy $H(X)$ is

$$D(\mathbf{p}) = \prod_{i \in \operatorname{supp}(\mathbf{p})} \left(\frac{1}{p_i}\right)^{p_i}.$$

This is simply the diversity of order 1 of the species distribution $\mathbf{p}$ of the whole metacommunity, ignoring its division into subcommunities. (Throughout this section, all diversities are of order 1.) We write $G(P) = D(\mathbf{p})$ and call it the **metacommunity gamma-diversity**.

The metacommunity gamma-diversity $G(P)$ is related to the subcommunity

Quantity	Name	Formula	Description
$\exp H(X)$	$G(P)$	$\prod_i \left(\frac{1}{p_i}\right)^{p_i}$	Effective number of species in metacommunity, ignoring division into subcommunities
$\exp H(Y)$	$D(\mathbf{w})$	$\prod_j \left(\frac{1}{w_j}\right)^{w_j}$	Effective number of subcommunities in metacommunity, ignoring division into species
$\exp H(X, Y)$	$A(P)$	$\prod_{i,j} \left(\frac{1}{P_{ij}}\right)^{P_{ij}}$	Effective number of (species, subcommunity) pairs
$\exp H(X \mid Y)$	$\overline{A}(P)$	$\prod_{i,j} \left(\frac{w_j}{P_{ij}}\right)^{P_{ij}}$	Average effective number of species per subcommunity
$\exp H(Y \mid X)$	$R(P)$	$\prod_{i,j} \left(\frac{p_i}{P_{ij}}\right)^{P_{ij}}$	Redundancy of subcommunities
$\exp I(X; Y)$	$\overline{B}(P)$	$\prod_{i,j} \left(\frac{P_{ij}}{p_i w_j}\right)^{P_{ij}}$	Effective number of isolated subcommunities

Table 8.1 Formulas for and descriptions of the metacommunity diversity measures. The first product is over the support of $\mathbf{p}$, the second is over the support of $\mathbf{w}$, and the others are over the support of P.

Name	Range	Minimized when	Maximized when
$G(P)$	$[1, S]$	Only one species in metacommunity	Species in metacommunity are balanced
$D(\mathbf{w})$	$[1, N]$	Only one subcommunity in metacommunity	Subcommunities are same size
$A(P)$	$[1, SN]$	Only one species and one subcommunity in metacommunity	Subcommunities are same size and all balanced
$\overline{A}(P)$	$[1, S]$	Each subcommunity contains only one species	Each subcommunity is balanced
$R(P)$	$[1, N]$	Subcommunities share no species	Subcommunities have same size and composition
$\overline{B}(P)$	$[1, N]$	Subcommunities have same composition	Subcommunities have same size and share no species

Table 8.2 Minimum and maximum values of the metacommunity diversity measures. The bounds shown depend only on S and N; tighter bounds are given in the text. **Balanced** means that all species have equal abundance.

gamma-diversities $\gamma_1(P), \ldots, \gamma_N(P)$ as follows:

$$\begin{aligned} G(P) &= \prod_{i,j:\ i \in \mathrm{supp}(\mathbf{p})} \left(\frac{1}{p_i} \right)^{P_{ij}} \\ &= \prod_j \gamma_j(P)^{w_j} \\ &= M_0\big(\mathbf{w}, (\gamma_1(P), \ldots, \gamma_N(P))\big). \end{aligned} \tag{8.9}$$

Here we have used the definition of p_i as $\sum_j P_{ij}$ and the formula (8.7) for $\gamma_j(P)$. So, the metacommunity gamma-diversity $G(P)$ is the geometric mean of the subcommunity gamma-diversities $\gamma_j(P)$, weighted by the sizes of the subcommunities.

In this sense, the subcommunity gamma-diversity $\gamma_j(P)$ is the mean contribution per individual of the jth subcommunity to the metacommunity diversity.

The metacommunity gamma-diversity is constrained by the bounds

$$1 \leq G(P) \leq S,$$

by Lemma 2.4.3. It attains the lower bound $G(P) = 1$ when the metacommunity consists of a single species, and the upper bound $G(P) = S$ when all S species have equal abundance in the metacommunity (regardless of how they are distributed across the subcommunities).

Now consider the random variable Y for subcommunities. The exponential of $H(Y)$ is

$$D(\mathbf{w}) = \prod_{j \in \mathrm{supp}(\mathbf{w})} \left(\frac{1}{w_j} \right)^{w_j}.$$

It measures how evenly the population is distributed across the subcommunities (regardless of how it is distributed across species). It is bounded by

$$1 \leq D(\mathbf{w}) \leq N$$

(by Lemma 2.4.3 again), with $D(\mathbf{w}) = 1$ when the metacommunity contains only one nonempty subcommunity and $D(\mathbf{w}) = N$ when the populations of the N subcommunities are of equal size.

Metacommunity Alpha-Diversities

The joint entropy $H(X, Y)$ has exponential

$$D(P) = \prod_{(i,j) \in \mathrm{supp}(P)} \left(\frac{1}{P_{ij}} \right)^{P_{ij}}.$$

Here we are treating the $S \times N$ matrix P as a probability distribution on the set $\{1, \ldots, S\} \times \{1, \ldots, N\}$. So, $D(P)$ is the effective number of (species, subcommunity) pairs, in the sense of Section 2.4. It is the species diversity that the metacommunity would have if individuals in different subcommunities were decreed to be of different species (as in the island scenario). We write $A(P) = D(P)$ and call it the **raw metacommunity alpha-diversity**.

Since $A(P)$ measures diversity as if no species were shared between subcommunities, it overestimates the true diversity. Indeed, taking exponentials in the inequalities

$$H(X) \leq H(X, Y) \leq H(X) + H(Y)$$

of Proposition 8.1.12(i) gives

$$G(P) \leq A(P) \leq D(\mathbf{w})G(P). \tag{8.10}$$

The upper bound states that the factor of overestimation is at most $D(\mathbf{w})$, the effective number of subcommunities.

The minimum value $A(P) = G(P)$ occurs when $H(X, Y) = H(X)$, which by Proposition 8.1.12(ii) means that Y is determined by X: the subcommunity is determined by the species. So, $A(P) = G(P)$ when no species are shared between subcommunities.

The maximum value $A(P) = D(\mathbf{w})G(P)$ occurs when $H(X, Y) = H(X) + H(Y)$. By Proposition 8.1.12(i), this is true just when X and Y are independent. Equivalently, $A(P)$ attains its maximum when the metacommunity is **well-mixed**, meaning that each of the subcommunity species distributions $P_{\bullet j}/w_j$ is equal to the metacommunity species distribution $\mathbf{p}$.

In summary, $A(P)$ does not overestimate $G(P)$ at all when the subcommunities share no species, whereas the overestimation is most pronounced when all of the subcommunities have identical composition.

Since $1 \leq G(P) \leq S$ and $1 \leq D(\mathbf{w}) \leq N$, the inequalities (8.10) imply the cruder bounds

$$1 \leq A(P) \leq SN.$$

This conforms with the interpretation of $A(P)$ as the effective number of (species, subcommunity) pairs. The minimum $A(P) = 1$ is attained when there is only one species present and only one nonempty subcommunity. The maximum $A(P) = SN$ is attained when the metacommunity is well-mixed and the subcommunities all have the same size.

Next consider conditional entropy. By Lemma 8.1.3(i), the conditional entropy $H(X \mid Y)$ is given by

$$H(X \mid Y) = \sum_{i,j:\, \mathbb{P}(i,j)>0} \mathbb{P}(i,j) \log \frac{\mathbb{P}(j)}{\mathbb{P}(i,j)}.$$

The **normalized metacommunity alpha-diversity** $\overline{A}(P)$ is its exponential:

$$\overline{A}(P) = \exp H(X \mid Y) = \prod_{i,j:\, P_{ij}>0} \left(\frac{w_j}{P_{ij}} \right)^{P_{ij}}. \tag{8.11}$$

To understand $\overline{A}$, we use one of the other formulas for conditional entropy:

$$H(X \mid Y) = \sum_{j:\, \mathbb{P}(j)>0} \mathbb{P}(j) H(X \mid j) \tag{8.12}$$

(Lemma 8.1.3(ii)). The random variable $X \mid j$ is the species of a random individual from the jth subcommunity, so taking exponentials throughout equation (8.12) gives

$$\begin{aligned} \overline{A}(P) &= \prod_{j:\, w_j>0} \overline{\alpha}_j(P)^{w_j} \\ &= M_0\big(\mathbf{w}, (\overline{\alpha}_1(P), \ldots, \overline{\alpha}_N(P))\big). \end{aligned} \tag{8.13}$$

Hence $\overline{A}(P)$ is the geometric mean of the individual subcommunity diversities $\overline{\alpha}_j(P)$, weighted by their sizes. It is therefore an alpha-diversity in the traditional sense (Remark 3.3.9 and p. 269).

To find the maximum and minimum values of $\overline{A}$, we take exponentials throughout the inequalities

$$0 \leq H(X \mid Y) \leq H(X)$$

(Proposition 8.1.12(ii)). This gives

$$1 \leq \overline{A}(P) \leq G(P), \tag{8.14}$$

with $\overline{A}(P) = 1$ when each subcommunity contains at most one species and $\overline{A}(P) = G(P)$ when the metacommunity is well-mixed. Since $G(P) \leq S$, we also have the cruder bounds

$$1 \leq \overline{A}(P) \leq S,$$

with $\overline{A}(P) = S$ when each subcommunity contains all S species in equal proportions.

The raw and normalized metacommunity alpha-diversities, $A(P)$ and $\overline{A}(P)$, are linked by the equation

$$\overline{A}(P) = A(P)/D(\mathbf{w}), \tag{8.15}$$

Quantity	Name	(i) Well-mixed metacommunity	(ii) Only one subcommunity	(iii) Subcommunities are species	(iv) Isolated subcommunities
$\exp H(X)$	$G(P)$	$D(\mathbf{p})$	$D(\mathbf{p})$	$D(\mathbf{p})$	$D(\mathbf{p})$
$\exp H(Y)$	$D(\mathbf{w})$	$D(\mathbf{w})$	1	$D(\mathbf{p})$	$D(\mathbf{w})$
$\exp H(X,Y)$	$A(P)$	$D(\mathbf{p})D(\mathbf{w})$	$D(\mathbf{p})$	$D(\mathbf{p})$	$D(\mathbf{w})\overline{A}(P)$
$\exp H(X \mid Y)$	$\overline{A}(P)$	$D(\mathbf{p})$	$D(\mathbf{p})$	1	$\overline{A}(P)$
$\exp H(Y \mid X)$	$R(P)$	$D(\mathbf{w})$	1	1	1
$\exp I(X;Y)$	$\overline{B}(P)$	1	1	$D(\mathbf{p})$	$D(\mathbf{w})$

Table 8.3 Summary of Examples 8.3.1 and 8.3.2.

which is the exponential of the definition

$$H(X \mid Y) = H(X,Y) - H(Y)$$

of conditional entropy.

Examples 8.3.1 Here we take the four running examples of Section 8.1 and translate them into ecological terms. The results below follow immediately from those examples, and are summarized in the first few rows of Table 8.3.

i. Suppose that the metacommunity is well-mixed. Then $P = \mathbf{p} \otimes \mathbf{w}$ and $A(P) = D(P) = D(\mathbf{p})D(\mathbf{w})$. Each subcommunity has the same species composition as the metacommunity, so the mean subcommunity diversity $\overline{A}(P)$ is the same as the metacommunity diversity $G(P)$.
ii. Suppose that the metacommunity consists of a single subcommunity. Then $N = 1$, $\mathbf{w} = (1)$, and $P = \mathbf{p}$. The effective number $A(P)$ of (species, subcommunity) pairs is just the effective number $D(\mathbf{p})$ of species, and since there is only one subcommunity, the average subcommunity diversity $\overline{A}(P)$ is also $D(\mathbf{p})$.
iii. Suppose that the subcommunities are exactly the species. Thus, $N = S$, $\mathbf{w} = \mathbf{p}$, and P is the diagonal matrix with entries $p_1, \ldots, p_S$. The effective number $A(P)$ of (species, subcommunity) pairs is again just $D(\mathbf{p})$, but since each subcommunity has a diversity of 1, the average subcommunity diversity $\overline{A}(P)$ is now 1.
iv. Finally, suppose that all subcommunities are isolated (share no species). Nothing special can be said about $\overline{A}(P)$, the mean subcommunity diversity, since it is unaffected by the degree of overlap of species between subcommunities. As always, $A(P) = D(\mathbf{w})\overline{A}(P)$.

The Redundancy of a Metacommunity

We have already considered one conditional entropy, $H(X \mid Y)$. The other,

$$H(Y \mid X) = \sum_{i,j:\, \mathbb{P}(i,j)>0} \mathbb{P}(i,j) \log \frac{\mathbb{P}(i)}{\mathbb{P}(i,j)},$$

has exponential

$$R(P) = \prod_{i,j:\, P_{ij}>0} \left(\frac{p_i}{P_{ij}} \right)^{P_{ij}}. \tag{8.16}$$

This is the **redundancy** of the metacommunity. The word is meant in the following sense: if some subcommunities were to be destroyed, how much of the diversity in the metacommunity would be preserved? High redundancy means that there is enough repetition of species across subcommunities that loss of some subcommunities would probably not cause great loss of diversity.

We now justify this interpretation. For each species i, consider the relative abundances $P_{i1}, \ldots, P_{iN}$ of that species within the N subcommunities, and normalize to obtain a probability distribution

$$P_{i\bullet}/p_i = (P_{i1}/p_i, \ldots, P_{iN}/p_i)$$

on the set $\{1, \ldots, N\}$ of subcommunities. (We assume in this explanation that $p_i > 0$.) Then $D(P_{i\bullet}/p_i)$ measures the extent to which the ith species is spread evenly across the subcommunities; for instance, it takes its maximum value N when there is the same amount of species i in every subcommunity. This is the 'redundancy' of the ith species.

To obtain a measure of the redundancy of the whole metacommunity, we take the geometric mean

$$\prod_{i \in \mathrm{supp}(\mathbf{p})} D(P_{i\bullet}/p_i)^{p_i} \tag{8.17}$$

of the redundancies of the species, weighted by their relative abundances. But this is exactly $R(P)$, since, using Lemma 8.1.3(ii),

$$\begin{aligned}
\prod_{i \in \mathrm{supp}(\mathbf{p})} D(P_{i\bullet}/p_i)^{p_i} &= \exp\Bigg(\sum_{i \in \mathrm{supp}(\mathbf{p})} p_i H(P_{i\bullet}/p_i) \Bigg) \\
&= \exp\Bigg(\sum_{i:\, \mathbb{P}(i)>0} \mathbb{P}(i) H(Y \mid i) \Bigg) \\
&= \exp H(Y \mid X) \\
&= R(P).
\end{aligned}$$

In conclusion, $R(P)$ is the average species redundancy (8.17): the effective number of subcommunities across which a typical species is spread.

A different way to understand redundancy is through the equation

$$R = A/G, \tag{8.18}$$

which is the exponential of the definition

$$H(Y \mid X) = H(X, Y) - H(X)$$

of conditional entropy. The gamma-diversity $G(P)$ is the effective number of species in the metacommunity, whereas $A(P)$ is the effective number of species if we pretend that individuals in different subcommunities are always of different species. The amount by which $A(P)$ overestimates $G(P)$ reflects the extent to which species are, in reality, shared across subcommunities: thus, it measures redundancy.

Bounds on the redundancy can be obtained by dividing inequalities (8.10) by $G(P)$. This gives

$$1 \leq R(P) \leq D(\mathbf{w}),$$

with the same extremal cases as for (8.10): redundancy takes its minimum value of 1 when no species are shared between subcommunities, and its maximum value of $D(\mathbf{w})$ when the species distributions in the subcommunities are all the same. It follows that

$$1 \leq R(P) \leq N,$$

with $R(P) = N$ when the subcommunities have not only the same composition, but also the same size.

Metacommunity Beta-Diversity

Finally, consider the mutual information $I(X; Y)$. By Lemma 8.1.7(i), its exponential $\overline{B}(P)$ is given by

$$\overline{B}(P) = \prod_{i,j:\, P_{ij}>0} \left(\frac{P_{ij}}{p_i w_j} \right)^{P_{ij}}.$$

This is the **metacommunity beta-diversity**. (In Reeve et al. [293], it is called the 'normalized' beta-diversity, and there is also a 'raw' beta-diversity $B(P)$, not treated here.) By the discussion of mutual information in Section 8.1, $\overline{B}(P)$ can be understood as the exponential of the amount of information that knowledge of an individual's species gives us about its subcommunity – or equivalently, vice versa.

Loosely, then, $\overline{B}(P)$ measures the alignment between subcommunity structure and species structure. It is a beta-diversity in the traditional sense of Remark 3.3.9 and p. 269.

By Proposition 8.1.12(iii) on mutual information,

$$1 \le \overline{B}(P) \le \min\{G(P), D(\mathbf{w})\}. \tag{8.19}$$

The minimum $\overline{B}(P) = 1$ is attained when X and Y are independent, that is, the metacommunity is well-mixed. In that case, knowing an individual's species does not help us to guess its subcommunity, nor vice versa. By Proposition 8.1.12(iii), there are two cases in which the maximum is attained. One is where X is determined by Y, that is, there is at most one species in each subcommunity. Then by Example 8.1.4(iv),

$$\overline{B}(P) = G(P) \le D(\mathbf{w}).$$

In this case, knowing the subcommunity to which an individual belongs enables us to infer its species with certainty. The other case in which the maximum is attained is where Y is determined by X, that is, the subcommunities are isolated. Then

$$\overline{B}(P) = D(\mathbf{w}) \le G(P)$$

by Example 8.1.4(iv) again, and knowing an individual's species enables us to infer its subcommunity with certainty.

We can also interpret $\overline{B}$ as the effective number of isolated subcommunities. Indeed, since $1 \le D(\mathbf{w}) \le N$, the inequalities (8.19) imply that

$$1 \le \overline{B}(P) \le N. \tag{8.20}$$

The maximum $\overline{B}(P) = N$ occurs when the N subcommunities are isolated and of equal size. We will see in Proposition 8.4.8 and Corollary 8.4.10 that $\overline{B}$ satisfies a chain rule and a replication principle, supporting the effective number interpretation.

For yet another viewpoint on $\overline{B}$, recall from equations (8.5) that

$$H(X \mid Y) + I(X; Y) = H(X).$$

Taking exponentials throughout gives

$$\overline{A}\,\overline{B} = G, \tag{8.21}$$

that is,

$$\text{alpha-diversity} \times \text{beta-diversity} = \text{gamma-diversity}.$$

This equation partitions the diversity of the metacommunity (gamma) into two components: the average diversity within subcommunities (alpha) and the variation between subcommunities (beta). The general principle has a long history, going back to the foundational work of Whittaker (p. 321 of [351] and p. 232 of [352]).

From equations (8.21), (8.15) and (8.18), it follows that

$$\overline{B}(P) = \frac{G(P)}{\overline{A}(P)} = \frac{G(P)D(\mathbf{w})}{A(P)} = \frac{D(\mathbf{w})}{R(P)},$$

giving

$$\overline{B}(P) = \frac{D(\mathbf{w})}{R(P)}.$$

So when the subcommunity sizes $\mathbf{w}$ are fixed, the effective number $\overline{B}(P)$ of isolated subcommunities is inversely proportional to the redundancy $R(P)$. This is reasonable: $R(P)$ measures overlap of species between subcommunities, whereas $\overline{B}(P)$ measures how disjoint the subcommunities are.

(Matters are more subtle outside the case $q = 1$, $Z = I$ to which we have confined ourselves. When $q \neq 1$ in the work of Reeve et al., the dependency between $\overline{B}$ and R breaks down, in the strong sense that the two quantities no longer determine one another; they convey different information.)

Examples 8.3.2 We return to the four scenarios of Examples 8.3.1, finding the redundancy $R(P)$ and metacommunity beta-diversity $\overline{B}(P)$. The results are summarized in Table 8.3.

i. A well-mixed metacommunity is maximally redundant ($R(P) = D(\mathbf{w})$), since all subcommunities are identical. For the same reason, the effective number $\overline{B}(P)$ of isolated subcommunities is just 1.
ii. If the metacommunity consists of a single subcommunity ($N = 1$) then the redundancy $R(P)$ and effective number $\overline{B}(P)$ of isolated subcommunities both take their minimum values of 1.
iii. Suppose that the subcommunities are exactly the species. Then the metacommunity is minimally redundant ($R(P) = 1$), reflecting the fact that each species is present in just one subcommunity: losing any of the subcommunities means losing a species. And since subcommunities are species, the effective number $\overline{B}(P)$ of isolated subcommunities is the effective number $D(\mathbf{p})$ of species.
iv. More generally, suppose that all subcommunities are isolated. The redundancy is minimal ($R(P) = 1$), since no species is repeated across subcommunities. The effective number $\overline{B}(P)$ of isolated subcommunities is simply the diversity $D(\mathbf{w})$ of the subcommunity distribution $\mathbf{w}$, which is reasonable since, in fact, the subcommunities *are* isolated.

Just as the metacommunity gamma-diversity $G(P)$ is the geometric mean of the subcommunity gamma-diversities $\gamma_j(P)$, and just as the metacommunity alpha-diversity $\overline{A}(P)$ is the geometric mean of the subcommunity alpha-diversities $\overline{\alpha}_j(P)$, the metacommunity beta-diversity $\overline{B}(P)$ is the geometric mean of the subcommunity beta-diversities $\overline{\beta}_j(P)$. Indeed, recall from Lemma 8.1.7(ii) that

$$I(X;Y) = \sum_{j\colon \mathbb{P}(j)>0} \mathbb{P}(j) H((X \mid j) \,\|\, X).$$

Taking exponentials throughout gives

$$\begin{aligned} \overline{B}(P) &= \prod_{j\colon w_j>0} \overline{\beta}_j(P)^{w_j} \\ &= M_0\Big(\mathbf{w}, (\overline{\beta}_1(P), \ldots, \overline{\beta}_N(P))\Big), \end{aligned} \tag{8.22}$$

as claimed.

We have seen that $\overline{\beta}_j(P)$ measures how unusual the jth subcommunity is in the context of the metacommunity. Taking the geometric mean over all subcommunities gives $\overline{B}(P)$, which is therefore an overall measure of the atypicality or isolation of the subcommunities within the metacommunity.

Further connections between beta-diversity and information-theoretic quantities are described in the first appendix of the paper [293] of Reeve et al.

8.4 Properties of the Metacommunity Measures

In this chapter so far, we have introduced a system of measures of the diversity and structure of a metacommunity, and explained their behaviour in a variety of hypothetical examples. But just as a measure of the diversity of a *single* community should not be accepted or used until it can be shown to behave logically (Section 2.4), the metacommunity measures should also be required to have sensible logical and algebraic properties. Here we show that they do.

Independence

We begin by showing that the alpha-diversity $\overline{A}$ and beta-diversity $\overline{B}$ are independent – not in the sense of probability theory, but in the sense of *certain* knowledge. An informal example illustrates the idea. Assume for simplicity that every person in the world is either dark-haired or fair-haired, and either brown-eyed or blue-eyed. These two variables, hair colour and eye colour, are not independent in the probabilistic sense: people with dark hair are more likely to have dark eyes. However, they are independent in a weaker sense: knowing an individual's hair colour gives no *certain* knowledge of their eye colour, nor vice versa. All four combinations occur.

The formal definition is as follows.

Definition 8.4.1 Let J, K and L be sets. Functions

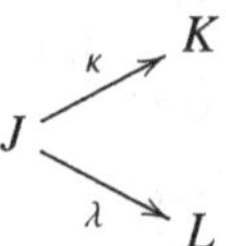

are **independent** if for all $k \in \kappa J$ and $\ell \in \lambda J$, there exists $j \in J$ such that $\kappa(j) = k$ and $\lambda(j) = \ell$.

For κ and λ to be independent means that if I choose in secret an element j of J, and tell you the value of $\kappa(j) \in K$, you gain no certain information about the value of $\lambda(j)$. (For by definition of independence, the image under λ of the fibre $\kappa^{-1}\{\kappa(j)\}$ is no smaller than the whole image λJ.) Of course, the same is also true with the roles of κ and λ reversed. In the informal example above, J is the set of all people, $K = \{\text{dark hair, fair hair}\}$, and $L = \{\text{brown eyes, blue eyes}\}$.

We have discussed the general goal of decomposing any measure of metacommunity diversity (any 'gamma-diversity') into within-group (alpha) and between-group (beta) components. The alpha-diversity and beta-diversity should be independent, otherwise the word 'decomposition' is not deserved: certain values of alpha would exclude certain values of beta, and vice versa. This requirement has been recognized in ecology since at least the 1984 work of Wilson and Shmida [360]. As Jost put it:

> Since [alpha- and beta-diversity] measure completely different aspects of regional diversity, they must be free to vary independently; alpha should not put mathematical constraints on the possible values of beta, and vice versa. If beta depended on alpha, it would be impossible to compare beta diversities of regions whose alpha diversities differed.

([167], p. 2428.)

We will show that the decomposition

$$\overline{A}\,\overline{B} = G$$

(equation (8.21)) passes this test. Since the number N of subcommunities is usually known in advance, but the number S of species may not be, we interpret independence as meaning that for each $N \geq 1$, the functions

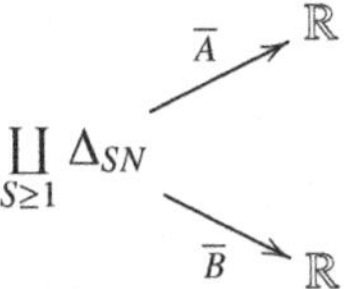

are independent. Here Δ_{SN} is understood as the set of $S \times N$ matrices P of nonnegative reals summing to 1, so that the disjoint union $\coprod_{S\geq 1} \Delta_{SN}$ is the set of all such matrices P with N columns and any number of rows.

Independence of these two functions means that given a metacommunity divided into a known number N of subcommunities, knowledge of the mean diversity $\overline{A}(P)$ of the subcommunities does not restrict the range of possible values of $\overline{B}(P)$, the effective number of isolated subcommunities. Thus, $\overline{B}(P)$ can still take all the values that it could have taken had we not known $\overline{A}(P)$. Equivalently, independence means that knowing the value of $\overline{B}(P)$ does not enable us to deduce anything about $\overline{A}(P)$. We prove this now.

Proposition 8.4.2 (Independence of alpha- and beta-diversities) *For each $N \geq 1$, the functions*

$$\coprod_{S\geq 1} \Delta_{SN} \underset{\overline{B}}{\overset{\overline{A}}{\rightrightarrows}} \mathbb{R}$$

are independent.

Proof We have already shown that $\overline{A}(P) \geq 1$ and $1 \leq \overline{B}(P) \leq N$ for all $P \in \coprod_{S\geq 1} \Delta_{SN}$ (inequalities (8.14) and (8.20)). So it suffices to show that given any $a \in [1, \infty)$ and $b \in [1, N]$, there exist some $S \geq 1$ and some $S \times N$ matrix $P \in \Delta_{SN}$ such that $\overline{A}(P) = a$ and $\overline{B}(P) = b$.

One way to do this is as follows. Choose an integer $T \geq a$. The diversity measure $D \colon \Delta_T \to \mathbb{R}$ is continuous (Lemma 2.4.4) with minimum 1 and maximum T (Lemma 2.4.3), so we can choose some $\mathbf{t} \in \Delta_T$ such that $D(\mathbf{t}) = a$. Similarly, we can choose some $\mathbf{w} \in \Delta_N$ such that $D(\mathbf{w}) = b$.

Now consider a metacommunity made up of N subcommunities of relative sizes $w_1, \ldots, w_N$, with no shared species, where each subcommunity contains

T species in proportions $t_1, \ldots, t_T$. Thus, there are TN species in all, and

$$P = \begin{pmatrix} w_1t_1 & 0 & & 0 \\ \vdots & \vdots & \cdots & \vdots \\ w_1t_T & 0 & & 0 \\ 0 & w_2t_1 & & 0 \\ \vdots & \vdots & \cdots & \vdots \\ 0 & w_2t_T & & 0 \\ & & & \\ \vdots & \vdots & & \vdots \\ & & & \\ 0 & 0 & & w_Nt_1 \\ \vdots & \vdots & \cdots & \vdots \\ 0 & 0 & & w_Nt_T \end{pmatrix}.$$

The species distribution $P_{\bullet i}/w_j$ in the jth subcommunity is

$$(0, \ldots, 0, t_1, \ldots, t_T, 0, \ldots, 0),$$

so its diversity $\overline{\alpha}_j(P)$ is $D(\mathbf{t}) = a$. But $\overline{A}(P)$ is an average of $\overline{\alpha}_1(P), \ldots, \overline{\alpha}_N(P)$ (equation (8.13)), so $\overline{A}(P) = a$. Moreover, the subcommunities are isolated, so by Example 8.3.2(iv), $\overline{B}(P) = D(\mathbf{w}) = b$. □

In the same sense, the average subcommunity diversity $\overline{A}$ and the redundancy R are independent.

Proposition 8.4.3 (Independence of alpha-diversity and redundancy) *For each $N \geq 1$, the functions*

$$\coprod_{S\geq 1} \Delta_{SN} \underset{R}{\overset{\overline{A}}{\rightrightarrows}} \mathbb{R}$$

are independent.

Proof The proof is similar to that of the last proposition. We have already seen that $\overline{A}(P) \in [1, \infty)$ and $R(P) \in [1, N]$ for all $P \in \coprod_{S\geq 1} \Delta_{SN}$. So it suffices to show that given $a \in [1, \infty)$ and $r \in [1, N]$, there exist an integer $S \geq 1$ and an $S \times N$ matrix $P \in \Delta_{SN}$ such that $\overline{A}(P) = a$ and $R(P) = r$.

To prove this, choose an integer $S \geq a$ and distributions $\mathbf{p} \in \Delta_S$, $\mathbf{w} \in \Delta_N$ such that $D(\mathbf{p}) = a$ and $D(\mathbf{w}) = r$. Consider a well-mixed metacommunity made up of N subcommunities of relative sizes $w_1, \ldots, w_N$, each with the same S species in proportions $p_1, \ldots, p_S$. Thus, $P = \mathbf{p} \otimes \mathbf{w}$. By Examples 8.3.1(i) and 8.3.2(i), $\overline{A}(P) = D(\mathbf{p}) = a$ and $R(P) = D(\mathbf{w}) = r$. □

Identical Subcommunities

When we were analysing the diversity of a single community, we argued that any similarity-sensitive diversity measure should be unchanged if a species is reclassified into two identical smaller species, and we proved that the diversity measures D_q^Z do indeed enjoy this property (Lemma 6.2.9, Example 6.2.10, and the text afterwards). It follows by continuity that if a species is divided into two nearly identical parts, the diversity increases only slightly. This is sensible behaviour, given that diversity is intended to measure the effective number of completely dissimilar species (p. 183).

The same principle applies to $\overline{B}$, the effective number of isolated subcommunities in a metacommunity. Dividing a subcommunity into two smaller subcommunities of identical composition should not change $\overline{B}$. In other words, the effective number of isolated subcommunities should not be changed by the presence or absence of boundaries between subcommunities that are identical. The average subcommunity diversity $\overline{A}$ should be similarly unaffected.

In summary, then, the decomposition of global diversity into within-subcommunity and between-subcommunity components,

$$\overline{A}\,\overline{B} = G, \tag{8.23}$$

should be unaffected by arbitrary decisions about where subcommunity boundaries lie. This is best explained by example.

Example 8.4.4 Suppose that we are interested in the tree diversity of a country that is divided into administrative districts with no ecological significance. Suppose further that in a particular pair of neighbouring districts, the distributions of tree species are identical. In that case, the partitioning (8.23) of the overall diversity into within- and between-district components should be the same as if the neighbouring districts had been merged into one. The effective number of isolated subcommunities should be invariant under the removal or addition of ecologically irrelevant boundaries.

Example 8.4.5 Suppose that we are studying the various species of grass on a hillside. To investigate the varying abundances of different species at different altitudes, we divide the hillside into height bands (0–10m, 10–20m, etc.) and regard them as subcommunities.

The beta-diversity $\overline{B}$ measures the effective number of isolated or disjoint subcommunities, so if it turns out that the bottom two height bands have the same species distribution, then $\overline{B}$ should be the same as if they were considered as a single band (0–20m).

In short, we require that the decomposition (8.23) of metacommunity diversity into alpha and beta components is ecologically meaningful, not an artefact of the particular subcommunity division chosen. As far as possible, the decomposition should be independent of resolution (that is, how fine or coarse the subcommunity division may be). In general, the finer the division one uses, the more variation between subcommunities one will observe. But if a subcommunity is ecologically uniform (has the same species distribution throughout) then dividing it further should make no difference to $\overline{B}$ or $\overline{A}$.

We now give a formal statement and proof of the desired invariance property of $\overline{A}$, $\overline{B}$ and G. To minimize notational overhead, we consider splitting a single subcommunity into two rather than splitting every subcommunity into an arbitrary number of smaller parts; but the general case follows by induction.

In the standard notation of this chapter, take an $S \times N$ matrix $P \in \Delta_{SN}$ with species distribution $\mathbf{p} \in \Delta_S$ and subcommunity size distribution $\mathbf{w} \in \Delta_N$, so that $p_i = \sum_j P_{ij}$ and $w_j = \sum_i P_{ij}$. Split the last subcommunity into two parts of relative sizes t and $1 - t$ (where $0 \leq t \leq 1$), and suppose that the two parts have the same species distribution. Then the new relative abundance matrix is the $S \times (N + 1)$ matrix P' given by

$$P'_{ij} = \begin{cases} P_{ij} & \text{if } 1 \leq j \leq N-1, \\ tP_{iN} & \text{if } j = N, \\ (1-t)P_{iN} & \text{if } j = N+1. \end{cases}$$

Proposition 8.4.6 (Identical subcommunities) *In the situation described,*

$$\overline{A}(P') = \overline{A}(P), \qquad \overline{B}(P') = \overline{B}(P), \qquad G(P') = G(P).$$

(In Reeve et al. [293], the splitting of a subcommunity into smaller parts with the same species distribution is called 'shattering', so the result is that $\overline{A}$, $\overline{B}$ and G are invariant under shattering.)

The idea behind the proof is that $\overline{A}$ is the average diversity of the subcommunity of an individual chosen at random, and this quantity is unchanged if a well-mixed subcommunity is split into smaller parts.

Proof Write $\mathbf{p}' \in \Delta_S$ and $\mathbf{w}' \in \Delta_{N+1}$ for the row- and column-sums of P'. Then for each $i \in \{1, \ldots, S\}$,

$$p'_i = \sum_{j=1}^{N-1} P_{ij} + tP_{iN} + (1-t)P_{iN} = \sum_{j=1}^{N} P_{ij} = p_i,$$

so $\mathbf{p}' = \mathbf{p}$, and for each $j \in \{1, \ldots, N+1\}$,

$$w'_j = \sum_{i=1}^{S} P'_{ij} = \begin{cases} w_j & \text{if } j \leq N-1, \\ tw_N & \text{if } j = N, \\ (1-t)w_N & \text{if } j = N+1. \end{cases}$$

First, $G(P') = D(\mathbf{p}') = D(\mathbf{p}) = G(P)$, so $G(P') = G(P)$. (This is also clear informally, since the definition of metacommunity gamma-diversity G does not refer to the division into subcommunities.)

Next, to calculate $\overline{A}(P')$, consider the subcommunity alpha-diversities $\overline{\alpha}_j(P')$. For each $j \in \{1, \ldots, N+1\}$ such that $w'_j > 0$, the species distribution of subcommunity j is

$$P'_{\bullet j}/w'_j = \begin{cases} P_{\bullet j}/w_j & \text{if } 1 \leq j \leq N-1, \\ P_{\bullet N}/w_N & \text{if } j \in \{N, N+1\}, \end{cases}$$

giving

$$\overline{\alpha}_j(P') = \begin{cases} \overline{\alpha}_j(P) & \text{if } 1 \leq j \leq N-1, \\ \overline{\alpha}_N(P) & \text{if } j \in \{N, N+1\}. \end{cases}$$

Equation (8.13) then gives

$$\begin{aligned} \overline{A}(P') &= M_0\big(\mathbf{w}', (\overline{\alpha}_1(P'), \ldots, \overline{\alpha}_N(P'), \overline{\alpha}_{N+1}(P'))\big) \\ &= M_0\big((w_1, \ldots, w_{N-1}, tw_N, (1-t)w_N), (\overline{\alpha}_1(P), \ldots, \overline{\alpha}_N(P), \overline{\alpha}_N(P))\big) \\ &= M_0\big((w_1, \ldots, w_{N-1}, w_N), (\overline{\alpha}_1(P), \ldots, \overline{\alpha}_{N-1}(P), \overline{\alpha}_N(P))\big) \qquad (8.24) \\ &= \overline{A}(P), \end{aligned}$$

where in equation (8.24) we used the repetition property of the power means (Lemma 4.2.11). Hence $\overline{A}(P') = \overline{A}(P)$.

Finally, by equation (8.23),

$$\overline{B}(P') = \frac{G(P')}{\overline{A}(P')} = \frac{G(P)}{\overline{A}(P)} = \overline{B}(P),$$

completing the proof. □

Although $\overline{A}$, $\overline{B}$ and G have the invariance property just established, the redundancy R and raw metacommunity alpha-diversity A do not:

Example 8.4.7 Consider a metacommunity consisting of a single subcommunity, as in Examples 8.3.1(ii) and 8.3.2(ii). Suppose that it is ecologically homogeneous, and split it arbitrarily into new subcommunities of relative sizes

$w_1, \ldots, w_N$. Then the species distributions in the new subcommunities are identical. As we see from columns (i) and (ii) of Table 8.3, the global diversity G, average subcommunity diversity $\overline{A}$ and effective number $\overline{B}$ of isolated subcommunities are the same before and after the division.

On the other hand, the effective number A of (species, subcommunity) pairs is greater by a factor of $D(\mathbf{w})$ in the newly divided metacommunity. This is because A counts individuals of the same species but different subcommunities as being in different groups, and therefore depends directly on the subcommunity divisions, however arbitrary they may be. The redundancy R is also greater by a factor of $D(\mathbf{w})$ in the divided metacommunity, because it too measures properties of the subcommunity division (namely, the effective number of subcommunities that a typical species is spread across). So it is reasonable that A and R increase when a subcommunity is split into smaller units, even when that subcommunity is well-mixed.

Chain Rule, Modularity and Replication Principle

For measures of the diversity of a single community, we have seen that the most important algebraic properties are the chain rule and the principles of modularity and replication. Here we show that versions of these properties also hold for the metacommunity measures.

Consider a group of islands, each divided into several regions (Figure 8.6). Each island can be considered as a metacommunity, and has associated with it all the metacommunity measures $A, \overline{A}, R, \overline{B}$, etc., discussed above. On the other hand, the whole island group can be considered as a metacommunity made up of regions, ignoring the intermediate level of islands. Can the redundancy of the whole island group be computed from the redundancies and relative sizes of the individual islands? If so, how? And the same questions can be asked for all the other metacommunity measures.

To give a precise statement of the problem and its solution, we need some notation and terminology.

We consider a **multicommunity** divided into m metacommunities, which have no species in common. The kth metacommunity ($1 \leq k \leq m$) is further divided into N_k subcommunities and S_k species; the subcommunities of each metacommunity *may* have species in common. There are $N_1 + \cdots + N_m$ subcommunities and $S_1 + \cdots + S_m$ species in the multicommunity as a whole. The relative sizes (that is, relative population abundances) of the metacommunities are denoted by $x_1, \ldots, x_m$, so that $\mathbf{x} = (x_1, \ldots, x_m) \in \Delta_m$.

Write P^k for the relative abundance matrix of the kth metacommunity divided into its subcommunities. Thus, P^k is an $S_k \times N_k$ matrix. Write $\mathbf{p}^k \in \Delta_{S_k}$ for

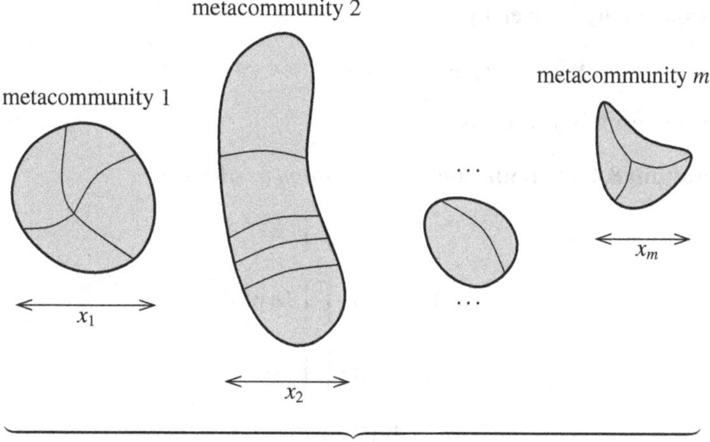

Figure 8.6 Terminology for the chain rule, modularity principle and replication principle. A multicommunity is divided into m metacommunities, with no species in common, of relative sizes $x_1, \ldots, x_m$. Each metacommunity is further divided into subcommunities, which *may* have species in common. In the example shown, there are $m = 4$ metacommunities, divided into $N_1 = 4$, $N_2 = 5$, $N_3 = 2$ and $N_4 = 3$ subcommunities, respectively. We may choose to ignore the metacommunity level and view the multicommunity as divided into $\sum_k N_k = 14$ subcommunities.

the relative abundance distribution of species in the kth metacommunity, and $\mathbf{w}^k \in \Delta_{N_k}$ for the relative sizes of its subcommunities. Since the metacommunities share no species, the relative abundance matrix $P^\star$ of the multicommunity (with respect to its division into subcommunities, ignoring the metacommunity level) is the matrix block sum

$$
\begin{aligned}
P^\star &= x_1 P^1 \oplus \cdots \oplus x_m P^m && (8.25) \\
&= \begin{pmatrix} x_1 P^1 & 0 & \cdots & 0 \\ 0 & x_2 P^2 & \ddots & \vdots \\ \vdots & \ddots & \ddots & 0 \\ 0 & \cdots & 0 & x_m P^m \end{pmatrix}.
\end{aligned}
$$

The relative abundance distribution $\mathbf{p}^\star$ of species in the multicommunity is given by

$$\mathbf{p}^\star = x_1 \mathbf{p}^1 \oplus \cdots \oplus x_m \mathbf{p}^m = \mathbf{x} \circ (\mathbf{p}^1, \ldots, \mathbf{p}^m) \in \Delta_{S_1 + \cdots + S_m},$$

and the relative size distribution $\mathbf{w}^\star$ of the $N_1 + \cdots + N_k$ subcommunities in the multicommunity is given by

$$\mathbf{w}^\star = x_1\mathbf{w}^1 \oplus \cdots \oplus x_m\mathbf{w}^m = \mathbf{x} \circ (\mathbf{w}^1, \ldots, \mathbf{w}^m).$$

The main result is as follows.

Proposition 8.4.8 (Chain rule) *With notation as above,*

$$G(P^\star) = D(\mathbf{x}) \cdot \prod_k G(P^k)^{x_k},$$

$$D(\mathbf{w}^\star) = D(\mathbf{x}) \cdot \prod_k D(\mathbf{w}^k)^{x_k},$$

$$A(P^\star) = D(\mathbf{x}) \cdot \prod_k A(P^k)^{x_k},$$

$$\overline{A}(P^\star) = \prod_k \overline{A}(P^k)^{x_k},$$

$$R(P^\star) = \prod_k R(P^k)^{x_k},$$

$$\overline{B}(P^\star) = D(\mathbf{x}) \cdot \prod_k \overline{B}(P^k)^{x_k},$$

where the products are over all $k \in \{1, \ldots, m\}$ such that $x_k > 0$.

Proof The statement on gamma-diversity is simply the chain rule for the diversity of a single community (Corollary 2.4.8):

$$\begin{aligned} G(P^\star) &= D(\mathbf{p}^\star) \\ &= D(\mathbf{x} \circ (\mathbf{p}^1, \ldots, \mathbf{p}^m)) \\ &= D(\mathbf{x}) \cdot \prod_k D(\mathbf{p}^k)^{x_k} \\ &= D(\mathbf{x}) \cdot \prod_k G(P^k)^{x_k}. \end{aligned}$$

The same argument gives the formula for $D(\mathbf{w}^\star)$. It also gives the formula for $A(P^\star)$, as follows. When the matrices $P^\star$ and $P^1, \ldots, P^m$ are regarded as finite probability distributions, equation (8.25) implies that $P^\star$ can be obtained from $\mathbf{x} \circ (P^1, \ldots, P^m)$ by permutation of its entries and insertion of zeros. By the symmetry and absence-invariance of D, it follows that

$$D(P^\star) = D(\mathbf{x} \circ (P^1, \ldots, P^m)).$$

The chain rule for D then gives

$$D(P^\star) = D(\mathbf{x}) \cdot \prod_k D(P^k)^{x_k},$$

or equivalently,

$$A(P^\star) = D(\mathbf{x}) \cdot \prod_k A(P^k)^{x_k}.$$

This proves the first three equations. Since the last three left-hand sides can be calculated from the first three (Figure 8.5), the rest of the proof is routine. Indeed, by equation (8.15),

$$\overline{A}(P^\star) = \frac{A(P^\star)}{D(\mathbf{w}^\star)} = \prod_k \left(\frac{A(P^k)}{D(\mathbf{w}^k)}\right)^{x_k} = \prod_k \overline{A}(P^k)^{x_k},$$

and similarly, by equation (8.18),

$$R(P^\star) = \frac{A(P^\star)}{G(P^\star)} = \prod_k \left(\frac{A(P^k)}{G(P^k)}\right)^{x_k} = \prod_k R(P^k)^{x_k}.$$

Finally, by equation (8.21),

$$\overline{B}(P^\star) = \frac{G(P^\star)}{\overline{A}(P^\star)} = D(\mathbf{x}) \cdot \prod_k \left(\frac{G(P^k)}{\overline{A}(P_k)}\right)^{x_k} = D(\mathbf{x}) \cdot \prod_k \overline{B}(P^k)^{x_k},$$

completing the proof. □

In particular, each multicommunity measure (such as $A(P^\star)$) is determined by the corresponding metacommunity measures (such as $A(P^1), \ldots, A(P^m)$) and the relative sizes of the metacommunities $(x_1, \ldots, x_m)$. This is the **modularity** property of the measures G, A, $\overline{A}$, R and $\overline{B}$.

The formulas in Proposition 8.4.8 can be restated more compactly in terms of the value measure σ_1 (defined in Section 7.1) and the geometric mean M_0. Write

$$A(P^\bullet) = (A(P^1), \ldots, A(P^m)) \in \mathbb{R}^m,$$

and similarly for the other measures. Then Proposition 8.4.8 states:

Corollary 8.4.9 *In the notation of Proposition 8.4.8,*

$$G(P^\star) = \sigma_1(\mathbf{x}, G(P^\bullet)), \quad D(\mathbf{w}^\star) = \sigma_1(\mathbf{x}, D(\mathbf{w}^\bullet)), \quad A(P^\star) = \sigma_1(\mathbf{x}, A(P^\bullet)),$$
$$\overline{A}(P^\star) = M_0(\mathbf{x}, \overline{A}(P^\bullet)), \quad R(P^\star) = M_0(\mathbf{x}, R(P^\bullet)),$$
$$\overline{B}(P^\star) = \sigma_1(\mathbf{x}, \overline{B}(P^\bullet)).$$

□

Here, the first row consists of exponentials of ordinary and joint entropies, the second of exponentials of conditional entropies, and the last of the exponential of mutual information.

An essential distinction is clear. The formulas in the first and third rows are value measures, meaning that the multicommunity measure $G(P^\star)$ *aggregates* the measures $G(P^1), \ldots, G(P^m)$ of the islands that make up the multicommunity, and similarly for $D(\mathbf{w}^\star)$, $A(P^\star)$ and $\overline{B}(P^\star)$. Those in the second row are means: $\overline{A}(P^\star)$ *averages* the island measures $\overline{A}(P^1), \ldots, \overline{A}(P^m)$, and similarly for $R(P^\star)$.

The point is clarified by considering a specific case. Suppose that the m islands are identical in almost every way: they have the same size, the same number S of species, the same division into subcommunities, and the same species distribution within each subcommunity. The only difference is that each island uses a disjoint set of species. Thus, $G(P^k)$, $D(\mathbf{w}^k)$, $A(P^k)$, $\overline{A}(P^k)$, $R(P^k)$ and $\overline{B}(P^k)$ are all independent of $k \in \{1, \ldots, m\}$, and $\mathbf{x} = \mathbf{u}_m$.

Corollary 8.4.10 (Replication) *In this situation,*

$$G(P^\star) = mG(P^1), \qquad D(\mathbf{w}^\star) = mD(\mathbf{w}^1), \qquad A(P^\star) = mA(P^1),$$
$$\overline{A}(P^\star) = \overline{A}(P^1), \qquad R(P^\star) = R(P^1),$$
$$\overline{B}(P^\star) = m\overline{B}(P^1).$$

□

Example 8.4.11 Take a single island divided into subcommunities, then make a copy of it, using a disjoint set of species for the copy. In the new, larger system consisting of both islands, four of the metacommunity measures are twice what they were for a single island: the effective number G of species, the diversity of the relative size distribution of the subcommunities, the effective number A of (species, subcommunity) pairs, and the effective number $\overline{B}$ of isolated subcommunities.

But the other two remain the same. The mean diversity of the subcommunities, $\overline{A}$, is unchanged, because the subcommunities on the second island have the same abundance distributions as those on the first. The redundancy, R, is also unchanged, because the two islands have no species in common. Put another way, the average spread of species across subcommunities is the same in the two-island system as on either island individually.

8.5 All Entropy Is Relative

The title of this section has two meanings. First, the definition of the entropy of a probability distribution on a finite set is implicitly relative to the uniform distribution. Hence on a general measurable space, ordinary entropy does not even make sense; only relative entropy does. This point was discussed in Section 3.4.

Here we explore a different meaning: that all of the entropies associated with a pair of random variables – cross, joint, conditional, and mutual information – can be reduced to relative entropy. This reduction sheds new light on the subcommunity and metacommunity diversity measures.

We examine each type of entropy in turn, beginning with ordinary Shannon entropy. Let X be a random variable taking values in a finite set $\mathcal{X}$, and let $U_{\mathcal{X}}$ denote a random variable uniformly distributed in $\mathcal{X}$. We have already seen that

$$H(X) = \log|\mathcal{X}| - H(X \parallel U_{\mathcal{X}}) \tag{8.26}$$

(Example 3.1.2), which expresses ordinary entropy in terms of relative entropy together with the cardinality $|\mathcal{X}|$ of the set $\mathcal{X}$.

For cross entropy, let X_1 and X_2 be random variables taking values in the same finite set $\mathcal{X}$. By equations (3.6) and (8.26),

$$\begin{aligned} H^{\times}(X_1 \parallel X_2) &= H(X_1 \parallel X_2) + H(X_1) \\ &= \log|\mathcal{X}| + H(X_1 \parallel X_2) - H(X_1 \parallel U_{\mathcal{X}}), \end{aligned} \tag{8.27}$$

expressing cross entropy in terms of relative entropy and $|\mathcal{X}|$.

Now let X and Y be random variables, not necessarily independent, taking values in finite sets $\mathcal{X}$ and $\mathcal{Y}$ respectively. Thus, the random variable (X, Y) takes values in $\mathcal{X} \times \mathcal{Y}$. By equation (8.26), the joint entropy of X and Y is

$$H(X, Y) = \log|\mathcal{X}| + \log|\mathcal{Y}| - H((X, Y) \parallel U_{\mathcal{X}} \otimes U_{\mathcal{Y}}), \tag{8.28}$$

where $\otimes$ denotes the independent coupling of random variables (as in Remark 8.1.13). Here we have used the observation that $U_{\mathcal{X}} \otimes U_{\mathcal{Y}}$ is uniformly distributed on $\mathcal{X} \times \mathcal{Y}$.

By the explicit formula for mutual information in Lemma 8.1.7(i),

$$I(X; Y) = H((X, Y) \parallel X \otimes Y). \tag{8.29}$$

Thus, mutual information is not merely expressible in terms of relative entropy; it is an *instance* of relative entropy. By Lemma 3.1.4 on relative entropy, $I(X; Y) \geq 0$ with equality if and only if (X, Y) and $X \otimes Y$ are identically distributed, that is, X and Y are independent. This gives another proof of the lower bound in Proposition 8.1.12(iii).

It remains to consider conditional entropy.

Lemma 8.5.1 *Take random variables V and W on the same sample space, with values in finite sets $\mathcal{V}$ and $\mathcal{W}$ respectively. Also take random variables V' and W', with values in $\mathcal{V}$ and $\mathcal{W}$ respectively. Then*

$$H((V, W) \parallel V' \otimes W') = H((V, W) \parallel V \otimes W') + H(V \parallel V').$$

Equations of this type are called **Pythagorean identities** (as in Theorem 4.2 of Csiszár and Shields [76]), because of the features that relative entropy shares with a squared distance (Section 3.4).

Proof By definition, the right-hand side is

$$\sum_{v,w:\, \mathbb{P}(V=v,W=w)>0} \mathbb{P}(V=v, W=w) \log \frac{\mathbb{P}(V=v, W=w)}{\mathbb{P}(V=v)\mathbb{P}(W'=w)} + \sum_{v:\, \mathbb{P}(V=v)>0} \mathbb{P}(V=v) \log \frac{\mathbb{P}(V=v)}{\mathbb{P}(V'=v)}.$$

But since $\mathbb{P}(V = v) = \sum_w \mathbb{P}(V = v, W = w)$, the second term is equal to

$$\sum_{v,w:\, \mathbb{P}(V=v,W=w)>0} \mathbb{P}(V=v, W=w) \log \frac{\mathbb{P}(V=v)}{\mathbb{P}(V'=v)}.$$

Collecting terms and cancelling gives the result. □

We return to our setting of random variables X and Y taking values in finite sets $\mathcal{X}$ and $\mathcal{Y}$. By equations (8.26) and (8.29), we can express the conditional entropy as

$$\begin{aligned} H(X \mid Y) &= H(X) - I(X;Y) \\ &= \log|\mathcal{X}| - \{H(X \,\|\, U_{\mathcal{X}}) + H((X,Y) \,\|\, X \otimes Y)\}, \end{aligned}$$

which by Lemma 8.5.1 gives a formula for conditional entropy in terms of relative entropy:

$$H(X \mid Y) = \log|\mathcal{X}| - H((X,Y) \,\|\, U_{\mathcal{X}} \otimes Y). \tag{8.30}$$

For example, if we fix $\mathcal{X}$, $\mathcal{Y}$ and Y but allow X to vary, the conditional entropy $H(X \mid Y)$ is greatest when the relative entropy $H((X,Y) \,\|\, U_{\mathcal{X}} \otimes Y)$ is least. This happens when (X, Y) has the same distribution as $U_{\mathcal{X}} \otimes Y$, that is, when X is independent of Y and uniformly distributed.

We have now reduced each of the various kinds of entropy to relative entropy. The purpose of this reduction is to illuminate the various measures of subcommunity and metacommunity diversity. In this setting, relative entropy is replaced by relative diversity (introduced in Section 3.3), and the concept of value (Chapter 7) also plays an important part. The results are summarized in Table 8.4.

As before in this chapter, let $P \in \Delta_{SN}$ be an $S \times N$ matrix representing the relative abundances of S species in N subcommunities, write $\mathbf{p} \in \Delta_S$ for the overall relative abundance vector of the species, and write $\mathbf{w} \in \Delta_N$ for the relative sizes of the subcommunities. We will often want to refer to the relative

$\overline{\alpha}_j(P) = \sigma(\overline{P}_{\bullet j}, \mathbf{1}_S)$	$\overline{\beta}_j(P) = D(\overline{P}_{\bullet j} \,\|\, \mathbf{p})$	$\gamma_j(P) = \sigma(\overline{P}_{\bullet j}, \overline{P}_{\bullet j}/\mathbf{p})$
$\overline{A}(P) = \sigma(P, \mathbf{1}_S \otimes \mathbf{w})$	$\overline{B}(P) = D(P \,\|\, \mathbf{p} \otimes \mathbf{w})$	$G(P) = \sigma(\mathbf{p}, \mathbf{1}_S)$
$A(P) = \sigma(P, \mathbf{1}_S \otimes \mathbf{1}_N)$	$R(P) = \sigma(P, \mathbf{p} \otimes \mathbf{1}_N)$	

Table 8.4 The subcommunity and metacommunity measures expressed in terms of relative diversity $D(- \,\|\, -)$ and value σ.

abundance distribution $P_{\bullet j}/w_j$ of species in the jth subcommunity, so let us write

$$\overline{P}_{\bullet j} = P_{\bullet j}/w_j = (P_{1j}/w_j, \ldots, P_{Sj}/w_j) \in \Delta_S$$

for each $j \in \{1, \ldots, N\}$ such that $w_j > 0$.

We begin with beta-diversity. The subcommunity measure $\overline{\beta}_j(P)$ is the relative diversity

$$\overline{\beta}_j(P) = D(\overline{P}_{\bullet j} \,\|\, \mathbf{p})$$

by definition (equation (8.6)); its interpretation was discussed in Sections 3.3 and 8.2. The metacommunity measure $\overline{B}(P)$, which is the effective number of isolated subcommunities, is given by

$$\overline{B}(P) = D(P \,\|\, \mathbf{p} \otimes \mathbf{w}).$$

This is simply the exponential of equation (8.29) (taking (X, Y) to have distribution P). Here P is the distribution of (species, subcommunity) pairs, $\mathbf{p} \otimes \mathbf{w}$ is the hypothetical distribution of (species, subcommunity) pairs in which the overall proportions of species and subcommunities are correct but the subcommunities all have the same composition, and $\overline{B}(P)$ is the diversity of the first relative to the second. It is the divergence of our metacommunity from being well-mixed.

The minimal value of $\overline{B}(P)$, which is 1, is taken when the metacommunity *is* well-mixed. Fixing $\mathbf{w}$ and letting P and $\mathbf{p}$ vary, the maximum value of $\overline{B}(P)$ is $D(\mathbf{w})$ (inequality (8.19)). This maximum is attained when the metacommunity is as far as possible from being well-mixed, that is, when the subcommunities share no species.

To interpret the other subcommunity and metacommunity measures in terms of relative diversity, we use the value measure

$$\begin{aligned} \sigma_1 : \quad \Delta_n \times [0, \infty)^n \quad &\to \quad [0, \infty) \\ (\mathbf{p}, \mathbf{v}) \quad &\mapsto \quad \prod_{i=1}^{n} \left(\frac{v_i}{p_i}\right)^{p_i}, \end{aligned}$$

defined in Section 7.1. Since we are currently abbreviating D_1 as D and working exclusively with the parameter value $q = 1$, we also abbreviate σ_1 as σ. Note that when $\mathbf{v}$ is a probability distribution on $\{1, \ldots, n\}$,

$$\sigma(\mathbf{p}, \mathbf{v}) = \frac{1}{D(\mathbf{p} \,\|\, \mathbf{v})}. \tag{8.31}$$

We now consider the metacommunity gamma-diversity $G(P)$. By equation (8.26) for ordinary entropy in terms of relative entropy,

$$G(P) = D(\mathbf{p}) = \frac{S}{D(\mathbf{p} \,\|\, \mathbf{u}_S)}.$$

But the reciprocal of $D(- \,\|\, -)$ is σ (equation (8.31)), which is homogeneous in its second argument, so

$$G(P) = \sigma(\mathbf{p}, \mathbf{1}_S)$$

where

$$\mathbf{1}_S = (1, \ldots, 1) \in [0, \infty)^S.$$

The same conclusion also follows from Example 7.1.6, where we showed that the diversity of a species distribution $\mathbf{p}$ is the value of the community when each species is given value 1.

The gamma-diversity of the jth subcommunity, $\gamma_j(P)$, is by definition the cross diversity

$$\gamma_j(P) = D^\times(\overline{P}_{\bullet j} \,\|\, \mathbf{p}).$$

Directly from the definition of σ, we also have

$$\gamma_j(P) = \sigma(\overline{P}_{\bullet j}, \overline{P}_{\bullet j}/\mathbf{p}). \tag{8.32}$$

In this expression, the value $\overline{P}_{ij}/p_i$ of species i is high if it is common in subcommunity j but rare in the metacommunity as a whole. Thus, $\gamma_j(P)$ is high if subcommunity j is rich in species that are globally rare. This supports the earlier interpretation of $\gamma_j(P)$ as the contribution of subcommunity j to metacommunity diversity (p. 275).

Taking the exponential of the formula (8.28) for joint entropy in terms of relative entropy gives

$$A(P) = \frac{SN}{D(P \,\|\, \mathbf{u}_S \otimes \mathbf{u}_N)} = \sigma(P, \mathbf{1}_S \otimes \mathbf{1}_N).$$

This is the effective number of (species, subcommunity) pairs. It takes no account of the extent to which the same species appear in different subcommunities, simply treating these SN pairs as separate classes. This formula is another instance of Example 7.1.6, which expressed the diversity of a *single* community in terms of value. So too is the value expression for the diversity $\overline{\alpha}_j(P)$ of subcommunity j in isolation:

$$\overline{\alpha}_j(P) = D(\overline{P}_{\bullet j}) = \sigma(\overline{P}_{\bullet j}, \mathbf{1}_S).$$

The average subcommunity diversity $\overline{A}(P)$ and the redundancy $R(P)$ are both exponentials of conditional entropies, so by equation (8.30),

$$\overline{A}(P) = \frac{S}{D(P \parallel \mathbf{u}_S \otimes \mathbf{w})} = \sigma(P, \mathbf{1}_S \otimes \mathbf{w}), \tag{8.33}$$

$$R(P) = \frac{N}{D(P \parallel \mathbf{p} \otimes \mathbf{u}_N)} = \sigma(P, \mathbf{p} \otimes \mathbf{1}_N). \tag{8.34}$$

Hence by Lemma 7.1.3, the average subcommunity diversity $\overline{A}(P)$ is greatest when P_{ij} is proportional to $(\mathbf{1}_S \otimes \mathbf{w})_{ij} = w_j$, that is, when each subcommunity has a uniform species distribution. Similarly, the redundancy $R(P)$ is greatest when P_{ij} is proportional to $(\mathbf{p} \otimes \mathbf{1}_N)_{ij} = p_i$, that is, when each species is distributed uniformly across subcommunities. These observations confirm the upper bounds on $\overline{A}(P)$ and $R(P)$ obtained in Section 8.3.

8.6 Beyond

The entropies and diversities discussed in this chapter so far are all situated in the case $q = 1$ and $Z = I$ (hence, not incorporating any notion of similarity or distance between species). In this short section, we sketch the definitions for a general q, omitting proofs and details. A more detailed development can be found in Reeve et al. [293], on which this section is based.

In generalizing from $q = 1$ to an arbitrary $q \in [0, \infty]$, we replace the Shannon entropy H by the Rényi entropy H_q, and its exponential D by the Hill number D_q. The Rényi analogue of relative entropy has already been discussed (Section 7.2), and Rényi-type analogues of conditional entropy and mutual information have appeared in other works such as Arimoto [17] and Csiszár [74].

In terms of diversity, q controls the comparative importance attached to rare and common species, and to smaller and larger subcommunities. (See the discussion at the end of Section 4.3.) We obtain the q-analogues of each of $\overline{\alpha}_j$, γ_j, A, $\overline{A}$, R and G by taking its expression in terms of value σ (Table 8.4) and replacing σ by σ_q. The q-analogue of $\overline{\beta}_j(P)$ is $1/\sigma_q(\overline{P}_{\bullet j}, \mathbf{p})$, as Table 8.4 and equation (8.31) would lead one to expect. But the situation for $\overline{B}$ is more subtle, and for this we refer to Reeve et al. [293].

The previously established relationships

$$\begin{aligned}
\overline{A} &= M_0(\mathbf{w}, (\overline{\alpha}_1, \ldots, \overline{\alpha}_N)),\\
\overline{B} &= M_0(\mathbf{w}, (\overline{\beta}_1, \ldots, \overline{\beta}_N)),\\
G &= M_0(\mathbf{w}, (\gamma_1, \ldots, \gamma_N))
\end{aligned}$$

(equations (8.13), (8.22) and (8.9)) continue to hold with M_{1-q} in place of M_0. Moreover, all of the bounds and extremal cases established in Section 8.3 and listed in Table 8.2 remain true without alteration for general q, as proved in the second appendix of Reeve et al. [293].

Example 8.6.1 In the $q = 1$ setting, we proved that

$$1 \leq \frac{A(P)}{G(P)} \leq D(\mathbf{w})$$

(equation (8.10)), or equivalently,

$$0 \leq \frac{A(P)}{G(P)} - 1 \leq D(\mathbf{w}) - 1.$$

These inequalities persist for arbitrary q, and in particular for $q = 0$, where they reduce to the elementary statement that

$$0 \leq \frac{|\mathrm{supp}(P)|}{|\mathrm{supp}(\mathbf{p})|} - 1 \leq N - 1. \tag{8.35}$$

Here, $|\mathrm{supp}(\mathbf{p})|$ is the number of species present in the metacommunity, $|\mathrm{supp}(P)|$ is the number of pairs (i, j) such that species i is present in subcommunity j, and we are assuming that no subcommunity is empty (so that $|\mathrm{supp}(\mathbf{w})| = N$).

For instance, suppose that our metacommunity is divided into just two subcommunities ($N = 2$), so that (8.35) reads

$$0 \leq \frac{|\mathrm{supp}(P)|}{|\mathrm{supp}(\mathbf{p})|} - 1 \leq 1. \tag{8.36}$$

The middle term in (8.36) is known as the **Jaccard index**, after the early twentieth-century botanist Paul Jaccard [153]. (For a modern reference, see pp. 172–173 of Magurran [240].) Traditionally, one writes a for the number of species present in both subcommunities, b for the number present in the first only, and c for the number present in the second only; then the middle term in (8.36) is

$$\frac{(a + b) + (a + c)}{a + b + c} - 1 = \frac{a}{a + b + c}.$$

In other words, the Jaccard index is the proportion of species in the metacommunity that are present in both subcommunities. It is, therefore, a simple measure of how much the two subcommunities overlap. The q-analogue $\frac{A(P)}{G(P)} - 1$ therefore functions as a generalization of Jaccard's index to an arbitrary number of subcommunities and an arbitrary degree q of emphasis on rare or common species. (I thank Richard Reeve for this observation.)

Several good properties of the metacommunity measures were proved in Section 8.4: independence of $\overline{A}$ and $\overline{B}$, independence of $\overline{A}$ and R, the identical subcommunities property, chain rules for the various metacommunity measures, and the consequent modularity and replication principles. All of these results extend without change to an arbitrary $q \in [0, \infty]$, as shown in the second appendix of Reeve et al. [293].

In contrast, the equations

$$\overline{\alpha}_j \overline{\beta}_j = \gamma_j, \qquad \overline{A}\,\overline{B} = G$$

are a special feature of the case $q = 1$. These relationships ultimately derive from the identity

$$M_0(\mathbf{p}, \mathbf{xy}) = M_0(\mathbf{p}, \mathbf{x}) M_0(\mathbf{p}, \mathbf{y})$$

($\mathbf{p} \in \Delta_n$, $\mathbf{x}, \mathbf{y} \in [0, \infty)^n$), which becomes false when M_0 is replaced by M_{1-q}. For arbitrary q, there appears to be *no* formula for G in terms of $\overline{A}$ and $\overline{B}$. That is, although $\overline{A}$ and $\overline{B}$ are canonical measures of average diversity within subcommunities and of variation between them, they do not together determine the diversity G of the metacommunity. As we have seen many times, and as Shannon himself recognized, entropy of order 1 has uniquely good properties.

The challenge of partitioning metacommunity diversity into within- and between-subcommunity components, for arbitrary q, was taken up by Jost [167, 170], who proposed formulas for alpha- and beta-diversities. When $q = 1$, they are equal to our $\overline{A}$ and $\overline{B}$, but for $q \neq 1$, they disagree. Jost's measures satisfy the relationship

$$\text{alpha} \times \text{beta} = \text{gamma}$$

for arbitrary q, but his beta-diversity does not have the 'identical subcommunities' property of Proposition 8.4.6. (The second appendix of Reeve et al. [293] gives a counterexample.) That is, an artificial division of a subcommunity into two identically composed smaller subcommunities can cause a change in the alpha- and beta-diversities that Jost proposed.

In summary, the generalization of the metacommunity and subcommunity diversity measures from $q = 1$ to an arbitrary $q \in [0, \infty]$ is mostly straightforward, as long as we abandon the idea that metacommunity gamma-diversity must be determined by metacommunity alpha- and beta-diversities.

However, to incorporate a species similarity matrix Z into the measures requires more care. We do not discuss this generalization here; again, the reader is referred to Reeve et al. [293].

9

Probabilistic Methods

Much of this book is about characterization theorems for entropies, diversities and means, and the conditions that characterize these quantities are mostly functional equations. In this chapter, we will see how to solve certain functional equations using results from probability theory, following the pioneering 2011 work of Aubrun and Nechita [20]. The technique is demonstrated first with their startlingly simple characterization of the ℓ^p norms, and then with a similar theorem for the power means, different from the characterizations in Chapter 5.

Functional equations are completely deterministic entities, with no stochastic element. How, then, can the power of probability theory be brought to bear?

A simple analogy demonstrates the general idea. Suppose that we want to multiply out the expression

$$(x+y)^{1000} = (x+y)(x+y)\cdots(x+y)$$

as a sum of terms $x^a y^b$. Which terms $x^a y^b$ appear, and how many of them are there?

The standard answer is, of course, that all the terms in the expansion satisfy $a + b = 1000$ with $a, b \geq 0$, and that the number of such terms is exactly $1000!/a!b!$. But there is a different kind of answer: that *most* of the terms are of the form $x^a y^b$ where a and b are each *about* 500. To see this, we can contemplate the process of multiplying out the brackets, in which one has to go through all 2^{1000} ways of making 1000 choices between x and y. If we flip a fair coin 1000 times, we usually obtain about 500 each of heads and tails, and this is the reason why most values of a and b are about 500.

This alternative answer has several distinguishing features. It is approximate, and the approximation is obtained by probabilistic reasoning. Depending on the degree of precision required for the purpose at hand, and depending on the meanings of 'most' and 'about', this approximation may be all that we need. It is also simpler than the first, precise, answer. All of these features are also displayed by the probabilistic method described in this chapter.

For us, the key theorem from probability theory is a variational formula for the moment generating function (Section 9.1). Conceptually, this formula can be understood as the convex conjugate of Cramér's large deviation theorem (Section 9.2). The probabilistic method is applied to characterize the ℓ^p norms in Section 9.3 and the power means in Section 9.4.

This chapter assumes some basic probability theory, but not much more than the language of random variables. The most technically sophisticated part, Section 9.2, is for context only and is not logically necessary for anything that follows.

9.1 Moment Generating Functions

In this short section, we give a variational formula for the moment generating function of any real random variable. It can be found in Cerf and Petit [64], who call it a 'dual equality', a name explained in the next section. The proof given here is different from theirs.

Let X be a real random variable. The **moment generating function** of X is the function

$$\begin{array}{rccc} m_X\colon & \mathbb{R} & \to & [0,\infty] \\ & \lambda & \mapsto & \mathbb{E}(e^{\lambda X}), \end{array}$$

where $\mathbb{E}$ denotes expected value.

Theorem 9.1.1 *Let $X, X_1, X_2, \ldots$ be independent identically distributed real random variables. Write*

$$\overline{X}_r = \tfrac{1}{r}(X_1 + \cdots + X_r)$$

($r \geq 1$). Then

$$m_X(\lambda) = \sup_{x\in\mathbb{R},\ r\geq 1} e^{\lambda x}\, \mathbb{P}(\overline{X}_r \geq x)^{1/r} \tag{9.1}$$

for all $\lambda \geq 0$, where the supremum is over all real x and positive integers r.

We allow infinite values on either side of equation (9.1).

The proof, given below, uses the elementary result of probability theory known as **Markov's inequality** (Grimmett and Stirzaker [130], Lemma 7.2(7)):

Lemma 9.1.2 (Markov) *Let Z be a random variable taking nonnegative real values. Then for all $z \in \mathbb{R}$,*

$$\mathbb{E}(Z) \geq z \cdot \mathbb{P}(Z \geq z).$$

This is intuitively clear: if one third of the people in a room are at least 60 years old, then the mean age is at least 20.

For the proof, we use some standard notation: given $S \subseteq \mathbb{R}$, let $I_S \colon \mathbb{R} \to \mathbb{R}$ denote the **indicator function** (or **characteristic function**) of S, defined by

$$I_S(x) = \begin{cases} 1 & \text{if } x \in S, \\ 0 & \text{otherwise.} \end{cases}$$

Proof We have

$$Z \geq z \cdot I_{[z,\infty)}(Z),$$

by considering the cases $Z \geq z$ and $Z < z$ separately. Hence

$$\mathbb{E}(Z) \geq \mathbb{E}(z \cdot I_{[z,\infty)}(Z)) = z \cdot \mathbb{P}(Z \geq z). \qquad \square$$

Proof of Theorem 9.1.1 Let $\lambda \geq 0$. We prove equation (9.1) by showing that each side is greater than or equal to the other.

First we show that

$$m_X(\lambda) \geq \sup_{x \in \mathbb{R},\, r \geq 1} e^{\lambda x}\, \mathbb{P}(\overline{X}_r \geq x)^{1/r}.$$

Let $x \in \mathbb{R}$ and $r \geq 1$; we must show that

$$\mathbb{E}(e^{\lambda X})^r \geq e^{r\lambda x}\, \mathbb{P}(\overline{X}_r \geq x).$$

And indeed,

$$\begin{aligned} \mathbb{E}(e^{\lambda X})^r &= \mathbb{E}(e^{\lambda(X_1+\cdots+X_r)}) && (9.2)\\ &= \mathbb{E}(e^{r\lambda \overline{X}_r}) \\ &\geq e^{r\lambda x}\, \mathbb{P}(e^{r\lambda \overline{X}_r} \geq e^{r\lambda x}) && (9.3)\\ &\geq e^{r\lambda x}\, \mathbb{P}(\overline{X}_r \geq x), && (9.4) \end{aligned}$$

where (9.2) holds because $X, X_1, X_2, \ldots$ are independent and identically distributed, (9.3) follows from Markov's inequality, and (9.4) holds because $e^{r\lambda y}$ is increasing in $y \in \mathbb{R}$.

Now we prove the opposite inequality,

$$m_X(\lambda) \leq \sup_{x \in \mathbb{R},\, r \geq 1} e^{\lambda x}\, \mathbb{P}(\overline{X}_r \geq x)^{1/r}. \qquad (9.5)$$

The strategy is to show that $\mathbb{E}(e^{\lambda X} I_{[-a,a]}(X))$ is bounded above by the right-hand side of (9.5) for each $a > 0$, then to deduce that the same is true of $\mathbb{E}(e^{\lambda X}) = m_X(\lambda)$ itself.

Let $a > 0$ and $\delta > 0$ be real numbers. We can choose an integer $d \geq 1$ and real numbers $v_0, \ldots, v_d$ such that

$$-a = v_0 < v_1 < \cdots < v_d = a$$

and $v_k \leq v_{k-1} + \delta$ for all $k \in \{1, \ldots, d\}$. Then for all integers $s \geq 1$,

$$\begin{aligned}
\mathbb{E}(e^{\lambda X} I_{[-a,a]}(X))^s &= \mathbb{E}\Big(e^{\lambda X_1} I_{[-a,a]}(X_1) \cdots e^{\lambda X_s} I_{[-a,a]}(X_s)\Big) && (9.6)\\
&\leq \mathbb{E}\Big(e^{s\lambda \overline{X}_s} I_{[-a,a]}(\overline{X}_s)\Big) && (9.7)\\
&\leq \sum_{k=1}^{d} \mathbb{P}(v_{k-1} \leq \overline{X}_s \leq v_k) e^{s\lambda v_k} && (9.8)\\
&\leq \sum_{k=1}^{d} \mathbb{P}(\overline{X}_s \geq v_{k-1}) e^{s\lambda v_{k-1}} e^{s\lambda\delta} && (9.9)\\
&\leq e^{s\lambda\delta} d \sup_{x \in \mathbb{R}} e^{s\lambda x}\, \mathbb{P}(\overline{X}_s \geq x),
\end{aligned}$$

where (9.6) holds because $X, X_1, X_2, \ldots$ are independent and identically distributed, (9.7) holds because the mean of a collection of numbers in the interval $[-a, a]$ also belongs to that interval, (9.8) follows from the definition of expected value, and (9.9) follows from the defining properties of $v_0, \ldots, v_d$. Hence

$$\begin{aligned}
\mathbb{E}(e^{\lambda X} I_{[-a,a]}(X)) &\leq e^{\lambda\delta} d^{1/s} \sup_{x \in \mathbb{R}} e^{\lambda x}\, \mathbb{P}(\overline{X}_s \geq x)^{1/s}\\
&\leq e^{\lambda\delta} d^{1/s} \sup_{x \in \mathbb{R},\, r \geq 1} e^{\lambda x}\, \mathbb{P}(\overline{X}_r \geq x)^{1/r}.
\end{aligned}$$

This holds for all real $\delta > 0$ and integers $s \geq 1$, so we can let $\delta \to 0$ and $s \to \infty$, which gives

$$\mathbb{E}(e^{\lambda X} I_{[-a,a]}(X)) \leq \sup_{x \in \mathbb{R},\, r \geq 1} e^{\lambda x}\, \mathbb{P}(\overline{X}_r \geq x)^{1/r}.$$

Finally, letting $a \to \infty$ and using the monotone convergence theorem gives the desired inequality (9.5). □

The following example is the only instance of Theorem 9.1.1 that we will need.

Example 9.1.3 Let $n \geq 1$ and $c_1, \ldots, c_n \in \mathbb{R}$. Let $X, X_1, X_2, \ldots$ be independent random variables with distribution $\frac{1}{n} \sum_{i=1}^{n} \delta_{c_i}$, where δ_c denotes the point mass at a real number c. Thus, the random variables take values $c_1, \ldots, c_n$ with probability $1/n$ each, adding probabilities if there are repeats among the c_i. For instance, if $n = 3$ and $(c_1, c_2, c_3) = (7, 7, 8)$, then X takes value 7 with probability $2/3$ and 8 with probability $1/3$.

The moment generating function of X is given by

$$m_X(\lambda) = \tfrac{1}{n}(e^{c_1\lambda} + \cdots + e^{c_n\lambda})$$

($\lambda \in \mathbb{R}$). On the other hand, for $r \geq 1$,

$$\mathbb{P}(\overline{X}_r \geq x) = \tfrac{1}{n^r}\big|\{(i_1, \ldots, i_r) : c_{i_1} + \cdots + c_{i_r} \geq rx\}\big|.$$

Hence by Theorem 9.1.1, for all $\lambda \geq 0$,

$$e^{c_1\lambda} + \cdots + e^{c_n\lambda} = \sup_{x\in\mathbb{R},\ r\geq 1} e^{\lambda x}\big|\{(i_1, \ldots, i_r) : c_{i_1} + \cdots + c_{i_r} \geq rx\}\big|^{1/r}. \tag{9.10}$$

This is a completely deterministic statement about real numbers $c_1, \ldots, c_n, \lambda$.

Theorem 9.1.1 gives the value of $m_X(\lambda)$ for $\lambda \geq 0$ only, but it is easy to deduce the value for negative λ:

Corollary 9.1.4 *In the context of Theorem 9.1.1,*

$$m_X(\lambda) = \sup_{x\in\mathbb{R},\ r\geq 1} e^{\lambda x}\,\mathbb{P}(\overline{X}_r \leq x)^{1/r}$$

for all $\lambda \leq 0$.

Proof Apply Theorem 9.1.1 to $-X$ and $-\lambda$, renaming x as $-x$. □

9.2 Large Deviations and Convex Duality

This section is not logically necessary for anything that follows, but places the moment generating function formula of Theorem 9.1.1 into a wider context. Briefly put, that formula is the convex conjugate of Cramér's large deviation theorem. Here we explain what this means and why it is true.

Cramér's Theorem

Let $X, X_1, X_2, \ldots$ be independent identically distributed real random variables, with mean μ, say. Given $x \in \mathbb{R}$, what can be said about $\mathbb{P}(\overline{X}_r \geq x)$ for large integers r?

The law of large numbers implies that

$$\mathbb{P}(\overline{X}_r \geq x) \to \begin{cases} 1 & \text{if } x < \mu, \\ 0 & \text{if } x > \mu \end{cases}$$

as $r \to \infty$ (assuming that $\mathbb{E}(|X|)$ is finite). However, it is silent on the question of how *fast* $\mathbb{P}(\overline{X}_r \geq x)$ converges as $r \to \infty$.

Consider, then, the central limit theorem. Loosely, this states that when r is large, the distribution of $\overline{X}_r$ is approximately normal. This enables us to estimate $\mathbb{P}(\overline{X}_r \geq x)$ for large r; but again, it does not help us with the *rate* of convergence.

More exactly, assume without loss of generality that $\mu = 0$. Then for each $r \geq 1$, the random variable

$$\sqrt{r}\,\overline{X}_r = \frac{1}{\sqrt{r}}(X_1 + \cdots + X_r)$$

has mean 0 and the same variance (σ^2, say) as X. The central limit theorem states that as $r \to \infty$, the distribution of $\sqrt{r}\,\overline{X}_r$ converges to the normal distribution with mean 0 and variance σ^2. This gives a way of estimating the probability

$$\mathbb{P}\Big(\tfrac{1}{\sqrt{r}}(X_1 + \cdots + X_r) \geq x\Big) = \mathbb{P}\Big(\overline{X}_r \geq \tfrac{x}{\sqrt{r}}\Big)$$

for any $x \in \mathbb{R}$ and large integer r. But the original question was about $\mathbb{P}(\overline{X}_r \geq x)$, not $\mathbb{P}(\overline{X}_r \geq x/\sqrt{r})$. In other words, we are interested in larger deviations from the mean than those addressed by the central limit theorem.

So, neither the law of large numbers nor the central limit theorem tells us the rate of convergence of $\mathbb{P}(\overline{X}_r \geq x)$ as $r \to \infty$. But large deviation theory does. Roughly speaking, the basic fact is that for each $x \in \mathbb{R}$ there is a constant $k(x) \in [0, 1]$ such that

$$\mathbb{P}(\overline{X}_r \geq x) \approx k(x)^r$$

when r is large. If $x > \mu$ then $k(x) < 1$, so the decay of $\mathbb{P}(\overline{X}_r \geq x)$ as $r \to \infty$ is exponential. The precise result is this.

Theorem 9.2.1 (Cramér) *Let $X, X_1, X_2, \ldots$ be independent identically distributed real random variables, and let $x \in \mathbb{R}$. Then the limit*

$$\lim_{r\to\infty} \mathbb{P}(\overline{X}_r \geq x)^{1/r}$$

exists and is equal to

$$\inf_{\lambda\geq 0} \frac{\mathbb{E}(e^{\lambda X})}{e^{\lambda x}}.$$

Part of this statement is an easy consequence of Markov's inequality. Indeed, we used Markov's inequality in equations (9.2)–(9.4) to show that

$$\mathbb{P}(\overline{X}_r \geq x)^{1/r} \leq \frac{\mathbb{E}(e^{\lambda X})}{e^{\lambda x}}$$

for each $r \geq 1$ and $\lambda \geq 0$, so if the limit in Cramér's theorem does exist then it is at most the stated infimum. We do not prove Cramér's theorem here, but a short proof can be found in Cerf and Petit [64] (who deduce it from Theorem 9.1.1 using the convex duality that we are about to discuss), or see standard probability texts such as Grimmett and Stirzaker ([130], Theorem 5.11(4)).

Example 9.2.2 When X is distributed normally with mean μ and variance σ^2, its moment generating function is

$$\mathbb{E}(e^{\lambda X}) = \exp(\lambda\mu + \tfrac{1}{2}\lambda^2\sigma^2).$$

Hence

$$\frac{\mathbb{E}(e^{\lambda X})}{e^{\lambda x}} = \exp(\tfrac{1}{2}\sigma^2 \cdot \lambda^2 - (x-\mu)\cdot\lambda).$$

Minimizing $\mathbb{E}(e^{\lambda X})/e^{\lambda x}$ over $\lambda \geq 0$ therefore reduces to the routine task of minimizing a quadratic. This done, Cramér's theorem gives

$$\lim_{r\to\infty} \mathbb{P}(\overline{X}_r \geq x)^{1/r} = \begin{cases} 1 & \text{if } x \leq \mu, \\ \exp\left(-\dfrac{(x-\mu)^2}{2\sigma^2}\right) & \text{if } x \geq \mu. \end{cases}$$

As one would expect, this is a decreasing function of x but an increasing function of both μ and σ.

As this example suggests, it is natural to split Cramér's theorem into two cases, according to whether x is greater than or less than $\mathbb{E}(X)$:

Corollary 9.2.3 *Let $X, X_1, X_2, \ldots$ be independent identically distributed real random variables.*

i. *For all $x \geq \mathbb{E}(X)$,*

$$\lim_{r\to\infty} \mathbb{P}(\overline{X}_r \geq x)^{1/r} = \inf_{\lambda\in\mathbb{R}} \frac{\mathbb{E}(e^{\lambda X})}{e^{\lambda x}},$$

and for all $x \leq \mathbb{E}(X)$,

$$\lim_{r\to\infty} \mathbb{P}(\overline{X}_r \leq x)^{1/r} = \inf_{\lambda\in\mathbb{R}} \frac{\mathbb{E}(e^{\lambda X})}{e^{\lambda x}}.$$

(Note that both infima are over all $\lambda \in \mathbb{R}$, *in contrast to Theorem 9.2.1.)*

ii. *For all $x \leq \mathbb{E}(X)$,*

$$\lim_{r\to\infty} \mathbb{P}(\overline{X}_r \geq x)^{1/r} = 1,$$

and for all $x \geq \mathbb{E}(X)$,

$$\lim_{r\to\infty} \mathbb{P}(\overline{X}_r \leq x)^{1/r} = 1.$$

Proof For both parts, we use the inequality $e^x \geq 1 + x$, which implies that

$$\frac{\mathbb{E}(e^{\lambda X})}{e^{\lambda x}} = \mathbb{E}(e^{\lambda(X-x)}) \geq \mathbb{E}(1 + \lambda(X - x)) = 1 + \lambda(\mathbb{E}(X) - x) \tag{9.11}$$

for all $\lambda, x \in \mathbb{R}$. We also use the fact that

$$\frac{\mathbb{E}(e^{0X})}{e^{0x}} = 1 \tag{9.12}$$

for all $x \in \mathbb{R}$.

For (i), let $x \geq \mathbb{E}(X)$. When $\lambda \leq 0$, (9.11) and (9.12) give

$$\frac{\mathbb{E}(e^{\lambda X})}{e^{\lambda x}} \geq 1 = \frac{\mathbb{E}(e^{0X})}{e^{0x}}, \tag{9.13}$$

so the infimum in Theorem 9.2.1 is unchanged if we allow λ to range over all of $\mathbb{R}$. This gives the first equation of (i). The second follows by applying the first to $-X$ and $-x$, renaming λ as $-\lambda$.

For (ii), let $x \leq \mathbb{E}(X)$. When $\lambda \geq 0$, (9.11) and (9.12) again imply (9.13), so the infimum in Theorem 9.2.1 is 1. This gives the first equation of (ii), and again, the second follows by applying the first to $-X$ and $-x$. □

Convex Duality

To relate the formula for moment generating functions in Theorem 9.1.1 to Cramér's theorem, we use the principle of convex duality.

Definition 9.2.4 Let $f: \mathbb{R} \to [-\infty, \infty]$ be a function. Its **convex conjugate** or **Legendre–Fenchel transform** is the function $f^*: \mathbb{R} \to [-\infty, \infty]$ defined by

$$f^*(\lambda) = \sup_{x\in\mathbb{R}}(\lambda x - f(x)). \tag{9.14}$$

The theory of convex conjugates is developed thoroughly in texts such as Borwein and Lewis [47] and Rockafellar [299]. Here we give a brief summary tailored to our needs.

Examples 9.2.5 i. Let $f: \mathbb{R} \to \mathbb{R}$ be a differentiable function such that $f': \mathbb{R} \to \mathbb{R}$ is an increasing bijection. Then for each $\lambda \in \mathbb{R}$, the function

$$x \mapsto \lambda x - f(x)$$

has a unique critical point $x_\lambda = f'^{-1}(\lambda)$, which is also the unique global maximum. Hence

$$f^*(\lambda) = \lambda x_\lambda - f(x_\lambda).$$

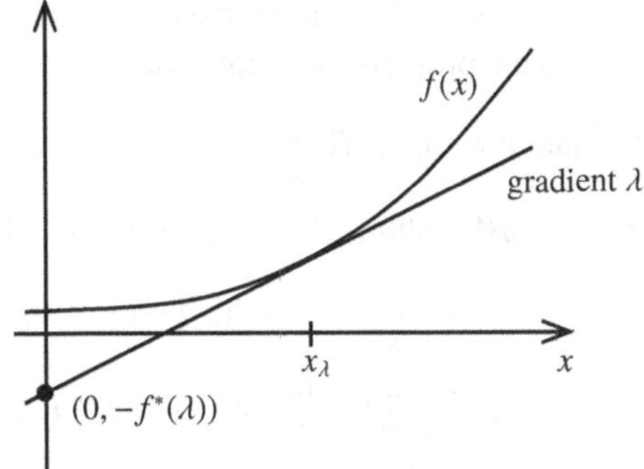

Figure 9.1 The relationship between a differentiable function f and its convex conjugate f^* (Example 9.2.5(i)).

In graphical terms, for each real number λ, there is a unique tangent line to the graph of f with gradient (slope) λ, and its equation is

$$y = \lambda x - f^*(\lambda)$$

(Figure 9.1). Thus, $f^*(\lambda)$ is the negative of the y-intercept of this tangent line. The convex conjugate f^* therefore describes f in terms of its envelope of tangent lines.

ii. Let $p, q \in (1, \infty)$ be conjugate exponents, that is, $1/p + 1/q = 1$. Then the functions $x \mapsto |x|^p/p$ and $x \mapsto |x|^q/q$ are convex conjugate to one another, as can be shown using (i).

iii. More generally, let $f, g\colon \mathbb{R} \to \mathbb{R}$ be differentiable functions such that f' and g' are increasing and $f'(0) = 0 = g'(0)$. It can be shown that if $f', g'\colon \mathbb{R} \to \mathbb{R}$ are mutually inverse then f and g are mutually convex conjugate (Section I.9 of Zygmund [363]). At this level of generality, convex duality has also been called Young complementarity or Young duality, as in [363] or Section 14D of Arnold [18].

Lemma 9.2.6 *For every function* $f\colon \mathbb{R} \to [-\infty, \infty]$, *the convex conjugate* $f^*\colon \mathbb{R} \to [-\infty, \infty]$ *is convex.*

Before we can prove this, we need to state what it means for a function into $[-\infty, \infty]$ to be convex. If the function takes only finite values, or takes ∞ but not $-\infty$ as a value, or vice versa, then the meaning is clear. If it takes both $-\infty$ and ∞ as values then the matter is more delicate; a careful treatment can be found in Section 2.2.2 of Willerton [356] (ultimately derived from Lawvere [205]). Fortunately, we can avoid the issue here. If $f \equiv \infty$ then $f^* \equiv -\infty$, in which case f^* is convex by any reasonable definition. Otherwise, f^* never takes the value $-\infty$, so the problem does not arise.

Proof Let $\lambda, \mu \in \mathbb{R}$ and $p \in [0, 1]$. Then

$$\begin{aligned} f^*(p\lambda + (1-p)\mu) &= \sup_{x\in\mathbb{R}}(p\lambda x + (1-p)\mu x - f(x)) \\ &= \sup_{x\in\mathbb{R}}(p[\lambda x - f(x)] + (1-p)[\mu x - f(x)]) \\ &\leq \sup_{y,z\in\mathbb{R}}(p[\lambda y - f(y)] + (1-p)[\mu z - f(z)]) \\ &= pf^*(\lambda) + (1-p)f^*(\mu), \end{aligned}$$

as required. □

The examples above suggest that often $f^{**} = f$. By Lemma 9.2.6, this cannot be true unless f is convex. For finite-valued f, that is the *only* restriction:

Theorem 9.2.7 (Legendre–Fenchel) *Let $f: \mathbb{R} \to \mathbb{R}$ be a convex function. Then $f^{**} = f$.*

Proof This standard result can be found in textbooks on convex analysis; see Theorem 4.2.1 of Borwein and Lewis [47] or Section 14C of Arnold [18], for instance. A proof is also included as Appendix A.7. □

Remarks 9.2.8 i. Theorem 9.2.7 is a very special case of the full Legendre–Fenchel theorem. For a start, we restricted to finite-valued functions, thus avoiding the semicontinuity requirement on f that is needed when values of $\pm\infty$ are allowed. But much more significantly, the duality can be generalized beyond functions on $\mathbb{R}$ to functions on a finite-dimensional real vector space X.

In that context, the convex conjugate of a function $f: X \to [-\infty, \infty]$ is a function $f^*: X^* \to [-\infty, \infty]$ on the dual vector space X^*. The function f^* is defined by the same formula (9.14) as before, now understanding the term λx to mean the functional $\lambda \in X^*$ evaluated at the vector $x \in X$. For the Legendre–Fenchel theorem at this level of generality, see Theorem 4.2.1 of Borwein and Lewis [47], Theorem 12.2 of Rockafellar [299], or Fenchel [100].

ii. The Legendre–Fenchel theorem for vector spaces is itself an instance of a more general duality still, recently discovered by Willerton [356]. It is framed in terms of enriched categories, as follows.

Let $\mathscr{V}$ be a complete symmetric monoidal closed category. In the rest of this remark, all categories, functors, adjunctions, etc., are taken to be enriched in $\mathscr{V}$. For any categories $\mathbf{A}$ and $\mathbf{B}$ and functor $M\colon \mathbf{A}^{\mathrm{op}} \times \mathbf{B} \to \mathscr{V}$, there is an induced adjunction

$$[\mathbf{A}^{\mathrm{op}}, \mathscr{V}] \rightleftarrows [\mathbf{B}, \mathscr{V}]^{\mathrm{op}}$$

between functor categories, in which both functors are defined by mapping into M. For instance, given $X \in [\mathbf{A}^{\mathrm{op}}, \mathscr{V}]$, the resulting functor $\mathbf{B} \to \mathscr{V}$ is

$$b \mapsto [\mathbf{A}^{\mathrm{op}}, \mathscr{V}](X, M(-, b))$$

($b \in \mathbf{B}$). On the other hand, any adjunction restricts canonically to an equivalence between full subcategories, consisting of its fixed points. Here, this gives a dual equivalence

$$\mathscr{C} \rightleftarrows \mathscr{D}^{\mathrm{op}} \tag{9.15}$$

between a full subcategory $\mathscr{C}$ of $[\mathbf{A}^{\mathrm{op}}, \mathscr{V}]$ and a full subcategory $\mathscr{D}$ of $[\mathbf{B}, \mathscr{V}]$. Pavlovic calls either of the categories $\mathscr{C}$ and $\mathscr{D}^{\mathrm{op}}$ the **nucleus** of M (Definition 3.9 of [274]).

Willerton showed that the Legendre–Fenchel theorem is a special case of this very general categorical construction. Let $\mathscr{V}$ be the ordered set $([-\infty, \infty], \geq)$, regarded as a category in the standard way, and with monoidal structure defined by addition. Any real vector space X gives rise to a category enriched in $\mathscr{V}$: the objects are the elements of X, and $\mathrm{Hom}(x, y) \in \mathscr{V}$ is 0 if $x = y$ and ∞ otherwise. The usual pairing between a vector space and its dual gives a canonical functor $M\colon (X^*)^{\mathrm{op}} \times X \to \mathscr{V}$. Applying the general construction above then gives a dual equivalence (9.15) between two enriched categories. As Willerton showed, this is precisely the convex duality established by the classical Legendre–Fenchel theorem for $[-\infty, \infty]$-valued functions on finite-dimensional vector spaces.

The Dual of Cramér's Theorem

As before, let $X, X_1, X_2, \ldots$ be independent identically distributed real random variables. In Corollary 9.2.3(i), Cramér's theorem was restated as

$$\inf_{\lambda \in \mathbb{R}} \frac{\mathbb{E}(e^{\lambda X})}{e^{\lambda x}} = \begin{cases} \lim_{r\to\infty} \mathbb{P}(\overline{X}_r \geq x)^{1/r} & \text{if } x \geq \mathbb{E}(X), \\ \lim_{r\to\infty} \mathbb{P}(\overline{X}_r \leq x)^{1/r} & \text{if } x \leq \mathbb{E}(X). \end{cases}$$

Taking logarithms and changing sign, an equivalent statement is that

$$(\log m_X)^*(x) = \begin{cases} -\lim_{r\to\infty} \frac{1}{r} \log \mathbb{P}(\overline{X}_r \geq x) & \text{if } x \geq \mathbb{E}(X), \\ -\lim_{r\to\infty} \frac{1}{r} \log \mathbb{P}(\overline{X}_r \leq x) & \text{if } x \leq \mathbb{E}(X). \end{cases} \tag{9.16}$$

It is a general fact that $\log m_X$, called the **cumulant generating function** of X, is a convex function (Appendix A.8). So by taking convex conjugates on each side of (9.16) and using the Legendre–Fenchel theorem, we will obtain an expression for $\log m_X$ and, therefore, the moment generating function m_X itself.

Specifically, equation (9.16) and the Legendre–Fenchel theorem imply that for all $\lambda \in \mathbb{R}$,

$$\log m_X(\lambda) =$$

$$\max\left\{ \sup_{x \geq \mathbb{E}(X)} \left(\lambda x + \lim_{r\to\infty} \tfrac{1}{r} \log \mathbb{P}(\overline{X}_r \geq x)\right), \sup_{x \leq \mathbb{E}(X)} \left(\lambda x + \lim_{r\to\infty} \tfrac{1}{r} \log \mathbb{P}(\overline{X}_r \leq x)\right)\right\},$$

or equivalently,

$$m_X(\lambda) = \max\left\{ \sup_{x \geq \mathbb{E}(X)} e^{\lambda x} \lim_{r\to\infty} \mathbb{P}(\overline{X}_r \geq x)^{1/r}, \sup_{x \leq \mathbb{E}(X)} e^{\lambda x} \lim_{r\to\infty} \mathbb{P}(\overline{X}_r \leq x)^{1/r}\right\}. \tag{9.17}$$

Let $\lambda \geq 0$. We analyse the second supremum in equation (9.17). The quantity $e^{\lambda x} \lim_{r\to\infty} \mathbb{P}(\overline{X}_r \leq x)^{1/r}$ is increasing in x, so the supremum is attained when $x = \mathbb{E}(X)$. But by Corollary 9.2.3(ii),

$$\lim_{r\to\infty} \mathbb{P}(\overline{X}_r \leq \mathbb{E}(X))^{1/r} = 1,$$

so the second supremum is just $e^{\lambda \mathbb{E}(X)}$. On the other hand, Corollary 9.2.3(ii) also states that for all $x \leq \mathbb{E}(X)$,

$$\lim_{r\to\infty} \mathbb{P}(\overline{X}_r \geq x)^{1/r} = 1,$$

so the second supremum can be expressed as

$$\sup_{x \leq \mathbb{E}(X)} e^{\lambda x} \lim_{r\to\infty} \mathbb{P}(\overline{X}_r \geq x)^{1/r}.$$

Hence by (9.17),

$$m_X(\lambda) = \sup_{x \in \mathbb{R}} e^{\lambda x} \lim_{r\to\infty} \mathbb{P}(\overline{X}_r \geq x)^{1/r}. \tag{9.18}$$

We have derived equation (9.18) as the convex dual of Cramér's theorem. It is very nearly the moment generating function formula of Theorem 9.1.1. The only difference is that where (9.18) has a limit as $r \to \infty$, Theorem 9.1.1 has a supremum over $r \geq 1$. However, Cerf and Petit showed that the two forms are equivalent:

$$\lim_{r\to\infty} \mathbb{P}(\overline{X}_r \geq x)^{1/r} = \sup_{r \geq 1} \mathbb{P}(\overline{X}_r \geq x)^{1/r} \tag{9.19}$$

([64], p. 928). In this sense, Theorem 9.1.1 can also be regarded as the dual of Cramér's theorem.

Remark 9.2.9 In their work, Cerf and Petit [64] travelled the opposite path from the one just described. They started by proving Theorem 9.1.1, took convex conjugates, and thus, with the aid of (9.19), deduced Cramér's theorem.

9.3 Multiplicative Characterization of the p-Norms

Here we show how probabilistic methods can be used to solve functional equations, following Aubrun and Nechita [20]. We give a version of their theorem that among all coherent ways of putting a norm on each of the vector spaces $\mathbb{R}^0, \mathbb{R}^1, \mathbb{R}^2, \ldots$, the only ones satisfying a certain multiplicativity condition are the p-norms.

Definition 9.3.1 Let $n \geq 0$. A **norm** $\|\cdot\|$ on $\mathbb{R}^n$ is a function $\mathbb{R}^n \to [0, \infty)$, written as $\mathbf{x} \mapsto \|\mathbf{x}\|$, with the following properties:

i. $\|\mathbf{x}\| = 0 \implies \mathbf{x} = 0$;
ii. $\|c\mathbf{x}\| = |c|\,\|\mathbf{x}\|$ for all $c \in \mathbb{R}$ and $\mathbf{x} \in \mathbb{R}^n$;
iii. $\|\mathbf{x} + \mathbf{y}\| \leq \|\mathbf{x}\| + \|\mathbf{y}\|$ for all $\mathbf{x}, \mathbf{y} \in \mathbb{R}^n$ (the **triangle inequality**).

Example 9.3.2 Let $n \geq 0$ and $p \in [1, \infty]$. The **p-norm** or **ℓ^p norm** $\|\cdot\|_p$ on $\mathbb{R}^n$ is defined by

$$\|\mathbf{x}\|_p = \left(\sum_{i=1}^{n} |x_i|^p\right)^{1/p}$$

for $p < \infty$, and for $p = \infty$ by

$$\|\mathbf{x}\|_\infty = \max_{1 \leq i \leq n} |x_i|$$

($\mathbf{x} \in \mathbb{R}^n$). Then $\|\mathbf{x}\|_\infty = \lim_{p\to\infty} \|\mathbf{x}\|_p$, by Lemma 4.2.7 on power means: writing $|\mathbf{x}| = (|x_1|, \ldots, |x_n|)$,

$$\|\mathbf{x}\|_p = n^{1/p} M_p(\mathbf{u}_n, |\mathbf{x}|) \to M_\infty(\mathbf{u}_n, |\mathbf{x}|) = \|\mathbf{x}\|_\infty$$

as $p \to \infty$.

Example 9.3.3 Let $\phi\colon [0, \infty) \to [0, \infty)$ be an increasing convex function such that $\phi^{-1}\{0\} = \{0\}$. For $n \geq 0$, put

$$K_n = \left\{\mathbf{x} \in \mathbb{R}^n : \sum_{i=1}^{n} \phi(|x_i|) \leq 1\right\},$$

which is a convex subset of $\mathbb{R}^n$. Then for $\mathbf{x} \in \mathbb{R}^n$, put

$$\|\mathbf{x}\| = \inf\{\lambda \geq 0 : \mathbf{x} \in \lambda K_n\}.$$

It can be shown that $\|\cdot\|$ is a norm on $\mathbb{R}^n$ (known as an **Orlicz norm**), whose unit ball $\{\mathbf{x} \in \mathbb{R}^n : \|\mathbf{x}\| \leq 1\}$ is K_n. For instance, taking $\phi(x) = x^p$ for some $p \in [1, \infty)$ gives the p-norm of Example 9.3.2.

Fix $p \in [1, \infty]$. The p-norms on the sequence of spaces $\mathbb{R}^0, \mathbb{R}^1, \mathbb{R}^2, \ldots$ are compatible with one another in the following two ways.

First, the p-norm of a vector is unchanged by permuting its entries or inserting zeros. For instance,

$$\|(x_1, x_2, x_3)\|_p = \|(x_2, 0, x_3, x_1)\|_p. \tag{9.20}$$

Generally, writing $\mathbf{n} = \{1, \ldots, n\}$, any injection $f\colon \mathbf{n} \to \mathbf{m}$ induces an injective linear map $f_*\colon \mathbb{R}^n \to \mathbb{R}^m$, defined by

$$(f_*\mathbf{x})_j = \begin{cases} x_i & \text{if } j = f(i) \text{ for some } i \in \{1, \ldots, n\}, \\ 0 & \text{otherwise} \end{cases}$$

($\mathbf{x} \in \mathbb{R}^n$, $j \in \{1, \ldots, m\}$). Then the p-norm has the property that

$$\|f_*\mathbf{x}\|_p = \|\mathbf{x}\|_p \tag{9.21}$$

for all injections $f\colon \mathbf{n} \to \mathbf{m}$ and all $\mathbf{x} \in \mathbb{R}^n$. For example, equation (9.20) is the instance of equation (9.21) where f is the map $\{1, 2, 3\} \to \{1, 2, 3, 4\}$ defined by $f(1) = 4$, $f(2) = 1$, and $f(3) = 3$.

Second, the p-norm satisfies a multiplicativity law. Let $\mathbf{x} \in \mathbb{R}^n$ and $\mathbf{y} \in \mathbb{R}^m$, and recall from equation (4.12) (p. 110) the definition of $\mathbf{x} \otimes \mathbf{y} \in \mathbb{R}^{nm}$. Then

$$\|\mathbf{x} \otimes \mathbf{y}\|_p = \|\mathbf{x}\|_p \, \|\mathbf{y}\|_p.$$

For instance,

$$\|(Ax, Ay, Az, Bx, By, Bz)\|_p = \|(A, B)\|_p \|(x, y, z)\|_p$$

for all $A, B, x, y, z \in \mathbb{R}$.

These two properties of the p-norms determine them completely, as we shall see.

Definition 9.3.4 i. A **system of norms** consists of a norm $\|\cdot\|$ on $\mathbb{R}^n$ for each $n \geq 0$, such that for each $n, m \geq 0$ and injection $f\colon \mathbf{n} \to \mathbf{m}$,

$$\|f_*\mathbf{x}\| = \|\mathbf{x}\|$$

for all $\mathbf{x} \in \mathbb{R}^n$.

ii. A system of norms $\|\cdot\|$ is **multiplicative** if

$$\|\mathbf{x} \otimes \mathbf{y}\| = \|\mathbf{x}\| \, \|\mathbf{y}\|$$

for all $n, m \geq 0$, $\mathbf{x} \in \mathbb{R}^n$, and $\mathbf{y} \in \mathbb{R}^m$.

Examples 9.3.5 i. For each $p \in [1, \infty]$, the p-norm $\|\cdot\|_p$ is a multiplicative system of norms.

ii. Fix a function ϕ as in Example 9.3.3. The norms $\|\cdot\|$ defined there always form a system of norms, but it is not in general multiplicative.

Remark 9.3.6 The notion of a system of norms can be recast in two equivalent ways. First, instead of only considering $\mathbb{R}^n$ for natural numbers n, we can consider

$$\mathbb{R}^I = \{\text{functions } I \to \mathbb{R}\} = \{\text{families } (x_i)_{i \in I} \text{ of reals}\}$$

for arbitrary finite sets I. (This was the approach taken in Leinster [214].) We then require the equation $\|f_* \mathbf{x}\| = \|\mathbf{x}\|$ to hold for every injection $f \colon I \to J$ between finite sets. In particular, taking f to be a bijection, the norm on $\mathbb{R}^J$ determines the norm on $\mathbb{R}^I$ for all sets I of the same cardinality as J. So, the norm on $\mathbb{R}^n$ determines the norm on $\mathbb{R}^I$ for all n-element sets I. It follows that this apparently more general notion of a system of norms is equivalent to the original one.

In the opposite direction, we can construe a system of norms as a norm on the single space c_{00} of infinite real sequences with only finitely many nonzero entries, subject to a symmetry axiom. (This was the approach taken in Aubrun and Nechita [20].) To state the multiplicativity property, we have to choose a bijection between the set of nonnegative integers and its cartesian square, but by symmetry, the definition of multiplicativity is unaffected by that choice.

We now come to the main theorem of this section. In its present form, it was first stated by Aubrun and Nechita [20]. The result also follows from Theorem 3.9 of an earlier paper of Fernández-González, Palazuelos and Pérez-García [102] (at least, putting aside some delicacies concerning $\|\cdot\|_\infty$). The arguments in [102] are very different, coming as they do from the theory of Banach spaces. We will consider only Aubrun and Nechita's method.

Theorem 9.3.7 *Every multiplicative system of norms is equal to $\|\cdot\|_p$ for some $p \in [1, \infty]$.*

The proof will rest on the moment generating function formula of Theorem 9.1.1. Specifically, we will need the following consequence of that theorem. Given $\mathbf{v} = (v_1, \ldots, v_n) \in \mathbb{R}^n$ and $t \in \mathbb{R}$, write

$$N(\mathbf{v}, t) = \bigl|\{i \in \{1, \ldots, n\} : v_i \geq t\}\bigr|.$$

Proposition 9.3.8 (Aubrun and Nechita) *Let $p \in [1, \infty)$, $n \geq 0$, and $\mathbf{x} \in (0, \infty)^n$. Then*

$$\|\mathbf{x}\|_p = \sup_{u>0,\ r\geq 1} u \cdot N(\mathbf{x}^{\otimes r}, u^r)^{1/rp},$$

where the supremum is over real $u > 0$ and integers $r \geq 1$.

This formula was central to Aubrun and Nechita's argument in [20], although not quite stated explicitly there.

Proof In equation (9.10) (Example 9.1.3), put $c_i = \log x_i$ and $\lambda = p$. Then

$$\begin{aligned} x_1^p + \cdots + x_n^p &= \sup_{y\in\mathbb{R},\ r\geq 1} e^{py} \bigl|\{(i_1, \ldots, i_r) : x_{i_1} \cdots x_{i_r} \geq e^{ry}\}\bigr|^{1/r} \\ &= \sup_{u>0,\ r\geq 1} u^p N(\mathbf{x}^{\otimes r}, u^r)^{1/r}, \end{aligned}$$

and the result follows by taking pth roots throughout. □

We now embark on the proof of Theorem 9.3.7, roughly following Aubrun and Nechita [20], but with some simplifications described in Remark 9.3.10. In the words of Aubrun and Nechita, the proof proceeds by 'examining the statistical distribution of large coordinates of the rth tensor power $\mathbf{x}^{\otimes r}$ (r large)' ([20], Section 1.1; notation adapted).

For the rest of this section, let $\|\cdot\|$ be a multiplicative system of norms.

Step 1: elementary results We begin by deriving some elementary properties of the norms $\|\cdot\|$. For $n \geq 0$, write $\mathbf{1}_n = (1, \ldots, 1) \in \mathbb{R}^n$.

Lemma 9.3.9 *Let $n \geq 0$ and $\mathbf{x}, \mathbf{y} \in \mathbb{R}^n$.*

i. *If $y_i = \pm x_i$ for each i then $\|\mathbf{x}\| = \|\mathbf{y}\|$.*
ii. *If $\mathbf{0} \leq \mathbf{x} \leq \mathbf{y}$ then $\|\mathbf{x}\| \leq \|\mathbf{y}\|$.*
iii. *$\|\mathbf{1}_m\| \leq \|\mathbf{1}_n\|$ whenever $0 \leq m \leq n$.*

Proof For (i), the vector $\mathbf{x} \otimes (1, -1)$ is a permutation of $\mathbf{y} \otimes (1, -1)$, so by definition of system of norms,

$$\|\mathbf{x} \otimes (1, -1)\| = \|\mathbf{y} \otimes (1, -1)\|.$$

But by multiplicativity, this equation is equivalent to

$$\|\mathbf{x}\|\, \|(1, -1)\| = \|\mathbf{y}\|\, \|(1, -1)\|.$$

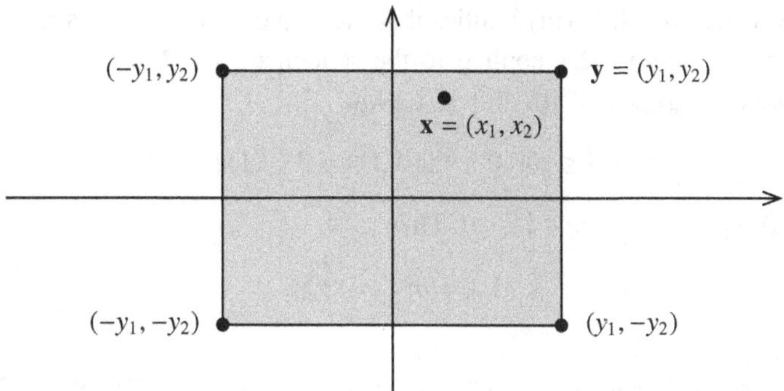

Figure 9.2 The vector $\mathbf{x}$ in the convex hull of the set S, as in the proof of Lemma 9.3.9(ii), shown for $n = 2$.

Hence $\|\mathbf{x}\| = \|\mathbf{y}\|$.

For (ii), let S be the set of vectors of the form $(\varepsilon_1 y_1, \ldots, \varepsilon_n y_n) \in \mathbb{R}^n$ with $\varepsilon_i = \pm 1$. Recall that the **convex hull** of S is the set of vectors expressible as $\sum_{\mathbf{s} \in S} \lambda_{\mathbf{s}} \mathbf{s}$ for some nonnegative reals $(\lambda_{\mathbf{s}})_{\mathbf{s} \in S}$ summing to 1. A straightforward induction shows that the convex hull of S is

$$\prod_{i=1}^{n} [-y_i, y_i] = [-y_1, y_1] \times \cdots \times [-y_n, y_n]$$

(Figure 9.2). But $\mathbf{x} \in \prod[-y_i, y_i]$, and $\|\mathbf{s}\| = \|\mathbf{y}\|$ for each $\mathbf{s} \in S$ by part (i). Hence, writing $\mathbf{x} = \sum \lambda_{\mathbf{s}} \mathbf{s}$ and using the triangle inequality,

$$\|\mathbf{x}\| \leq \sum_{\mathbf{s} \in S} \lambda_{\mathbf{s}} \|\mathbf{s}\| = \sum_{\mathbf{s} \in S} \lambda_{\mathbf{s}} \|\mathbf{y}\| = \|\mathbf{y}\|.$$

For (iii), let $0 \leq m \leq n$. We have

$$\|\mathbf{1}_m\| = \|(\overbrace{\underbrace{1, \ldots, 1}_{m}, 0, \ldots, 0}^{n})\| \leq \|\mathbf{1}_n\|,$$

where the equality follows from the definition of system of norms and the inequality follows from part (ii). □

Step 2: finding p The idea now is that since $\|\mathbf{1}_n\|_p = n^{1/p}$ for all $p \in [1, \infty]$ and $n \geq 1$, we should be able to recover p from $\|\cdot\|$ by examining the sequence $(\|\mathbf{1}_n\|)_{n \geq 1}$.

Indeed, for all $m, n \geq 1$, multiplicativity gives

$$\|\mathbf{1}_{mn}\| = \|\mathbf{1}_m \otimes \mathbf{1}_n\| = \|\mathbf{1}_m\| \, \|\mathbf{1}_n\|.$$

Moreover, Lemma 9.3.9(iii) implies that the sequence $(\|\mathbf{1}_n\|)_{n\geq 1}$ is increasing. Hence by Theorem 1.2.1 applied to the sequence $(\log\|\mathbf{1}_n\|)_{n\geq 1}$, there exists $c \geq 0$ such that $\|\mathbf{1}_n\| = n^c$ for all $n \geq 1$. Now

$$2^c = \|(1,1)\| \leq \|(1,0)\| + \|(0,1)\| = 2\cdot\|(1)\| = 2\cdot 1^c = 2,$$

so $c \in [0,1]$. Put $p = 1/c \in [1,\infty]$. Then

$$\|\mathbf{1}_n\| = n^{1/p} = \|\mathbf{1}_n\|_p$$

for all $n \geq 1$.

We will show that $\|\mathbf{x}\| = \|\mathbf{x}\|_p$ for all $n \geq 0$ and $\mathbf{x} \in \mathbb{R}^n$. By definition of system of norms and Lemma 9.3.9(i), it is enough to prove this when $\mathbf{x} \in (0,\infty)^n$. The case $n = 0$ is trivial, so we can also restrict to $n \geq 1$.

Step 3: the case $p = \infty$ This case needs separate handling, and is straightforward anyway. We show directly that if $p = \infty$ (that is, if $\|\mathbf{1}_n\| = 1$ for all $n \geq 1$) then $\|\cdot\| = \|\cdot\|_\infty$.

Let $\mathbf{x} \in (0,\infty)^n$, and choose j such that $x_j = \|\mathbf{x}\|_\infty$. Then by Lemma 9.3.9(ii),

$$\|\mathbf{x}\| \leq \|(x_j,\ldots,x_j)\| = x_j\|\mathbf{1}_n\| = x_j.$$

But also

$$\|\mathbf{x}\| \geq \|(\underbrace{0,\ldots,0}_{j-1}, x_j, \underbrace{0,\ldots,0}_{n-j})\| = \|(x_j)\| = x_j\|\mathbf{1}_1\| = x_j.$$

Hence $\|\mathbf{x}\| = x_j = \|\mathbf{x}\|_\infty$, as required.

So, we may assume henceforth that $p \in [1,\infty)$.

Step 4: exploiting the variational formula for p-norms We now use the formula for p-norms in Proposition 9.3.8: for $\mathbf{x} \in (0,\infty)^n$,

$$\|\mathbf{x}\|_p = \sup_{u>0,\ r\geq 1} \left(u^r N(\mathbf{x}^{\otimes r}, u^r)^{1/p}\right)^{1/r}.$$

(This is where the probability theory is used, as Proposition 9.3.8 was derived from the variational formula for moment generating functions.) Since $m^{1/p} = \|\mathbf{1}_m\|$ for all m, an equivalent statement is that

$$\|\mathbf{x}\|_p = \sup_{u>0,\ r\geq 1} \left\|(N(\mathbf{x}^{\otimes r}, u^r) * u^r)\right\|^{1/r}. \tag{9.22}$$

Here we have used the notation $*$ introduced after Definition 5.2.9.

The expression (9.22) for $\|\mathbf{x}\|_p$ has the feature that it makes no mention of p. We will use it to prove first that $\|\mathbf{x}\| \geq \|\mathbf{x}\|_p$, then that $\|\mathbf{x}\| \leq \|\mathbf{x}\|_p$.

Step 5: the lower bound Let $\mathbf{x} \in (0, \infty)^n$. We show that $\|\mathbf{x}\| \geq \|\mathbf{x}\|_p$. By (9.22) and multiplicativity, it is equivalent to show that

$$\|\mathbf{x}^{\otimes r}\| \geq \left\|(N(\mathbf{x}^{\otimes r}, u^r) * u^r)\right\|$$

for all real $u > 0$ and integers $r \geq 1$. But this is clear, since by Lemma 9.3.9(ii) and the definition of system of norms,

$$\|\mathbf{x}^{\otimes r}\| \geq \Big\|\overbrace{(\underbrace{u^r, \ldots, u^r}_{N(\mathbf{x}^{\otimes r}, u^r)}, 0, \ldots, 0)}^{n^r}\Big\| = \left\|(N(\mathbf{x}^{\otimes r}, u^r) * u^r)\right\|.$$

Step 6: the upper bound Let $\mathbf{x} \in (0, \infty)^n$. We show that $\|\mathbf{x}\| \leq \|\mathbf{x}\|_p$. The argument is structurally very similar to the second part of the proof of Theorem 9.1.1, and uses the tensor power trick (Tao [325], Section 1.9).

Let $\theta \in (1, \infty)$. We will prove that $\|\mathbf{x}\| \leq \theta\|\mathbf{x}\|_p$. Since $\min_i x_i > 0$, we can choose an integer $d \geq 1$ and real numbers $u_0, \ldots, u_d$ such that

$$\min_i x_i = u_0 < u_1 < \cdots < u_d = \max_i x_i$$

and $u_k/u_{k-1} < \theta$ for all $k \in \{1, \ldots, d\}$.

Let $r \geq 1$. We have the vector $\mathbf{x}^{\otimes r} \in \mathbb{R}^{n^r}$, and we define a new vector $\mathbf{y}_r \in \mathbb{R}^{n^r}$ by rounding up each coordinate of $\mathbf{x}^{\otimes r}$ to the next element of the set $\{u_1^r, \ldots, u_d^r\}$. (Formally, define a map $f_r\colon [u_0^r, u_d^r] \to [u_0^r, u_d^r]$ by $f_r(w) = u_k^r$, where $k \in \{1, \ldots, d\}$ is least such that $w \leq u_k^r$. Then $\mathbf{y}_r$ is obtained from $\mathbf{x}^{\otimes r}$ by applying f_r in each coordinate.)

By construction, $\mathbf{x}^{\otimes r} \leq \mathbf{y}_r$, and the number $n_{k,r}$ of coordinates of $\mathbf{y}_r$ equal to u_k^r is at most $N(\mathbf{x}^{\otimes r}, u_{k-1}^r)$. Hence

$$\begin{aligned}
\|\mathbf{x}^{\otimes r}\| &\leq \|\mathbf{y}_r\| && (9.23)\\
&= \left\|(n_{1,r} * u_1^r, \ldots, n_{d,r} * u_d^r)\right\| && (9.24)\\
&\leq \sum_{k=1}^{d} \left\|((n_{1,r} + \cdots + n_{k-1,r}) * 0, n_{k,r} * u_k^r, (n_{k+1,r} + \cdots + n_{d,r}) * 0)\right\| && (9.25)\\
&= \sum_{k=1}^{d} \|(n_{k,r} * u_k^r)\| && (9.26)\\
&\leq d \max_{1 \leq k \leq d} \|(n_{k,r} * u_k^r)\| && (9.27)\\
&\leq d\theta^r \max_{1 \leq k \leq d} \|(n_{k,r} * u_{k-1}^r)\| && (9.28)\\
&\leq d\theta^r \max_{1 \leq k \leq d} \left\|(N(\mathbf{x}^{\otimes r}, u_{k-1}^r) * u_{k-1}^r)\right\| && (9.29)\\
&\leq d\theta^r \|\mathbf{x}\|_p^r, && (9.30)
\end{aligned}$$

where (9.23) is by Lemma 9.3.9(ii), (9.24) is by symmetry and definition of $n_{k,r}$, (9.25) is by the triangle inequality, (9.26) is by definition of system of norms, (9.27) is elementary, (9.28) is by hypothesis on $u_0, \ldots, u_d$ and Lemma 9.3.9(ii), (9.29) uses Lemma 9.3.9(iii), and (9.30) follows from (9.22). Hence by multiplicativity,

$$\|\mathbf{x}\| = \|\mathbf{x}^{\otimes r}\|^{1/r} \leq d^{1/r}\theta\|\mathbf{x}\|_p.$$

This holds for all integers $r \geq 1$ and real numbers $\theta > 1$. Letting $r \to \infty$ and $\theta \to 1$ gives $\|\mathbf{x}\| \leq \|\mathbf{x}\|_p$, completing the proof of Theorem 9.3.7.

Remark 9.3.10 The proof of Theorem 9.3.7 originally given by Aubrun and Nechita relied on both Cramér's theorem and the Legendre–Fenchel theorem. Effectively, they used Cramér's theorem and convex duality to derive the moment generating function formula of Theorem 9.1.1 in the specific case required.

However, Cerf and Petit [64] showed how the moment generating function formula can be proved without these tools. (In fact, they used it as part of their proof of Cramér's theorem.) The proof of the moment generating function formula given in Section 9.1 is similarly elementary. Our proof of Theorem 9.3.7 works directly from the moment generating function formula, and does not, therefore, need Cramér's theorem, the Legendre–Fenchel theorem, or even the notion of convex conjugate.

Aubrun and Nechita went on to prove similar characterizations of the L^p norms (Theorem 1.2 of [20]) and the Schatten p-norms (their Theorem 4.2). The main focus of the article of Fernández-González, Palazuelos and Pérez-García [102] was also the L^p norms (their Theorem 3.1). We do not discuss these results further.

9.4 Multiplicative Characterization of the Power Means

From the multiplicative characterization of the p-norms, we derive a multiplicative characterization of the power means of order at least 1. It differs from the characterizations of power means in Section 5.5 in that it does not assume modularity. Instead, it uses the multiplicativity condition of Definition 4.2.27, as well as a convexity axiom that provides the connection with norms.

Definition 9.4.1 A sequence of functions $(M\colon \Delta_n \times [0,\infty)^n \to [0,\infty))_{n\geq 1}$ is **convex** if

$$M(\mathbf{p}, \tfrac{1}{2}(\mathbf{x}+\mathbf{y})) \leq \max\{M(\mathbf{p},\mathbf{x}), M(\mathbf{p},\mathbf{y})\}$$

for all $n \geq 1$, $\mathbf{p} \in \Delta_n$, and $\mathbf{x}, \mathbf{y} \in [0, \infty)^n$.

Example 9.4.2 The power mean M_t is multiplicative for all $t \in [-\infty, \infty]$ (Corollary 4.2.28). If $t \in [1, \infty]$ then M_t is also convex. To show this, it suffices to prove the inequality in Definition 9.4.1 in the case where $\mathbf{p}$ has full support. In that case, M_t can be expressed in terms of $\|\cdot\|_t$ by the formula

$$M_t(\mathbf{p}, \mathbf{x}) = \|\mathbf{p}^{1/t}\mathbf{x}\|_t$$

($\mathbf{x} \in [0, \infty)^n$), where both the power and the product of vectors are defined coordinatewise. Now, for $\mathbf{x}, \mathbf{y} \in [0, \infty)^n$,

$$\begin{aligned} M_t(\mathbf{p}, \tfrac{1}{2}(\mathbf{x} + \mathbf{y})) &= \left\|\tfrac{1}{2}\mathbf{p}^{1/t}\mathbf{x} + \tfrac{1}{2}\mathbf{p}^{1/t}\mathbf{y}\right\|_t \\ &\leq \tfrac{1}{2}\left\|\mathbf{p}^{1/t}\mathbf{x}\right\|_t + \tfrac{1}{2}\left\|\mathbf{p}^{1/t}\mathbf{y}\right\|_t \\ &= \tfrac{1}{2}(M_t(\mathbf{p}, \mathbf{x}) + M_t(\mathbf{p}, \mathbf{y})) \\ &\leq \max\{M_t(\mathbf{p}, \mathbf{x}), M_t(\mathbf{p}, \mathbf{y})\}, \end{aligned}$$

by the triangle inequality for $\|\cdot\|_t$. Thus, M_t is convex for $t \in [1, \infty]$.

On the other hand, M_t is not convex for $t \in [-\infty, 1)$, since then

$$M_t\big((\tfrac{1}{2}, \tfrac{1}{2}), \tfrac{1}{2}((1, 0) + (0, 1))\big) = \tfrac{1}{2}$$

but

$$\begin{aligned} \max\big\{M_t((\tfrac{1}{2}, \tfrac{1}{2}), (1, 0)), M_t((\tfrac{1}{2}, \tfrac{1}{2}), (0, 1))\big\} &= M_t((\tfrac{1}{2}, \tfrac{1}{2}), (1, 0)) \\ &= \begin{cases} (\tfrac{1}{2})^{1/t} & \text{if } t \in (0, 1), \\ 0 & \text{if } t \in [-\infty, 0], \end{cases} \end{aligned}$$

which is strictly less than $1/2$.

The multiplicative characterization of the power means is as follows. For a review of the terminology used in (i), see Appendix B.

Theorem 9.4.3 *Let* $(M \colon \Delta_n \times [0, \infty)^n \to [0, \infty))_{n\geq 1}$ *be a sequence of functions. The following are equivalent:*

i. M is natural, consistent, increasing, multiplicative, and convex;
ii. $M = M_t$ *for some* $t \in [1, \infty]$.

The proof follows shortly.

Remarks 9.4.4 i. We have already made some elementary inferences from combinations of the properties in part (i) of the theorem. In the proof of Lemma 4.2.11 (p. 107), we showed that naturality implies symmetry, absence-invariance and repetition. Since M is increasing, Lemma 5.5.3 then implies that M also has the transfer property. Moreover, M is homogeneous, since for all $\mathbf{p} \in \Delta_n$, $\mathbf{x} \in [0, \infty)^n$, and $c \in [0, \infty)$,

$$\begin{aligned} M(\mathbf{p}, c\mathbf{x}) &= M(\mathbf{u}_1 \otimes \mathbf{p}, (c) \otimes \mathbf{x}) \\ &= M(\mathbf{u}_1, (c)) M(\mathbf{p}, \mathbf{x}) \\ &= cM(\mathbf{p}, \mathbf{x}), \end{aligned}$$

by definition of $\otimes$, multiplicativity, and consistency.

ii. Theorem 9.4.3 first appeared as Theorem 1.3 of Leinster [214]. There, the result was stated with a superficially weaker consistency axiom: that $M(\mathbf{u}_1, (x)) = x$ for all $x \in [0, \infty)$. But in the presence of the naturality property, this easily implies full consistency: for naturality implies repetition (as just noted), which in turn implies that

$$M(\mathbf{p}, (x, \ldots, x)) = M((p_1 + \cdots + p_n), (x)) = M(\mathbf{u}_1, (x)) = x$$

for all $\mathbf{p} \in \Delta_n$ and $x \in [0, \infty)$.

We now embark on the proof of Theorem 9.4.3. Certainly (ii) implies (i), by Lemmas 4.2.14, 4.2.17 and 4.2.19, Corollary 4.2.28, and Example 9.4.2. For the converse, and ***for the rest of this section***, let M be a sequence of functions satisfying the conditions in Theorem 9.4.3(i). We will prove that $M = M_t$ for some $t \in [1, \infty]$.

Step 1: finding t The observation behind this step is that

$$M_t((p, 1 - p), (1, 0)) = p^{1/t}$$

for all $p \in (0, 1)$.

Define a function $f : (0, 1) \to [0, \infty)$ by

$$f(p) = M((p, 1 - p), (1, 0)).$$

By multiplicativity and repetition (proved in Remark 9.4.4(i)),

$$\begin{aligned} f(p)f(r) &= M((p, 1 - p), (1, 0)) \cdot M((r, 1 - r), (1, 0)) \\ &= M((p, 1 - p) \otimes (r, 1 - r), \ (1, 0) \otimes (1, 0)) \\ &= M\big((pr, p(1 - r), (1 - p)r, (1 - p)(1 - r)), \ (1, 0, 0, 0)\big) \\ &= M((pr, 1 - pr), \ (1, 0)) \\ &= f(pr) \end{aligned}$$

for all $p, r \in (0, 1)$. By transfer (Remark 9.4.4(i)), f is increasing. If $f(r) = 0$ for some $r \in (0, 1)$ then for all $p \in (0, 1)$,

$$f(p) = f(p/r)f(r) = 0 = p^{\infty}.$$

If not, then f defines an increasing multiplicative function $(0, 1) \to (0, \infty)$, so by Corollary 1.1.16, there is some constant $c \in [0, \infty)$ such that $f(p) = p^c$ for all $p \in (0, 1)$. So in either case, there is a constant $c \in [0, \infty]$ such that $f(p) = p^c$ for all $p \in (0, 1)$. But

$$f(\tfrac{1}{2}) = M(\mathbf{u}_2, (1, 0)) = M(\mathbf{u}_2, (0, 1))$$

by symmetry, so

$$\begin{aligned}(\tfrac{1}{2})^c &= f(\tfrac{1}{2}) \\ &= \max\bigl\{M(\mathbf{u}_2, (1, 0)), M(\mathbf{u}_2, (0, 1))\bigr\} \\ &\geq M(\mathbf{u}_2, (\tfrac{1}{2}, \tfrac{1}{2})) = \tfrac{1}{2}\end{aligned}$$

by convexity and consistency. It follows that $c \in [0, 1]$. Put $t = 1/c \in [1, \infty]$. Then

$$M((p, 1 - p), (1, 0)) = p^{1/t} = M_t((p, 1 - p), (1, 0)) \tag{9.31}$$

for all $p \in (0, 1)$.

Step 2: constructing a system of norms Here we take our inspiration from the relationship

$$\|\mathbf{x}\|_t = n^{1/t} M_t(\mathbf{u}_n, (|x_1|, \ldots, |x_n|))$$

($\mathbf{x} \in \mathbb{R}^n$) between the t-norm and the power mean of order t.

For each $n \geq 1$, define a function $\|\cdot\|: \mathbb{R}^n \to [0, \infty)$ by

$$\|\mathbf{x}\| = n^{1/t} M(\mathbf{u}_n, (|x_1|, \ldots, |x_n|))$$

($\mathbf{x} \in \mathbb{R}^n$). To cover the case $n = 0$, let $\|\cdot\|: \mathbb{R}^0 \to [0, \infty)$ be the function whose single value is 0. The next few lemmas show that $\|\cdot\|$ is a multiplicative system of norms.

Lemma 9.4.5 $n^{-1/t} = M(\mathbf{u}_n, (1, 0, \ldots, 0))$ *for all* $n \geq 1$.

Proof By the defining property of t (equation (9.31)) and the repetition property of M, both sides are equal to $M((1/n, 1 - 1/n), (1, 0))$. □

Lemma 9.4.6 *For each* $n \geq 0$, *the function* $\|\cdot\|: \mathbb{R}^n \to [0, \infty)$ *is a norm.*

Proof This is trivial when $n = 0$; suppose that $n \geq 1$. We verify the three conditions in the definition of norm (Definition 9.3.1).

First, we have to prove that if $\mathbf{0} \neq \mathbf{x} \in \mathbb{R}^n$ then $||\mathbf{x}|| \neq 0$. We may assume by symmetry that $x_1 \neq 0$, and then

$$\begin{aligned}||\mathbf{x}|| &\geq n^{1/t} M(\mathbf{u}_n, (|x_1|, 0, \ldots, 0)) \\ &= n^{1/t} |x_1| M(\mathbf{u}_n, (1, 0, \ldots, 0)) \\ &= |x_1| > 0\end{aligned}$$

by definition of $||\mathbf{x}||$, the increasing and homogeneity properties of M, and Lemma 9.4.5.

The homogeneity of M implies that $||c\mathbf{x}|| = |c|\,||\mathbf{x}||$ for all $\mathbf{x} \in \mathbb{R}^n$ and $c \in \mathbb{R}$.

It remains to prove the triangle inequality, which we do in stages. First let $\mathbf{x}, \mathbf{y} \in \mathbb{R}^n$ with $||\mathbf{x}||, ||\mathbf{y}|| \leq 1$ and $x_i, y_i \geq 0$ for all i. Using the convexity of M,

$$\begin{aligned}\left\|\tfrac{1}{2}\mathbf{x} + \tfrac{1}{2}\mathbf{y}\right\| &= n^{1/t} M(\mathbf{u}_n, \tfrac{1}{2}(\mathbf{x} + \mathbf{y})) \\ &\leq n^{1/t} \max\{M(\mathbf{u}_n, \mathbf{x}), M(\mathbf{u}_n, \mathbf{y})\} \\ &= \max\{||\mathbf{x}||, ||\mathbf{y}||\} \\ &\leq 1.\end{aligned}$$

It follows that

$$||\lambda\mathbf{x} + (1 - \lambda)\mathbf{y}|| \leq 1 \tag{9.32}$$

for all dyadic rationals $\lambda = k/2^\ell \in [0, 1]$, by induction on ℓ. We now show that (9.32) holds for all $\lambda \in [0, 1]$. Indeed, given $\lambda \in [0, 1]$ and $\varepsilon > 0$, we can choose a dyadic rational $\lambda' \in [0, 1]$ such that

$$\lambda \leq (1 + \varepsilon)\lambda', \qquad 1 - \lambda \leq (1 + \varepsilon)(1 - \lambda'),$$

and then

$$\begin{aligned}||\lambda\mathbf{x} + (1 - \lambda)\mathbf{y}|| &\leq ||(1 + \varepsilon)\lambda'\mathbf{x} + (1 + \varepsilon)(1 - \lambda')\mathbf{y}|| \\ &= (1 + \varepsilon)||\lambda'\mathbf{x} + (1 - \lambda')\mathbf{y}|| \\ &\leq 1 + \varepsilon,\end{aligned}$$

where in the first inequality, we used the assumptions that M is increasing and $x_i, y_i \geq 0$. This holds for all $\varepsilon > 0$, proving the claimed inequality (9.32).

Now take any $\mathbf{x}, \mathbf{y} \in \mathbb{R}^n$ with $x_i, y_i \geq 0$ for all i. We will prove that

$$||\mathbf{x} + \mathbf{y}|| \leq ||\mathbf{x}|| + ||\mathbf{y}||. \tag{9.33}$$

This is immediate if $\mathbf{x} = \mathbf{0}$ or $\mathbf{y} = \mathbf{0}$. Supposing otherwise, put

$$\hat{\mathbf{x}} = \frac{\mathbf{x}}{||\mathbf{x}||}, \qquad \hat{\mathbf{y}} = \frac{\mathbf{y}}{||\mathbf{y}||}, \qquad \lambda = \frac{||\mathbf{x}||}{||\mathbf{x}|| + ||\mathbf{y}||}.$$

Then $\|\hat{\mathbf{x}}\| = \|\hat{\mathbf{y}}\| = 1$, so by inequality (9.32) applied to $\hat{\mathbf{x}}$, $\hat{\mathbf{y}}$ and λ,

$$\|\mathbf{x} + \mathbf{y}\| = (\|\mathbf{x}\| + \|\mathbf{y}\|)\, \|\lambda\hat{\mathbf{x}} + (1 - \lambda)\hat{\mathbf{y}}\| \leq \|\mathbf{x}\| + \|\mathbf{y}\|.$$

Finally, take any $\mathbf{x}, \mathbf{y} \in \mathbb{R}^n$. To prove the triangle inequality (9.33), put $\mathbf{x}' = (|x_1|, \dots, |x_n|)$ and $\mathbf{y}' = (|y_1|, \dots, |y_n|)$. Then $\|\mathbf{x}\| = \|\mathbf{x}'\|$ and $\|\mathbf{y}\| = \|\mathbf{y}'\|$ by definition of $\|\cdot\|$, and

$$\|\mathbf{x} + \mathbf{y}\| \leq \|\mathbf{x}' + \mathbf{y}'\|$$

since M is increasing. By the inequality proved in the previous paragraph,

$$\|\mathbf{x}' + \mathbf{y}'\| \leq \|\mathbf{x}'\| + \|\mathbf{y}'\|,$$

and the triangle inequality (9.33) follows. □

Lemma 9.4.7 $\|\cdot\|$ *is a multiplicative system of norms.*

Proof We have just shown that $\|\cdot\|$ is a norm on $\mathbb{R}^n$ for each individual n. Symmetry of M implies symmetry of $\|\cdot\|$, so to show that $\|\cdot\|$ is a system of norms, it suffices to prove that

$$\|(x_1, \dots, x_n)\| = \|(x_1, \dots, x_n, 0)\| \tag{9.34}$$

for all $n \geq 1$ and $\mathbf{x} \in \mathbb{R}^n$. By definition of $\|\cdot\|$ and Lemma 9.4.5, equation (9.34) is equivalent to

$$\frac{M(\mathbf{u}_n, (|x_1|, \dots, |x_n|))}{M(\mathbf{u}_n, (1, 0, \dots, 0))} = \frac{M(\mathbf{u}_{n+1}, (|x_1|, \dots, |x_n|, 0))}{M(\mathbf{u}_{n+1}, (1, 0, \dots, 0, 0))},$$

or equivalently,

$$\begin{aligned} &M(\mathbf{u}_{n+1}, (1, 0, \dots, 0, 0)) \cdot M(\mathbf{u}_n, (|x_1|, \dots, |x_n|)) \\ &\quad = M(\mathbf{u}_n, (1, 0, \dots, 0)) \cdot M(\mathbf{u}_{n+1}, (|x_1|, \dots, |x_n|, 0)). \end{aligned}$$

But by multiplicativity and symmetry, both sides are equal to

$$M(\mathbf{u}_{n(n+1)}, (|x_1|, \dots, |x_n|, \underbrace{0, \dots, 0}_{n^2})),$$

proving (9.34).

Finally, the system of norms $\|\cdot\|$ is multiplicative, by multiplicativity of M. □

Step 3: using the norm theorem It now follows from Theorem 9.3.7 that $\|\cdot\| = \|\cdot\|_s$ for some $s \in [1, \infty]$. Thus, $\|\mathbf{x}\|_s = n^{1/t} M(\mathbf{u}_n, \mathbf{x})$ for all $n \geq 1$ and $\mathbf{x} \in [0, \infty)^n$. But also, $\|\mathbf{x}\|_s = n^{1/s} M_s(\mathbf{u}_n, \mathbf{x})$, so

$$n^{1/t} M(\mathbf{u}_n, \mathbf{x}) = n^{1/s} M_s(\mathbf{u}_n, \mathbf{x})$$

for all $n \geq 1$ and $\mathbf{x} \in [0, \infty)$. Putting $n = 2$ and $\mathbf{x} = (1, 1)$, and using the consistency of both M and M_s, gives $s = t$. Hence for all $n \geq 1$ and $\mathbf{x} \in [0, \infty)^n$,

$$M(\mathbf{u}_n, \mathbf{x}) = M_t(\mathbf{u}_n, \mathbf{x}).$$

Step 4: arbitrary weights We have now shown that $M(\mathbf{u}_n, -) = M_t(\mathbf{u}_n, -)$ for all $n \geq 1$. To extend the equality to arbitrary weights, we use Proposition 5.5.7, taking $M' = M_t$ there. The hypotheses of that proposition are satisfied, by Remark 9.4.4(i) and Lemma 4.2.6(i). Hence $M = M_t$.

This completes the proof of Theorem 9.4.3, the multiplicative characterization of the power means.

10

Information Loss

Grothendieck came along and said, 'No, the Riemann–Roch theorem is *not* a theorem about varieties, it's a theorem about morphisms between varieties.' – Nicholas Katz (quoted in [154], p. 1046).

This short chapter tells the following story. A measure-preserving map between finite probability spaces can be regarded as a deterministic process. As such, it loses information. We can attempt to quantify how much information is lost. It turns out that as soon as we impose a few reasonable requirements on this quantity, it is highly constrained: up to a constant factor, it must be the difference between the entropies of the domain and the codomain. That is our main theorem.

This result is essentially another characterization of Shannon entropy, and first appeared in a 2011 paper of Baez, Fritz and Leinster [25]. The broad idea is to shift the focus from *objects* (finite probability spaces) to *maps* between objects (measure-preserving maps). Entropy is an invariant of finite probability spaces; information loss is an invariant of measure-preserving maps. The shift of emphasis from objects to maps is integral to category theory, and has borne fruit such as the Grothendieck–Riemann–Roch theorem alluded to in the opening quotation, as well as the considerably more humble characterization of information loss described here.

In full categorical generality, a map $\mathcal{X} \xrightarrow{f} \mathcal{Y}$ of any kind can be viewed as an object $\mathcal{X}$ parametrized by another object $\mathcal{Y}$. An object $\mathcal{X}$ can be viewed as a map of a special kind, namely, the unique map $\mathcal{X} \xrightarrow{!_{\mathcal{X}}} 1$ to the terminal object 1 of the category concerned. In the case at hand, we associate with any probability space $\mathcal{X}$ the unique measure-preserving map $\mathcal{X} \xrightarrow{!_{\mathcal{X}}} 1$ to the one-point space 1, and the information loss of the map $!_{\mathcal{X}}$ is equal to the entropy of the space $\mathcal{X}$. Thus, entropy is a special case of information loss.

An advantage of working with information loss rather than entropy (that is, maps rather than objects) is that the characterization theorems take on a new simplicity. For instance, the conditions in our main result (Theorem 10.2.1) look just like the linearity or homomorphism conditions that appear throughout mathematics. In contrast, the chain rule for entropy, while justifiable in many other ways, has a more complicated algebraic form.

We begin with a review of measure-preserving maps, then define information loss (Section 10.1). After recording a few simple properties of information loss, we prove that they characterize it uniquely (Section 10.2). An analogous and even simpler result is then proved for q-logarithmic information loss ($q \neq 1$). Both of these theorems first appeared in the 2011 paper of Baez, Fritz and Leinster [25].

10.1 Measure-Preserving Maps

So far in this text, we have focused on probability distributions on finite sets of the special form $\{1, \ldots, n\}$. Here, it is convenient to use arbitrary finite sets. The difference is cosmetic, but does cause some shifts in notation, as follows.

Definition 10.1.1 i. Let $\mathcal{X}$ be a finite set. A **probability distribution p** on $\mathcal{X}$ is a family $(p_i)_{i \in \mathcal{X}}$ of nonnegative real numbers such that $\sum_{i \in \mathcal{X}} p_i = 1$. We write $\Delta_{\mathcal{X}}$ for the set of probability distributions on $\mathcal{X}$.

ii. A **finite probability space** is a pair $(\mathcal{X}, \mathbf{p})$ where $\mathcal{X}$ is a finite set and $\mathbf{p} \in \Delta_{\mathcal{X}}$.

The set $\Delta_{\mathcal{X}}$ is topologized as a subspace of the product space $\mathbb{R}^{\mathcal{X}}$.

Definition 10.1.2 Let $(\mathcal{Y}, \mathbf{s})$ and $(\mathcal{X}, \mathbf{p})$ be finite probability spaces. A **measure-preserving map** $(\mathcal{Y}, \mathbf{s}) \to (\mathcal{X}, \mathbf{p})$ is a function $f \colon \mathcal{Y} \to \mathcal{X}$ such that

$$p_i = \sum_{j \in f^{-1}(i)} s_j \tag{10.1}$$

for all $i \in \mathcal{X}$.

An equivalent statement is that $f \colon (\mathcal{Y}, \mathbf{s}) \to (\mathcal{X}, \mathbf{p})$ is measure-preserving if and only if

$$\sum_{i \in \mathcal{V}} p_i = \sum_{j \in f^{-1}\mathcal{V}} s_j \tag{10.2}$$

for all $\mathcal{V} \subseteq \mathcal{X}$. Indeed, (10.1) is the case of (10.2) where $\mathcal{V} = \{i\}$, and (10.2) follows from (10.1) by summing over all $i \in \mathcal{V}$.

Remarks 10.1.3 i. For any finite probability space $(\mathcal{Y}, \mathbf{s})$ and function f from $\mathcal{Y}$ to another finite set $\mathcal{X}$, there is an induced probability distribution $f\mathbf{s}$ on

$\mathcal{X}$, the **pushforward** of $\mathbf{s}$ along f. It is defined by the obvious generalization of Definition 2.1.10:

$$(f\mathbf{s})_i = \sum_{j \in f^{-1}(i)} s_j$$

($i \in \mathcal{X}$). In these terms, a function $f\colon (\mathcal{Y}, \mathbf{s}) \to (\mathcal{X}, \mathbf{p})$ is measure-preserving if and only if $f\mathbf{s} = \mathbf{p}$.

ii. Finite probability spaces and measure-preserving maps form a category **FinProb**. We note in passing that by (i), the forgetful functor **FinProb** $\to$ **FinSet** is a discrete opfibration. In fact, **FinProb** is the category of elements of the functor **FinSet** $\to$ **Set** defined on objects by $\mathcal{X} \mapsto \Delta_{\mathcal{X}}$ and on maps by pushforward. (For the categorical terminology used here, see for instance Riehl [297], Definition 2.4.1 and Exercise 2.4.viii.)

Although a measure-preserving map need not be literally surjective, it is essentially so, in the sense that all elements not in the image have probability zero.

Example 10.1.4 Let $\mathcal{Y} = \{\mathsf{a}, \mathsf{à}, \mathsf{â}, \mathsf{b}, \mathsf{c}, \mathsf{ç}, \ldots\}$ be the set of symbols in the French language, and let $\mathbf{s} \in \Delta_{\mathcal{Y}}$ be their frequency distribution (as in Example 2.1.5). Let $\mathcal{X} = \{\mathsf{a}, \mathsf{b}, \mathsf{c}, \ldots\}$ be the 26-element set of letters, and $\mathbf{p} \in \Delta_{\mathcal{X}}$ their frequency distribution. There is a function $f\colon \mathcal{Y} \to \mathcal{X}$ that forgets accents; for instance, $f(\mathsf{a}) = f(\mathsf{à}) = f(\mathsf{â}) = \mathsf{a}$. Then $f\colon (\mathcal{Y}, \mathbf{s}) \to (\mathcal{X}, \mathbf{p})$ is measure-preserving and surjective.

Example 10.1.5 Let ℓ be the inclusion function $\{1\} \hookrightarrow \{1, 2\}$. Give $\{1\}$ its unique probability distribution $(1) = \mathbf{u}_1$, and give $\{1, 2\}$ the distribution $(1, 0)$. Then ℓ is measure-preserving but not surjective.

Any measure-preserving map between finite probability spaces can be factorized canonically into maps of the two types in these two examples: a surjection followed by a subset inclusion, where the subset concerned has total probability 1. Specifically, $f\colon (\mathcal{Y}, \mathbf{s}) \to (\mathcal{X}, \mathbf{p})$ factorizes as

$$(\mathcal{Y}, \mathbf{s}) \xrightarrow{f'} (f\mathcal{Y}, \mathbf{p}') \xrightarrow{\ell} (\mathcal{X}, \mathbf{p}),$$

where $\mathbf{p}'$ is the probability distribution on $f\mathcal{Y}$ defined by $p'_i = p_i$ for all $i \in f\mathcal{Y}$, the surjection f' is defined by $f'(j) = f(j)$ for all $j \in \mathcal{Y}$, and ℓ is inclusion.

A measure-preserving surjection simply discards information (such as the accents in Example 10.1.4). It is a coarse-graining, in the sense of taking finely grained information (such as letters with accents) and converting it into more coarsely grained information (such as mere letters). A measure-preserving inclusion is essentially trivial, simply appending some events of probability zero.

For any measure-preserving bijection $f\colon (\mathcal{Y}, \mathbf{s}) \to (\mathcal{X}, \mathbf{p})$ between finite probability spaces, the inverse f^{-1} is also measure-preserving. We call such an f an **isomorphism**, and write $(\mathcal{Y}, \mathbf{s}) \cong (\mathcal{X}, \mathbf{p})$.

An important feature of probability spaces is that we can take convex combinations of them. Given $\mathbf{w} \in \Delta_n$ and finite probability spaces $(\mathcal{X}_1, \mathbf{p}^1), \ldots, (\mathcal{X}_n, \mathbf{p}^n)$, we obtain a new probability space

$$\left(\coprod_{i=1}^n \mathcal{X}_i, \coprod_{i=1}^n w_i \mathbf{p}^i \right),$$

where $\coprod \mathcal{X}_i$ is the disjoint union of sets $\mathcal{X}_1 \sqcup \cdots \sqcup \mathcal{X}_n$ and $\coprod w_i \mathbf{p}^i$ is the probability distribution on $\coprod \mathcal{X}_i$ that gives probability $w_i p^i_j$ to an element $j \in \mathcal{X}_i$.

Convex combination of probability spaces is just composition of probability distributions, translated into different notation. More exactly, if $\mathcal{X}_i = \{1, \ldots, k_i\}$ then $\coprod \mathcal{X}_i$ is in canonical bijection with $\{1, \ldots, k_1 + \cdots + k_n\}$, and under this bijection, $\coprod w_i \mathbf{p}^i$ corresponds to the composite distribution $\mathbf{w} \circ (\mathbf{p}^1, \ldots, \mathbf{p}^n)$.

The construction of convex combinations is functorial, that is, applies not only to probability spaces but also to maps between them. Indeed, take measure-preserving maps

$$\begin{array}{ccc} (\mathcal{Y}_1, \mathbf{s}^1) & \xrightarrow{f_1} & (\mathcal{X}_1, \mathbf{p}^1) \\ \vdots & & \vdots \\ (\mathcal{Y}_n, \mathbf{s}^n) & \xrightarrow{f_n} & (\mathcal{X}_n, \mathbf{p}^n) \end{array}$$

between finite probability spaces, and a probability distribution $\mathbf{w} \in \Delta_n$. There is a function

$$\coprod_{i=1}^n \mathcal{Y}_i \xrightarrow{\coprod_{i=1}^n f_i} \coprod_{i=1}^n \mathcal{X}_i$$

that maps $j \in \mathcal{Y}_i$ to $f_i(j) \in \mathcal{X}_i$, and it is easily checked that $\coprod f_i$ is a measure-preserving map

$$\left(\coprod_{i=1}^n \mathcal{Y}_i, \coprod_{i=1}^n w_i \mathbf{s}^i \right) \xrightarrow{\coprod_{i=1}^n f_i} \left(\coprod_{i=1}^n \mathcal{X}_i, \coprod_{i=1}^n w_i \mathbf{p}^i \right). \tag{10.3}$$

It will be convenient to use the alternative notation

$$\coprod_{i=1}^n w_i f_i \qquad \text{or} \qquad w_1 f_1 \sqcup \cdots \sqcup w_n f_n$$

for the measure-preserving map $\coprod_{i=1}^n f_i$ of (10.3).

We defined Shannon entropy only for probability distributions on sets of the form $\{1, \ldots, n\}$, but, of course, the definition for general finite probability spaces $(\mathcal{X}, \mathbf{p})$ is

$$H(\mathbf{p}) = - \sum_{i \in \operatorname{supp}(\mathbf{p})} p_i \log p_i,$$

where $\operatorname{supp}(\mathbf{p}) = \{i \in \mathcal{X} : p_i > 0\}$. Shannon entropy is **isomorphism-invariant**, meaning that $H(\mathbf{p}) = H(\mathbf{s})$ whenever $(\mathcal{X}, \mathbf{p})$ and $(\mathcal{Y}, \mathbf{s})$ are isomorphic finite probability spaces.

Translated into this notation, the chain rule for Shannon entropy states that

$$H\Bigl(\coprod_{i=1}^{n} w_i \mathbf{p}^i\Bigr) = H(\mathbf{w}) + \sum_{i=1}^{n} w_i H(\mathbf{p}^i) \tag{10.4}$$

for all $\mathbf{w} \in \Delta_n$ and finite probability spaces $(\mathcal{X}_1, \mathbf{p}^1), \ldots, (\mathcal{X}_n, \mathbf{p}^n)$. The continuity property of entropy is that for each finite set $\mathcal{X}$, the function

$$\begin{array}{ccc} \Delta_{\mathcal{X}} & \to & \mathbb{R} \\ \mathbf{p} & \mapsto & H(\mathbf{p}) \end{array} \tag{10.5}$$

is continuous.

We now set out to quantify the information lost by a measure-preserving map f, first exploring through examples how a reasonable definition of information loss ought to behave.

Example 10.1.6 If f is an isomorphism then f should lose no information at all. More generally, the same should be true if f is injective.

Example 10.1.7 The unique measure-preserving map $(\{1, 2\}, \mathbf{u}_2) \to (\{1\}, \mathbf{u}_1)$ forgets the result of a fair coin toss. Intuitively, then, it loses one bit of information.

Example 10.1.8 More generally, for any finite probability space $(\mathcal{X}, \mathbf{p})$, consider the unique measure-preserving map

$$f : (\mathcal{X}, \mathbf{p}) \to (\{1\}, \mathbf{u}_1),$$

which forgets the result of an observation drawn from the distribution $\mathbf{p}$. Such an observation contains $H^{(2)}(\mathbf{p})$ bits of information (in the sense of Section 2.3), so the information lost by f should be $H^{(2)}(\mathbf{p})$ bits.

Example 10.1.9 Suppose that I draw fairly from a pack of playing cards, and tell you only the rank (number) of the card chosen. The information that I am withholding is the suit, which needs $\log_2 4 = 2$ bits to encode. Thus, if $f : \mathcal{Y} \to \mathcal{X}$ is a four-to-one map from a 52-element set $\mathcal{Y}$ to a 13-element set

$\mathcal{X}$, and if we equip $\mathcal{Y}$ and $\mathcal{X}$ with their uniform distributions $\mathbf{u}_{\mathcal{Y}}$ and $\mathbf{u}_{\mathcal{X}}$, then the information loss of the measure-preserving map $f\colon (\mathcal{Y}, \mathbf{u}_{\mathcal{Y}}) \to (\mathcal{X}, \mathbf{u}_{\mathcal{X}})$ should be 2 bits.

Example 10.1.10 Take the measure-preserving map

$$f\colon (\{\mathsf{a}, \mathsf{à}, \mathsf{â}, \mathsf{b}, \ldots\}, \mathbf{s}) \to (\{\mathsf{a}, \mathsf{b}, \ldots\}, \mathbf{p})$$

of Example 10.1.4, representing the process of forgetting the accent on a letter in the French language. There are two quantities that we could reasonably call the 'amount of information lost' by the process f.

First, we could condition on the underlying letter. To do this, we go through the 26 letters, we take for each letter the amount of information lost by forgetting the accent on that letter, and we form the weighted mean. Write

$$\mathbf{r}^1 \in \Delta_3, \ \mathbf{r}^2 \in \Delta_1, \ \ldots, \ \mathbf{r}^{26} \in \Delta_1$$

for the accent distributions on each letter, so that $\mathbf{s} = \coprod_{i=1}^{26} p_i \mathbf{r}^i$. As in Example 10.1.8, the amount of information lost by forgetting the accent on an a (for instance) should be $H^{(2)}(\mathbf{r}^1)$ bits. So, the expected amount of information lost by forgetting the accent on a random letter should be

$$\sum_{i=1}^{26} p_i H^{(2)}(\mathbf{r}^i). \tag{10.6}$$

This is one possible definition of the amount of information lost by f.

Alternatively, we could define the information loss to be the amount of information we had at the start of the process minus the amount of information that remains at the end. This is

$$H^{(2)}(\mathbf{s}) - H^{(2)}(\mathbf{p}). \tag{10.7}$$

But since $\mathbf{s} = \coprod p_i \mathbf{r}^i$, the chain rule (10.4) tells us that the two quantities (10.6) and (10.7) are equal. So, our two ways of quantifying information loss are equivalent.

Motivated by these examples, we make the following definition.

Definition 10.1.11 Let

$$f\colon (\mathcal{Y}, \mathbf{s}) \to (\mathcal{X}, \mathbf{p})$$

be a measure-preserving map of finite probability spaces. The **information loss** of f is

$$L(f) = H(\mathbf{s}) - H(\mathbf{p}).$$

As with other entropic quantities that we have encountered, the definition of information loss depends on a choice of logarithmic base, and changing that base scales the quantity by a constant factor.

A deterministic process cannot create new information, and correspondingly, information loss is always nonnegative:

Lemma 10.1.12 *Let $f: (\mathcal{Y}, \mathbf{s}) \to (\mathcal{X}, \mathbf{p})$ be a measure-preserving map of finite probability spaces. Then:*

i. $L(f) = \sum_{j \in \mathrm{supp}(\mathbf{s})} s_j \log \frac{p_{f(j)}}{s_j}$;

ii. $L(f) \geq 0$.

Proof By definition of measure-preserving map (Definition 10.1.2), $p_{f(j)} \geq s_j$ for all $j \in \mathcal{Y}$. It follows that

$$j \in \mathrm{supp}(\mathbf{s}) \implies f(j) \in \mathrm{supp}(\mathbf{p}). \tag{10.8}$$

It also follows that $\log(p_{f(j)}/s_j) \geq 0$ for all $j \in \mathrm{supp}(\mathbf{s})$, so part (ii) will follow once we have proved (i).

To prove (i), first note that by definition of measure-preserving map,

$$\begin{aligned} H(\mathbf{p}) &= \sum_{i \in \mathrm{supp}(\mathbf{p})} p_i \log \frac{1}{p_i} \\ &= \sum_{i \in \mathrm{supp}(\mathbf{p}),\ j \in \mathcal{Y}:\ f(j)=i} s_j \log \frac{1}{p_i} \\ &= \sum_{j:\ f(j) \in \mathrm{supp}(\mathbf{p})} s_j \log \frac{1}{p_{f(j)}}. \end{aligned}$$

By (10.8), this sum is unchanged if we take j to range over supp(s) instead. Hence

$$\begin{aligned} L(f) &= H(\mathbf{s}) - H(\mathbf{p}) \\ &= \sum_{j \in \mathrm{supp}(\mathbf{s})} s_j \log \frac{1}{s_j} - \sum_{j \in \mathrm{supp}(\mathbf{s})} s_j \log \frac{1}{p_{f(j)}} \\ &= \sum_{j \in \mathrm{supp}(\mathbf{s})} s_j \log \frac{p_{f(j)}}{s_j}, \end{aligned}$$

as claimed. □

Remark 10.1.13 This result is also an instance of Lemma 8.1.3(i) on conditional entropy, as follows. Let V be a random variable taking values in $\mathcal{Y}$, with distribution $\mathbf{s}$. Put $U = f(V)$, which is a random variable taking values in $\mathcal{X}$, with distribution $f\mathbf{s} = \mathbf{p}$. Then U is determined by V, so by Example 8.1.4(iv),

$$0 \leq H(V \mid U) = H(V) - H(U) = H(\mathbf{s}) - H(\mathbf{p}) = L(f).$$

On the other hand, by Lemma 8.1.3(i),

$$H(V \mid U) = \sum_{j,i:\, \mathbb{P}(j,i)>0} \mathbb{P}(j,i) \log \frac{\mathbb{P}(i)}{\mathbb{P}(j,i)} = \sum_{j:\, s_j>0} s_j \log \frac{p_{f(j)}}{s_j}.$$

Comparing the two expressions for $H(V \mid U)$ gives another proof of Lemma 10.1.12.

This argument shows that information loss is a special case of conditional entropy. But conditional entropy is also a special case of information loss. Indeed, let U and V be random variables with the same sample space, taking values in finite sets $\mathcal{X}$ and $\mathcal{Y}$ respectively. Equip $\mathcal{X} \times \mathcal{Y}$ with the distribution of (U, V) and $\mathcal{X}$ with the distribution of U. Then the projection map

$$\begin{array}{rccc} \mathrm{pr}_1 : & \mathcal{X} \times \mathcal{Y} & \to & \mathcal{X} \\ & (i, j) & \mapsto & i \end{array}$$

is measure-preserving. By definition, its information loss is

$$L(\mathrm{pr}_1) = H(U, V) - H(U) = H(V \mid U).$$

Hence $H(V \mid U) = L(\mathrm{pr}_1)$, expressing conditional entropy in terms of information loss.

10.2 Characterization of Information Loss

In this section, we prove that information loss is uniquely characterized (up to a constant factor) by four basic properties.

First, a reversible process loses no information: $L(f) = 0$ for all isomorphisms f. This follows from the definition of L and the isomorphism-invariance of H.

Second, the amount of information lost by two processes in series is the sum of the amounts of information lost by each individually. Formally,

$$L(g \circ f) = L(g) + L(f) \tag{10.9}$$

whenever

$$(\mathcal{Y}, \mathbf{s}) \xrightarrow{f} (\mathcal{X}, \mathbf{p}) \xrightarrow{g} (W, \mathbf{t})$$

are measure-preserving maps of finite probability spaces. This is immediate from the definition of information loss.

Third, given n measure-preserving maps

$$\begin{array}{ccc} (\mathcal{Y}_1, \mathbf{s}^1) & \xrightarrow{f_1} & (\mathcal{X}_1, \mathbf{p}^1) \\ \vdots & & \vdots \\ (\mathcal{Y}_n, \mathbf{s}^n) & \xrightarrow{f_n} & (\mathcal{X}_n, \mathbf{p}^n) \end{array}$$

and a distribution $\mathbf{w} \in \Delta_n$, the amount of information lost by the convex combination $\coprod w_i f_i$ is given by

$$L\left(\coprod_{i=1}^{n} w_i f_i\right) = \sum_{i=1}^{n} w_i L(f_i). \tag{10.10}$$

This follows from the chain rule (10.4):

$$\begin{aligned} L\left(\coprod w_i f_i\right) &= H\left(\coprod w_i \mathbf{s}^i\right) - H\left(\coprod w_i \mathbf{p}^i\right) \\ &= \left\{H(\mathbf{w}) + \sum w_i H(\mathbf{s}^i)\right\} - \left\{H(\mathbf{w}) + \sum w_i H(\mathbf{p}^i)\right\} \\ &= \sum w_i L(f_i). \end{aligned}$$

In particular, given measure-preserving maps

$$\begin{aligned} (\mathcal{Y}, \mathbf{s}) &\xrightarrow{f} (\mathcal{X}, \mathbf{p}), \\ (\mathcal{Y}', \mathbf{s}') &\xrightarrow{f'} (\mathcal{X}', \mathbf{p}') \end{aligned}$$

and a constant $\lambda \in [0, 1]$,

$$L(\lambda f \sqcup (1 - \lambda) f') = \lambda L(f) + (1 - \lambda) L(f').$$

Intuitively, this means that if we flip a probability-λ coin and, depending on the outcome, do either the process f or the process f', then the expected information loss is λ times the information loss of f plus $1 - \lambda$ times the information loss of f'. So, while the previous property of L (equation (10.9)) concerned the information lost by two processes in *series*, this property (equation (10.10)) concerns the information lost by two or more processes in *parallel*.

Fourth and finally, information loss is continuous, in the following sense. Let $f: \mathcal{Y} \to \mathcal{X}$ be a map of finite sets. For each probability distribution $\mathbf{s}$ on $\mathcal{Y}$, we have the pushforward distribution $f\mathbf{s}$ on $\mathcal{X}$, and f defines a measure-preserving map

$$f: (\mathcal{Y}, \mathbf{s}) \to (\mathcal{X}, f\mathbf{s})$$

(Remark 10.1.3(i)). The statement is that the map

$$\begin{array}{ccc} \Delta_{\mathcal{Y}} & \to & \mathbb{R} \\ \mathbf{s} & \mapsto & L\big((\mathcal{Y}, \mathbf{s}) \xrightarrow{f} (\mathcal{X}, f\mathbf{s})\big) \end{array}$$

is continuous. This follows from the fact that all the maps in the (noncommutative) triangle

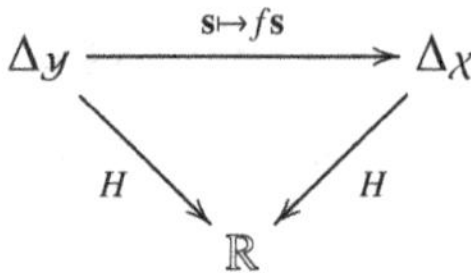

are continuous.

An equivalent way to state continuity is as follows. Let us say that an infinite sequence

$$\big((\mathcal{Y}_m, \mathbf{s}^m) \xrightarrow{f_m} (\mathcal{X}_m, \mathbf{p}^m)\big)_{m \geq 1}$$

of measure-preserving maps of finite probability spaces **converges** to a map

$$(\mathcal{Y}, \mathbf{s}) \xrightarrow{f} (\mathcal{X}, \mathbf{p})$$

if

$$\big(\mathcal{Y}_m \xrightarrow{f_m} \mathcal{X}_m\big) = \big(\mathcal{Y} \xrightarrow{f} \mathcal{X}\big)$$

for all sufficiently large m, and $\mathbf{s}^m \to \mathbf{s}$ and $\mathbf{p}^m \to \mathbf{p}$ as $m \to \infty$. Then continuity of L is equivalent to the statement that for any such convergent sequence,

$$L\big((\mathcal{Y}_m, \mathbf{s}^m) \xrightarrow{f_m} (\mathcal{X}_m, \mathbf{p}^m)\big) \to L\big((\mathcal{Y}, \mathbf{s}) \xrightarrow{f} (\mathcal{X}, \mathbf{p})\big) \text{ as } m \to \infty.$$

The equivalence between these two formulations of continuity follows from the elementary fact that a map of metrizable spaces is continuous if and only if it preserves convergence of sequences.

We now state the main theorem, which first appeared as Theorem 2 of Baez, Fritz and Leinster [25].

Theorem 10.2.1 (Baez, Fritz and Leinster) *Let K be a function assigning a real number $K(f)$ to each measure-preserving map f of finite probability spaces. The following are equivalent.*

i. *K has these four properties:*
   a. *$K(f) = 0$ for all isomorphisms f;*
   b. *$K(g \circ f) = K(g) + K(f)$ for all composable pairs (f, g) of measure-preserving maps;*

c. $K(\lambda f \sqcup (1-\lambda)f') = \lambda K(f) + (1-\lambda)K(f')$ *for all measure-preserving maps* f *and* f' *and all* $\lambda \in [0,1]$*;*

d. K *is continuous.*

ii. $K = cL$ *for some* $c \in \mathbb{R}$.

The proof, given below, will use a version of Faddeev's theorem.

Theorem 10.2.2 (Faddeev, version 2) *Let* I *be a function assigning a real number* $I(\mathbf{p})$ *to each finite probability space* $(\mathcal{X}, \mathbf{p})$*. The following are equivalent:*

i. I *is isomorphism-invariant, satisfies the chain rule* (10.4)*, and is continuous in the sense of* (10.5) *(with* I *in place of* H*);*

ii. $I = cH$ *for some* $c \in \mathbb{R}$.

Proof We have already observed that H satisfies the conditions in (i), and it follows that (ii) implies (i).

Conversely, take a function I satisfying (i). Restricting I to finite sets of the form $\{1, \ldots, n\}$ defines, for each $n \geq 1$, a continuous function $I\colon \Delta_n \to \mathbb{R}$ satisfying the chain rule. Hence by Faddeev's Theorem 2.5.1, there is some constant $c \in \mathbb{R}$ such that $I(\mathbf{p}) = cH(\mathbf{p})$ for all $n \geq 1$ and $\mathbf{p} \in \Delta_n$. Next, take any finite probability space $(\mathcal{Y}, \mathbf{s})$. We have

$$(\mathcal{Y}, \mathbf{s}) \cong (\{1, \ldots, n\}, \mathbf{p})$$

for some $n \geq 1$ and $\mathbf{p} \in \Delta_n$, and then by isomorphism-invariance of both I and H,

$$I(\mathbf{s}) = I(\mathbf{p}) = cH(\mathbf{p}) = cH(\mathbf{s}),$$

as required. □

Remark 10.2.3 The version of Faddeev's theorem just stated is slightly weaker than the earlier version, Theorem 2.5.1. To see this, take $\mathbf{p} \in \Delta_n$ and a permutation σ of $\{1, \ldots, n\}$. Then σ defines a measure-preserving bijection

$$\sigma\colon (\{1, \ldots, n\}, \mathbf{p}\sigma) \to (\{1, \ldots, n\}, \mathbf{p}).$$

In Theorem 10.2.2, therefore, the isomorphism-invariance axiom on I includes as a special case that $I(\mathbf{p}\sigma) = I(\mathbf{p})$ for all $\mathbf{p} \in \Delta_n$ and permutations σ. This is the symmetry axiom that is traditionally included in statements of Faddeev's theorem, but is not in fact necessary, as observed in Remark 2.5.2(ii). So, Theorem 10.2.2 is a restatement of that traditional, weaker form of Faddeev's theorem. The analogous restatement of the stronger Theorem 2.5.1 would involve *ordered* probability spaces.

We can now prove the characterization theorem for information loss.

Proof of Theorem 10.2.1 We have already shown that information loss L satisfies the four conditions of (i), and it follows that (ii) implies (i).

For the converse, suppose that K satisfies (i). Given a finite probability space $(\mathcal{X}, \mathbf{p})$, write $!_{\mathbf{p}}$ for the unique measure-preserving map

$$!_{\mathbf{p}}: (\mathcal{X}, \mathbf{p}) \to (\{1\}, \mathbf{u}_1),$$

and define $I(\mathbf{p}) = K(!_{\mathbf{p}})$. For any measure-preserving map $f: (\mathcal{Y}, \mathbf{s}) \to (\mathcal{X}, \mathbf{p})$, the triangle

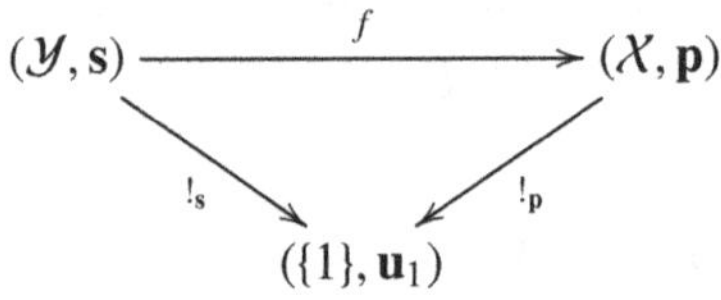

commutes, so by the composition condition on K,

$$K(!_{\mathbf{s}}) = K(!_{\mathbf{p}}) + K(f).$$

Equivalently,

$$K(f) = I(\mathbf{s}) - I(\mathbf{p}). \tag{10.11}$$

So in order to prove the theorem, it suffices to show that $I = cH$ for some constant c; and for this, it is enough to prove that I satisfies the hypotheses of Theorem 10.2.2.

First, I is isomorphism-invariant, since if $f: (\mathcal{Y}, \mathbf{s}) \to (\mathcal{X}, \mathbf{p})$ is an isomorphism then $K(f) = 0$, so $I(\mathbf{s}) = I(\mathbf{p})$ by (10.11).

Second, I satisfies the chain rule (10.4); that is,

$$I\left(\coprod_{i=1}^{n} w_i \mathbf{p}^i\right) = I(\mathbf{w}) + \sum_{i=1}^{n} w_i I(\mathbf{p}^i) \tag{10.12}$$

for all $\mathbf{w} \in \Delta_n$ and finite probability spaces $(\mathcal{X}_1, \mathbf{p}^1), \ldots, (\mathcal{X}_n, \mathbf{p}^n)$. To see this, write

$$f: \coprod_{i=1}^{n} \mathcal{X}_i \to \{1, \ldots, n\}$$

for the function defined by $f(j) = i$ whenever $j \in \mathcal{X}_i$. Then f defines a measure-preserving map

$$f: \left(\coprod \mathcal{X}_i, \coprod w_i \mathbf{p}^i\right) \to (\{1, \ldots, n\}, \mathbf{w}).$$

We now evaluate $K(f)$ in two ways. On the one hand, by equation (10.11),

$$K(f) = I\Big(\coprod w_i \mathbf{p}^i\Big) - I(\mathbf{w}).$$

On the other,

$$f = \coprod w_i \, !_{\mathbf{p}^i},$$

so by hypothesis on K and induction,

$$K(f) = \sum w_i K(!_{\mathbf{p}^i}) = \sum w_i I(\mathbf{p}^i).$$

Comparing the two expressions for $K(f)$ gives the chain rule (10.12) for I.

Third and finally, for each finite set $\mathcal{X}$, the function $I\colon \Delta_{\mathcal{X}} \to \mathbb{R}$ is continuous, by continuity of K.

Theorem 10.2.2 can therefore be applied, giving $I = cH$ for some $c \in \mathbb{R}$. It follows from equation (10.11) that $K = cL$. □

As observed in [25] (p. 1947), the charm of Theorem 10.2.1 is that the axioms on the information loss function K are entirely linear. They give no hint of any special role for the function

$$p \mapsto -p \log p.$$

And yet, this function emerges in the conclusion.

Another striking feature of Theorem 10.2.1 is that the natural conditions imposed on K force $K(f)$ to depend only on the domain and codomain of f. This is a consequence of condition (b) alone (on the information lost by a composite process), as can be seen from the argument leading up to equation (10.11). It is an instance of a general categorical fact: for any functor K from a category $\mathscr{P}$ with a terminal object to a groupoid, $K(f) = K(f')$ whenever f and f' are maps in $\mathscr{P}$ with the same domain and the same codomain.

Theorem 10.2.1 has several variants. We can drop the condition that $K(f) = 0$ for isomorphisms f if we instead require that $K(f) \geq 0$ for all f. (This was the version stated in Baez, Fritz and Leinster [25].) There is another version of Theorem 10.2.1 for finite sets equipped with arbitrary finite measures instead of probability measures (Corollary 4 of [25]). And there is a further variant for the q-logarithmic entropies S_q, which we give now.

For a measure-preserving map

$$f\colon (\mathcal{Y}, \mathbf{s}) \to (\mathcal{X}, \mathbf{p})$$

between finite probability spaces, define the **q-logarithmic information loss** of f as

$$L_q(f) = S_q(\mathbf{s}) - S_q(\mathbf{p}).$$

The following characterization of L_q is identical to Theorem 10.2.1 except for a change in the rule for the information lost by two processes in parallel (condition (c) below) and the absence of a continuity condition. With some minor differences, it first appeared as Theorem 7 of Baez, Fritz and Leinster [25].

Theorem 10.2.4 (Baez, Fritz and Leinster) *Let $1 \neq q \in \mathbb{R}$. Let K be a function assigning a real number $K(f)$ to each measure-preserving map f of finite probability spaces. The following are equivalent.*

i. *K has these three properties:*
   a. *$K(f) = 0$ for all isomorphisms f;*
   b. *$K(g \circ f) = K(g) + K(f)$ for all composable pairs (f, g) of measure-preserving maps;*
   c. *$K(\lambda f \sqcup (1 - \lambda)f') = \lambda^q K(f) + (1 - \lambda)^q K(f')$ for all measure-preserving maps f and f' and all $\lambda \in (0, 1)$.*

ii. *$K = cL_q$ for some $c \in \mathbb{R}$.*

No continuity or other regularity condition is needed, in contrast to Theorem 10.2.1.

Proof As for the proof of Theorem 10.2.1, but using the characterization theorem for S_q (Theorem 4.1.5) instead of Faddeev's characterization of H (Theorem 2.5.1). □

11

Entropy Modulo a Prime

> *Conclusion:* If we have a random variable ξ which takes finitely many values with all probabilities in $\mathbb{Q}$ then we can define not only the transcendental number $H(\xi)$ but also its 'residues modulo p' for almost all primes p ! – Maxim Kontsevich [195].

In this chapter, we define the entropy of any probability distribution whose 'probabilities' are not real numbers, but integers modulo a prime p. Its entropy, too, is an integer mod p. We justify the definition by proving a characterization theorem very similar to Faddeev's theorem on real entropy (Theorem 2.5.1), and by a characterization theorem for information loss mod p that is also closely analogous to the real case.

In earlier chapters, we reached our axiomatic characterization of real information loss in three steps:

(I) characterize the sequence $(\log n)_{n\geq 1}$ (Theorem 1.2.2);
(II) using (I), characterize entropy (Theorem 2.5.1);
(III) using (II), characterize information loss (Theorem 10.2.1).

Here, we follow three analogous steps to characterize entropy and information loss modulo p (Sections 11.1 and 11.2). The analytic subtleties disappear, but instead we encounter a number-theoretic obstacle.

With the definition of entropy mod p in place, we implement the idea proposed by Kontsevich in the quotation above. That is, we define a sense in which certain real numbers can be said to have residues mod p (Section 11.3). The residue map establishes a direct relationship between entropy over $\mathbb{R}$ and entropy over $\mathbb{Z}/p\mathbb{Z}$, supplementing the analogy between the Faddeev-type theorems over $\mathbb{R}$ and $\mathbb{Z}/p\mathbb{Z}$.

We finish by developing an alternative but equivalent approach to entropy modulo a prime (Section 11.4). It takes place in the ring of polynomials over the field of p elements. It is related more closely than the rest of this chapter to the subject of polylogarithms, which formed the context of Kontsevich's note [195] and of subsequent related work such as that of Elbaz-Vincent and Gangl [88, 89].

The results of this chapter first appeared in [219]. While [219] seems to have been the first place where the theory of entropy mod p was developed in detail, many of the ideas had been sketched or at least hinted at in Kontsevich's note [195], which itself was preceded by related work of Cathelineau [60, 61]. The introduction to Elbaz-Vincent and Gangl [88] relates some of the history, including the connection with polylogarithms; see also Remark 11.4.8 below.

11.1 Fermat Quotients and the Definition of Entropy

For the whole of this chapter, fix a prime p. To avoid confusion between the prime p and a probability distribution $\mathbf{p}$, we now denote a typical probability distribution by $\boldsymbol{\pi} = (\pi_1, \ldots, \pi_n)$.

Our first task is to formulate the correct definition of the entropy of a probability distribution $\boldsymbol{\pi}$ in which $\pi_1, \ldots, \pi_n$ are not real numbers, but elements of the field $\mathbb{Z}/p\mathbb{Z}$ of integers modulo p.

A problem arises immediately. Real probabilities are ordinarily required to be nonnegative, and the logarithms in the definition of entropy over $\mathbb{R}$ would be undefined if any probability were negative. So in the familiar real setting, the notion of positivity seems to be needed in order to state a definition of entropy. But in $\mathbb{Z}/p\mathbb{Z}$, there is no sense of positive or negative. How, then, are we to imitate the definition of entropy in $\mathbb{Z}/p\mathbb{Z}$?

This problem is solved by a simple observation. Although Shannon entropy is usually only defined for sequences $\boldsymbol{\pi} = (\pi_1, \ldots, \pi_n)$ of *nonnegative* reals summing to 1, it can just as easily be defined for sequences $\boldsymbol{\pi}$ of *arbitrary* reals summing to 1. One simply puts

$$H(\boldsymbol{\pi}) = -\sum_{i \in \operatorname{supp}(\boldsymbol{\pi})} \pi_i \log |\pi_i|, \tag{11.1}$$

where $\operatorname{supp}(\boldsymbol{\pi}) = \{i : \pi_i \neq 0\}$. (See Kontsevich [195], for instance.) This extended entropy is still continuous and symmetric, and still satisfies the chain rule. So, real entropy can in fact be defined without reference to the notion of positivity. (And generally speaking, negative probabilities are not as outlandish as they might seem; see Feynman [103] and Blass and Gurevich [41, 42].)

Thus, writing

$$\Pi_n = \{\boldsymbol{\pi} \in (\mathbb{Z}/p\mathbb{Z})^n : \pi_1 + \cdots + \pi_n = 1\},$$

it is reasonable to attempt to define the entropy of any element of Π_n. We will refer to elements $\boldsymbol{\pi} = (\pi_1, \ldots, \pi_n)$ of Π_n as **probability distributions mod** p, or simply **distributions**. Geometrically, the set Π_n of distributions on n elements is a hyperplane in the n-dimensional vector space $(\mathbb{Z}/p\mathbb{Z})^n$ over the field $\mathbb{Z}/p\mathbb{Z}$.

The function $x \mapsto \log|x|$ is a homomorphism from the multiplicative group $\mathbb{R}^\times$ of nonzero reals to the additive group $\mathbb{R}$. But when we look for an analogue over $\mathbb{Z}/p\mathbb{Z}$, we run into an obstacle.

Lemma 11.1.1 *There is no nontrivial homomorphism from the multiplicative group $(\mathbb{Z}/p\mathbb{Z})^\times$ of nonzero integers modulo p to the additive group $\mathbb{Z}/p\mathbb{Z}$.*

Proof Let $\phi\colon (\mathbb{Z}/p\mathbb{Z})^\times \to \mathbb{Z}/p\mathbb{Z}$ be a homomorphism. The image of ϕ is a subgroup of $\mathbb{Z}/p\mathbb{Z}$, which by Lagrange's theorem has order 1 or p. Since $(\mathbb{Z}/p\mathbb{Z})^\times$ has order $p - 1$, the image of ϕ has order at most $p - 1$. It therefore has order 1; that is, $\phi = 0$. □

In this sense, there is no logarithm for the integers modulo p. Nevertheless, there is an acceptable substitute. For integers n not divisible by p, Fermat's little theorem implies that p divides $n^{p-1} - 1$. The **Fermat quotient** of n modulo p is defined as

$$q_p(n) = \frac{n^{p-1} - 1}{p} \in \mathbb{Z}/p\mathbb{Z}.$$

The resemblance between the formulas for the Fermat quotient and the q-logarithm (equation (1.17)) hints that the Fermat quotient might function as some kind of logarithm, and part (i) of the following lemma confirms that this is so.

Lemma 11.1.2 *The map $q_p\colon \{n \in \mathbb{Z} : p \nmid n\} \to \mathbb{Z}/p\mathbb{Z}$ has the following properties:*

i. $q_p(mn) = q_p(m) + q_p(n)$ for all $m, n \in \mathbb{Z}$ not divisible by p, and $q_p(1) = 0$;
ii. $q_p(n + rp) = q_p(n) - r/n$ for all $n, r \in \mathbb{Z}$ such that n is not divisible by p;
iii. $q_p(n + p^2) = q_p(n)$ for all $n \in \mathbb{Z}$ not divisible by p.

Proof For (i), certainly $q_p(1) = 0$. We now have to show that

$$m^{p-1}n^{p-1} - 1 \equiv (m^{p-1} - 1) + (n^{p-1} - 1) \pmod{p^2},$$

or equivalently,

$$(m^{p-1} - 1)(n^{p-1} - 1) \equiv 0 \pmod{p^2}.$$

Since both $m^{p-1} - 1$ and $n^{p-1} - 1$ are integer multiples of p, this is true.

For (ii), we have

$$\begin{aligned}(n+rp)^{p-1} &= n^{p-1} + (p-1)n^{p-2}rp + \sum_{i=2}^{p-1}\binom{p-1}{i}n^{p-i-1}r^i p^i \\ &\equiv n^{p-1} + p(p-1)rn^{p-2} \pmod{p^2}.\end{aligned}$$

Subtracting 1 from each side and dividing by p gives

$$q_p(n+rp) \equiv q_p(n) + (p-1)rn^{p-2} \pmod{p},$$

and (ii) then follows from the fact that $n^{p-1} \equiv 1 \pmod{p}$. Taking $r = p$ in (ii) gives (iii). □

It follows that q_p defines a group homomorphism

$$q_p\colon (\mathbb{Z}/p^2\mathbb{Z})^\times \to \mathbb{Z}/p\mathbb{Z},$$

where $(\mathbb{Z}/p^2\mathbb{Z})^\times$ is the multiplicative group of integers modulo p^2. (The elements of $(\mathbb{Z}/p^2\mathbb{Z})^\times$ are the congruence classes modulo p^2 of the integers not divisible by p.) Moreover, the homomorphism q_p is surjective, since the lemma implies that

$$q_p(1-rp) = q_p(1) + r \equiv r \pmod{p}$$

for all integers r.

Lemma 11.1.1 states that there is no logarithm mod p, in the sense that there is no nontrivial group homomorphism $(\mathbb{Z}/p\mathbb{Z})^\times \to \mathbb{Z}/p\mathbb{Z}$. But the Fermat quotient is the next best thing, being a homomorphism $(\mathbb{Z}/p^2\mathbb{Z})^\times \to \mathbb{Z}/p\mathbb{Z}$. It is essentially the only such homomorphism:

Proposition 11.1.3 *Every group homomorphism $(\mathbb{Z}/p^2\mathbb{Z})^\times \to \mathbb{Z}/p\mathbb{Z}$ is a scalar multiple of the Fermat quotient.*

Proof It is a standard fact that the group $(\mathbb{Z}/p^2\mathbb{Z})^\times$ is cyclic (Theorem 10.6 of Apostol [16], for instance). Choose a generator e. Since q_p is surjective, it is not identically zero, so $q_p(e) \neq 0$.

Let $\phi\colon (\mathbb{Z}/p^2\mathbb{Z})^\times \to \mathbb{Z}/p\mathbb{Z}$ be a group homomorphism. Put $c = \phi(e)/q_p(e) \in \mathbb{Z}/p\mathbb{Z}$. Then for all $n \in \mathbb{Z}$,

$$\phi(e^n) = n\phi(e) = ncq_p(e) = cq_p(e^n).$$

Since e is a generator, it follows that $\phi = cq_p$. □

In Section 1.2, we proved characterization theorems for the sequence $(\log(n))_{n\geq 1}$. The next result plays a similar role for $(q_p(n))_{n\geq 1,\, p\nmid n}$.

Theorem 11.1.4 *Let $f\colon \{n \in \mathbb{N} : p \nmid n\} \to \mathbb{Z}/p\mathbb{Z}$ be a function. The following are equivalent:*

i. $f(mn) = f(m) + f(n)$ and $f(n + p^2) = f(n)$ for all $m, n \in \mathbb{N}$ not divisible by p;
ii. $f = cq_p$ for some $c \in \mathbb{Z}/p\mathbb{Z}$.

Proof We have already shown that q_p satisfies the conditions in (i), so (ii) implies (i). For the converse, suppose that f satisfies the conditions in (i). Then f induces a group homomorphism $(\mathbb{Z}/p^2\mathbb{Z})^\times \to \mathbb{Z}/p\mathbb{Z}$, which by Proposition 11.1.3 is a scalar multiple of q_p. The result follows. □

In terms of the three-step plan in the introduction to this chapter, we have now completed step (I): defining and characterizing the appropriate notion of logarithm. We now begin step (II): defining and characterizing the appropriate notion of entropy.

To state the definition, we will need an elementary lemma.

Lemma 11.1.5 *Let $a, b \in \mathbb{Z}$. If $a \equiv b \pmod{p}$ then $a^p \equiv b^p \pmod{p^2}$.*

Proof If $b = a + rp$ with $r \in \mathbb{Z}$, then

$$\begin{aligned} b^p &= (a + rp)^p \\ &= a^p + pa^{p-1}rp + \sum_{i=2}^{p} \binom{p}{i} a^{p-i} r^i p^i \\ &\equiv a^p \pmod{p^2}. \end{aligned}$$

□

Definition 11.1.6 Let $n \geq 1$ and $\boldsymbol{\pi} \in \Pi_n$. The **entropy** of $\boldsymbol{\pi}$ is

$$H(\boldsymbol{\pi}) = \frac{1}{p}\left(1 - \sum_{i=1}^{n} a_i^p\right) \in \mathbb{Z}/p\mathbb{Z},$$

where $a_i \in \mathbb{Z}$ represents $\pi_i \in \mathbb{Z}/p\mathbb{Z}$ for each $i \in \{1, \ldots, n\}$.

Lemma 11.1.5 guarantees that the definition is independent of the choice of $a_1, \ldots, a_n$.

We now explain and justify the definition of entropy mod p. In particular, we will prove a theorem characterizing the sequence of functions $(H\colon \Pi_n \to \mathbb{Z}/p\mathbb{Z})$ uniquely up to a scalar multiple. This result is plainly analogous to Faddeev's theorem for real entropy, and as such, is the strongest justification for the definition. But the analogy with the real case can also be seen in terms of derivations, as follows.

The entropy of a real probability distribution $\boldsymbol{\pi}$ is equal to $\sum_i \partial(\pi_i)$, where

$$\partial(x) = \begin{cases} -x \log x & \text{if } x > 0, \\ 0 & \text{if } x = 0 \end{cases} \tag{11.2}$$

(as in Section 2.2). What is the analogue of ∂ over $\mathbb{Z}/p\mathbb{Z}$? Given the analogy between the logarithm and the Fermat quotient, it is natural to consider $-nq_p(n)$ as a candidate. For integers n not divisible by p,

$$-nq_p(n) = \frac{n - n^p}{p}.$$

The right-hand side is a well-defined integer even if n is divisible by p. We therefore define a map $\partial \colon \mathbb{Z} \to \mathbb{Z}/p\mathbb{Z}$ by

$$\partial(n) = \frac{n - n^p}{p} \in \mathbb{Z}/p\mathbb{Z}. \tag{11.3}$$

Thus,

$$\partial(n) = \begin{cases} -nq_p(n) & \text{if } p \nmid n, \\ n/p & \text{if } p \mid n. \end{cases}$$

If $n \equiv m \pmod{p^2}$ then $\partial(n) = \partial(m)$, so ∂ can also be regarded as a map $\mathbb{Z}/p^2\mathbb{Z} \to \mathbb{Z}/p\mathbb{Z}$. Like its real counterpart, ∂ satisfies a version of the Leibniz rule (and for this reason, is essentially what is called a p-**derivation** [54, 55]):

Lemma 11.1.7 *$\partial(mn) = m\partial(n) + \partial(m)n$ for all $m, n \in \mathbb{Z}$.*

Proof The proof is similar to that of Lemma 11.1.2(i). The statement to be proved is equivalent to

$$mn - m^p n^p \equiv m(n - n^p) + (m - m^p)n \pmod{p^2}.$$

Rearranging, this in turn is equivalent to

$$0 \equiv (m - m^p)(n - n^p) \pmod{p^2},$$

which is true since $m \equiv m^p \pmod{p}$ and $n \equiv n^p \pmod{p}$. □

Using this lemma, we derive an equivalent expression for entropy mod p.

Lemma 11.1.8 *For all $n \geq 1$ and $\boldsymbol{\pi} \in \Pi_n$,*

$$H(\boldsymbol{\pi}) = \sum_{i=1}^{n} \partial(a_i) - \partial\left(\sum_{i=1}^{n} a_i\right)$$

whenever $a_i \in \mathbb{Z}$ represents $\pi_i \in \mathbb{Z}/p\mathbb{Z}$ for each $i \in \{1, \ldots, n\}$.

Proof An equivalent statement is that

$$1 - \sum a_i^p \equiv \sum (a_i - a_i^p) - \Big\{ \sum a_i - \big(\sum a_i\big)^p \Big\} \pmod{p^2}.$$

Cancelling, this reduces to

$$1 \equiv \big(\sum a_i\big)^p \pmod{p^2}.$$

But $\sum \pi_i = 1$ in $\mathbb{Z}/p\mathbb{Z}$ by definition of Π_n, so $\sum a_i \equiv 1 \pmod{p}$, so $(\sum a_i)^p \equiv 1 \pmod{p^2}$ by Lemma 11.1.5. □

Thus, $H(\boldsymbol{\pi})$ measures the extent to which the nonlinear derivation ∂ fails to preserve the sum $\sum a_i$.

The analogy with entropy over $\mathbb{R}$ is now evident. For a real probability distribution $\boldsymbol{\pi}$, and defining $\partial\colon [0, \infty) \to \mathbb{R}$ as in equation (11.2), we also have

$$H(\boldsymbol{\pi}) = \sum \partial(\pi_i) - \partial\big(\sum \pi_i\big).$$

In the real case, since $\sum \pi_i = 1$, the second term on the right-hand side vanishes. But over $\mathbb{Z}/p\mathbb{Z}$, it is not true in general that $\partial(\sum a_i) = 0$, so it is not true either that $H(\boldsymbol{\pi}) = \sum \partial(a_i)$. (Indeed, $\sum \partial(a_i)$, unlike $H(\boldsymbol{\pi})$, depends on the choice of representatives a_i.) So in the formula

$$H(\boldsymbol{\pi}) = \sum \partial(a_i) - \partial\big(\sum a_i\big)$$

for entropy mod p, the second summand is indispensable.

Example 11.1.9 Let $n \geq 1$ with $p \nmid n$. Since n is invertible mod p, there is a **uniform distribution**

$$\mathbf{u}_n = \underbrace{(1/n, \ldots, 1/n)}_{n} \in \Pi_n.$$

Choose $a \in \mathbb{Z}$ representing $1/n \in \mathbb{Z}/p\mathbb{Z}$. By Lemma 11.1.8 and then the derivation property of ∂,

$$H(\mathbf{u}_n) = n\partial(a) - \partial(na) = -a\partial(n).$$

But $\partial(n) = -nq_p(n)$, so $H(\mathbf{u}_n) = q_p(n)$. This result over $\mathbb{Z}/p\mathbb{Z}$ is analogous to the formula $H(\mathbf{u}_n) = \log n$ for the real entropy of a uniform distribution.

Example 11.1.10 Let $p = 2$. Any distribution $\boldsymbol{\pi} \in \Pi_n$ has an odd number of elements in its support, since $\sum \pi_i = 1$. Directly from the definition of entropy, $H(\boldsymbol{\pi}) \in \mathbb{Z}/2\mathbb{Z}$ is given by

$$H(\boldsymbol{\pi}) = \tfrac{1}{2}(|\mathrm{supp}(\boldsymbol{\pi})| - 1) = \begin{cases} 0 & \text{if } |\mathrm{supp}(\boldsymbol{\pi})| \equiv 1 \pmod{4}, \\ 1 & \text{if } |\mathrm{supp}(\boldsymbol{\pi})| \equiv 3 \pmod{4}. \end{cases}$$

In preparation for the next example, we record a useful standard lemma.

Lemma 11.1.11 $\binom{p-1}{s} \equiv (-1)^s \pmod{p}$ *for all* $s \in \{0, \ldots, p-1\}$.

Proof In $\mathbb{Z}/p\mathbb{Z}$, we have equalities

$$\begin{aligned}\binom{p-1}{s} &= \frac{(p-1)(p-2)\cdots(p-s)}{s!} \\ &= \frac{(-1)(-2)\cdots(-s)}{s!} \\ &= (-1)^s.\end{aligned}$$

□

Example 11.1.12 Here we find the entropy of a distribution $(\pi, 1-\pi)$ on two elements. Choose an integer a representing $\pi \in \mathbb{Z}/p\mathbb{Z}$. From the definition of entropy, assuming that $p \neq 2$,

$$H(\pi, 1-\pi) = \frac{1}{p}(1 - a^p - (1-a)^p) = \sum_{r=1}^{p-1} (-1)^{r+1} \frac{1}{p}\binom{p}{r} a^r.$$

But $\frac{1}{p}\binom{p}{r} = \frac{1}{r}\binom{p-1}{r-1}$, so by Lemma 11.1.11, the coefficient of a^r in the sum is simply $\frac{1}{r}$. We can now replace a by π, giving

$$H(\pi, 1-\pi) = \sum_{r=1}^{p-1} \frac{\pi^r}{r}.$$

The function on the right-hand side was the starting point of Kontsevich's note [195], and we return to it in Section 11.4.

In the case $p = 2$, we have $H(\pi, 1-\pi) = 0$ for both values of $\pi \in \mathbb{Z}/2\mathbb{Z}$.

Example 11.1.13 Appending zero probabilities to a distribution does not change its entropy:

$$H(\pi_1, \ldots, \pi_n, 0, \ldots, 0) = H(\pi_1, \ldots, \pi_n).$$

This is immediate from the definition. But a subtlety of distributions mod p, absent in the standard real setting, is that nonzero probabilities can sum to zero. So, one might ask whether

$$H(\pi_1, \ldots, \pi_n, \tau_1, \ldots, \tau_m) = H(\pi_1, \ldots, \pi_n)$$

whenever $\tau_1, \ldots, \tau_m \in \mathbb{Z}/p\mathbb{Z}$ with $\sum \tau_j = 0$. The answer is trivially yes for $m = 0$ and $m = 1$, and it is also yes for $m = 2$ as long as $p \neq 2$. (For if we choose an integer a to represent τ_1 then $-a$ represents τ_2, and $a^p + (-a)^p = 0$.) But the answer is no for $m \geq 3$. For instance, when $p = 3$, we have

$$H(1, 1, 1, 1) = H(\mathbf{u}_4) = q_3(4) = \tfrac{1}{3}(4^2 - 1) = -1$$

by Example 11.1.9, which is not equal to $H(1) = 0$, even though $1 + 1 + 1 = 0$.

Distributions over $\mathbb{Z}/p\mathbb{Z}$ can be composed, using the same formula as in the real case (Definition 2.1.3). As in the real case, entropy mod p satisfies the chain rule.

Proposition 11.1.14 (Chain rule) *We have*

$$H(\boldsymbol{\gamma} \circ (\boldsymbol{\pi}^1, \ldots, \boldsymbol{\pi}^n)) = H(\boldsymbol{\gamma}) + \sum_{i=1}^{n} \gamma_i H(\boldsymbol{\pi}^i)$$

for all $n, k_1, \ldots, k_n \geq 1$, *all* $\boldsymbol{\gamma} = (\gamma_1, \ldots, \gamma_n) \in \Pi_n$, *and all* $\boldsymbol{\pi}^i \in \Pi_{k_i}$.

Proof Write $\boldsymbol{\pi}^i = (\pi^i_1, \ldots, \pi^i_{k_i})$. Choose $b_i \in \mathbb{Z}$ representing $\gamma_i \in \mathbb{Z}/p\mathbb{Z}$ and $a^i_j \in \mathbb{Z}$ representing $\pi^i_j \in \mathbb{Z}/p\mathbb{Z}$, for each i and j. Write $A^i = a^i_1 + \cdots + a^i_{k_i}$.

We evaluate in turn the three terms $H(\boldsymbol{\gamma} \circ (\boldsymbol{\pi}^1, \ldots, \boldsymbol{\pi}^n))$, $H(\boldsymbol{\gamma})$, and $\sum \gamma_i H(\boldsymbol{\pi}^i)$. First, by Lemma 11.1.8 and the derivation property of ∂ (Lemma 11.1.7),

$$\begin{aligned} H(\boldsymbol{\gamma} \circ (\boldsymbol{\pi}^1, \ldots, \boldsymbol{\pi}^n)) &= \sum_{i=1}^{n} \sum_{j=1}^{k_i} \partial(b_i a^i_j) - \partial\Bigg(\sum_{i=1}^{n} \sum_{j=1}^{k_i} b_i a^i_j\Bigg) \\ &= \sum_{i=1}^{n} \sum_{j=1}^{k_i} (\partial(b_i) a^i_j + b_i \partial(a^i_j)) - \partial\Bigg(\sum_{i=1}^{n} b_i A^i\Bigg) \\ &= \sum_{i=1}^{n} \partial(b_i) A^i + \sum_{i=1}^{n} b_i \sum_{j=1}^{k_i} \partial(a^i_j) - \partial\Bigg(\sum_{i=1}^{n} b_i A^i\Bigg). \end{aligned}$$

Second, $A^i \equiv 1 \pmod{p}$ since $\boldsymbol{\pi}^i \in \Pi_{k_i}$, so $b_i A^i \in \mathbb{Z}$ represents $\gamma_i \in \mathbb{Z}/p\mathbb{Z}$. Hence

$$\begin{aligned} H(\boldsymbol{\gamma}) &= \sum_{i=1}^{n} \partial(b_i A^i) - \partial\Bigg(\sum_{i=1}^{n} b_i A^i\Bigg) \\ &= \sum_{i=1}^{n} \partial(b_i) A^i + \sum_{i=1}^{n} b_i \partial(A^i) - \partial\Bigg(\sum_{i=1}^{n} b_i A^i\Bigg). \end{aligned}$$

Third,

$$\sum_{i=1}^{n} \gamma_i H(\boldsymbol{\pi}^i) = \sum_{i=1}^{n} b_i \sum_{j=1}^{k_i} \partial(a^i_j) - \sum_{i=1}^{n} b_i \partial(A^i).$$

The result follows. □

There is a tensor product for distributions mod p, defined as in the real case (p. 38), and entropy mod p has the familiar logarithmic property:

Corollary 11.1.15 $H(\boldsymbol{\gamma} \otimes \boldsymbol{\pi}) = H(\boldsymbol{\gamma}) + H(\boldsymbol{\pi})$ *for all* $\boldsymbol{\gamma} \in \Pi_n$ *and* $\boldsymbol{\pi} \in \Pi_m$. □

11.2 Characterizations of Entropy and Information Loss

We now state our characterization theorem for entropy mod p, whose close resemblance to the characterization theorem for real entropy (Theorem 2.5.1) is the main justification for the definition.

Theorem 11.2.1 *Let $(I\colon \Pi_n \to \mathbb{Z}/p\mathbb{Z})_{n\geq 1}$ be a sequence of functions. The following are equivalent:*

i. I satisfies the chain rule (that is, satisfies the conclusion of Proposition 11.1.14 with I in place of H);
ii. $I = cH$ for some $c \in \mathbb{Z}/p\mathbb{Z}$.

As in our sharper version of Faddeev's theorem over $\mathbb{R}$ (Theorem 2.5.1), no symmetry condition is needed.

Since H satisfies the chain rule, so does any constant multiple of H. Hence (ii) implies (i). We now begin the proof of the converse. ***For the rest of the proof***, let $(I\colon \Pi_n \to \mathbb{Z}/p\mathbb{Z})_{n\geq 1}$ be a sequence of functions satisfying the chain rule.

Lemma 11.2.2 *i. $I(\mathbf{u}_{mn}) = I(\mathbf{u}_m) + I(\mathbf{u}_n)$ for all $m, n \in \mathbb{N}$ not divisible by p;*
ii. $I(\mathbf{u}_1) = 0$.

Proof Both parts are proved exactly as in the real case (Lemma 2.5.3). □

Lemma 11.2.3 $I(1, 0) = I(0, 1) = 0$.

Proof The proof that $I(1, 0) = 0$ is identical to the proof in the real case (Lemma 2.5.4), and $I(0, 1) = 0$ is proved similarly. □

Lemma 11.2.4 *For all $\boldsymbol{\pi} \in \Pi_n$ and $i \in \{0, \ldots, n\}$,*

$$I(\pi_1, \ldots, \pi_n) = I(\pi_1, \ldots, \pi_i, 0, \pi_{i+1}, \ldots, \pi_n).$$

Proof First suppose that $i \neq 0$. Then

$$(\pi_1, \ldots, \pi_i, 0, \pi_{i+1}, \ldots, \pi_n) = \boldsymbol{\pi} \circ (\underbrace{\mathbf{u}_1, \ldots, \mathbf{u}_1}_{i-1}, (1, 0), \underbrace{\mathbf{u}_1, \ldots, \mathbf{u}_1}_{n-i}).$$

Applying I to both sides, then using the chain rule and that $I(\mathbf{u}_1) = I(1, 0) = 0$, gives the result. The case $i = 0$ is proved similarly, now using $I(0, 1) = 0$. □

As in the real case, we will prove the characterization theorem by analysing $I(\mathbf{u}_n)$ as n varies. And as in the real case, the chain rule will allow us to deduce the value of $I(\boldsymbol{\pi})$ for more general distributions $\boldsymbol{\pi}$:

Lemma 11.2.5 *Let $\pi \in \Pi_n$ with $\pi_i \neq 0$ for all i. For each i, let $k_i \geq 1$ be an integer representing $\pi_i \in \mathbb{Z}/p\mathbb{Z}$, and write $k = \sum_{i=1}^n k_i$. Then*

$$I(\pi) = I(\mathbf{u}_k) - \sum_{i=1}^{n} k_i I(\mathbf{u}_{k_i}).$$

Proof First note that none of $k_1, \ldots, k_n$ is a multiple of p, and since k represents $\sum \pi_i = 1 \in \mathbb{Z}/p\mathbb{Z}$, neither is k. Hence $\mathbf{u}_{k_i}$ and $\mathbf{u}_k$ are well defined. By definition of composition,

$$\pi \circ (\mathbf{u}_{k_1}, \ldots, \mathbf{u}_{k_n}) = (\underbrace{1, \ldots, 1}_{k}) = \mathbf{u}_k.$$

Applying I and using the chain rule gives the result. □

We now come to the most delicate part of the argument. Since $H(\mathbf{u}_n) = q_p(n)$, and since $q_p(n)$ is p^2-periodic in n, if I is to be a constant multiple of H then $I(\mathbf{u}_n)$ must also be p^2-periodic in n. We show this directly.

Lemma 11.2.6 $I(\mathbf{u}_{n+p^2}) = I(\mathbf{u}_n)$ *for all natural numbers n not divisible by p.*

Proof First we prove the existence of a constant $c \in \mathbb{Z}/p\mathbb{Z}$ such that for all $n \in \mathbb{N}$ not divisible by p,

$$I(\mathbf{u}_{n+p}) = I(\mathbf{u}_n) - c/n. \tag{11.4}$$

(Compare Lemma 11.1.2(ii).) An equivalent statement is that $n(I(\mathbf{u}_{n+p}) - I(\mathbf{u}_n))$ is independent of n. For any n_1 and n_2, we can choose some $m \geq \max\{n_1, n_2\}$ with $m \equiv 1 \pmod p$, so it is enough to show that whenever $0 \leq n \leq m$ with $n \not\equiv 0 \pmod p$ and $m \equiv 1 \pmod p$,

$$n(I(\mathbf{u}_{n+p}) - I(\mathbf{u}_n)) = I(\mathbf{u}_{m+p}) - I(\mathbf{u}_m). \tag{11.5}$$

To prove this, consider the distribution

$$\pi = (n, \underbrace{1, \ldots, 1}_{m-n}).$$

By Lemma 11.2.5 and the fact that $I(\mathbf{u}_1) = 0$,

$$I(\pi) = I(\mathbf{u}_m) - nI(\mathbf{u}_n).$$

But also

$$\pi = (n + p, \underbrace{1, \ldots, 1}_{m-n}),$$

so by the same argument,

$$I(\boldsymbol{\pi}) = I(\mathbf{u}_{m+p}) - (n+p)I(\mathbf{u}_{n+p})$$
$$= I(\mathbf{u}_{m+p}) - nI(\mathbf{u}_{n+p}).$$

Comparing the two expressions for $I(\boldsymbol{\pi})$ gives equation (11.5), thus proving the initial claim.

By induction on equation (11.4),

$$I(\mathbf{u}_{n+rp}) = I(\mathbf{u}_n) - cr/n$$

for all $n, r \in \mathbb{N}$ with n not divisible by p. The result follows by setting $r = p$. □

We can now prove the characterization theorem for entropy modulo p.

Proof of Theorem 11.2.1 Define $f\colon \{n \in \mathbb{N} : p \nmid n\} \to \mathbb{Z}/p\mathbb{Z}$ by $f(n) = I(\mathbf{u}_n)$. By Lemma 11.2.2, $f(mn) = f(m) + f(n)$ for all m, n not divisible by p. By Lemma 11.2.6, $f(n + p^2) = f(n)$ for all n not divisible by p. Hence by Theorem 11.1.4, $f = cq_p$ for some $c \in \mathbb{Z}/p\mathbb{Z}$. It follows from Example 11.1.9 that $I(\mathbf{u}_n) = cH(\mathbf{u}_n)$ for all n not divisible by p.

Since both I and cH satisfy the chain rule, Lemma 11.2.5 applies to both; and since I and cH are equal on uniform distributions, they are also equal on all distributions $\boldsymbol{\pi}$ such that $\pi_i \neq 0$ for all i. Finally, applying Lemma 11.2.4 to both I and cH, we deduce by induction that $I(\boldsymbol{\pi}) = cH(\boldsymbol{\pi})$ for all $\boldsymbol{\pi} \in \Pi_n$. □

In the real case, the characterization theorem for entropy leads to a characterization of information loss involving only linear conditions (Theorem 10.2.1). The same holds for entropy mod p, and the argument can be copied over from the real case nearly verbatim.

Thus, given a finite set $\mathcal{X}$, we write $\Pi_{\mathcal{X}}$ for the set of families $\boldsymbol{\pi} = (\pi_i)_{i \in \mathcal{X}}$ of elements of $\mathbb{Z}/p\mathbb{Z}$ such that $\sum_{i \in \mathcal{X}} \pi_i = 1$. A **finite probability space mod** p is a finite set $\mathcal{X}$ together with an element $\boldsymbol{\pi} \in \Pi_{\mathcal{X}}$. A **measure-preserving map** $f\colon (\mathcal{Y}, \boldsymbol{\sigma}) \to (\mathcal{X}, \boldsymbol{\pi})$ between such spaces is a function $f\colon \mathcal{Y} \to \mathcal{X}$ such that

$$\pi_i = \sum_{j \in f^{-1}(i)} s_j$$

for all $i \in \mathcal{X}$.

As in the real case, we can take convex combinations of both probability spaces and maps between them. Given two finite probability spaces mod p, say $(\mathcal{X}, \boldsymbol{\pi})$ and $(\mathcal{X}', \boldsymbol{\pi}')$, and given also a scalar $\lambda \in \mathbb{Z}/p\mathbb{Z}$, we obtain another such space, $(\mathcal{X} \sqcup \mathcal{X}', \lambda\boldsymbol{\pi} \sqcup (1-\lambda)\boldsymbol{\pi}')$. Given two measure-preserving maps

$$f\colon (\mathcal{Y}, \boldsymbol{\sigma}) \to (\mathcal{X}, \boldsymbol{\pi}),$$
$$f'\colon (\mathcal{Y}', \boldsymbol{\sigma}') \to (\mathcal{X}', \boldsymbol{\pi}')$$

and an element $\lambda \in \mathbb{Z}/p\mathbb{Z}$, we obtain a new measure-preserving map

$$\lambda f \sqcup (1-\lambda)f' : (\mathcal{Y} \sqcup \mathcal{Y}', \lambda\sigma \sqcup (1-\lambda)\sigma') \to (\mathcal{X} \sqcup \mathcal{X}', \lambda\pi \sqcup (1-\lambda)\pi'),$$

exactly as in Section 10.1.

The **entropy** of $\pi \in \Pi_{\mathcal{X}}$ is, naturally,

$$H(\pi) = \frac{1}{p}\left(1 - \sum_{i \in \mathcal{X}} a_i^p\right),$$

where $a_i \in \mathbb{Z}$ represents $\pi_i \in \mathbb{Z}/p\mathbb{Z}$ for each $i \in \mathcal{X}$. The **information loss** of a measure-preserving map $f: (\mathcal{Y}, \sigma) \to (\mathcal{X}, \pi)$ between finite probability spaces mod p is

$$L(f) = H(\sigma) - H(\pi) \in \mathbb{Z}/p\mathbb{Z}.$$

Theorem 11.2.7 *Let K be a function assigning an element $K(f) \in \mathbb{Z}/p\mathbb{Z}$ to each measure-preserving map f of finite probability spaces mod p. The following are equivalent.*

i. K has these three properties:

a. $K(f) = 0$ for all isomorphisms f;

b. $K(g \circ f) = K(g) + K(f)$ for all composable pairs (f, g) of measure-preserving maps;

c. $K(\lambda f \sqcup (1-\lambda)f') = \lambda K(f) + (1-\lambda)K(f')$ for all measure-preserving maps f and f' and all $\lambda \in \mathbb{Z}/p\mathbb{Z}$.

ii. $K = cL$ for some $c \in \mathbb{Z}/p\mathbb{Z}$.

Proof The proof is identical to that of the real case, Theorem 10.2.1, but with $\mathbb{Z}/p\mathbb{Z}$ in place of $\mathbb{R}$, Theorem 11.2.1 in place of Faddeev's theorem, and all mention of continuity removed. □

11.3 The Residues of Real Entropy

Having found a satisfactory definition of the entropy of a probability distribution mod p, we are now in a position to develop Kontsevich's suggestion about the residues mod p of real entropy, quoted at the start of this chapter. (That quotation was the sum total of what he wrote on the subject.)

Let $\pi \in \Delta_n$ be a probability distribution with rational probabilities, say $\pi = (a_1/b_1, \ldots, a_n/b_n)$ with $a_i, b_i \in \mathbb{Z}$. There are only finitely many primes that divide one or more of the denominators b_i. If p is not in that exceptional set then π defines an element of Π_n, and therefore has a mod p entropy $H(\pi) \in$

$\mathbb{Z}/p\mathbb{Z}$. Kontsevich invites us to think of this as the residue class modulo p of the real entropy $H(\boldsymbol{\pi}) \in \mathbb{R}$.

Kontsevich phrased his suggestion playfully, but there is more to it than meets the eye. To explain it, let us write $H_{\mathbb{R}}$ for entropy over the reals, H_p for entropy mod p, and $\Delta_n^{(p)}$ for the set of real probability distributions $\boldsymbol{\pi} \in \Delta_n$ such that each π_i can be expressed as a rational number with denominator not divisible by p. The proposal is that given $\boldsymbol{\pi} \in \Delta_n^{(p)}$, we regard $H_p(\boldsymbol{\pi}) \in \mathbb{Z}/p\mathbb{Z}$ as the residue mod p of $H_{\mathbb{R}}(\boldsymbol{\pi}) \in \mathbb{R}$.

Now, different distributions can have the same entropy over $\mathbb{R}$. For instance,

$$H_{\mathbb{R}}(\tfrac{1}{2}, \tfrac{1}{8}, \tfrac{1}{8}, \tfrac{1}{8}, \tfrac{1}{8}) = H_{\mathbb{R}}(\tfrac{1}{4}, \tfrac{1}{4}, \tfrac{1}{4}, \tfrac{1}{4}).$$

There is, therefore, a question of consistency: Kontsevich's proposal only makes sense if

$$H_{\mathbb{R}}(\boldsymbol{\pi}) = H_{\mathbb{R}}(\boldsymbol{\gamma}) \implies H_p(\boldsymbol{\pi}) = H_p(\boldsymbol{\gamma})$$

for all $\boldsymbol{\pi} \in \Delta_n^{(p)}$ and $\boldsymbol{\gamma} \in \Delta_m^{(p)}$. We now show that this is true.

Lemma 11.3.1 *Let $n, m \geq 1$ and let $a_1, \ldots, a_n, b_1, \ldots, b_m \geq 0$ be integers. Then*

$$\prod_{i=1}^{n} a_i^{a_i} = \prod_{j=1}^{m} b_j^{b_j} \implies \sum_{i=1}^{n} \partial(a_i) = \sum_{j=1}^{m} \partial(b_j),$$

where the first equality is in $\mathbb{Z}$, the second is in $\mathbb{Z}/p\mathbb{Z}$, and we use the convention that $0^0 = 1$.

Here ∂ is the map $\mathbb{Z} \to \mathbb{Z}/p\mathbb{Z}$ defined in equation (11.3). The analogue of this lemma for the real-valued map ∂ of equation (11.2) is trivial: simply discard the factors in the products for which a_i or b_j is 0, then take logarithms. But over $\mathbb{Z}/p\mathbb{Z}$, it is not so simple. The subtlety arises from the possibility that some a_i or b_j is not zero but is divisible by p. In that case, $\prod a_i^{a_i} = \prod b_j^{b_j}$ is divisible by p, so its Fermat quotient – the analogue of the logarithm – is undefined. A more detailed analysis is therefore required.

Proof Since $0^0 = 1$ and $\partial(0) = 0$, it is enough to prove the result in the case where each of the integers a_i and b_j is strictly positive. We may then write $a_i = p^{\alpha_i} A_i$ with $\alpha_i \geq 0$ and $p \nmid A_i$, and similarly $b_j = p^{\beta_j} B_j$. We adopt the convention that unless mentioned otherwise, the index i ranges over $1, \ldots, n$ and the index j over $1, \ldots, m$.

Assume that $\prod a_i^{a_i} = \prod b_j^{b_j}$. We have

$$\prod a_i^{a_i} = p^{\sum \alpha_i a_i} \prod A_i^{a_i}$$

with $p \nmid \prod A_i^{a_i}$, and similarly for $\prod b_j^{b_j}$. It follows that

$$\prod A_i^{a_i} = \prod B_j^{b_j}, \tag{11.6}$$

$$\sum \alpha_i a_i = \sum \beta_j b_j \tag{11.7}$$

in $\mathbb{Z}$. We consider each of these equations in turn.

First, since $p \nmid \prod A_i^{a_i}$, the Fermat quotient $q_p(\prod A_i^{a_i})$ is well defined, and the logarithmic property of q_p (Lemma 11.1.2(i)) gives

$$-q_p\Big(\prod A_i^{a_i}\Big) = \sum -a_i q_p(A_i).$$

Consider the right-hand side as an element of $\mathbb{Z}/p\mathbb{Z}$. When $p \mid a_i$, the i-summand vanishes. When $p \nmid a_i$, the i-summand is $-a_i q_p(a_i) = \partial(a_i)$. Hence

$$-q_p\Big(\prod A_i^{a_i}\Big) = \sum_{i:\, \alpha_i=0} \partial(a_i)$$

in $\mathbb{Z}/p\mathbb{Z}$. A similar result holds for $\prod B_j^{b_j}$, so equation (11.6) gives

$$\sum_{i:\, \alpha_i=0} \partial(a_i) = \sum_{j:\, \beta_j=0} \partial(b_j). \tag{11.8}$$

Second,

$$\sum_{i=1}^{n} \alpha_i a_i = \sum_{i:\, \alpha_i \geq 1} \alpha_i a_i,$$

so $p \mid \sum \alpha_i a_i$. Now

$$\frac{1}{p} \sum \alpha_i a_i = \sum_{i:\, \alpha_i \geq 1} \alpha_i p^{\alpha_i - 1} A_i \equiv \sum_{i:\, \alpha_i = 1} A_i \pmod{p},$$

and if $\alpha_i = 1$ then $A_i = a_i/p = \partial(a_i)$. A similar result holds for $\sum \beta_j b_j$, so equation (11.7) gives

$$\sum_{i:\, \alpha_i=1} \partial(a_i) = \sum_{j:\, \beta_j=1} \partial(b_j) \tag{11.9}$$

in $\mathbb{Z}/p\mathbb{Z}$.

Finally, for each i such that $\alpha_i \geq 2$, we have $p^2 \mid a_i$ and so $\partial(a_i) = 0$ in $\mathbb{Z}/p\mathbb{Z}$. The same holds for b_j, so

$$\sum_{i:\, \alpha_i \geq 2} \partial(a_i) = \sum_{j:\, \beta_j \geq 2} \partial(b_j), \tag{11.10}$$

both sides being 0. Adding equations (11.8), (11.9) and (11.10) gives the result. □

We deduce that the real entropy of a rational distribution determines its entropy mod p.

Theorem 11.3.2 *Let $n, m \geq 1$, $\boldsymbol{\pi} \in \Delta_n^{(p)}$, and $\boldsymbol{\gamma} \in \Delta_m^{(p)}$. Then*

$$H_{\mathbb{R}}(\boldsymbol{\pi}) = H_{\mathbb{R}}(\boldsymbol{\gamma}) \implies H_p(\boldsymbol{\pi}) = H_p(\boldsymbol{\gamma}).$$

Proof We can write

$$\boldsymbol{\pi} = (r_1/t, \ldots, r_n/t), \qquad \boldsymbol{\gamma} = (s_1/t, \ldots, s_m/t),$$

where r_i, s_j and t are nonnegative integers such that $p \nmid t$ and

$$r_1 + \cdots + r_n = t = s_1 + \cdots + s_m.$$

By multiplying all of these integers by a constant if necessary, we may assume that $t \equiv 1 \pmod{p}$.

By definition,

$$e^{-H_{\mathbb{R}}(\boldsymbol{\pi})} = \prod_i (r_i/t)^{r_i/t},$$

with the convention that $0^0 = 1$. Multiplying both sides by t and then raising to the power of t gives

$$t^t e^{-tH_{\mathbb{R}}(\boldsymbol{\pi})} = \prod_i r_i^{r_i}.$$

By the analogous equation for $\boldsymbol{\gamma}$ and the assumption that $H_{\mathbb{R}}(\boldsymbol{\pi}) = H_{\mathbb{R}}(\boldsymbol{\gamma})$, it follows that

$$\prod_i r_i^{r_i} = \prod_j s_j^{s_j}.$$

Lemma 11.3.1 now gives

$$\sum_i \partial(r_i) = \sum_j \partial(s_j)$$

in $\mathbb{Z}/p\mathbb{Z}$. Since $\sum r_i = t = \sum s_j$, it follows that

$$\sum_i \partial(r_i) - \partial\left(\sum_i r_i\right) = \sum_j \partial(s_j) - \partial\left(\sum_j s_j\right).$$

But $t \equiv 1 \pmod{p}$, so r_i represents the element $r_i/t = \pi_i$ of $\mathbb{Z}/p\mathbb{Z}$, so by Lemma 11.1.8, the left-hand side of this equation is $H_p(\boldsymbol{\pi})$. Similarly, the right-hand side is $H_p(\boldsymbol{\gamma})$. Hence $H_p(\boldsymbol{\pi}) = H_p(\boldsymbol{\gamma})$. □

It follows that Kontsevich's residue classes of real entropies are well defined. That is, writing

$$\mathbb{E}^{(p)} = \bigcup_{n=1}^{\infty} \left\{ H_{\mathbb{R}}(\boldsymbol{\pi}) : \boldsymbol{\pi} \in \Delta_n^{(p)} \right\} \subseteq \mathbb{R},$$

there is a unique map of sets

$$[\cdot]: \mathbb{E}^{(p)} \to \mathbb{Z}/p\mathbb{Z}$$

such that $[H_{\mathbb{R}}(\boldsymbol{\pi})] = H_p(\boldsymbol{\pi})$ for all $\boldsymbol{\pi} \in \Delta_n^{(p)}$ and $n \geq 1$. We now show that this map is additive, as the word 'residue' leads one to expect.

Proposition 11.3.3 *The set $\mathbb{E}^{(p)}$ is closed under addition, and the residue map*

$$\begin{array}{rccc} [\cdot]: & \mathbb{E}^{(p)} & \to & \mathbb{Z}/p\mathbb{Z} \\ & H_{\mathbb{R}}(\boldsymbol{\pi}) & \mapsto & H_p(\boldsymbol{\pi}) \end{array}$$

preserves addition.

Proof Let $\boldsymbol{\pi} \in \Delta_n^{(p)}$ and $\boldsymbol{\gamma} \in \Delta_m^{(p)}$. We must show that $H_{\mathbb{R}}(\boldsymbol{\pi}) + H_{\mathbb{R}}(\boldsymbol{\gamma}) \in \mathbb{E}^{(p)}$ and

$$[H_{\mathbb{R}}(\boldsymbol{\pi}) + H_{\mathbb{R}}(\boldsymbol{\gamma})] = [H_{\mathbb{R}}(\boldsymbol{\pi})] + [H_{\mathbb{R}}(\boldsymbol{\gamma})].$$

Evidently $\boldsymbol{\pi} \otimes \boldsymbol{\gamma} \in \Delta_{nm}^{(p)}$, so by the logarithmic property of $H_{\mathbb{R}}$,

$$H_{\mathbb{R}}(\boldsymbol{\pi}) + H_{\mathbb{R}}(\boldsymbol{\gamma}) = H_{\mathbb{R}}(\boldsymbol{\pi} \otimes \boldsymbol{\gamma}) \in \mathbb{E}^{(p)}.$$

Now also using the logarithmic property of H_p (Corollary 11.1.15),

$$\begin{aligned} [H_{\mathbb{R}}(\boldsymbol{\pi}) + H_{\mathbb{R}}(\boldsymbol{\gamma})] &= [H_{\mathbb{R}}(\boldsymbol{\pi} \otimes \boldsymbol{\gamma})] \\ &= H_p(\boldsymbol{\pi} \otimes \boldsymbol{\gamma}) \\ &= H_p(\boldsymbol{\pi}) + H_p(\boldsymbol{\gamma}) \\ &= [H_{\mathbb{R}}(\boldsymbol{\pi})] + [H_{\mathbb{R}}(\boldsymbol{\gamma})], \end{aligned}$$

as required. □

11.4 Polynomial Approach

There is an alternative approach to entropy modulo a prime. It repairs a defect of the approach above: that in order to define the entropy of a distribution $\boldsymbol{\pi}$ over $\mathbb{Z}/p\mathbb{Z}$, we had to step outside $\mathbb{Z}/p\mathbb{Z}$ to make arbitrary choices of integers representing the probabilities π_i, then show that the definition was independent of those choices. We now show how to define $H(\boldsymbol{\pi})$ directly as a function of $\pi_1, \ldots, \pi_n$.

Inevitably, that function is a polynomial, by the classical fact that every function $K^n \to K$ on a finite field K is induced by some polynomial in n variables. Indeed, there is a *unique* such polynomial whose degree in each variable is strictly less than the order of the field:

Lemma 11.4.1 *Let K be a finite field with q elements, let $n \geq 0$, and let $F\colon K^n \to K$ be a function. Then there is a unique polynomial f of the form*

$$f(x_1, \ldots, x_n) = \sum_{0 \leq r_1, \ldots, r_n < q} c_{r_1, \ldots, r_n} x_1^{r_1} \cdots x_n^{r_n}$$

($c_{r_1, \ldots, r_n} \in K$) such that

$$f(\pi_1, \ldots, \pi_n) = F(\pi_1, \ldots, \pi_n)$$

for all $\pi_1, \ldots, \pi_n \in K$.

Proof See Appendix A.9. □

In particular, taking $K = \mathbb{Z}/p\mathbb{Z}$, entropy modulo p can be expressed as a polynomial of degree less than p in each variable. We now identify such a polynomial.

For each $n \geq 1$, define $h(x_1, \ldots, x_n) \in (\mathbb{Z}/p\mathbb{Z})[x_1, \ldots, x_n]$ by

$$h(x_1, \ldots, x_n) = -\sum_{\substack{0 \leq r_1, \ldots, r_n < p \\ r_1 + \cdots + r_n = p}} \frac{x_1^{r_1} \cdots x_n^{r_n}}{r_1! \cdots r_n!}.$$

Proposition 11.4.2 *For all $n \geq 1$ and $(\pi_1, \ldots, \pi_n) \in \Pi_n$,*

$$H(\pi_1, \ldots, \pi_n) = h(\pi_1, \ldots, \pi_n).$$

Proof Let $\pi_1, \ldots, \pi_n \in \mathbb{Z}/p\mathbb{Z}$. We will show that whenever $a_1, \ldots, a_n$ are integers representing $\pi_1, \ldots, \pi_n$, then

$$\frac{1}{p}\left(\left(\sum_{i=1}^{n} a_i\right)^p - \sum_{i=1}^{n} a_i^p\right) \tag{11.11}$$

is an integer representing $h(\pi_1, \ldots, \pi_n)$. The result will follow, since if $\boldsymbol{\pi} \in \Pi_n$ then $\sum \pi_i = 1$, so $(\sum a_i)^p \equiv 1 \pmod{p^2}$ by Lemma 11.1.5.

We have to prove that

$$\left(\sum_{i=1}^{n} a_i\right)^p - \sum_{i=1}^{n} a_i^p \equiv -p \sum_{\substack{0 \leq r_1, \ldots, r_n < p \\ r_1 + \cdots + r_n = p}} \frac{a_1^{r_1} \cdots a_n^{r_n}}{r_1! \cdots r_n!} \pmod{p^2}.$$

Since $(p-1)!$ is invertible in $\mathbb{Z}/p^2\mathbb{Z}$, an equivalent statement is that

$$(p-1)!\left(\sum_{i=1}^{n} a_i^p - \left(\sum_{i=1}^{n} a_i\right)^p\right) \equiv \sum_{\substack{0\le r_1,\dots,r_n<p\\ r_1+\cdots+r_n=p}} \frac{p!}{r_1!\cdots r_n!} a_1^{r_1}\cdots a_n^{r_n} \pmod{p^2}. \tag{11.12}$$

The right-hand side is $(\sum a_i)^p - \sum a_i^p$, so equation (11.12) reduces to

$$((p-1)!+1)\left(\sum_{i=1}^{n} a_i^p - \left(\sum_{i=1}^{n} a_i\right)^p\right) \equiv 0 \pmod{p^2}.$$

And since $(p-1)! \equiv -1 \pmod{p}$ and $\sum a_i^p \equiv \sum a_i \equiv (\sum a_i)^p \pmod{p}$, this is true. □

We have shown that $h(\pi_1,\dots,\pi_n) = H(\pi_1,\dots,\pi_n)$ whenever $\sum \pi_i = 1$. Lemma 11.4.1 does not imply that h is the unique such polynomial of degree less than p in each variable, since this equation is only stated (and H is only defined) in the case where the arguments sum to 1. However, h has further good properties. It is homogeneous of degree p, which implies that the polynomial function $\bar{H}\colon (\mathbb{Z}/p\mathbb{Z})^n \to \mathbb{Z}/p\mathbb{Z}$ induced by h is homogeneous of degree 1. In fact,

$$\bar{H}(\boldsymbol{\pi}) = \sum_{i=1}^{n} \partial(a_i) - \partial\left(\sum_{i=1}^{n} a_i\right).$$

(This follows from the fact that the integer (11.11) represents $h(\pi_1,\dots,\pi_n)$.) So in the light of Lemma 11.1.8 and the explanation that follows it, h is the natural choice of polynomial representing entropy mod p.

We now establish several polynomial identities satisfied by h, which are stronger than the functional equations previously proved for H. The first is closely related to the chain rule, as we shall see.

Theorem 11.4.3 *Let $n, k_1, \dots, k_n \ge 0$. Then h satisfies the following identity of polynomials in commuting variables x_{ij} over $\mathbb{Z}/p\mathbb{Z}$:*

$$\begin{aligned} &h(x_{11},\dots,x_{1k_1},\ \dots,\ x_{n1},\dots,x_{nk_n}) \\ &\quad = h(x_{11}+\cdots+x_{1k_1},\ \dots,\ x_{n1}+\cdots+x_{nk_n}) + \sum_{i=1}^{n} h(x_{i1},\dots,x_{ik_i}). \end{aligned}$$

Proof The left-hand side of this equation is equal to

$$-\sum_{\substack{0\le s_1,\dots,s_n\le p\\ s_1+\cdots+s_n=p}} \sum \frac{x_{11}^{r_{11}}\cdots x_{1k_1}^{r_{1k_1}}\ \cdots\ x_{n1}^{r_{n1}}\cdots x_{nk_n}^{r_{nk_n}}}{r_{11}!\cdots r_{1k_1}!\ \cdots\ r_{n1}!\cdots r_{nk_n}!}, \tag{11.13}$$

where the inner sum is over all $r_{11}, \ldots, r_{nk_n}$ such that $0 \le r_{ij} < p$ and

$$r_{11} + \cdots + r_{1k_1} = s_1, \ \ldots, \ r_{n1} + \cdots + r_{nk_n} = s_n.$$

Split the outer sum into two parts, the first consisting of the summands in which none of $s_1, \ldots, s_n$ is equal to p, and the second consisting of the summands in which one s_i is equal to p and the others are zero. Then the polynomial (11.13) is equal to $A + B$, where

$$A = - \sum_{\substack{0 \le s_1, \ldots, s_n < p \\ s_1 + \cdots + s_n = p}} \prod_{i=1}^{n} \sum_{\substack{r_{i1}, \ldots, r_{ik_i} \ge 0 \\ r_{i1} + \cdots + r_{ik_i} = s_i}} \frac{x_{i1}^{r_{i1}} \cdots x_{ik_i}^{r_{ik_i}}}{r_{i1}! \cdots r_{ik_i}!},$$

$$B = - \sum_{i=1}^{n} \sum_{\substack{0 \le r_{i1}, \ldots, r_{ik_i} < p \\ r_{i1} + \cdots + r_{ik_i} = p}} \frac{x_{i1}^{r_{i1}} \cdots x_{ik_i}^{r_{ik_i}}}{r_{i1}! \cdots r_{ik_i}!}.$$

We have

$$\begin{aligned} A &= - \sum_{\substack{0 \le s_1, \ldots, s_n < p \\ s_1 + \cdots + s_n = p}} \frac{1}{s_1! \cdots s_n!} \prod_{i=1}^{n} \sum_{\substack{r_{i1}, \ldots, r_{ik_i} \ge 0 \\ r_{i1} + \cdots + r_{ik_i} = s_i}} \frac{s_i!}{r_{i1}! \cdots r_{ik_i}!} x_{i1}^{r_{i1}} \cdots x_{ik_i}^{r_{ik_i}} \\ &= - \sum_{\substack{0 \le s_1, \ldots, s_n < p \\ s_1 + \cdots + s_n = p}} \frac{1}{s_1! \cdots s_n!} \prod_{i=1}^{n} (x_{i1} + \cdots + x_{ik_i})^{s_i} \\ &= h(x_{11} + \cdots + x_{1k_1}, \ \ldots, \ x_{n1} + \cdots + x_{nk_n}) \end{aligned}$$

and

$$B = \sum_{i=1}^{n} h(x_{i1}, \ldots, x_{ik_i}).$$

The result follows. □

We easily deduce the polynomial form of the chain rule.

Corollary 11.4.4 (Chain rule) *Let $n, k_1, \ldots, k_n \ge 0$. Then h satisfies the following identity of polynomials in commuting variables y_i, x_{ij} over $\mathbb{Z}/p\mathbb{Z}$:*

$$\begin{aligned} &h(y_1 x_{11}, \ldots, y_1 x_{1k_1}, \ \ldots, \ y_n x_{n1}, \ldots, y_n x_{nk_n}) \\ &\quad = h(y_1(x_{11} + \cdots + x_{1k_1}), \ \ldots, \ y_n(x_{n1} + \cdots + x_{nk_n})) + \sum_{i=1}^{n} y_i^p h(x_{i1}, \ldots, x_{ik_i}). \end{aligned}$$

Proof This follows from Theorem 11.4.3 by substituting $y_i x_{ij}$ for x_{ij} then using the degree p homogeneity of h. □

This polynomial identity provides another proof of the chain rule for entropy mod p: given $\boldsymbol{\gamma} \in \Pi_n$ and $\boldsymbol{\pi}^i \in \Pi_{k_i}$ as in Proposition 11.1.14, substitute $y_i = \gamma_i$ and $x_{ij} = \pi^i_j$, then use the facts that $\sum_j \pi^i_j = 1$ and $\gamma_i^p = \gamma_i$ for each i. (Here i is a superscript and p is a power.)

The entropy polynomial $h(x)$ in a single variable is 0, by definition. But the entropy polynomial in two variables has important properties.

Corollary 11.4.5 *The two-variable entropy polynomial h satisfies the cocycle condition*

$$h(x, y) - h(x, y + z) + h(x + y, z) - h(y, z) = 0$$

as a polynomial identity.

Similar results appear in Cathelineau [60] (pp. 58–59), Kontsevich [195], and Elbaz-Vincent and Gangl [89] (Section 2.3), and can be understood through the information cohomology of Baudot, Bennequin and Vigneaux [32, 345].

Proof Theorem 11.4.3 with $n = 2$ and $(k_1, k_2) = (2, 1)$ gives

$$h(x, y, z) = h(x + y, z) + h(x, y)$$

(since $h(z)$ is the zero polynomial), and similarly,

$$h(x, y, z) = h(x, y + z) + h(y, z).$$

The result follows. □

We are especially interested in the case where the arguments of the entropy function sum to 1, and under that restriction, $h(x, y)$ reduces to a simple form:

Lemma 11.4.6 *If $p \neq 2$, there is an identity of polynomials*

$$h(x, 1 - x) = \sum_{r=1}^{p-1} \frac{x^r}{r},$$

and if $p = 2$, there is an identity of polynomials

$$h(x, 1 - x) = x + x^2.$$

Proof The case $p = 2$ is trivial. Suppose, then, that $p > 2$. The result can be proved by direct calculation, but we shorten the proof using Example 11.1.12, which implies that

$$h(\pi, 1 - \pi) = \sum_{r=1}^{p-1} \frac{\pi^r}{r}$$

for all $\pi \in \mathbb{Z}/p\mathbb{Z}$. We now want to prove that this is a *polynomial* identity, not just an equality of functions. By Lemma 11.4.1, it suffices to show that the polynomial

$$h(x, 1-x) = -\sum_{r=1}^{p-1} \frac{x^r(1-x)^{p-r}}{r!(p-r)!}$$

has degree strictly less than p. Since it plainly has degree at most p, we need only show that the coefficient of x^p vanishes. That coefficient is

$$-\sum_{r=1}^{p-1} \frac{(-1)^{p-r}}{r!(p-r)!}.$$

For $1 \leq r \leq p-1$,

$$-\frac{(-1)^{p-r}}{r!(p-r)!} = (-1)^{p-r}\frac{(p-1)!}{r!(p-r)!} = (-1)^{p-r}\frac{1}{r}\binom{p-1}{r-1} = (-1)^{p-1}\frac{1}{r}$$

in $\mathbb{Z}/p\mathbb{Z}$, using first the fact that $(p-1)! = -1$ and then Lemma 11.1.11. Hence the coefficient of x^p in $h(x, 1-x)$ is

$$(-1)^{p-1} \sum_{r \in (\mathbb{Z}/p\mathbb{Z})^\times} \frac{1}{r}.$$

But $r \mapsto 1/r$ defines a permutation of $(\mathbb{Z}/p\mathbb{Z})^\times$, so the sum here is equal to $\sum_{r=1}^{p-1} r$, which is 0 since p is odd. □

Following Elbaz-Vincent and Gangl [88], we write

$$£_1(x) = h(x, 1-x) = \begin{cases} \sum_{r=1}^{p-1} x^r/r & \text{if } p \neq 2, \\ x + x^2 & \text{if } p = 2. \end{cases} \tag{11.14}$$

(Elbaz-Vincent and Gangl omitted the case $p = 2$.) The function $£_1$ is the mod p analogue of the real function

$$x \mapsto H_{\mathbb{R}}(x, 1-x) = -x \log x - (1-x)\log(1-x). \tag{11.15}$$

This may be a surprise, given the lack of formal resemblance between the expressions (11.14) and (11.15). Indeed, the polynomial $\sum_{r=1}^{p-1} x^r/r$ is the truncation of the power series of $-\log(1-x)$, not (11.15). Nevertheless, the Faddeev theorem and its mod p counterpart (Theorems 2.5.1 and 11.2.1) establish a tight analogy between the entropy functions over $\mathbb{R}$ and $\mathbb{Z}/p\mathbb{Z}$.

It is immediate from the definition of h that it is a symmetric polynomial, so there is a polynomial identity

$$£_1(x) = £_1(1-x). \tag{11.16}$$

The polynomial $£_1$ also satisfies a more complicated identity, whose significance will be explained shortly. Following Kontsevich [195], Elbaz-Vincent and Gangl proved the following.

Proposition 11.4.7 (Elbaz-Vincent and Gangl) *There is a polynomial identity*

$$£_1(x) + (1-x)^p £_1\left(\frac{y}{1-x}\right) = £_1(y) + (1-y)^p £_1\left(\frac{x}{1-y}\right).$$

Both sides of this equation are indeed polynomials, since $£_1$ has degree at most p. Elbaz-Vincent and Gangl proved it using differential equations (Proposition 5.9(2) of [88]), but it also follows easily from the cocycle identity.

Proof We work in the field of rational expressions over $\mathbb{Z}/p\mathbb{Z}$ in commuting variables x and y. Since h is homogeneous of degree p,

$$h(x,y) = (x+y)^p £_1\left(\frac{x}{x+y}\right).$$

The identity to be proved is, therefore, equivalent to

$$h(x, 1-x) + h(y, 1-x-y) = h(y, 1-y) + h(x, 1-x-y).$$

Since h is symmetric, this in turn is equivalent to

$$h(x, 1-x-y) - h(x, 1-x) + h(1-y, y) - h(1-x-y, y) = 0,$$

which is an instance of the cocycle identity of Corollary 11.4.5. □

Proposition 11.4.7 can be understood as follows. Any probability distribution mod p can be expressed as an iterated composite of distributions on two elements. Hence, the entropy of any distribution can be computed in terms of entropies $H(\pi, 1-\pi)$ of distributions on two elements, using the chain rule. In this sense, the sequence of functions

$$(H\colon \Pi_n \to \mathbb{Z}/p\mathbb{Z})_{n\geq 1}$$

reduces to the single function $H\colon \Pi_2 \to \mathbb{Z}/p\mathbb{Z}$, which is effectively a function in one variable:

$$\begin{array}{rccc} F\colon & \mathbb{Z}/p\mathbb{Z} & \to & \mathbb{Z}/p\mathbb{Z} \\ & \pi & \mapsto & H(\pi, 1-\pi). \end{array}$$

The same is true over $\mathbb{R}$: the sequence of functions $(H\colon \Delta_n \to \mathbb{R})_{n\geq 1}$ reduces to the single function $F\colon [0,1] \to \mathbb{R}$ defined by $F(\pi) = H(\pi, 1-\pi)$.

On the other hand, given an arbitrary function $F\colon \mathbb{Z}/p\mathbb{Z} \to \mathbb{Z}/p\mathbb{Z}$, one cannot generally extend it to a sequence of functions $(\Pi_n \to \mathbb{Z}/p\mathbb{Z})_{n\geq 1}$ satisfying the chain rule (nor, similarly, in the real case). Indeed, by expressing a distribution $(\pi, 1-\pi-\tau, \tau)$ as a composite in two different ways, we obtain an equation that F must satisfy if such an extension is to exist. Assuming the symmetry property $F(\pi) = F(1-\pi)$, that equation is

$$F(\pi) + (1-\pi)F\left(\frac{\tau}{1-\pi}\right) = F(\tau) + (1-\tau)F\left(\frac{\pi}{1-\tau}\right) \tag{11.17}$$

$(\pi, \tau \neq 1)$. When the function F is $\pi \mapsto H(\pi, 1-\pi)$, equation (11.17) also follows from Proposition 11.4.7.

Equation (11.17) is sometimes called the **fundamental equation of information theory**. (Over $\mathbb{R}$, this functional equation has been studied since at least the 1958 work of Tverberg [337]. The name seems to have come later, and appears in Aczél and Daróczy's 1975 book [3].) Assuming that F is symmetric, it is the only obstacle to the extension problem, in the sense that if F satisfies the fundamental equation then the extension can be performed.

In the real case, the function (11.15) is a solution of the fundamental equation. In fact, up to a scalar multiple, it is the *only* measurable solution F of the fundamental equation such that $F(0) = F(1)$ (Corollary 3.4.22 of Aczél and Daróczy [3]). It can be deduced that up to a constant factor, Shannon entropy for finite real probability distributions is characterized uniquely by measurability, symmetry and the chain rule. This is the 1964 theorem of Lee mentioned in Remark 2.5.2(iii), proofs of which can be found in Lee [206] and Aczél and Daróczy [3] (Corollary 3.4.23).

In the mod p case, we know that the function $F = \pounds_1$ is symmetric and satisfies the fundamental equation. Since any such function F can be extended to a sequence of functions $\Pi_n \to \mathbb{Z}/p\mathbb{Z}$ satisfying the chain rule, it follows from Theorem 11.2.1 that up to a constant factor, $\pounds_1$ is unique with these properties.

Kontsevich [195] proposed calling $\pounds_1$ the $1\frac{1}{2}$**-logarithm**, because the ordinary logarithm satisfies a three-term functional equation ($\log(xy) = \log x + \log y$), the dilogarithm satisfies a five-term functional equation (as in Section 2 of Zagier [362] or Proposition 3.5 of Elbaz-Vincent and Gangl [88]), and the $1\frac{1}{2}$-logarithm satisfies the four-term functional equation (11.17).

Remark 11.4.8 In his seminal note [195], Kontsevich unified the real and mod p cases with a homological argument, using a cocycle identity equivalent to the one in Corollary 11.4.5. In doing so, he established that $\sum_{0<r<p} \pi^r/r$ is the correct formula for the entropy of a distribution $(\pi, 1-\pi)$ mod p on two elements (assuming, as he did, that $p \neq 2$). Although he gave no definition of the entropy of a distribution mod p on an arbitrary number of elements, his arguments showed that a unique reasonable such definition must exist.

In this chapter, we have developed the framework hinted at by Kontsevich, and also provided the definition and characterization of information loss mod p. Two further features of this theory are apparent. The first is the streamlined inclusion of the case $p = 2$. The second is the dropping of all symmetry requirements. In axiomatic approaches to entropy based on the fundamental equation of information theory (11.17), such as those of Lee [206] and Kontsevich, the symmetry axiom $F(\pi) = F(1 - \pi)$ is essential. Indeed, $F(\pi) = \pi$ is also a solution of (11.17), and the polynomial identity of Proposition 11.4.7 is also satisfied by x^p in place of $\pounds_1(x)$. The symmetry axiom is used to rule out these and other undesired solutions. This is why Lee's characterization of real entropy H needed the assumption that H is a symmetric function. In contrast, symmetry is needed nowhere in the approach taken here.

12

The Categorical Origins of Entropy

In this chapter, we describe a general category-theoretic construction which, when given as input the real line and the notion of finite probability distribution, automatically produces as output the notion of Shannon entropy (Figure 12.1).

The moral of this result is that even in the pure-mathematical heartlands of algebra and topology, entropy is inescapable. This may come as a surprise: for although entropy is a major concept in many branches of science, an algebraist, topologist or category theorist can easily go a lifetime without encountering entropy of any kind.

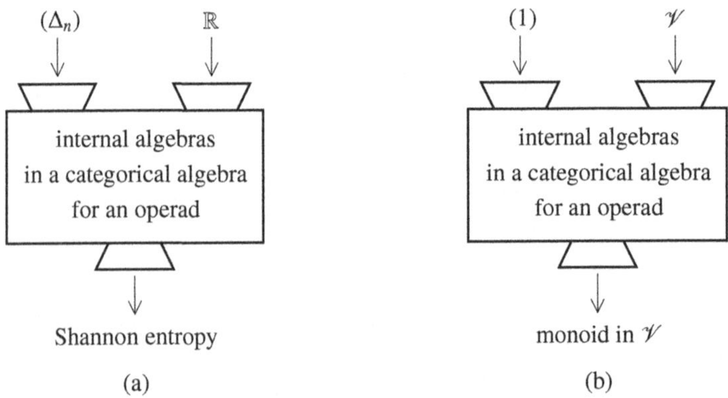

Figure 12.1 Schematic illustration of the main result of this chapter, Theorem 12.3.1. There is a general categorical machine which (a) when given as input the simplices (Δ_n) and the real line, produces as output the notion of Shannon entropy, and (b) when given as input the one-point set 1 and a monoidal category $\mathscr{V}$, produces as output the notion of monoid in $\mathscr{V}$.

Yet the categorical construction described here is entirely general and natural. It is not tailor-made for this particular purpose. Other familiar inputs produce familiar outputs. And the inputs that we give to the construction here, the real line $\mathbb{R}$ and the standard topological simplices Δ_n, are fundamental objects of pure mathematics. So, we are all but forced to accept Shannon entropy as a natural concept in pure mathematics too – quite independently of any motivation in terms of information, diversity, thermodynamics, and so on.

The categorical construction involves operads and their algebras. The first two sections set out some standard definitions, beginning with operads and algebras themselves in Section 12.1. For an operad P, there are notions of categorical P-algebra $\mathbf{A}$ (a category acted on by P) and of internal algebra in $\mathbf{A}$. With these definitions in place (Section 12.2), we can fulfil the promise of the first paragraph above (Theorem 12.3.1). Specifically, we show that for the operad Δ of simplices and the categorical Δ-algebra $\mathbb{R}$, the internal algebras in $\mathbb{R}$ are precisely the scalar multiples of Shannon entropy.

In the final section, we describe the free categorical Δ-algebra containing an internal algebra. The result proved is analogous to the classical theorem that the free monoidal category containing a monoid is the category of finite totally ordered sets. To reach this result involves a further climb up the mountain of categorical abstraction. But at the end of the path is a characterization of information loss that is entirely concrete. It is almost exactly the characterization theorem of Chapter 10.

This chapter assumes some knowledge of category theory, including the concepts of product in a category, monoid in a monoidal category, and internal category in a category with finite limits.

12.1 Operads and Their Algebras

An operad is a system of abstract operations, somewhat like an algebraic theory in the sense of universal algebra, but more restricted in nature. Operads first emerged in algebraic topology (Boardman and Vogt [43]; May [247]), while independently, the more general notion of multicategory was being developed in categorical logic (Lambek [201]). Nowadays, operads (like many other categorical structures) have found application in a very wide range of subjects, from algebra to theoretical physics. Some samples of such applications can be found in Kontsevich [196], Loday and Vallette [234], and Markl, Shnider and Stasheff [244].

Many introductions to operads are available, such as [234], [244], and Chap-

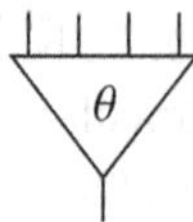

Figure 12.2 An element $\theta \in P_4$ of an operad P.

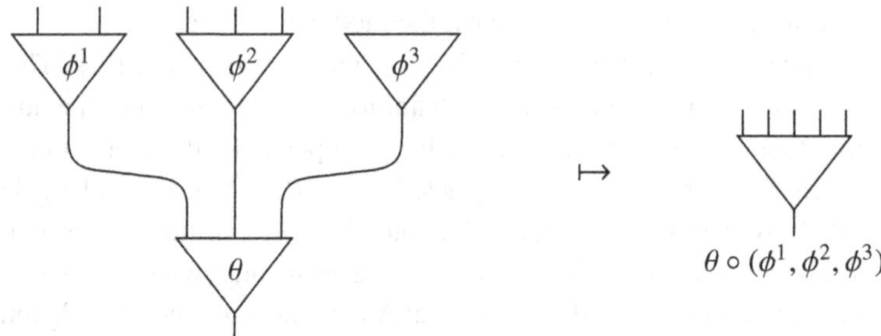

Figure 12.3 Composition in an operad: $\theta \in P_3$ composes with $\phi^1 \in P_2$, $\phi^2 \in P_3$ and $\phi^3 \in P_0$ to give $\theta \circ (\phi^1, \phi^2, \phi^3) \in P_5$.

ter 2 of [207]. Here we give only the definitions and results that are needed in order to reach our goal.

An operad consists of a sequence $(P_n)_{n \geq 0}$ of sets equipped with certain algebraic structure obeying certain laws. It is useful to view the elements θ of P_n as abstract operations with n inputs and one output, as in Figure 12.2. A typical example will be given by $P_n = \mathscr{A}(A^{\otimes n}, A)$, for any object A of a monoidal category $\mathscr{A}$. The algebraic structure on the sequence of sets $(P_n)_{n \geq 0}$, and the equational laws that this structure obeys, are exactly those suggested by this example.

Definition 12.1.1 An **operad** P consists of:

- a sequence $(P_n)_{n \geq 0}$ of sets;
- for each $n, k_1, \ldots, k_n \geq 0$, a function

$$P_n \times P_{k_1} \times \cdots \times P_{k_n} \to P_{k_1 + \cdots + k_n} \tag{12.1}$$

(Figure 12.3), called **composition** and written as

$$(\theta, \phi^1, \ldots, \phi^n) \mapsto \theta \circ (\phi^1, \ldots, \phi^n);$$

- an element $1_P \in P_1$, called the **identity**,

satisfying the following axioms:

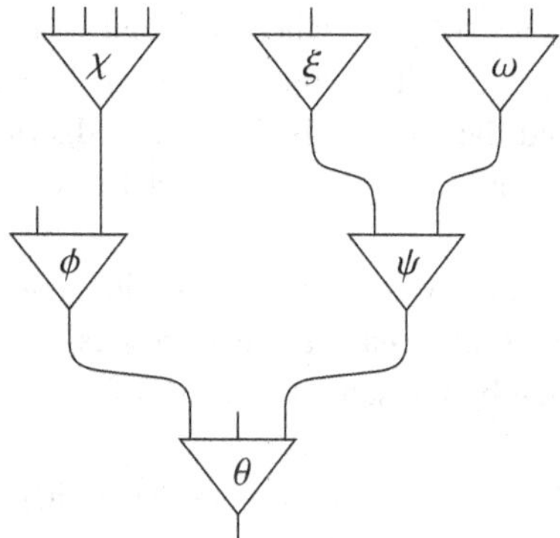

Figure 12.4 Every tree of operations in an operad P has a well-defined composite (in this case, an element of P_9).

- **associativity**: for each $n, k_i, \ell_{ij} \geq 0$ and $\theta \in P_n$, $\phi^i \in P_{k_i}$, $\psi^{ij} \in P_{\ell_{ij}}$,

$$\big(\theta \circ (\phi^1, \ldots, \phi^n)\big) \circ (\psi^{11}, \ldots, \psi^{1k_1}, \ \ldots, \ \psi^{n1}, \ldots, \psi^{nk_n})$$
$$= \theta \circ \big(\phi^1 \circ (\psi^{11}, \ldots, \psi^{1k_1}), \ \ldots, \ \phi^n \circ (\psi^{n1}, \ldots, \psi^{nk_n})\big);$$

- **identity**: for each $n \geq 0$ and $\theta \in P_n$,

$$\theta \circ (\underbrace{1_P, \ldots, 1_P}_{n}) = \theta = 1_P \circ (\theta).$$

Every tree of operations such as that shown in Figure 12.4 has an unambiguous composite, obtained by repeatedly using the composition and identity of the operad. The associativity and identity axioms guarantee that the order in which this is done makes no difference to the outcome.

Examples 12.1.2 i. There is an operad $\mathbf{1}$ in which $\mathbf{1}_n$ is the one-element set for each $n \geq 0$. The composition and identities are uniquely determined. With the obvious notion of map of operads, $\mathbf{1}$ is the terminal operad.

ii. Fix a monoid M. There is an operad $P(M)$ given by

$$P(M)_n = \begin{cases} M & \text{if } n = 1, \\ \varnothing & \text{otherwise} \end{cases}$$

($n \geq 0$). There is no choice in how to define the composition of $P(M)$ except when $n = k_1 = 1$ (in the notation of (12.1)), and in that case it is defined to be the multiplication of M. Similarly, the identity of the operad $P(M)$ is the identity of the monoid M.

iii. There is an operad $\Delta = (\Delta_n)_{n\geq 0}$, where as usual Δ_n is the set of probability distributions on $\{1, \ldots, n\}$. The composition of the operad is composition of distributions, and the identity is the unique distribution $\mathbf{u}_1$ on $\{1\}$. We already noted in Remark 2.1.8 that the associativity and identity axioms are satisfied.

iv. There is a larger operad Λ consisting of not just the *probability* measures on finite sets, but all finite measures on finite sets. Thus, $\Lambda_n = [0, \infty)^n$. The composition is given by the same formula as for Δ (Definition 2.1.3), and the identity is $(1) \in \Lambda_1$.

v. Let $\mathscr{A}$ be a monoidal category and $A \in \mathscr{A}$. Then there is an operad $\mathrm{End}(A)$ with

$$\mathrm{End}(A)_n = \mathscr{A}(A^{\otimes n}, A),$$

and with composition and identities defined using the composition, identities and monoidal structure of $\mathscr{A}$. For a general operad P, we have suggested that elements of P_n be thought of as operations, but when $P = \mathrm{End}(A)$, this is true in a concrete sense: $\mathrm{End}(A)_n$ is the set of maps $A^{\otimes n} \to A$.

vi. Fix a field k, and let $P_n = k[x_1, \ldots, x_n]$ be the set of polynomials over k in n variables. Then $P = (P_n)_{n\geq 0}$ has the structure of an operad, with composition given by substitution and reindexing of variable names. For instance, if

$$\theta = x_1^2 + x_2^3 \in P_2, \quad \phi = 2x_1x_3 - x_2 \in P_3, \quad \psi = x_1 + x_2x_3x_4 \in P_4,$$

then

$$\theta \circ (\phi, \psi) = (2x_1x_3 - x_2)^2 + (x_4 + x_5x_6x_7)^3 \in P_7.$$

This example of an operad is just one of a large family. In this case, P_n is the free k-algebra on n generators. There are similar examples where P_n is the free group, free Lie algebra, free distributive lattice, etc., on n generators. In all cases, composition is by substitution and reindexing.

vii. Fix $d \geq 1$. The **little d-discs operad** P is defined as follows. Let P_n be the set of configurations of n d-dimensional discs inside the unit disc, numbered in order and with disjoint interiors (Figure 12.5). Composition is by substitution (using affine transformations) and reindexing, as suggested by the figure.

The little discs operad and its close relative, the little cubes operad, were some of the very first operads to be defined (Boardman and Vogt [43] and Section 4 of May [247]).

In more precise terminology, operads as defined in Definition 12.1.1 are

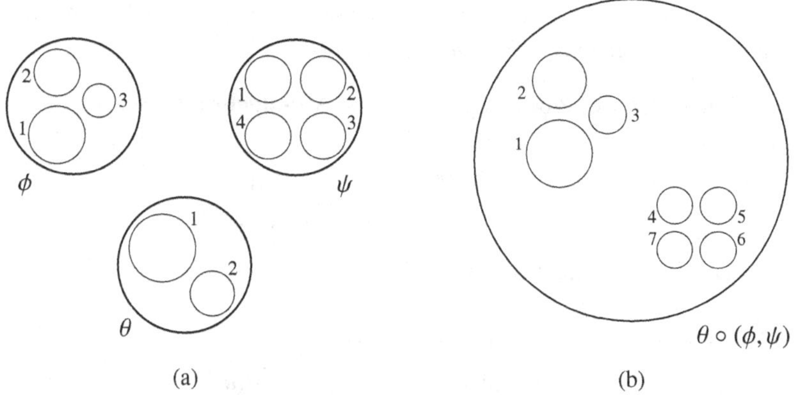

Figure 12.5 (a) Operations $\theta \in P_2$, $\phi \in P_3$ and $\psi \in P_4$ in the little 2-discs operad P (Example 12.1.2(vii)); (b) the composite operation $\theta \circ (\phi, \psi) \in P_7$.

called nonsymmetric operads of sets. Just as the definition of monoidal category has symmetric and nonsymmetric variants, so too does the definition of operad. We will concentrate on the nonsymmetric variant.

However, we will not only need operads *of sets*. Let $\mathscr{E}$ be any category with finite products (or indeed, any symmetric monoidal category, a level of generality that we will not need). An **operad in** $\mathscr{E}$ is a sequence $(P_n)_{n \geq 0}$ of objects of $\mathscr{E}$ together with maps (12.1) in $\mathscr{E}$ (encoding the composition) and a map $1 \to P_1$ in $\mathscr{E}$ (encoding the identity), all subject to commutative diagrams expressing the associativity and identity equations of Definition 12.1.1.

Details of this more general definition can be found in May [248], for instance, but we will need only two cases. The first is $\mathscr{E} = \mathbf{Set}$, the category of sets. In that case, an operad in $\mathscr{E}$ is just an operad as in Definition 12.1.1. The second is $\mathscr{E} = \mathbf{Top}$, the category of topological spaces. An operad in **Top** is just an operad P of sets in which each set P_n is equipped with a topology and the composition maps (12.1) are continuous. We will refer to operads in **Top** as **topological operads**.

Examples 12.1.3 i. The terminal operad **1** is a topological operad in a unique way.

ii. For a topological monoid M, the operad $P(M)$ of Example 12.1.2(ii) is a topological operad in an evident way.

iii. Putting the standard topology on the simplices Δ_n gives Δ the structure of a topological operad.

iv. The little discs operad is also naturally a topological operad.

An operad P is a system of abstract operations. An algebra for P is an interpretation of the elements of P as *actual* operations:

Definition 12.1.4 Let P be an operad of sets. A **P-algebra** is a set A together with a map

$$\begin{array}{cccc} \alpha_n\colon & P_n \times A^n & \to & A \\ & (\theta, (a^1, \ldots, a^n)) & \mapsto & \overline{\theta}(a^1, \ldots, a^n) \end{array}$$

for each $n \geq 0$, satisfying two axioms:

$$\begin{aligned} &\overline{\theta \circ (\phi^1, \ldots, \phi^n)}(a^{11}, \ldots, a^{1k_1}, \ \ldots, \ a^{n1}, \ldots, a^{nk_n}) \\ &\qquad = \overline{\theta}\big(\overline{\phi^1}(a^{11}, \ldots, a^{1k_1}), \ \ldots, \ \overline{\phi^n}(a^{n1}, \ldots, a^{nk_n})\big) \end{aligned} \tag{12.2}$$

for all $\theta \in P_n$, $\phi^i \in P_{k_i}$ and $a^{ij} \in A$; and

$$\overline{1_P}(a) = a \tag{12.3}$$

for all $a \in A$.

The definition of algebra extends easily from operads of sets to operads in any category $\mathscr{E}$ with finite products: then A is an object of $\mathscr{E}$ and α_n is a map in $\mathscr{E}$, while the equations (12.2) and (12.3) are expressed as commutative diagrams in $\mathscr{E}$ (May [248]). In the only other case that concerns us here, $\mathscr{E}$ = **Top**, an algebra for a topological operad P is a topological space A together with a sequence of continuous maps

$$\left(P_n \times A^n \xrightarrow{\alpha_n} A\right)_{n \geq 0}$$

satisfying equations (12.2) and (12.3).

Examples 12.1.5 i. Consider the terminal operad $\mathbf{1}$ of sets. A $\mathbf{1}$-algebra is a set A together with a map $\alpha_n\colon A^n \to A$ for each $n \geq 0$, satisfying equations (12.2) and (12.3). One easily deduces that a $\mathbf{1}$-algebra is exactly a monoid, with α_n as its n-fold multiplication. If $\mathbf{1}$ is regarded as a topological operad then a $\mathbf{1}$-algebra is exactly a topological monoid.

ii. Fix a monoid M. An algebra for the operad $P(M)$ is simply a set with a left M-action. If M is a topological monoid then a $P(M)$-algebra is a topological space with a continuous left M-action.

iii. Now consider the topological operad Δ of simplices. Any convex subset A of $\mathbb{R}^d$, for any $d \geq 0$, is a Δ-algebra in a natural way: given $\mathbf{p} \in \Delta_n$ and $\mathbf{a}^1, \ldots, \mathbf{a}^n \in A$, put

$$\overline{\mathbf{p}}(\mathbf{a}^1, \ldots, \mathbf{a}^n) = p_1\mathbf{a}^1 + \cdots + p_n\mathbf{a}^n \in A.$$

Equations (12.2) and (12.3) express elementary facts about convex combinations.

We refer to this as the **standard** Δ-algebra structure on A.

iv. The previous example admits a family of deformations, at least when A is a linear subspace of $\mathbb{R}^d$. For each $q \in \mathbb{R}$, there is a Δ-algebra structure on A given by

$$\overline{\mathbf{p}}(\mathbf{a}^1, \ldots, \mathbf{a}^n) = \sum_{i \in \operatorname{supp} \mathbf{p}} p_i^q \mathbf{a}^i.$$

(Here the superscript q is a power but the superscript i is an index.) The previous example is the case $q = 1$.

v. The chain rule for a weighted mean M on an interval I nearly states that the maps

$$(M \colon \Delta_n \times I^n \to I)_{n \geq 0}$$

give I the structure of an algebra for the operad Δ. More exactly, the chain rule is the composition axiom (12.2) for a Δ-algebra. For I to be a Δ-algebra, it must also satisfy the identity axiom (12.3), which is a special case of consistency: $M(\mathbf{u}_1, (x)) = x$ for each $x \in I$.

vi. For any operad P of sets, a P-algebra amounts to a set A together with a map $P \to \operatorname{End}(A)$ of operads. Here, $\operatorname{End}(A)$ is the operad defined in Example 12.1.2(v), with $\mathscr{A} = \mathbf{Set}$. This makes precise the earlier assertion that an algebra for P is an interpretation of the elements of P as actual operations.

vii. Any d-fold loop space is an algebra for the little d-discs operad in a natural way. This was one of the first examples of an algebra for an operad, the details of which can be found in Section 5 of May [247].

Let P be an operad in a finite product category $\mathscr{E}$, and let $A = (A, \alpha)$ and $B = (B, \beta)$ be P-algebras. A **map of P-algebras** from A to B is a map $f \colon A \to B$ in $\mathscr{E}$ such that the square

$$\begin{array}{ccc} P_n \times A^n & \xrightarrow{1 \times f^n} & P_n \times B^n \\ {\scriptstyle \alpha_n} \downarrow & & \downarrow {\scriptstyle \beta_n} \\ A & \xrightarrow[f]{} & B \end{array}$$

commutes for each $n \geq 0$. This defines a category $\mathbf{Alg}(P)$ of P-algebras.

Remarks 12.1.6 The language of operations and algebras invites comparison with other categorical formulations of the concept of algebraic theory. Systematic comparisons can be found in Kelly [185], Section 2.8 of Gould [124], and Chapter 3 of Avery [21]. Here, we just make the following observations.

i. Let $\mathscr{E}$ be a finite product category satisfying the further mild condition that it has countable coproducts over which the product distributes. (**Set** and **Top** are examples.) Then any operad P in $\mathscr{E}$ induces a monad T_P on $\mathscr{E}$, with functor part given by

$$T_P(A) = \coprod_{n\geq 0} P_n \times A^n$$

($A \in \mathscr{E}$). The category of algebras for the operad P is exactly the category of algebras for the monad T_P. Non-isomorphic operads P sometimes induce the same monad T_P [208], although many aspects of an operad can still be understood through its induced monad.

ii. Remark (i) provides a semantic connection between operads and a different conception of algebraic theory, monads. On the syntactic side, the definition of operad can easily be adapted to give a definition of finitary algebraic theory, equivalent to any of the usual definitions (as given in Manes [242], for instance). Indeed, a finitary algebraic theory can be defined as an operad P together with, for each map of sets

$$f: \{1, \ldots, m\} \to \{1, \ldots, n\},$$

a map $f_*: P_m \to P_n$, subject to equations expressing compatibility between the operad structure and these maps f_* (Tronin [329, 330]). The idea is that f_* transforms an m-ary operation into an n-ary operation by reindexing the variables according to f. For instance, in the theory P of groups, P_n is the underlying set of the free group on n generators (which can be regarded as the set of n-ary operations defined on any group), and if f is the unique map $\{1, 2\} \to \{1\}$ then $f_*: P_2 \to P_1$ sends the operation of multiplication to the operation of squaring.

If we take the definition of finitary algebraic theory sketched in the previous paragraph but restrict f to be a bijection, we obtain the definition of **symmetric operad**. (In much of the literature, 'operad' is taken to mean 'symmetric operad' by default.) If we further restrict f to be an identity, we recover the definition of nonsymmetric operad.

iii. As the previous remark suggests, most algebraic theories cannot be described by an operad. For instance, there is no operad P of sets such that **Alg**(P) is equivalent to the category of groups. For a proof of a strong version of this statement, see Lin [228].

12.2 Categorical Algebras and Internal Algebras

Let P be an operad of sets. An algebra for P is a set acted on by P, but more generally, we can consider *categories* acted on by P. Such a structure is called a categorical P-algebra.

More generally still, let $\mathscr{E}$ be a category with finite limits. Then there is the notion of internal category in $\mathscr{E}$ (as in Chapter 2 of Johnstone [163]), and when P is an operad in $\mathscr{E}$, we can consider actions of P on such an internal category.

Definition 12.2.1 Let $\mathscr{E}$ be a category with finite limits and let P be an operad in $\mathscr{E}$. A **categorical P-algebra** is an internal category in $\mathbf{Alg}(P)$.

It is straightforward to verify that $\mathbf{Alg}(P)$ has finite limits, computed as in $\mathscr{E}$, so this definition does make sense. But it is also helpful to have at hand a more explicit form, as follows.

A categorical P-algebra $\mathbf{A}$ can be described as a pair of ordinary P-algebras, $\mathbf{A}_0$ and $\mathbf{A}_1$, together with domain and codomain maps

$$\mathbf{A}_1 \rightrightarrows \mathbf{A}_0$$

and composition and identity maps

$$\mathbf{A}_1 \times_{\mathbf{A}_0} \mathbf{A}_1 \to \mathbf{A}_1, \qquad 1 \to \mathbf{A}_1,$$

all of which are required to be maps of P-algebras, as well as obeying the usual axioms for an internal category. Here $\mathbf{A}_0$ is to be thought of as the object of objects of $\mathbf{A}$, and $\mathbf{A}_1$ as the object of maps in $\mathbf{A}$.

Equivalently, a categorical P-algebra is an internal category in $\mathscr{E}$ on which P acts functorially. To see this, first note that for any object X of $\mathscr{E}$ and internal category $\mathbf{A}$ in $\mathscr{E}$, we can define another internal category

$$X \times \mathbf{A}$$

in $\mathscr{E}$. This is the product $D(X) \times \mathbf{A}$, where $D(X)$ is the discrete internal category on X. Thus, its object of objects and object of maps are given by

$$(X \times \mathbf{A})_0 = X \times \mathbf{A}_0, \qquad (X \times \mathbf{A})_1 = X \times \mathbf{A}_1.$$

In this notation, a categorical P-algebra consists of an internal category $\mathbf{A}$ in $\mathscr{E}$ together with internal functors

$$\alpha_n \colon P_n \times \mathbf{A}^n \to \mathbf{A} \tag{12.4}$$

($n \geq 0$), satisfying analogues of the usual algebra axioms.

As usual, we are principally concerned with the cases $\mathscr{E} = \mathbf{Set}$ and $\mathscr{E} = \mathbf{Top}$, where categorical algebras for an operad can be understood as follows.

Examples 12.2.2 i. Let P be an operad in $\mathscr{E} = \mathbf{Set}$. A categorical P-algebra consists of a small category $\mathbf{A}$ together with a functor

$$\overline{\theta}\colon \mathbf{A}^n \to \mathbf{A}$$

for each $n \geq 0$ and $\theta \in P_n$. These functors are required to satisfy equations (12.2) and (12.3) for objects a^{ij} and a of $\mathbf{A}$, and analogous equations for maps in $\mathbf{A}$.

ii. Let P be an operad in $\mathscr{E} = \mathbf{Top}$. The categorical P-algebras can be described as in (i), but with the following additions. $\mathbf{A}$ is now a topological category, so that $\mathbf{A}_0$ and $\mathbf{A}_1$ carry topologies with the property that the domain, codomain, composition and identity operations are continuous. Moreover, the structure maps

$$P_n \times \mathbf{A}_0^n \to \mathbf{A}_0, \quad P_n \times \mathbf{A}_1^n \to \mathbf{A}_1$$

of the P-algebras $\mathbf{A}_0$ and $\mathbf{A}_1$ are required to be continuous.

iii. Here we consider the special case of categorical algebras with only one object, over both **Set** and **Top**.

First, let P be an operad of sets. Let A be a monoid, viewed as a one-object category $\mathbf{A}$. To give $\mathbf{A}$ the structure of a categorical P-algebra is to give the set A the structure of an ordinary P-algebra in such a way that for each $n \geq 0$ and $\theta \in P_n$, the structure map

$$\overline{\theta}\colon A^n \to A$$

is a monoid homomorphism. In short, a one-object categorical P-algebra is a monoid on which P acts by homomorphisms.

Similarly, when P is a topological operad, a one-object categorical P-algebra is a topological monoid on which P acts by continuous homomorphisms.

Some specific examples now follow.

Examples 12.2.3 i. Consider the terminal operad of sets, **1**. By the description preceding equation (12.4), a categorical **1**-algebra is a category on which **1** acts functorially, that is, a category $\mathbf{A}$ together with a functor $\mathbf{A}^n \to \mathbf{A}$, subject to certain axioms. These axioms give $\mathbf{A}$ the structure of a monoid in **Cat**. Thus, a categorical **1**-algebra is exactly a strict monoidal category.

Alternatively, working directly from Definition 12.2.1, a categorical **1**-algebra is an internal category in **Mon**, the category of monoids. Again, this is just a strict monoidal category.

ii. Let M be a monoid, and form the operad $P(M)$ of sets (Example 12.1.2(ii)). A categorical $P(M)$-algebra is a category equipped with a left M-action.
iii. Consider the topological operad Δ of simplices. Let A be a linear subspace (or more generally, a convex additive submonoid) of $\mathbb{R}^d$. Then A is a topological monoid under addition. We have already considered the standard Δ-algebra structure on the topological space A, given for $\mathbf{p} \in \Delta_n$ by

$$\begin{array}{cccc} \overline{\mathbf{p}}\colon & A^n & \to & A \\ & (\mathbf{a}^1, \ldots, \mathbf{a}^n) & \mapsto & \sum_{i=1}^n p_i \mathbf{a}^i \end{array}$$

(Example 12.1.5(iii)). Each of these maps $\overline{\mathbf{p}}$ is a monoid homomorphism. Hence by Example 12.2.2(iii), A is a one-object categorical Δ-algebra.
iv. The same is true for the q-deformed algebra structure

$$(\mathbf{a}^1, \ldots, \mathbf{a}^n) \mapsto \sum_{i \in \mathrm{supp}(\mathbf{p})} p_i^q \mathbf{a}^i$$

of Example 12.1.5(iv), for any $q \in \mathbb{R}$.

For ordinary algebras for an operad, there is only one sensible notion of map between algebras, but for *categorical* algebras, there are several. Indeed, let $\mathscr{E}$ be a category with finite limits, let P be an operad in $\mathscr{E}$, and let $\mathbf{B}$ and $\mathbf{A}$ be categorical P-algebras. Then $\mathbf{B}$ and $\mathbf{A}$ are, by definition, internal categories in $\mathbf{Alg}(P)$, and a **strict map** from $\mathbf{B}$ to $\mathbf{A}$ is an internal functor $\mathbf{B} \to \mathbf{A}$ in $\mathbf{Alg}(P)$. Equivalently, it is an internal functor

$$G\colon \mathbf{B} \to \mathbf{A}$$

in $\mathscr{E}$ such that for all $n \geq 0$, the square

$$\begin{array}{ccc} P_n \times \mathbf{B}^n & \xrightarrow{1 \times G^n} & P_n \times \mathbf{A}^n \\ {\scriptstyle \beta_n} \downarrow & & \downarrow {\scriptstyle \alpha_n} \\ \mathbf{B} & \xrightarrow[G]{} & \mathbf{A} \end{array}$$

commutes, where β_n and α_n are the structure maps of $\mathbf{B}$ and $\mathbf{A}$ (as in equation (12.4)).

However, this is a square of (internal) categories and functors, so we can also consider variants in which the square is only required to commute up to a specified natural isomorphism, or a natural transformation in one direction or the other – subject, as usual, to coherence axioms. The particular variant that will concern us is the following.

Definition 12.2.4 Let $\mathscr{E}$ be a category with finite limits and let P be an operad in $\mathscr{E}$. Let $\mathbf{B}$ and $\mathbf{A}$ be categorical P-algebras, with structure maps (β_n) and (α_n)

respectively. A **lax map B → A** of categorical P-algebras consists of a functor $G\colon \mathbf{B} \to \mathbf{A}$ (internal to $\mathscr{E}$) together with a natural transformation

$$\begin{array}{ccc} P_n \times \mathbf{B}^n & \xrightarrow{1\times G^n} & P_n \times \mathbf{A}^n \\ {\scriptstyle\beta_n}\downarrow & \swArrow_{\gamma_n} & \downarrow{\scriptstyle\alpha_n} \\ \mathbf{B} & \xrightarrow[G]{} & \mathbf{A} \end{array}$$

(again internal to $\mathscr{E}$) for each $n \geq 0$, satisfying the following two axioms.

i. For each $n, k_1, \ldots, k_n \geq 0$, writing $k = \sum k_i$, the composite natural transformation

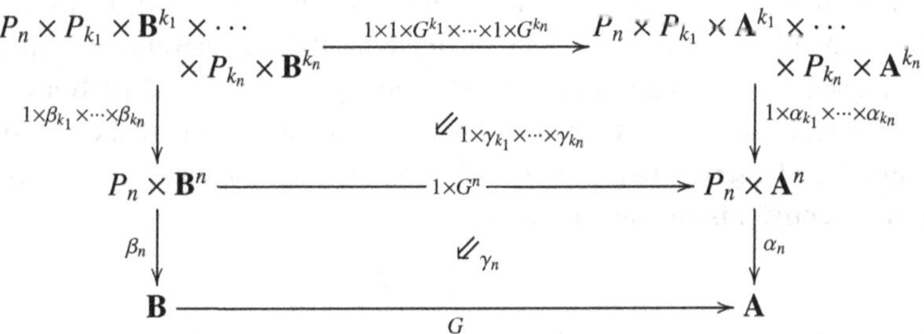

is equal to

$$\begin{array}{ccc} P_n \times P_{k_1} \times \mathbf{B}^{k_1} \times \cdots \times P_{k_n} \times \mathbf{B}^{k_n} & \xrightarrow{1\times 1\times G^{k_1}\times\cdots\times 1\times G^{k_n}} & P_n \times P_{k_1} \times \mathbf{A}^{k_1} \times \cdots \times P_{k_n} \times \mathbf{A}^{k_n} \\ \cong\downarrow & & \downarrow\cong \\ P_n \times P_{k_1} \times \cdots \times P_{k_n} \times \mathbf{B}^k & // & P_n \times P_{k_1} \times \cdots \times P_{k_n} \times \mathbf{A}^k \\ {\scriptstyle\circ\times 1}\downarrow & & \downarrow{\scriptstyle\circ\times 1} \\ P_k \times \mathbf{B}^k & \xrightarrow{1\times G^k} & P_k \times \mathbf{A}^k \\ {\scriptstyle\beta_k}\downarrow & \swArrow_{\gamma_k} & \downarrow{\scriptstyle\alpha_k} \\ \mathbf{B} & \xrightarrow[G]{} & \mathbf{A} \end{array}$$

ii. The composite natural transformation

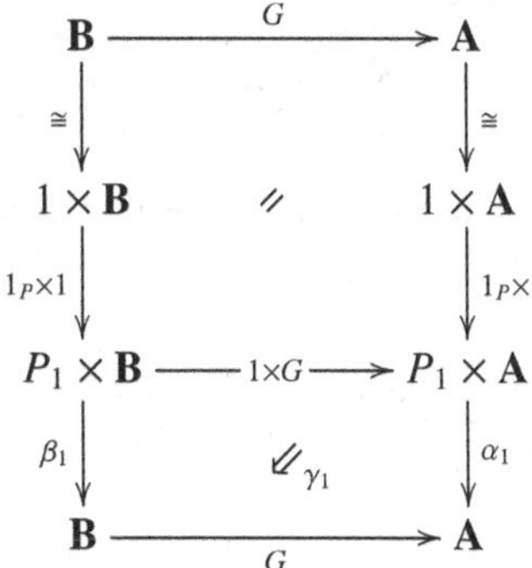

is equal to the identity. (Here, $1_P : 1 \to P_1$ denotes the map encoding the identity of the operad P.)

A strict map of P-algebras can equivalently be viewed as a lax map (G, γ) in which each of the maps γ_n is an identity.

Remark 12.2.5 Definition 12.2.4 can also be derived from the theory of 2-monads, as follows. We observed in Remark 12.1.6(i) that any operad P in $\mathscr{E}$ induces a monad T_P on $\mathscr{E}$ (under mild hypotheses on the category $\mathscr{E}$). In the same way, it induces a 2-monad on $\mathbf{Cat}(\mathscr{E})$, the 2-category of internal categories in $\mathscr{E}$. An algebra for that 2-monad is exactly a categorical P-algebra, and a lax map of algebras for the 2-monad (in the sense of Blackwell, Kelly and Power [40]) is exactly a lax map of categorical P-algebras.

As a general categorical principle, it is often worth considering the maps into an object from the terminal object. (In categories of spaces, this gives the notion of point.) For any operad P in any category $\mathscr{E}$ with finite limits, there is a terminal categorical P-algebra $\mathbf{1}$. We consider the lax maps from $\mathbf{1}$ to other categorical P-algebras.

Definition 12.2.6 Let $\mathscr{E}$ be a category with finite limits, let P be an operad in $\mathscr{E}$, and let $\mathbf{A}$ be a categorical P-algebra. An **internal algebra** in $\mathbf{A}$ is a lax map $\mathbf{1} \to \mathbf{A}$ of categorical P-algebras.

This definition is due to Batanin ([31], Definition 7.2). We will see that it generalizes the notion of internal monoid in a monoidal category. But first, we give an explicit description of internal algebras in the cases $\mathscr{E} = \mathbf{Set}$ and $\mathscr{E} = \mathbf{Top}$.

Examples 12.2.7 i. Let $\mathscr{E} = \mathbf{Set}$. Take an operad P of sets and a categorical P-algebra $\mathbf{A}$. An internal algebra in $\mathbf{A}$ consists of, first of all, a functor

$G: \mathbf{1} \to \mathbf{A}$. This simply picks out an object a of $\mathbf{A}$. Next, the natural transformations γ_n in Definition 12.2.4 amount to a family of maps

$$\gamma_\theta : \overline{\theta}(\underbrace{a, \ldots, a}_{n}) \to a,$$

one for each $n \geq 0$ and $\theta \in P_n$. The first coherence axiom in Definition 12.2.4 states that the diagram

$$\begin{array}{ccc} \overline{\theta}\big(\overline{\phi^1}(a,\ldots,a),\ldots,\overline{\phi^n}(a,\ldots,a)\big) & \xrightarrow{\overline{\theta}(\gamma_{\phi^1},\ldots,\gamma_{\phi^n})} & \overline{\theta}(a,\ldots,a) \\ \Big\| & & \Big\downarrow \gamma_\theta \\ \overline{\theta \circ (\phi^1,\ldots,\phi^n)}(a,\ldots,a) & \xrightarrow[\gamma_{\theta\circ(\phi^1,\ldots,\phi^n)}]{} & a \end{array}$$

commutes for all $\theta \in P_n$ and $\phi^i \in P_{k_i}$, and the second states that

$$\gamma_{1_P} : \overline{1_P}(a) \to a$$

is equal to the identity on a.

ii. An identical description of internal algebras applies when $\mathscr{E} = \mathbf{Top}$, with the additional condition that for each $n \geq 0$, the function

$$\begin{array}{ccc} P_n & \to & \mathbf{A}_1 \\ \theta & \mapsto & \gamma_\theta \end{array}$$

is continuous. (Here $\mathbf{A}_1$ denotes the space of maps in the topological category $\mathbf{A}$, as in Example 12.2.2(ii).)

iii. Now let $\mathscr{E} = \mathbf{Set}$, and let $\mathbf{A}$ be a one-object categorical P-algebra. As we saw in Example 12.2.2(iii), $\mathbf{A}$ amounts to a monoid A on which P acts by homomorphisms. An internal P-algebra in $\mathbf{A}$ consists of an element $\gamma_\theta \in A$ for each $n \geq 0$ and $\theta \in P_n$, satisfying the coherence axioms in (i).

   Writing γ_θ as $\gamma(\theta)$, we have a sequence of functions

$$(\gamma : P_n \to A)_{n \geq 0}.$$

The coherence axioms in (i) state that

$$\gamma(\theta) \cdot \overline{\theta}(\gamma(\phi^1), \ldots, \gamma(\phi^n)) = \gamma(\theta \circ (\phi^1, \ldots, \phi^n)) \tag{12.5}$$

for all $n, k_1, \ldots, k_n \geq 0$, $\theta \in P_n$ and $\phi^i \in P_{k_i}$, and that

$$\gamma(1_P) = 1. \tag{12.6}$$

In summary, when $\mathbf{A}$ is a one-object categorical P-algebra corresponding to a monoid A, an internal P-algebra in $\mathbf{A}$ amounts to a sequence of maps $\gamma : P_n \to A$ satisfying equations (12.5) and (12.6).

iv. For a topological operad P, internal algebras in a one-object categorical P-algebra admit exactly the same explicit description as in the previous example, with the added requirement that the maps $\gamma\colon P_n \to A$ are continuous.

We now give some specific examples.

Examples 12.2.8 i. Let $\mathbf{1}$ be the terminal operad of sets. As we have seen, a categorical $\mathbf{1}$-algebra is just a monoidal category. By the explicit description in Example 12.2.7(i), an internal algebra in a categorical $\mathbf{1}$-algebra $\mathbf{A}$ consists of an object $a \in \mathbf{A}$ together with a map

$$\gamma_n\colon a^{\otimes n} \to a$$

for each $n \geq 0$, satisfying the equations given there. It follows easily that an internal algebra in $\mathbf{A}$ is exactly a monoid in the monoidal category $\mathbf{A}$.

As an alternative proof, note that for strict monoidal categories $\mathbf{B}$ and $\mathbf{A}$, a lax map $\mathbf{B} \to \mathbf{A}$ of categorical $\mathbf{1}$-algebras is precisely a lax monoidal functor. This is immediate from the definitions. Hence an internal algebra in a strict monoidal category $\mathbf{A}$ is a lax monoidal functor $\mathbf{1} \to \mathbf{A}$, and it is well known that such functors correspond naturally to monoids in $\mathbf{A}$ (paragraph (5.4.1) of Bénabou [34]).

An algebra for $\mathbf{1}$ is exactly a monoid (Example 12.1.5(i)), so it is logical terminology that an internal algebra is exactly an internal monoid.

ii. Fix a monoid M. We saw in Example 12.2.3(ii) that a categorical $P(M)$-algebra is a category $\mathbf{A}$ with a left M-action; let us write the action as

$$\begin{array}{ccc} M \times \mathbf{A} & \to & \mathbf{A} \\ (m, a) & \mapsto & m \cdot a. \end{array}$$

By the explicit description in Example 12.2.7(i), an internal $P(M)$-algebra in $\mathbf{A}$ consists of an object $a \in \mathbf{A}$ together with a map

$$\gamma_m\colon m \cdot a \to a$$

for each $m \in M$, satisfying natural coherence axioms.

Missing from this list of examples is the case of internal algebras in a categorical algebra for Δ, the operad of simplices. This is the subject of the next section, and will transport us directly to the concept of entropy.

12.3 Entropy as an Internal Algebra

In this chapter so far, we have reviewed some established general concepts in the theory of operads. We now apply them to the topological operad Δ of simplices.

We saw in Example 12.2.3(iii) that the real line $\mathbb{R}$, as a topological monoid under addition, is a categorical Δ-algebra in a standard way:

$$(\mathbf{p}, (x_1, \dots, x_n)) \mapsto \sum_{i=1}^{n} p_i x_i \tag{12.7}$$

($\mathbf{p} \in \Delta_n$, $x_1, \dots, x_n \in \mathbb{R}$). What are the internal algebras in the categorical Δ-algebra $\mathbb{R}$?

By Example 12.2.7(iv), an internal Δ-algebra in $\mathbb{R}$ amounts to a sequence of functions $(\gamma\colon \Delta_n \to \mathbb{R})_{n \geq 0}$ satisfying certain axioms. It is in this sense that the following theorem holds.

Theorem 12.3.1 *Let Δ be the topological operad of simplices, and equip $\mathbb{R}$ with its standard categorical Δ-algebra structure* (12.7). *Then the internal algebras in $\mathbb{R}$ are precisely the real scalar multiples of Shannon entropy.*

In other words, a sequence of functions $(\gamma\colon \Delta_n \to \mathbb{R})_{n \geq 0}$ defines an internal algebra in $\mathbb{R}$ if and only if $\gamma = cH$ for some $c \in \mathbb{R}$.

Proof By Example 12.2.7(iv), an internal algebra in $\mathbb{R}$ is a sequence of functions $(\gamma\colon \Delta_n \to \mathbb{R})_{n \geq 0}$ with the following properties:

i. for all $n, k_1, \dots, k_n \geq 0$ and $\mathbf{w} \in \Delta_n$, $\mathbf{p}^1 \in \Delta_{k_1}, \dots, \mathbf{p}^n \in \Delta_{k_n}$,

$$\gamma(\mathbf{w}) + \sum_{i=1}^{n} w_i \gamma(\mathbf{p}^i) = \gamma(\mathbf{w} \circ (\mathbf{p}^1, \dots, \mathbf{p}^n));$$

ii. $\gamma(\mathbf{u}_1) = 0$;

iii. $\gamma\colon \Delta_n \to \mathbb{R}$ is continuous for each $n \geq 0$.

Condition (ii) is redundant, since it follows from (i) by taking $n = k_1 = 1$ and $\mathbf{w} = \mathbf{p}^1 = \mathbf{u}_1$. Hence by Faddeev's Theorem 2.5.1, γ defines an internal algebra if and only if $\gamma = cH$ for some $c \in \mathbb{R}$. □

This theorem can be deformed. In Example 12.2.3(iv), we defined a one-parameter family of categorical Δ-algebra structures on $\mathbb{R}$, where for a real parameter q, the action of Δ on $\mathbb{R}$ is

$$(\mathbf{p}, (x_1, \dots, x_n)) \mapsto \sum_{i \in \mathrm{supp}(\mathbf{p})} p_i^q x_i \tag{12.8}$$

($\mathbf{p} \in \Delta_n$, $x_1, \dots, x_n \in \mathbb{R}$).

Theorem 12.3.2 *Let* $1 \neq q \in \mathbb{R}$. *Let* Δ *be the operad of simplices, considered as an operad of sets, and equip* $\mathbb{R}$ *with its q-deformed categorical* Δ*-algebra structure* (12.8). *Then the internal algebras in* $\mathbb{R}$ *are precisely the real scalar multiples of q-logarithmic entropy.*

Proof By Example 12.2.7(iii), an internal algebra in $\mathbb{R}$ is a sequence of functions $(\gamma\colon \Delta_n \to \mathbb{R})_{n\geq 0}$ with the following properties:

i. for all $n, k_1, \ldots, k_n \geq 0$ and all $\mathbf{w} \in \Delta_n$, $\mathbf{p}^1 \in \Delta_{k_1}, \ldots, \mathbf{p}^n \in \Delta_{k_n}$,
$$\gamma(\mathbf{w}) + \sum_{i \in \mathrm{supp}(\mathbf{w})} w_i^q \gamma(\mathbf{p}^i) = \gamma(\mathbf{w} \circ (\mathbf{p}^1, \ldots, \mathbf{p}^n));$$

ii. $\gamma(\mathbf{u}_1) = 0$.

Condition (ii) is redundant, for the same reason as in the proof of Theorem 12.3.1. Condition (i) is satisfied if $\gamma = cS_q$, by the chain rule (4.2) for q-logarithmic entropies. Conversely, condition (i) implies that

$$\gamma(\mathbf{w} \otimes \mathbf{p}) = \gamma(\mathbf{w}) + \left(\sum_{i \in \mathrm{supp}(\mathbf{w})} w_i^q \right) \gamma(\mathbf{p})$$

for all $\mathbf{w} \in \Delta_n$ and $\mathbf{p} \in \Delta_k$, by taking $\mathbf{p}^1 = \cdots = \mathbf{p}^n = \mathbf{p}$. Hence by Theorem 4.1.5, if γ defines an internal algebra then $\gamma = cS_q$ for some $c \in \mathbb{R}$. □

Continuity was not needed in this theorem, and in fact the structure maps $\Delta_n \times \mathbb{R}^n \to \mathbb{R}$ of the q-deformed Δ-algebra $\mathbb{R}$ are discontinuous when $q \leq 0$. But they are evidently continuous when $q > 0$, so we have:

Corollary 12.3.3 *Let* $q \in (0, \infty)$. *Let* Δ *be the topological operad of simplices, and equip* $\mathbb{R}$ *with its q-deformed categorical* Δ*-algebra structure* (12.8). *Then the internal algebras in* $\mathbb{R}$ *are precisely the real scalar multiples of q-logarithmic entropy.*

Proof The case $q = 1$ is Theorem 12.3.1, and all other cases follow from Theorem 12.3.2. □

12.4 The Universal Internal Algebra

In algebra, an important role is played by free algebraic structures (groups, modules, etc.). But since one forms the free algebraic structure on a *set*, and a set is merely a cardinality (for these purposes at least), the possibilities are in a sense limited. Greater riches are to be found one categorical level up, where one can speak of the free categorical structure containing some specified internal algebraic structure. This leads to categorical characterizations of some important mathematical objects.

Examples 12.4.1 i. The free monoidal category containing a monoid is equivalent to the category of finite totally ordered sets (Mac Lane [236], Proposition VII.5.1). We will return to this example shortly. Informally, the statement is that if we build a monoidal category by starting from nothing, putting in an internal monoid, then adjoining no more other objects and maps than are forced by the definitions, and making no unnecessary identifications, then the result is the category of finite totally ordered sets.

ii. The free monoidal category containing an object A and an isomorphism $A \otimes A \to A$ is equivalent to the disjoint union of the terminal category and Thompson's group F, viewed as a one-object category (Fiore and Leinster [104]).

(Thompson's group is an infinite group with remarkable properties; it has been rediscovered multiple times in diverse contexts. Cannon, Floyd and Parry [58] provide a survey. A major open question, which has attracted an exceptional number of opposing claims and retractions, is whether F is amenable. Cannon and Floyd [57] report that even among experts, opinion is evenly split.)

iii. The free symmetric monoidal category containing a commutative Frobenius algebra is the category of compact oriented 1-manifolds and 2-dimensional cobordisms between them (Theorem 3.6.19 of Kock [191], for instance). This result lies at the foundations of topological quantum field theory.

iv. The free finite product category containing a group is the Lawvere theory of groups. The same statement holds for any other algebraic structure in place of groups (Lawvere [203]). This is essentially a tautology, but expresses a fundamental insight of categorical universal algebra: an algebraic theory can be understood as a finite product category, and a model of a theory as a finite-product-preserving functor.

In this section, we construct the free categorical P-algebra containing an internal algebra, where P is any given operad. We proceed as follows. First, we construct a certain categorical P-algebra FP. Then, we make precise what it means for a categorical P-algebra to be 'free containing an internal algebra'. Next, we prove that FP has that property. This last result, applied in the case $P = \Delta$, leads to a characterization of information loss.

We begin by constructing the categorical P-algebra FP, for an operad P of sets.

The objects of FP are the pairs (n, θ) with $n \geq 0$ and $\theta \in P_n$. Where confusion will not arise, we write (n, θ) as just θ. For objects $\psi = (k, \psi)$ and

$\theta = (n, \theta)$, a map $\psi \to \theta$ in FP consists of integers $k_1, \ldots, k_n \geq 0$ and operations $\phi^1 \in P_{k_1}, \ldots, \phi^n \in P_{k_n}$ such that

$$k = k_1 + \cdots + k_n, \qquad \psi = \theta \circ (\phi^1, \ldots, \phi^n).$$

We write this map as

$$\langle \phi^1, \ldots, \phi^n \rangle_\theta \colon \psi \to \theta. \tag{12.9}$$

Thus, the set of objects of the category FP and the set of maps in FP are, respectively,

$$\coprod_{n \geq 0} P_n, \qquad \coprod_{n, k_1, \ldots, k_n \geq 0} P_n \times P_{k_1} \times \cdots \times P_{k_n}. \tag{12.10}$$

Composition and identities in the category FP are defined using the composition and identity of the operad P.

To give the category FP the structure of a categorical P-algebra, we must construct from each operation $\pi \in P_m$ a functor

$$\overline{\pi} \colon (FP)^m \to FP.$$

On objects, $\overline{\pi}$ is defined by

$$\overline{\pi}(\theta^1, \ldots, \theta^m) = \pi \circ (\theta^1, \ldots, \theta^m).$$

To define the action of $\overline{\pi}$ on maps, take an m-tuple of maps

$$\begin{gathered} \langle \phi^{11}, \ldots, \phi^{1n_1} \rangle_{\theta^1} \colon \psi^1 \to \theta^1 \\ \vdots \\ \langle \phi^{m1}, \ldots, \phi^{mn_m} \rangle_{\theta^m} \colon \psi^m \to \theta^m \end{gathered}$$

in FP. Then

$$\begin{aligned} &\overline{\pi}\big(\langle \phi^{11}, \ldots, \phi^{1n_1} \rangle_{\theta^1}, \ \ldots, \ \langle \phi^{m1}, \ldots, \phi^{mn_m} \rangle_{\theta^m}\big) \\ &\qquad = \langle \phi^{11}, \ldots, \phi^{1n_1}, \ \ldots, \ \phi^{m1}, \ldots, \phi^{mn_m} \rangle_{\pi \circ (\theta^1, \ldots, \theta^m)}, \end{aligned} \tag{12.11}$$

which is a map $\overline{\pi}(\psi^1, \ldots, \psi^m) \to \overline{\pi}(\theta^1, \ldots, \theta^m)$ in FP.

Verifying that FP satisfies the axioms for a categorical P-algebra is routine.

Lemma 12.4.2 *Let P be an operad of sets.*

i. *The object 1_P of FP is terminal.*
ii. *Write $!_\phi \colon \phi \to 1_P$ for the unique map from an object ϕ of FP to 1_P. Then for any map*

$$\langle \phi^1, \ldots, \phi^n \rangle_\theta \colon \psi \to \theta$$

in FP, we have

$$\langle \phi^1, \ldots, \phi^n \rangle_\theta = \overline{\theta}(!_{\phi^1}, \ldots, !_{\phi^n}).$$

The notation in (i) refers to the identity element $1_P \in P_1$ of the operad P, which corresponds to the object $1_P = (1, 1_P)$ of the category FP. It is this object that is terminal.

Proof For (i), given any object ϕ of FP, it is immediate from the definition of FP that there is a unique map $\phi \to 1_P$, namely,

$$!_\phi = \langle \phi \rangle_{1_P} : \phi \to 1_P.$$

For (ii), take a map

$$\langle \phi^1, \ldots, \phi^n \rangle_\theta : \psi \to \theta$$

in FP. Since $!_{\phi^i}$ is a map $\phi^i \to 1_P$, the map $\overline{\theta}(!_{\phi^1}, \ldots, !_{\phi^n})$ has domain

$$\overline{\theta}(\phi^1, \ldots, \phi^n) = \theta \circ (\phi^1, \ldots, \phi^n) = \psi$$

and codomain

$$\overline{\theta}(1_P, \ldots, 1_P) = \theta \circ (1_P, \ldots, 1_P) = \theta,$$

matching the domain and codomain of $\langle \phi^1, \ldots, \phi^n \rangle_\theta$. Now by definition of $!_{\phi^i}$ and by definition (12.11) of the P-action on maps in FP,

$$\begin{aligned} \overline{\theta}(!_{\phi^1}, \ldots, !_{\phi^n}) &= \overline{\theta}(\langle \phi^1 \rangle_{1_P}, \ldots, \langle \phi^n \rangle_{1_P}) \\ &= \langle \phi^1, \ldots, \phi^n \rangle_{\theta \circ (1_P, \ldots, 1_P)} \\ &= \langle \phi^1, \ldots, \phi^n \rangle_\theta, \end{aligned}$$

as required. □

The categorical P-algebra FP contains a canonical internal algebra. To specify it, we use the description of internal algebras in Example 12.2.7(i). Its underlying object is the terminal object 1_P. To give 1_P the structure of an internal algebra, we have to specify, for each $n \geq 0$ and $\theta \in P_n$, a map

$$\overline{\theta}(\underbrace{1_P, \ldots, 1_P}_{n}) \to 1_P.$$

The domain here is θ, and the codomain is terminal, so the only possible choice is the unique map $!_\theta : \theta \to 1_P$. This gives 1_P the structure of an internal algebra in the categorical P-algebra FP. We refer to this internal algebra as $(1_P, !)$.

When P is a topological operad, the set of objects of FP and the set of maps in FP (both given in (12.10)) each carry a natural topology. For instance, the set of maps in FP is a coproduct of product spaces. In this way, FP is an internal category in **Top**. Indeed, FP is a categorical P-algebra in the topological sense (by the description in Example 12.2.2(ii)) and $(1_P, !)$ is an internal algebra in FP in the topological sense (by the description in Example 12.2.7(iv)).

Remark 12.4.3 As for all of the operadic definitions and constructions in this chapter, the construction of FP can be generalized to an operad P in an arbitrary category $\mathscr{E}$ with suitable properties (in this case, finite products and countable coproducts over which the products distribute). The general definition is exactly as suggested by the case $\mathscr{E} = \mathbf{Top}$.

Examples 12.4.4 i. Consider the terminal operad $\mathbf{1}$ of sets. The objects of the category $\mathbf{D} = F\mathbf{1}$ are the natural numbers $0, 1, \ldots$ A map $k \to n$ in $\mathbf{D}$ is an ordered n-tuple of natural numbers summing to k, or equivalently, an order-preserving map $\{1, \ldots, k\} \to \{1, \ldots, n\}$. Thus, $\mathbf{D}$ is equivalent to the category of finite totally ordered sets. It is almost the same as the category usually denoted by Δ in algebraic topology, the only difference being that it also contains the object 0 (corresponding to the empty ordered set).

By construction, $\mathbf{D}$ is a categorical $\mathbf{1}$-algebra, that is, a strict monoidal category. The monoidal structure is defined on objects by addition and on maps by disjoint union. Moreover, $\mathbf{D}$ contains a canonical internal algebra, that is, internal monoid. It is the object $1 \in \mathbf{D}$ with its unique monoid structure: the multiplication is the unique map $1 + 1 = 2 \to 1$ in $\mathbf{D}$, and the identity is the unique map $0 \to 1$.

ii. Fix a monoid M and consider the operad $P(M)$. Since $P(M)_n$ is empty for all $n \neq 1$, the objects of the category $FP(M)$ are just the elements $\theta \in M$. A map $\psi \to \theta$ in $FP(M)$ is an element $\phi \in M$ such that $\psi = \theta\phi$. In other words, regarding the monoid M as a category with a single object $\star$, the category $FP(M)$ is the slice $M/\star$. For instance, when the monoid M is cancellative, $FP(M)$ is the poset of elements of M ordered by divisibility.

iii. Now take the topological operad Δ. The objects of the category $F\Delta$ are the pairs $(n, \mathbf{p})$ with $n \geq 0$ and $\mathbf{p} \in \Delta_n$. A map $(k, \mathbf{s}) \to (n, \mathbf{p})$ consists of natural numbers $k_1, \ldots, k_n$ summing to k together with probability distributions $\mathbf{r}^i \in \Delta_{k_i}$ satisfying

$$\mathbf{s} = \mathbf{p} \circ (\mathbf{r}^1, \ldots, \mathbf{r}^n). \tag{12.12}$$

This category has a more familiar description. As in (i) above, the n-tuple $(k_1, \ldots, k_n)$ amounts to an order-preserving map

$$f\colon \{1, \ldots, k\} \to \{1, \ldots, n\}.$$

Then $\mathbf{p}$ is equal to the pushforward $f\mathbf{s}$ of the probability measure $\mathbf{s}$ along f.

(See Definition 2.1.10.) Thus, f is a measure-preserving map

$$(\{1,\ldots,k\},\mathbf{s}) \to (\{1,\ldots,n\},\mathbf{p}).$$

In Lemma 2.1.9, we showed that given $\mathbf{s}$, $\mathbf{p}$ and $k_1,\ldots,k_n$ (or equivalently $\mathbf{s}$, $\mathbf{p}$ and f), it is always possible to find distributions $\mathbf{r}^i$ satisfying equation (12.12). Furthermore, we showed that for $i \in \operatorname{supp}(\mathbf{p})$, the distribution $\mathbf{r}^i$ is uniquely determined, and for $i \notin \operatorname{supp}(\mathbf{p})$, we can choose $\mathbf{r}^i$ freely in Δ_{k_i}.

These observations together imply that up to equivalence, $F\Delta$ is the category whose objects are finite totally ordered probability spaces $(\mathcal{X},\mathbf{p})$, in which a map $(\mathcal{Y},\mathbf{s}) \to (\mathcal{X},\mathbf{p})$ is an order-preserving, measure-preserving map f together with a probability distribution on $f^{-1}(i)$ for each $i \in \mathcal{X}$ such that $p_i = 0$.

By construction, $F\Delta$ has the structure of a categorical Δ-algebra. On objects, the Δ-action takes convex combinations of finite probability spaces, as in Section 10.1. The one-element probability space $(1,\mathbf{u}_1)$ has a unique internal algebra structure in $F\Delta$.

Remark 12.4.5 The category $F\Delta$ just described is nearly the category **FinOrdProb** of finite totally ordered probability spaces. There is a forgetful functor $F\Delta \to$ **FinOrdProb**, but it is not an equivalence, because of the complication associated with zero probabilities.

From the point of view of Bayesian inference, it is broadly unsurprising that such a complication arises. In that subject, special caution is reserved for probabilities of exactly zero. The Bayesian statistician Dennis Lindley wrote:

> leave a little probability for the moon being made of green cheese; it can be as small as 1 in a million, but have it there since otherwise an army of astronauts returning with samples of the said cheese will leave you unmoved. [...] So never believe in anything absolutely, leave some room for doubt.

([230], p. 104.) He named this principle **Cromwell's rule**, after the English Lord Protector Oliver Cromwell, who wrote to the Church of Scotland in 1650:

> I beseech you, in the bowels of Christ, think it possible you may be mistaken.

Further discussion can be found in Section 6.8 of Lindley [231].

We now make precise, and prove, the statement that FP is the 'free categorical P-algebra containing an internal algebra'.

Let P be an operad of sets or topological spaces, and let $E\colon \mathbf{B} \to \mathbf{A}$ be a strict map of categorical P-algebras. An internal algebra in $\mathbf{B}$ is a lax map $\mathbf{1} \to \mathbf{B}$, and can be composed with E to obtain a lax map $\mathbf{1} \to \mathbf{A}$. In this way, E maps internal algebras in $\mathbf{B}$ to internal algebras in $\mathbf{A}$.

It will be convenient to use the explicit description of internal algebras derived in Example 12.2.7(i). There, we showed that an internal algebra (b, δ) in $\mathbf{B}$ consists of an object b and a family of maps $\delta_\theta \colon \overline{\theta}(b, \ldots, b) \to b$ subject to certain equations. In these terms, the induced internal algebra $E(b, \delta)$ in $\mathbf{A}$ consists of the object $E(b)$ and the maps $E(\delta_\theta)$.

We now state and prove the universal property of the categorical P-algebra FP equipped with its internal algebra $(1_P, !)$.

Theorem 12.4.6 *Let P be an operad of either sets or topological spaces, let $\mathbf{A}$ be a categorical P-algebra, and let (a, γ) be an internal algebra in $\mathbf{A}$. Then there is a unique strict map $E \colon FP \to \mathbf{A}$ of categorical P-algebras such that $E(1_P, !) = (a, \gamma)$.*

This is a universal property of FP together with its internal algebra, and therefore determines them uniquely up to isomorphism.

Proof To prove uniqueness, let E be a map with the properties stated. Let $\theta = (n, \theta)$ be an object of FP; thus, $n \geq 0$ and $\theta \in P_n$. By definition of the categorical P-algebra structure on FP,

$$\theta = \overline{\theta}(1_P, \ldots, 1_P).$$

Applying E to both sides gives

$$E(\theta) = E(\overline{\theta}(1_P, \ldots, 1_P)) = \overline{\theta}(E(1_P), \ldots, E(1_P)) = \overline{\theta}(a, \ldots, a),$$

where the second equality holds because E is a strict map of categorical P-algebras, and the last is by hypothesis. Hence

$$E(\theta) = \overline{\theta}(a, \ldots, a), \tag{12.13}$$

which determines E uniquely on the objects of FP.

To show the same for maps, take a map

$$\langle \phi^1, \ldots, \phi^n \rangle_\theta \colon \psi \to \theta$$

in FP. By Lemma 12.4.2(ii),

$$\langle \phi^1, \ldots, \phi^n \rangle_\theta = \overline{\theta}(!_{\phi^1}, \ldots, !_{\phi^n}).$$

Applying E to both sides gives

$$\begin{aligned} E(\langle\phi^1,\ldots,\phi^n\rangle_\theta) &= E(\overline{\theta}(!_{\phi^1},\ldots,!_{\phi^n})) \\ &= \overline{\theta}(E(!_{\phi^1}),\ldots,E(!_{\phi^n})) \\ &= \overline{\theta}(\gamma_{\phi^1},\ldots,\gamma_{\phi^n}), \end{aligned}$$

for the same reasons as in the argument for objects. Hence

$$E(\langle\phi^1,\ldots,\phi^n\rangle_\theta) = \overline{\theta}(\gamma_{\phi^1},\ldots,\gamma_{\phi^n}), \tag{12.14}$$

which determines E uniquely on the maps in FP. We have therefore proved uniqueness.

To prove existence, we define E on objects by equation (12.13) and on maps by equation (12.14). Verifying that E satisfies the stated conditions (including continuity in the topological case) is a series of routine checks. □

Corollary 12.4.7 *Let P be an operad of sets or topological spaces. Let* **A** *be a categorical P-algebra. Then there is a canonical bijection between internal algebras in* **A** *and strict maps $FP \to$* **A** *of categorical P-algebras.* □

Thus, an internal algebra in **A** can be described as either a lax map **1** → **A** or a strict map $FP \to$ **A**.

Example 12.4.8 In the case $P = \mathbf{1}$, Theorem 12.4.6 states that for any strict monoidal category **A** and monoid a in **A**, there is exactly one strict monoidal functor $E : \mathbf{D} \to \mathbf{A}$ that maps the trivial monoid 1 in **D** to the given monoid a in **A**.

Hence, Corollary 12.4.7 implies that given just a monoidal category **A**, the monoids in **A** correspond naturally to the strict monoidal functors **D** → **A**. We have therefore recovered the classical fact that a monoid in **A** can be described as either a lax monoidal functor **1** → **A** or a strict monoidal functor **D** → **A** (paragraph (5.4.1) of Bénabou [34] and Proposition VII.5.1 of Mac Lane [236], for instance).

Now consider Theorem 12.4.6 in the case where P is the topological operad Δ and **A** is the topological monoid $\mathbb{R}$. By Corollary 12.4.7, the strict maps $F\Delta \to$ **A** of categorical Δ-algebras are in natural bijection with the internal Δ-algebras in **A**. By Theorem 12.3.1, these in turn correspond to real scalar multiples of Shannon entropy. Together, these results imply that the strict maps $F\Delta \to$ **A** are naturally parametrized by $\mathbb{R}$.

We now make this parametrization explicit. Since **A** has only one object, a strict map $F\Delta \to$ **A** of categorical Δ-algebras amounts to a function

$$E\colon \{\text{maps in } F\Delta\} \to \mathbb{R}$$

satisfying certain conditions. Our final theorem classifies such functions.

Theorem 12.4.9 *Let E be a function $\{\text{maps in } F\Delta\} \to \mathbb{R}$. The following are equivalent:*

i. E defines a strict map $F\Delta \to \mathbb{R}$ of categorical Δ-algebras in **Top** *(with respect to the standard categorical Δ-algebra structure on $\mathbb{R}$);*
ii. there is some $c \in \mathbb{R}$ such that for all maps $f : \mathbf{s} \to \mathbf{p}$ in $F\Delta$,

$$E(f) = c(H(\mathbf{s}) - H(\mathbf{p})).$$

Proof First assume (i). Applying E to the internal algebra $(\mathbf{u}_1, !)$ in $F\Delta$ gives an internal algebra $E(\mathbf{u}_1, !)$ in $\mathbb{R}$ (whose underlying object is necessarily the unique object of the category $\mathbb{R}$). So by Theorem 12.3.1, there is some constant $c \in \mathbb{R}$ such that $E(!_{\mathbf{p}}) = cH(\mathbf{p})$ for all $n \geq 1$ and $\mathbf{p} \in \Delta_n$.

Now take any map

$$\langle \mathbf{r}^1, \ldots, \mathbf{r}^n \rangle_{\mathbf{p}} : \mathbf{s} \to \mathbf{p} \tag{12.15}$$

in $F\Delta$. Since $\mathbf{u}_1$ is terminal in $F\Delta$, there is a commutative triangle

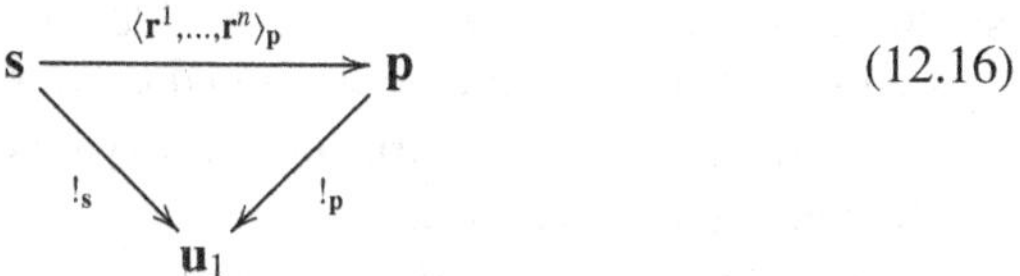

(12.16)

in $F\Delta$. Applying the functor E to this triangle gives

$$E(!_{\mathbf{s}}) = E(!_{\mathbf{p}}) + E(\langle \mathbf{r}^1, \ldots, \mathbf{r}^n \rangle_{\mathbf{p}}), \tag{12.17}$$

which by the result of the last paragraph gives

$$cH(\mathbf{s}) = cH(\mathbf{p}) + E(\langle \mathbf{r}^1, \ldots, \mathbf{r}^n \rangle_{\mathbf{p}}),$$

proving (ii).

To show that (ii) implies (i), let $c \in \mathbb{R}$. By Theorem 12.3.1, cH defines an internal algebra structure on the unique object of the category $\mathbb{R}$. Now take a map (12.15) in $F\Delta$. By definition of E,

$$E(\langle \mathbf{r}^1, \ldots, \mathbf{r}^n \rangle_{\mathbf{p}}) = c(H(\mathbf{s}) - H(\mathbf{p})).$$

But $\mathbf{s} = \mathbf{p} \circ (\mathbf{r}^1, \ldots, \mathbf{r}^n)$ by definition of the maps in $F\Delta$, so by the chain rule,

$$E(\langle \mathbf{r}^1, \ldots, \mathbf{r}^n \rangle_{\mathbf{p}}) = c \sum_{i=1}^{n} p_i H(\mathbf{r}^i) = \overline{\mathbf{p}}(cH(\mathbf{r}^1), \ldots, cH(\mathbf{r}^n)).$$

It follows from the proof of Theorem 12.4.6 that E is a strict map $F\Delta \to \mathbb{R}$ of categorical Δ-algebras. □

A result similar to Theorem 12.4.9 can also be proved for the q-logarithmic entropies, using the q-deformed categorical Δ-algebra structure on $\mathbb{R}$ and Theorem 12.3.2.

Theorem 12.4.9 bears a striking resemblance to the characterization of information loss in Theorem 10.2.1. It states that the strict maps $F\Delta \to \mathbb{R}$ are the scalar multiples of the information loss function. But where one theorem uses the category $F\Delta$, the other uses the category **FinProb** of finite probability spaces. The explicit description of $F\Delta$ in Example 12.4.4(iii) shows that there are three differences between $F\Delta$ and **FinProb**. First, the maps in $F\Delta$ are required to be order-preserving, whereas in **FinProb** there is no notion of ordering at all. Second, the category $F\Delta$ is skeletal (isomorphic objects are equal), but **FinProb** is not. Third, the maps in the category $F\Delta$ are not merely measure-preserving maps; they also come equipped with a probability distribution on the fibre over each zero-probability element of the codomain.

There is an analogue of Theorem 12.4.9 that comes close to Theorem 10.2.1; we sketch it now. It uses *symmetric* operads. As indicated in Remark 12.1.6(ii), a symmetric operad is an operad P together with an action of the symmetric group S_n on P_n for each $n \geq 0$, satisfying suitable axioms. For example, if A is an object of a symmetric monoidal category then the operad End(A) of Example 12.1.2(v) has the structure of a symmetric operad. The operad Δ is also symmetric in a natural way.

At the cost of some further complications, the notions of categorical P-algebra and internal algebra, and the construction of the free categorical P-algebra on an internal algebra, can be extended to symmetric operads P. The free categorical Δ-algebra $F_{\text{sym}}\Delta$ on an internal algebra is much like $F\Delta$, but the maps are no longer required to be order-preserving. In other words, the first of the three differences between $F\Delta$ and **FinProb** vanishes for $F_{\text{sym}}\Delta$. The second, skeletality, is categorically unimportant. So, the only substantial difference between $F_{\text{sym}}\Delta$ and **FinProb** is the third: a map in $F_{\text{sym}}\Delta$ between finite probability spaces is a measure-preserving map *together with* a probability distribution on each fibre over an element of probability zero.

The symmetric analogue of Theorem 12.4.9 states that the strict maps $F_{\text{sym}}\Delta \to \mathbb{R}$ of symmetric categorical Δ-algebras are precisely the scalar multiples of information loss. Translated into explicit terms, this theorem is nearly the same as the characterization of information loss in Theorem 10.2.1. The only difference is in the handling of zero probabilities. But the result can easily be adapted in an ad hoc way to discard the extra data associated with elements of probability zero, and it then becomes exactly Theorem 10.2.1. Historically, this categorical argument was, in fact, how the wholly elementary and concrete Theorem 10.2.1 was first obtained.

Appendix A

Proofs of Background Facts

This appendix consists of proofs deferred from the main text.

A.1 Forms of the Chain Rule for Entropy

In Remark 2.2.11, it was asserted that although the chain rule

$$H(\mathbf{w} \circ (\mathbf{p}^1, \ldots, \mathbf{p}^n)) = H(\mathbf{w}) + \sum_{i=1}^{n} w_i H(\mathbf{p}^i)$$

for Shannon entropy appears to be more general (that is, stronger) than the versions used by some previous authors, straightforward inductive arguments show that it is equivalent to those special cases. Remark 4.1.6 made a similar assertion for the q-logarithmic entropies S_q, where the equation becomes

$$S_q(\mathbf{w} \circ (\mathbf{p}^1, \ldots, \mathbf{p}^n)) = S_q(\mathbf{w}) + \sum_{i \in \mathrm{supp}(\mathbf{w})} w_i^q S_q(\mathbf{p}^i).$$

Here we prove those claims.

In Lemma A.1.1 below, part (i) is the general form of the chain rule, parts (ii) and (iv) are the special cases used by other authors, and part (iii) is an intermediate case that is helpful for the proof. Each of the four parts corresponds to a certain type of composition of probability distributions, depicted as a tree in Figure A.1.

Rather than working with sums over the support of $\mathbf{w}$, in this lemma we adopt the convention that $0^q = 0$ for all $q \in \mathbb{R}$.

Lemma A.1.1 *Let $q \in \mathbb{R}$. Let $(I\colon \Delta_n \to \mathbb{R})_{n \geq 1}$ be a sequence of symmetric functions. The following are equivalent:*

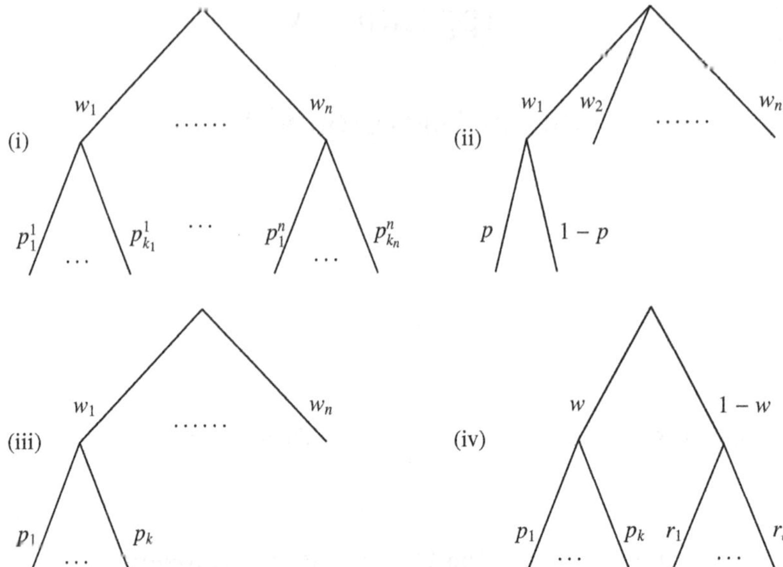

Figure A.1 Shapes of composites used in the four parts of Lemma A.1.1.

i. *for all* $n, k_1, \ldots, k_n \geq 1$, $\mathbf{w} \in \Delta_n$, *and* $\mathbf{p}^i \in \Delta_{k_i}$,

$$I(\mathbf{w} \circ (\mathbf{p}^1, \ldots, \mathbf{p}^n)) = I(\mathbf{w}) + \sum_{i=1}^{n} w_i^q I(\mathbf{p}^i);$$

ii. *for all* $n \geq 1$, $\mathbf{w} \in \Delta_n$, *and* $p \in [0, 1]$,

$$I(w_1 p, w_1(1-p), w_2, \ldots, w_n) = I(\mathbf{w}) + w_1^q I(p, 1-p);$$

iii. *for all* $n, k \geq 1$, $\mathbf{w} \in \Delta_n$, *and* $\mathbf{p} \in \Delta_k$,

$$I(w_1 p_1, \ldots, w_1 p_k, w_2, \ldots, w_n) = I(\mathbf{w}) + w_1^q I(\mathbf{p});$$

iv. *for all* $k, \ell \geq 1$, $\mathbf{p} \in \Delta_k$, $\mathbf{r} \in \Delta_\ell$, *and* $w \in [0, 1]$,

$$\begin{aligned} &I(wp_1, \ldots, wp_k, (1-w)r_1, \ldots, (1-w)r_\ell) \\ &\qquad = I(w, 1-w) + w^q I(\mathbf{p}) + (1-w)^q I(\mathbf{r}). \end{aligned}$$

Much of the following argument goes back to Feinstein ([99], pp. 5–6).

Proof Trivially, (i) implies (ii).

Assuming (ii), we prove (iii) by induction on k. The case $k = 1$ reduces to the statement that $I(\mathbf{u}_1) = 0$, which follows by taking $n = 1$ in (ii). Now let $k \geq 2$, and assume the result for $k - 1$.

Let $n \geq 1$, $\mathbf{w} \in \Delta_n$, and $\mathbf{p} \in \Delta_k$. By symmetry, we can assume that $p_k < 1$. Using the inductive hypothesis, we have

$$\begin{aligned}
&I(w_1p_1, \ldots, w_1p_k, w_2, \ldots, w_n) \\
&= I\Big(w_1(1-p_k)\cdot\frac{p_1}{1-p_k}, \ldots, w_1(1-p_k)\cdot\frac{p_{k-1}}{1-p_k}, w_1p_k, w_2, \ldots, w_n\Big) \\
&= I(w_1(1-p_k), w_1p_k, w_2, \ldots, w_n) + (w_1(1-p_k))^q I\Big(\frac{p_1}{1-p_k}, \ldots, \frac{p_{k-1}}{1-p_k}\Big),
\end{aligned}$$

which by (ii) is equal to

$$I(\mathbf{w}) + w_1^q\Big\{I(1-p_k, p_k) + (1-p_k)^q I\Big(\frac{p_1}{1-p_k}, \ldots, \frac{p_{k-1}}{1-p_k}\Big)\Big\}.$$

But by the inductive hypothesis again, the term $\{\cdots\}$ is equal to

$$I\Big((1-p_k)\cdot\frac{p_1}{1-p_k}, \ldots, (1-p_k)\cdot\frac{p_{k-1}}{1-p_k}, p_k\Big) = I(\mathbf{p}),$$

completing the induction.

Now assuming (iii), we prove (iv). Let $k, \ell \geq 1$, $\mathbf{p} \in \Delta_k$, $\mathbf{r} \in \Delta_\ell$, and $w \in [0,1]$. Using (iii), we have

$$\begin{aligned}
&I(wp_1, \ldots, wp_k, (1-w)r_1, \ldots, (1-w)r_\ell) \\
&\qquad = I(w, (1-w)r_1, \ldots, (1-w)r_\ell) + w^q I(\mathbf{p}).
\end{aligned}$$

By symmetry and (iii) again, this in turn is equal to

$$I(w, 1-w) + (1-w)^q I(\mathbf{r}) + w^q I(\mathbf{p}),$$

proving (iv).

Finally, assume (iv). We prove (i) by induction on n. The case $n = 1$ just states that $I(\mathbf{u}_1) = 0$, which follows from (iv) by taking $k = \ell = 1$. Now let $n \geq 2$, and assume the result for $n - 1$.

Let $k_1, \ldots, k_n \geq 1$, $\mathbf{w} \in \Delta_n$, and $\mathbf{p}^i \in \Delta_{k_i}$. By symmetry, we can assume that $w_1 > 0$. Write

$$\mathbf{p}^{12} = \Big(\frac{w_1}{w_1+w_2}p_1^1, \ldots, \frac{w_1}{w_1+w_2}p_{k_1}^1, \frac{w_2}{w_1+w_2}p_1^2, \ldots, \frac{w_2}{w_1+w_2}p_{k_2}^2\Big) \in \Delta_{k_1+k_2}.$$

Then

$$\mathbf{w} \circ (\mathbf{p}^1, \ldots, \mathbf{p}^n) = (w_1 + w_2, w_3, \ldots, w_n) \circ (\mathbf{p}^{12}, \mathbf{p}^3, \ldots, \mathbf{p}^n),$$

so by inductive hypothesis,

$$I(\mathbf{w} \circ (\mathbf{p}^1, \dots, \mathbf{p}^n)) \\ = I(w_1 + w_2, w_3, \dots, w_n) + (w_1 + w_2)^q I(\mathbf{p}^{12}) + \sum_{i=3}^{n} w_i^q I(\mathbf{p}^i). \quad \text{(A.1)}$$

On the other hand, by (iv),

$$I(\mathbf{p}^{12}) = I\left(\frac{w_1}{w_1 + w_2}, \frac{w_2}{w_1 + w_2}\right) + \left(\frac{w_1}{w_1 + w_2}\right)^q I(\mathbf{p}^1) + \left(\frac{w_2}{w_1 + w_2}\right)^q I(\mathbf{p}^2).$$

Substituting this into (A.1), we deduce that $I(\mathbf{w} \circ (\mathbf{p}^1, \dots, \mathbf{p}^n))$ is equal to

$$I(w_1 + w_2, w_3, \dots, w_n) + (w_1 + w_2)^q I\left(\frac{w_1}{w_1 + w_2}, \frac{w_2}{w_1 + w_2}\right) + \sum_{i=1}^{n} w_i^q I(\mathbf{p}^i). \quad \text{(A.2)}$$

But applying the inductive hypothesis to the composite

$$\mathbf{w} = (w_1 + w_2, w_3, \dots, w_n) \circ \left(\left(\frac{w_1}{w_1 + w_2}, \frac{w_2}{w_1 + w_2}\right), \mathbf{u}_1, \dots, \mathbf{u}_1\right)$$

gives

$$I(\mathbf{w}) = I(w_1 + w_2, w_3, \dots, w_n) + (w_1 + w_2)^q I\left(\frac{w_1}{w_1 + w_2}, \frac{w_2}{w_1 + w_2}\right)$$

(recalling that $I(\mathbf{u}_1) = 0$). Hence the expression (A.2) reduces to

$$I(\mathbf{w}) + \sum_{i=1}^{n} w_i^q I(\mathbf{p}^i),$$

proving (i). □

A.2 The Expected Number of Species in a Random Sample

Here we prove the result stated in Example 4.3.6, which expresses the diversity index of Hurlbert, Smith and Grassle in terms of the Hill numbers $D_q(\mathbf{p})$.

Recall that we are modelling an ecological community with n species via its relative abundance distribution, and that $H_m^{\text{HSG}}(\mathbf{p})$ denotes the expected number of different species represented in a random sample with replacement of m individuals. The claim is that

$$H_m^{\text{HSG}}(\mathbf{p}) = \sum_{q=1}^{m} (-1)^{q-1} \binom{m}{q} D_q(\mathbf{p})^{1-q}.$$

Define random variables $X_1, \ldots, X_n$ by

$$X_i = \begin{cases} 1 & \text{if species } i \text{ is present in the sample,} \\ 0 & \text{otherwise.} \end{cases}$$

Then $\sum_{i=1}^n X_i$ is the number of different species in the sample, so

$$\begin{aligned} H_m^{\text{HSG}}(\mathbf{p}) &= \mathbb{E}\left(\sum_{i=1}^n X_i\right) = \sum_{i=1}^n \mathbb{E}(X_i) \\ &= \sum_{i=1}^n \mathbb{P}(\text{species } i \text{ is present in the sample}) \\ &= \sum_{i=1}^n (1 - (1 - p_i)^m), \end{aligned}$$

as Hurlbert observed (equation (14) of [150]). It follows that

$$\begin{aligned} H_m^{\text{HSG}}(\mathbf{p}) &= n - \sum_{i=1}^n \sum_{q=0}^m \binom{m}{q} (-p_i)^q \\ &= n - \sum_{q=0}^m (-1)^q \binom{m}{q} \sum_{i=1}^n p_i^q \\ &= n - \left\{ \binom{m}{0} n - \binom{m}{1} 1 + \sum_{q=2}^m (-1)^q \binom{m}{q} D_q(\mathbf{p})^{1-q} \right\} \\ &= m - \sum_{q=2}^m (-1)^q \binom{m}{q} D_q(\mathbf{p})^{1-q} \\ &= \sum_{q=1}^m (-1)^{q-1} \binom{m}{q} D_q(\mathbf{p})^{1-q}, \end{aligned}$$

as claimed.

A.3 The Diversity Profile Determines the Distribution

Here we prove the result claimed in Remark 4.4.9: two probability distributions on the same finite set have the same diversity profile if and only if one is a permutation of the other. The formal statement is as follows.

Lemma A.3.1 *Let $n \geq 1$ and $\mathbf{p}, \mathbf{r} \in \Delta_n$. The following are equivalent:*

i. $D_q(\mathbf{p}) = D_q(\mathbf{r})$ *for all* $q \in [-\infty, \infty]$;

ii. there exists a subset $Q \subseteq [-\infty, \infty)$, *unbounded above, such that* $D_q(\mathbf{p}) = D_q(\mathbf{r})$ *for all* $q \in Q$;

iii. $\mathbf{p} = \mathbf{r}\sigma$ *for some permutation* σ *of* $\{1, \ldots, n\}$.

This result first appeared as Proposition A22 of the appendix to Leinster and Cobbold [220].

Proof (iii) implies (i) by the symmetry of the Hill numbers (Lemma 4.4.8), and (i) implies (ii) trivially. Now assuming (ii), we prove (iii) by induction on n. It is trivial for $n = 1$. Let $n \geq 2$, assume the result for $n-1$, and take $\mathbf{p}, \mathbf{r} \in \Delta_n$ such that $D_q(\mathbf{p}) = D_q(\mathbf{r})$ for all elements q of some set $Q \subseteq [-\infty, \infty)$ that is unbounded above. We may assume that $-\infty \notin Q$ and $1 \notin Q$ (for if not, remove them).

We know that $D_q(\mathbf{p})$ is continuous in $q \in [-\infty, \infty]$, by Lemma 4.2.7 or Lemma 6.2.4(i). Since Q is unbounded above,

$$\lim_{q \in Q, q \to \infty} D_q(\mathbf{p}) = D_\infty(\mathbf{p}) = 1 \Big/ \max_{1 \leq i \leq n} p_i.$$

The same is true for $D_q(\mathbf{r})$. Hence by assumption, $\max_i p_i = \max_i r_i$. Choose k and ℓ such that $p_k = \max_i p_i$ and $r_\ell = \max_i r_i$. Then $p_k = r_\ell$.

If $p_k = r_\ell = 1$ then $\mathbf{p}$ and $\mathbf{r}$ are both of the form $(0, \ldots, 0, 1, 0, \ldots, 0)$, so one is a permutation of the other. Assuming otherwise, define $\mathbf{p}', \mathbf{r}' \in \Delta_{n-1}$ by

$$\mathbf{p}' = \left(\frac{p_1}{1 - p_k}, \ldots, \frac{p_{k-1}}{1 - p_k}, \frac{p_{k+1}}{1 - p_k}, \ldots, \frac{p_n}{1 - p_k} \right)$$

and similarly for $\mathbf{r}'$. Then for all $q \in Q$,

$$\begin{aligned} D_q(\mathbf{p}') &= (1 - p_k)^{q/(q-1)} \left(\sum_{i \neq k} p_i^q \right)^{1/(1-q)} \\ &= (1 - p_k)^{q/(q-1)} (D_q(\mathbf{p})^{1-q} - p_k^q)^{1/(1-q)}. \end{aligned}$$

Similarly,

$$D_q(\mathbf{r}') = (1 - r_\ell)^{q/(q-1)} (D_q(\mathbf{r})^{1-q} - r_\ell^q)^{1/(1-q)}.$$

But $p_k = r_\ell$ and $D_q(\mathbf{p}) = D_q(\mathbf{r})$, so $D_q(\mathbf{p}') = D_q(\mathbf{r}')$. This holds for all $q \in Q$, so by inductive hypothesis, $\mathbf{p}'$ is a permutation of $\mathbf{r}'$. It follows that $\mathbf{p}$ is a permutation of $\mathbf{r}$, completing the induction. □

A.4 Affine Functions

Here we prove Lemma 5.1.7, which is restated here for convenience.

Lemma 5.1.7 *Let $\alpha\colon I \to J$ be a function between real intervals. The following are equivalent:*

i. α is affine;
ii. $\alpha(\sum \lambda_i x_i) = \sum \lambda_i \alpha(x_i)$ for all $n \geq 1$, $x_1, \ldots, x_n \in I$ and $\lambda_1, \ldots, \lambda_n \in \mathbb{R}$ such that $\sum \lambda_i = 1$ and $\sum \lambda_i x_i \in I$;
iii. there exist constants $a, b \in \mathbb{R}$ such that $\alpha(x) = ax + b$ for all $x \in I$;
iv. α is continuous and $\alpha(\frac{1}{2}(x_1 + x_2)) = \frac{1}{2}(\alpha(x_1) + \alpha(x_2))$ for all $x_1, x_2 \in I$.

Proof First we assume (i) and prove (ii). By induction,

$$\alpha\left(\sum_{i=1}^{n} p_i x_i\right) = \sum_{i=1}^{n} p_i \alpha(x_i) \tag{A.3}$$

for all $n \geq 1$, $\mathbf{p} \in \Delta_n$, and $\mathbf{x} \in I^n$. Now let $n \geq 1$, $x_1, \ldots, x_n \in I$ and $\lambda_1, \ldots, \lambda_n \in \mathbb{R}$ with $\sum \lambda_i = 1$ and $\sum \lambda_i x_i \in I$. Assume without loss of generality that

$$\lambda_1, \ldots, \lambda_k \geq 0, \quad \lambda_{k+1}, \ldots, \lambda_n < 0$$

for some $k \in \{1, \ldots, n\}$. Write

$$\mu = \sum_{i=1}^{k} \lambda_i = 1 - \sum_{i=k+1}^{n} \lambda_i \geq 1, \qquad w = \sum_{i=1}^{n} \lambda_i x_i \in I.$$

Then

$$\sum_{i=1}^{k} \frac{\lambda_i}{\mu} x_i = \frac{1}{\mu} w + \sum_{i=k+1}^{n} \frac{-\lambda_i}{\mu} x_i.$$

The coefficients $\lambda_1/\mu, \ldots, \lambda_k/\mu$ on the left-hand side are nonnegative and sum to 1, and the same is true of the coefficients $1/\mu, -\lambda_{k+1}/\mu, \ldots, -\lambda_n/\mu$ on the right-hand side. Hence we can apply α throughout and use equation (A.3) on both sides, giving

$$\sum_{i=1}^{k} \frac{\lambda_i}{\mu} \alpha(x_i) = \frac{1}{\mu} \alpha(w) + \sum_{i=k+1}^{n} \frac{-\lambda_i}{\mu} \alpha(x_i).$$

Rearranging gives

$$\alpha(w) = \sum_{i=1}^{n} \lambda_i \alpha(x_i),$$

proving (ii).

Next we assume (ii) and prove (iii). If I is trivial, the result is trivial. Otherwise, we can choose distinct $x_1, x_2 \in I$. Put

$$a = \frac{\alpha(x_2) - \alpha(x_1)}{x_2 - x_1}, \qquad b = \frac{\alpha(x_1)x_2 - \alpha(x_2)x_1}{x_2 - x_1},$$

and define $\alpha' : \mathbb{R} \to \mathbb{R}$ by $\alpha'(x) = ax + b$. We show that $\alpha(x) = \alpha'(x)$ for all $x \in I$. First, this is true when $x \in \{x_1, x_2\}$, by direct calculation. Second, every element of I can be written as $\lambda_1 x_1 + \lambda_2 x_2$ for some $\lambda_1, \lambda_2 \in \mathbb{R}$ with $\lambda_1 + \lambda_2 = 1$. Since both α and α' satisfy (ii), the result follows.

Trivially, (iii) implies (iv).

Finally, assuming (iv), we prove (i). By continuity, it is enough to prove that

$$\alpha(px_1 + (1-p)x_2) = p\alpha(x_1) + (1-p)\alpha(x_2)$$

whenever $x_1, x_2 \in I$ and $p \in [0, 1]$ is a dyadic rational, that is, $p = m/2^n$ for some integers $n \geq 0$ and $0 \leq m \leq 2^n$. We do this by induction on n. It is trivial for $n = 0$. Now let $n \geq 1$ and assume the result for $n - 1$. Let $x_1, x_2 \in I$, let $0 \leq m \leq 2^n$, and assume without loss of generality that $m \leq 2^{n-1}$ (otherwise we can reverse the roles of x_1 and x_2). Then

$$\alpha\left(\frac{m}{2^n}x_1 + \left(1 - \frac{m}{2^n}\right)x_2\right) = \alpha\left(\frac{m}{2^{n-1}} \cdot \frac{1}{2}(x_1 + x_2) + \left(1 - \frac{m}{2^{n-1}}\right)x_2\right) \tag{A.4}$$

$$= \frac{m}{2^{n-1}}\alpha\left(\frac{1}{2}(x_1 + x_2)\right) + \left(1 - \frac{m}{2^{n-1}}\right)\alpha(x_2) \tag{A.5}$$

$$= \frac{m}{2^{n-1}} \cdot \frac{1}{2}(\alpha(x_1) + \alpha(x_2)) + \left(1 - \frac{m}{2^{n-1}}\right)\alpha(x_2) \tag{A.6}$$

$$= \frac{m}{2^n}\alpha(x_1) + \left(1 - \frac{m}{2^n}\right)\alpha(x_2), \tag{A.7}$$

where (A.4) and (A.7) are elementary, (A.5) is by inductive hypothesis, and (A.6) is by (iv). This completes the induction and, therefore, the proof. □

A.5 Diversity of Integer Orders

Here we prove the statement made in Example 6.1.7 on computation of the diversity $D_q^Z(\mathbf{p})$ for integers $q \geq 2$: that in the notation defined there,

$$D_q^Z(\mathbf{p}) = \mu_q^{1/(1-q)}.$$

Indeed, adopting the convention that all sums run over $1, \ldots, n$,

$$\begin{aligned} D_q^Z(\mathbf{p})^{1-q} &= \sum_i p_i \Big(\sum_j Z_{ij} p_j \Big)^{q-1} \\ &= \sum_{i, j_1, \ldots, j_{q-1}} p_i Z_{ij_1} p_{j_1} Z_{ij_2} p_{j_2} \cdots Z_{ij_{q-1}} p_{j_{q-1}} \\ &= \sum_{i_1, i_2, \ldots, i_q} p_{i_1} p_{i_2} \cdots p_{i_q} Z_{i_1 i_2} Z_{i_1 i_3} \cdots Z_{i_1 i_q} \\ &= \mu_q, \end{aligned}$$

as required.

A.6 The Maximum Entropy of a Coupling

Let $\mathbf{p}$ and $\mathbf{r}$ be probability distributions on finite sets $\mathcal{X}$ and $\mathcal{Y}$, respectively. We showed in Remark 8.1.13 that among all distributions on $\mathcal{X} \times \mathcal{Y}$ with marginals $\mathbf{p}$ and $\mathbf{r}$, none has greater entropy than $\mathbf{p} \otimes \mathbf{r}$. In other words,

$$H(P) \leq H(\mathbf{p} \otimes \mathbf{r}) \tag{A.8}$$

for all probability distributions P on $\mathcal{X} \times \mathcal{Y}$ whose marginal distributions are $\mathbf{p}$ and $\mathbf{r}$. It was also claimed there that unless $q = 0$ or $q = 1$, the inequality (A.8) fails when H is replaced by the Rényi entropy H_q or the q-logarithmic entropy S_q. Here we prove this claim.

Since H_q and S_q are increasing, invertible transformations of one another, it suffices to prove it for H_q. And since Rényi entropy is logarithmic (equation (4.14)), the inequality in question can be restated as

$$H_q(P) \leq H_q(\mathbf{p}) + H_q(\mathbf{r}). \tag{A.9}$$

This is true for $q = 0$:

$$\operatorname{supp}(P) \subseteq \operatorname{supp}(\mathbf{p}) \times \operatorname{supp}(\mathbf{r}),$$

so

$$|\operatorname{supp}(P)| \leq |\operatorname{supp}(\mathbf{p})| \cdot |\operatorname{supp}(\mathbf{r})|,$$

giving

$$H_0(P) = \log|\operatorname{supp}(P)| \leq \log|\operatorname{supp}(\mathbf{p})| + \log|\operatorname{supp}(\mathbf{r})| = H_0(\mathbf{p}) + H_0(\mathbf{r}).$$

Our task now is to show that except in the cases $q = 0$ and $q = 1$, the inequality (A.9) is false. Thus, we prove that for each $q \in (0, 1) \cup (1, \infty]$, there exist finite sets $\mathcal{X}$ and $\mathcal{Y}$ and a probability distribution P on $\mathcal{X} \times \mathcal{Y}$ such that

$$H_q(P) > H_q(\mathbf{p}) + H_q(\mathbf{r}),$$

where $\mathbf{p}$ and $\mathbf{r}$ are the marginal distributions of P.

We will treat separately the cases $q \in (0,1)$, $q \in (1,\infty)$, and $q = \infty$. In all cases, we will take $\mathcal{X} = \mathcal{Y} = \{1,\ldots,N\}$ for some N. A probability distribution P on $\mathcal{X}\times\mathcal{Y}$ is then an $N\times N$ matrix of nonnegative real numbers whose entries sum to 1, and its marginals $\mathbf{p}$ and $\mathbf{r}$ are given by the row-sums and column-sums:

$$p_i = \sum_{j=1}^{N} P_{ij}, \qquad r_j = \sum_{i=1}^{N} P_{ij}$$

$(i, j \in \{1,\ldots,N\})$.

First let $q \in (0,1)$. For each $N \geq 2$, define an $N \times N$ matrix P by

$$P = \begin{pmatrix} 1-(N-1)^{(q-1)/q} & 0 & \cdots & 0 \\ 0 & (N-1)^{(-q-1)/q} & \cdots & (N-1)^{(-q-1)/q} \\ \vdots & \vdots & & \vdots \\ 0 & (N-1)^{(-q-1)/q} & \cdots & (N-1)^{(-q-1)/q} \end{pmatrix}.$$

The entries of P sum to 1, and $1-(N-1)^{(q-1)/q} \geq 0$ since $q \in (0,1)$, so $P \in \Delta_{N^2}$. We have

$$\begin{aligned} H_q(P) &= \frac{1}{1-q}\log\Big((1-(N-1)^{(q-1)/q})^q + (N-1)^2(N-1)^{-q-1}\Big) \\ &\geq \frac{1}{1-q}\log((N-1)^{-q+1}) \\ &= \log(N-1). \end{aligned}$$

The marginals of P are

$$\mathbf{p} = \mathbf{r} = \Big(1-(N-1)^{(q-1)/q}, \underbrace{(N-1)^{-1/q},\ldots,(N-1)^{-1/q}}_{N-1}\Big),$$

so

$$\begin{aligned} H_q(\mathbf{p}) = H_q(\mathbf{r}) &= \frac{1}{1-q}\log\Big((1-(N-1)^{(q-1)/q})^q + (N-1)\cdot(N-1)^{-1}\Big) \\ &< \frac{1}{1-q}\log 2. \end{aligned}$$

Hence

$$H_q(P) - (H_q(\mathbf{p}) + H_q(\mathbf{r})) > \log(N-1) - \frac{2}{1-q}\log 2 \to \infty$$

as $N \to \infty$. In particular, $H_q(P) > H_q(\mathbf{p}) + H_q(\mathbf{r})$ when N is sufficiently large.

Now let $q \in (1, \infty)$. For each $N \geq 2$, define an $N \times N$ matrix P by

$$P = \begin{pmatrix} 0 & 1/2(N-1) & \cdots & 1/2(N-1) \\ 1/2(N-1) & 0 & \cdots & 0 \\ \vdots & \vdots & & \vdots \\ 1/2(N-1) & 0 & \cdots & 0 \end{pmatrix}.$$

The entries of P are nonnegative and sum to 1, and

$$H_q(P) = H_q(\mathbf{u}_{2(N-1)}) = \log(2(N-1)) \to \infty$$

as $N \to \infty$. The marginals of P are

$$\mathbf{p} = \mathbf{r} = (1/2, \underbrace{1/2(N-1), \ldots, 1/2(N-1)}_{N-1}),$$

and

$$\begin{aligned} H_q(\mathbf{p}) = H_q(\mathbf{r}) &= \frac{1}{1-q} \log\Big((1/2)^q + (N-1) \cdot (1/2(N-1))^q\Big) \\ &= \frac{1}{1-q} \log((1/2)^q) + \frac{1}{1-q} \log(1 + (N-1)^{1-q}) \\ &\to \frac{1}{1-q} \log((1/2)^q) \end{aligned}$$

as $N \to \infty$, since $q > 1$. Hence

$$H_q(P) - (H_q(\mathbf{p}) + H_q(\mathbf{r})) \to \infty$$

as $N \to \infty$, which again implies that $H_q(P) > H_q(\mathbf{p}) + H_q(\mathbf{r})$ when N is sufficiently large.

Finally, let $q = \infty$. The same matrix P as in the previous case has

$$\begin{aligned} H_\infty(P) &= \log(2(N-1)), \\ H_\infty(\mathbf{p}) = H_\infty(\mathbf{r}) &= \log 2. \end{aligned}$$

Hence

$$H_\infty(P) - (H_\infty(\mathbf{p}) + H_\infty(\mathbf{r})) = \log(2(N-1)) - 2\log 2 \to \infty$$

as $N \to \infty$. Once again, this implies that $H_\infty(P) > H_\infty(\mathbf{p}) + H_\infty(\mathbf{r})$ for sufficiently large N.

A.7 Convex Duality

Here we prove Theorem 9.2.7, which is restated here for convenience.

Theorem 9.2.7 (Legendre–Fenchel) *Let $f\colon \mathbb{R} \to \mathbb{R}$ be a convex function. Then $f^{**} = f$.*

Proof Let $x \in \mathbb{R}$. By definition of convex conjugate,

$$\begin{aligned} f^{**}(x) &= \sup_{\lambda \in \mathbb{R}} (\lambda x - f^*(\lambda)) \\ &= \sup_{\lambda \in \mathbb{R}} \inf_{y \in \mathbb{R}} (\lambda(x-y) + f(y)). \end{aligned} \tag{A.10}$$

In particular,

$$f^{**}(x) \leq \sup_{\lambda \in \mathbb{R}} (\lambda(x-x) + f(x)) = f(x),$$

so it remains to prove that $f^{**}(x) \geq f(x)$. In fact, we will show that there exists $\lambda \in \mathbb{R}$ such that

$$\lambda(x-y) + f(y) \geq f(x) \quad \text{for all } y \in \mathbb{R}. \tag{A.11}$$

By (A.10), this will suffice. Now, a real number λ satisfies (A.11) if and only if

$$\sup_{y \in (-\infty, x)} \frac{f(x) - f(y)}{x - y} \leq \lambda \leq \inf_{z \in (x, \infty)} \frac{f(z) - f(x)}{z - x},$$

so such a λ exists if and only if

$$\frac{f(x) - f(y)}{x - y} \leq \frac{f(z) - f(x)}{z - x} \tag{A.12}$$

for all $y < x < z$. We now prove this. Take y and z such that $y < x < z$. Then $x = py + (1-p)z$ for some $p \in (0,1)$, and the inequality (A.12) to be proved states that

$$\frac{f(x) - f(y)}{(1-p)(z-y)} \leq \frac{f(z) - f(x)}{p(z-y)},$$

or equivalently,

$$f(x) \leq pf(y) + (1-p)f(z).$$

This is true by convexity of f. □

A.8 Cumulant Generating Functions Are Convex

In Section 9.2, we used the fact that the cumulant generating function of any real random variable is convex. Here we prove this.

If we are willing to assume that the cumulant generating function is twice differentiable, then the result can be deduced from the Cauchy–Schwarz inequality, as in Section 5.11 of Grimmett and Stirzaker [130]. But there is no need to make this assumption. Instead, we use a more general standard inequality.

Theorem A.8.1 (Hölder's inequality) *Let Ω be a measure space, let $p, q \in (1, \infty)$ with $1/p + 1/q = 1$, and let $f, g \colon \Omega \to [0, \infty)$ be measurable functions. Then*

$$\int_\Omega fg \leq \left(\int_\Omega f^p\right)^{1/p} \left(\int_\Omega g^q\right)^{1/q}.$$

Here we allow the possibility that one or more of the integrals is ∞.

Proof This is Theorem 6.2 of Folland [108], for instance. □

Corollary A.8.2 *Let X be a real random variable. Then the function*

$$\begin{aligned} \mathbb{R} &\to [0, \infty] \\ \lambda &\mapsto \log \mathbb{E}(e^{\lambda X}) \end{aligned}$$

is convex.

Proof We have to prove that for all $\lambda, \mu \in \mathbb{R}$ and $t \in [0, 1]$,

$$\log \mathbb{E}\left(e^{(t\lambda + (1-t)\mu)X}\right) \leq t \log \mathbb{E}(e^{\lambda X}) + (1 - t) \log \mathbb{E}(e^{\mu X}),$$

or equivalently,

$$\mathbb{E}\left(e^{t\lambda X} e^{(1-t)\mu X}\right) \leq \mathbb{E}(e^{\lambda X})^t \, \mathbb{E}(e^{\mu X})^{1-t}.$$

This is trivial if $t = 0$ or $t = 1$. Supposing otherwise, write $p = 1/t$, $q = 1/(1 - t)$, $U = e^{t\lambda X}$, and $V = e^{(1-t)\mu X}$. Thus, $p, q \in (1, \infty)$ with $1/p + 1/q = 1$, and U and V are nonnegative real random variables on the same sample space. The inequality to be proved is that

$$\mathbb{E}(UV) \leq \mathbb{E}(U^p)^{1/p} \, \mathbb{E}(V^q)^{1/q},$$

which is just Hölder's inequality in probabilistic notation. □

A.9 Functions on a Finite Field

Here we prove Lemma 11.4.1, which is restated here for convenience.

Lemma 11.4.1 *Let K be a finite field with q elements, let $n \geq 0$, and let $F\colon K^n \to K$ be a function. Then there is a unique polynomial f of the form*

$$f(x_1, \ldots, x_n) = \sum_{0 \leq r_1, \ldots, r_n < q} c_{r_1, \ldots, r_n} x_1^{r_1} \cdots x_n^{r_n} \tag{A.13}$$

($c_{r_1, \ldots, r_n} \in K$) such that

$$f(\pi_1, \ldots, \pi_n) = F(\pi_1, \ldots, \pi_n)$$

for all $\pi_1, \ldots, \pi_n \in K$.

This result is standard. For instance, Section 10.3 of Roman [300] gives a proof in the case $n = 1$.

Proof Write $K^{<q}[x_1, \ldots, x_n]$ for the set of polynomials of degree less than q in each variable, that is, of the form (A.13). Write $R(f)\colon K^n \to K$ for the function induced by a polynomial f in n variables. Then R defines a map

$$R\colon K^{<q}[x_1, \ldots, x_n] \to \{\text{functions } K^n \to K\}.$$

We have to prove that R is bijective. Both domain and codomain have q^{q^n} elements, so it suffices to prove that R is surjective.

First define a polynomial δ by

$$\delta(x_1, \ldots, x_n) = (1 - x_1^{q-1}) \cdots (1 - x_n^{q-1}).$$

Then δ has degree $q - 1$ in each variable, and for $a_1, \ldots, a_n \in K$,

$$R(\delta)(a_1, \ldots, a_n) = \begin{cases} 1 & \text{if } a_1 = \cdots = a_n = 0, \\ 0 & \text{otherwise.} \end{cases}$$

Now, given a function $F\colon K^n \to K$, define a polynomial f by

$$f(x_1, \ldots, x_n) = \sum_{a_1, \ldots, a_n \in K} F(a_1, \ldots, a_n) \delta(x_1 - a_1, \ldots, x_n - a_n).$$

Then f has degree at most $q - 1$ in each variable and $R(f) = F$, as required. □

There are other proofs. For instance, one can prove that R is injective rather than surjective, showing by induction on n that its kernel is trivial. I thank Todd Trimble for pointing out to me the Lagrange interpolation argument above.

Appendix B

Summary of Conditions

Here we list the main conditions on means, diversity measures and value measures used in the text. For each condition, we give an abbreviated form of the definition and a reference to the point(s) in the text where it is defined in full.

Weighted Means

The following conditions apply to a sequence $(M\colon \Delta_n \times I^n \to I)_{n\geq 1}$ of functions, where I is a real interval. For the homogeneity and multiplicativity conditions, I is assumed to be closed under multiplication.

Name	Abbreviated form	Definition
Absence-invariant	$M((\ldots, p_{i-1}, 0, p_{i+1}), (\ldots, x_{i-1}, x_i, x_{i+1}, \ldots))$ $= M((\ldots, p_{i-1}, p_{i+1}, \ldots), (\ldots, x_{i-1}, x_{i+1}, \ldots))$	4.2.10
Chain rule	$M(\mathbf{w} \circ (\mathbf{p}^1, \ldots, \mathbf{p}^n), \mathbf{x}^1 \oplus \cdots \oplus \mathbf{x}^n)$ $= M(\mathbf{w}, (M(\mathbf{p}^1, \mathbf{x}^1), \ldots, M(\mathbf{p}^n, \mathbf{x}^n)))$	4.2.23
Consistent	$M(\mathbf{p}, (x, \ldots, x)) = x$	4.2.16
Convex	$M(\mathbf{p}, \frac{1}{2}(\mathbf{x} + \mathbf{y})) \leq \max\{M(\mathbf{p}, \mathbf{x}), M(\mathbf{p}, \mathbf{y})\}$	9.4.1
Homogeneous	$M(\mathbf{p}, c\mathbf{x}) = cM(\mathbf{p}, \mathbf{x})$	4.2.21
Increasing	$\mathbf{x} \leq \mathbf{y} \implies M(\mathbf{p}, \mathbf{x}) \leq M(\mathbf{p}, \mathbf{y})$	4.2.18
Modular	$M(\mathbf{w} \circ (\mathbf{p}^1, \ldots, \mathbf{p}^n), \mathbf{x}^1 \oplus \cdots \oplus \mathbf{x}^n)$ depends only on $\mathbf{w}$ and $M(\mathbf{p}^1, \mathbf{x}^1), \ldots, M(\mathbf{p}^n, \mathbf{x}^n)$	4.2.25
Multiplicative	$M(\mathbf{p} \otimes \mathbf{p}', \mathbf{x} \otimes \mathbf{x}') = M(\mathbf{p}, \mathbf{x}) M(\mathbf{p}', \mathbf{x}')$	4.2.27
Natural	$M(f\mathbf{p}, \mathbf{x}) = M(\mathbf{p}, \mathbf{x}f)$	4.2.12
Quasiarithmetic	$M(\mathbf{p}, \mathbf{x}) = \phi^{-1}(\sum p_i \phi(x_i))$ for some ϕ	5.1.1

Repetition	$M((\dots, p_i, p_{i+1}, \dots), (\dots, x_i, x_i, \dots))$ $= M((\dots, p_i + p_{i+1}, \dots), (\dots, x_i, \dots))$	4.2.10
Strictly increasing	$\mathbf{x} \leq \mathbf{y}$ and $x_i < y_i$ for some $i \in \mathrm{supp}(\mathbf{p})$ $\implies M(\mathbf{p}, \mathbf{x}) < M(\mathbf{p}, \mathbf{y})$	4.2.18
Symmetric	$M(\mathbf{p}, \mathbf{x}) = M(\mathbf{p}\sigma, \mathbf{x}\sigma)$	4.2.10

Unweighted Means

The following conditions apply to a sequence $(M\colon I^n \to I)_{n\geq 1}$ of functions, where I is a real interval. For the homogeneity and multiplicativity conditions, I is assumed to be closed under multiplication.

Name	Abbreviated form	Definition
Consistent	$M(x, \dots, x) = x$	5.2.3
Decomposable	$M(x_1^1, \dots, x_{k_1}^1, \dots, x_1^n, \dots, x_{k_n}^n)$ $= M(a_1, \dots, a_1, \dots, a_n, \dots, a_n)$ where $a_i = M(x_1^i, \dots, x_{k_i}^i)$	5.2.9
Homogeneous	$M(c\mathbf{x}) = cM(\mathbf{x})$	5.2.13
Increasing	$\mathbf{x} \leq \mathbf{y} \implies M(\mathbf{x}) \leq M(\mathbf{y})$	5.2.5
Modular	$M(x_1^1, \dots, x_{k_1}^1, \dots, x_1^n, \dots, x_{k_n}^n)$ depends only on $k_1, \dots, k_n$ and $M(x_1^1, \dots, x_{k_1}^1), \dots, M(x_1^n, \dots, x_{k_n}^n)$	5.2.12
Multiplicative	$M(\mathbf{x} \otimes \mathbf{y}) = M(\mathbf{x})M(\mathbf{y})$	5.2.18
Quasiarithmetic	$M(\mathbf{x}) = \phi^{-1}(\sum \frac{1}{n}\phi(x_i))$ for some ϕ	p. 140
Strictly increasing	$\mathbf{x} \leq \mathbf{y} \neq \mathbf{x} \implies M(\mathbf{x}) < M(\mathbf{y})$	5.2.5
Symmetric	$M(\mathbf{x}) = M(\mathbf{x}\sigma)$	5.2.1

Diversity Measures

The following conditions apply to a sequence $(D\colon \Delta_n \to (0, \infty))_{n\geq 1}$ of functions, that is, a diversity measure for communities modelled as finite probability distributions (without incorporating species similarity).

Name	Abbreviated form	Definition
Absence-invariant	$D(\ldots, p_{i-1}, 0, p_{i+1}, \ldots)$ $= D(\ldots, p_{i-1}, p_{i+1}, \ldots)$	4.4.7
Continuous	$D\colon \Delta_n \to (0,\infty)$ is continuous	4.4.5
Continuous in positive probabilities	$D\colon \Delta_n^\circ \to (0,\infty)$ is continuous	4.4.5
Effective number	$D(\mathbf{u}_n) = n$	2.4.5
Modular	$D(\mathbf{w} \circ (\mathbf{p}^1, \ldots, \mathbf{p}^n))$ depends only on $\mathbf{w}$ and $D(\mathbf{p}^1), \ldots, D(\mathbf{p}^n)$	p. 56, 4.4.14
Modular-monotone	$D(\mathbf{p}^i) \leq D(\widetilde{\mathbf{p}}^i)$ for all $i \implies$ $D(\mathbf{w}\circ(\mathbf{p}^1, \ldots, \mathbf{p}^n)) \leq D(\mathbf{w}\circ(\widetilde{\mathbf{p}}^1, \ldots, \widetilde{\mathbf{p}}^n))$	7.4.1
Multiplicative	$D(\mathbf{p} \otimes \mathbf{r}) = D(\mathbf{p})D(\mathbf{r})$	4.4.16
Normalized	$D(\mathbf{u}_1) = 1$	p. 246
Replication principle	$D(\mathbf{u}_n \otimes \mathbf{p}) = nD(\mathbf{p})$	p. 56, 4.4.18
Symmetric	$D(\mathbf{p}) = D(\mathbf{p}\sigma)$	p. 96

Value Measures

The following conditions apply to a sequence $(\sigma\colon \Delta_n \times (0,\infty)^n \to (0,\infty))_{n\geq 1}$ of functions. Such a sequence is of the same type as a weighted mean on $(0,\infty)$, and the same terminology applies. We also use two further conditions.

Name	Abbreviated form	Definition
Continuous in positive probabilities	$\sigma(-, \mathbf{v})\colon \Delta_n^\circ \to (0,\infty)$ is continuous	7.3.1
Effective number	$\sigma(\mathbf{u}_n, (1, \ldots, 1)) = n$	7.3.2

References

[1] J. Aczél. On mean values. *Bulletin of the American Mathematical Society*, 54(4):392–400, 1948.

[2] J. Aczél. *Lectures on Functional Equations and their Applications*. Academic Press, New York, 1966.

[3] J. Aczél and Z. Daróczy. *On Measures of Information and Their Characterizations*, volume 115 of *Mathematics in Science and Engineering*. Academic Press, New York, 1975.

[4] R. L. Adler, A. G. Konheim, and M. H. McAndrew. Topological entropy. *Transactions of the American Mathematical Society*, 114:309–319, 1965.

[5] J. Aitchison. Simplicial inference. In M. A. G. Viana and D. S. P. Richards, editors, *Algebraic Methods in Statistics and Probability*, volume 287 of *Contemporary Mathematics*, pages 1–22. American Mathematical Society, Providence, RI, 2001.

[6] S. Alesker. Theory of valuations on manifolds: a survey. *Geometric and Functional Analysis*, 17:1321–1341, 2007.

[7] S. Alesker, S. Artstein-Avidan, and V. Milman. Characterization of the Fourier transform and related topics. *Comptes Rendus de l'Académie des Sciences, Paris, Series I, Mathématique*, 346:625–628, 2008.

[8] S. Alesker and J. H. G. Fu. *Integral Geometry and Valuations*. Advanced Courses in Mathematics CRM Barcelona. Birkhäuser, Basel, 2014.

[9] L. Alsedà, J. Llibre, and M. Misiurewicz. *Combinatorial Dynamics and Entropy in Dimension One*, volume 5 of *Advanced Series in Nonlinear Dynamics*. World Scientific, Singapore, 2nd edition, 2000.

[10] S. Amari. Differential geometry of curved exponential families – curvatures and information loss. *Annals of Statistics*, 10(2):357–385, 1982.

[11] S. Amari. A foundation of information geometry. *Electronics and Communications in Japan*, 66-A(6):1–10, 1983.

[12] S. Amari. *Information Geometry and its Applications*, volume 194 of *Applied Mathematical Sciences*. Springer, Tokyo, 2016.

[13] S. Amari and H. Nagaoka. *Methods of Information Geometry*, volume 191 of *Translations of Mathematical Monographs*. Oxford University Press, Oxford, 1993.

[14] M. Ancrenaz, M. Gumal, A. J. Marshall, E. Meijaard, S. A. Wich, and S. Husson. *Pongo pygmaeus. IUCN Red List of Threatened Species* e.T17975A17966347, 2016.

[15] T. M. Apostol. *Mathematical Analysis*. Addison-Wesley, Reading, MA, 1957.

[16] T. M. Apostol. *Introduction to Analytic Number Theory*. Undergraduate Texts in Mathematics. Springer, New York, 1976.

[17] S. Arimoto. Information measures and capacity of order α for discrete memoryless channels. In *Topics in Information Theory: 2nd Colloquium, Keszthely, Hungary, 1975*, pages 41–52. North-Holland, Amsterdam, 1977.

[18] V. I. Arnold. *Mathematical Methods of Classical Mechanics*, volume 60 of *Graduate Texts in Mathematics*. Springer, New York, 2nd edition, 1989.

[19] Y. Asao. Magnitude homology of geodesic metric spaces with an upper curvature bound. *Algebraic & Geometric Topology*, to appear, 2021.

[20] G. Aubrun and I. Nechita. The multiplicative property characterizes ℓ_p and L_p norms. *Confluentes Mathematici*, 3:637–647, 2011.

[21] T. Avery. *Structure and Semantics*. PhD thesis, University of Edinburgh, 2017.

[22] N. Ay, J. Jost, H. V. Lê, and L. Schwachhöfer. *Information Geometry*, volume 64 of *Ergebnisse der Mathematik und ihrer Grenzgebiete*. Springer, Cham, 2017.

[23] J. Baez and J. Dolan. From finite sets to Feynman diagrams. In B. Engquist and W. Schmid, editors, *Mathematics Unlimited – 2001 and Beyond*, pages 29–50. Springer, Berlin, 2001.

[24] J. Baez and T. Fritz. A Bayesian characterization of relative entropy. *Theory and Applications of Categories*, 29:421–456, 2014.

[25] J. Baez, T. Fritz, and T. Leinster. A characterization of entropy in terms of information loss. *Entropy*, 13:1945–1957, 2011.

[26] M. G. Bakker, J. M. Chaparro, D. K. Manter, and J. M. Vivanco. Impacts of bulk soil microbial community structure on rhizosphere microbiomes of *Zea mays*. *Plant Soil*, 392:115–126, 2015.

[27] S. Banach. Sur l'équation fonctionnelle $f(x + y) = f(x) + f(y)$. *Fundamenta Mathematicae*, 1:123–124, 1920.

[28] J. A. Barceló and A. Carbery. On the magnitudes of compact sets in Euclidean spaces. *American Journal of Mathematics*, 140(2):449–494, 2018.

[29] A. R. Barron. Entropy and the central limit theorem. *Annals of Probability*, 14(1):336–342, 1986.

[30] B. H. Barton and E. Moran. Measuring diversity on the Supreme Court with biodiversity statistics. *Journal of Empirical Legal Studies*, 10:1–34, 2013.

[31] M. A. Batanin. The Eckmann–Hilton argument and higher operads. *Advances in Mathematics*, 217:334–385, 2008.

[32] P. Baudot and D. Bennequin. The homological nature of entropy. *Entropy*, 17:3253–3318, 2015.

[33] C. Beck and F. Schlögl. *Thermodynamics of Chaotic Systems: An Introduction*, volume 4 of *Cambridge Nonlinear Science Series*. Cambridge University Press, Cambridge, 1993.

[34] J. Bénabou. Introduction to bicategories. In J. Bénabou, R. Davis, A. Dold, S. Mac Lane, J. Isbell, U. Oberst, and J.-E. Roos, editors, *Reports of the Midwest Category Seminar*, Lecture Notes in Mathematics 47. Springer, Berlin, 1967.

[35] P. Bereziński, B. Jasiul, and M. Szpyrka. An entropy-based network anomaly detection method. *Entropy*, 17:2367–2408, 2015.

[36] C. Berger and T. Leinster. The Euler characteristic of a category as the sum of a divergent series. *Homology, Homotopy and Applications*, 10(1):41–51, 2008.

[37] W. H. Berger and F. L. Parker. Diversity of planktonic Foraminifera in deep-sea sediments. *Science*, 168:1345–1347, 1970.

[38] A. Bhattacharyya. On a measure of divergence between two statistical populations defined by their probability distribution. *Bulletin of the Calcutta Mathematical Society*, 35(1):99–109, 1943.

[39] R. L. Bishop. Elasticities, cross-elasticities, and market relationships. *American Economic Review*, 42(5):780–803, 1952.

[40] R. Blackwell, G. M. Kelly, and A. J. Power. Two-dimensional monad theory. *Journal of Pure and Applied Algebra*, 59:1–41, 1989.

[41] A. Blass and Y. Gurevich. Negative probability. *Bulletin of the European Association for Theoretical Computer Science*, 115:126–142, 2015.

[42] A. Blass and Y. Gurevich. Negative probabilities, II: What they are and what they are for. *Bulletin of the European Association for Theoretical Computer Science*, 125:152–168, 2018.

[43] J. M. Boardman and R. Vogt. *Homotopy Invariant Algebraic Structures on Topological Spaces*, volume 347 of *Lecture Notes in Mathematics*. Springer, Berlin, 1973.

[44] F. Borceux. *Handbook of Categorical Algebra 2: Categories and Structures*, volume 51 of *Encyclopedia of Mathematics and its Applications*. Cambridge University Press, Cambridge, 1994.

[45] E. P. Borges. On a q-generalization of circular and hyperbolic functions. *Journal of Mathematical Physics A: Mathematical and General*, 31:5281–5288, 1998.

[46] A. Borgoo, P. Jaque, A. Toro-Labbé, C. V. Alsenoy, and P. Geerlings. Analyzing Kullback–Leibler information profiles: an indication of their chemical relevance. *Physical Chemistry Chemical Physics*, 11:476–482, 2009.

[47] J. M. Borwein and A. S. Lewis. *Convex Analysis and Nonlinear Optimization: Theory and Examples*. Canadian Mathematical Society Books in Mathematics. Springer, New York, 2000.

[48] A. Boularias, J. Kober, and J. Peters. Relative entropy inverse reinforcement learning. In G. Gordon, D. Dunson, and M. Dudík, editors, *Proceedings of the Fourteenth International Conference on Artificial Intelligence and Statistics, Fort Lauderdale, FL, USA, 11–13 April 2011*, volume 15 of *Proceedings of Machine Learning Research*, pages 182–189. PMLR, Fort Lauderdale, FL, 2011.

[49] G. E. P. Box and D. R. Cox. An analysis of transformations. *Journal of the Royal Statistical Society, Series B (Methodological)*, 26:211–252, 1964.

[50] J. F. Bromaghin, K. D. Rode, S. M. Budge, and G. W. Thiemann. Distance measures and optimization spaces in quantitative fatty acid signature analysis. *Ecology and Evolution*, 5:1249–1262, 2015.

[51] B. Buck and V. A. Macaulay, editors. *Maximum Entropy in Action*. Oxford University Press, Oxford, 1991.

[52] D. Buck and E. Flapan. Predicting knot or catenane type of site-specific recombination products. *Journal of Molecular Biology*, 374:1186–1199, 2007.

[53] D. Buck and E. Flapan. A topological characterization of knots and links arising from site-specific recombination. *Journal of Physics A: Mathematical and Theoretical*, 40:12377–12395, 2007.

[54] A. Buium. Differential characters of abelian varieties over p-adic fields. *Inventiones Mathematicae*, 122:309–340, 1995.

[55] A. Buium. Arithmetic analogues of derivations. *Journal of Algebra*, 198:290–299, 1997.

[56] M. A. Buzas and T. G. Gibson. Species diversity: benthonic Foraminifera in western North Atlantic. *Science*, 163:72–75, 1969.

[57] J. W. Cannon and W. J. Floyd. What is... Thompson's group? *Notices of the American Mathematical Society*, 58(8):1112–1113, 2011.

[58] J. W. Cannon, W. J. Floyd, and W. R. Parry. Introductory notes on Richard Thompson's groups. *L'Enseignement Mathématique*, 42:215–256, 1996.

[59] G. Carlsson. Topology and data. *Bulletin of the American Mathematical Society*, 46(2):255–308, 2009.

[60] J.-L. Cathelineau. Sur l'homologie de SL_2 à coefficients dans l'action adjointe. *Mathematica Scandinavica*, 63:51–86, 1988.

[61] J.-L. Cathelineau. Remarques sur les différentielles des polylogarithmes uniformes. *Annales de l'Institut Fourier*, 46:1327–1347, 1996.

[62] N. N. Čencov. Geometry of the "manifold" of a probability distribution (in Russian). *Doklady Akademii Nauk SSSR*, 158:543–546, 1964.

[63] N. N. Čencov. *Statistical Decision Rules and Optimal Inference*, volume 53 of *Translations of Mathematical Monographs*. American Mathematical Society, Providence, RI, 1982. Translated from the 1972 Russian edition.

[64] R. Cerf and P. Petit. Short proof of Cramér's theorem in $\mathbb{R}$. *American Mathematical Monthly*, pages 925–931, December 2011.

[65] S. R. Chakravarty and W. Eichhorn. An axiomatic characterization of a generalized index of concentration. *Journal of Productivity Analysis*, 2:103–112, 1991.

[66] L. Chalmandrier, T. Münkemüller, S. Lavergne, and W. Thuiller. Effects of species' similarity and dominance on the functional and phylogenetic structure of a plant meta-community. *Ecology*, 96:143–153, 2015.

[67] A. Chao, C.-H. Chiu, and L. Jost. Phylogenetic diversity measures based on Hill numbers. *Philosophical Transactions of the Royal Society B*, 365:3599–3609, 2010.

[68] S. Cho. Quantales, persistence, and magnitude homology. Preprint arXiv:1910.02905, available at arXiv.org, 2019.

[69] J. Chuang, A. King, and T. Leinster. On the magnitude of a finite dimensional algebra. *Theory and Applications of Categories*, 31:63–72, 2016.

[70] K.-S. Chung, W.-S. Chung, S.-T. Nam, and H.-J. Kang. New q-derivative and q-logarithm. *International Journal of Theoretical Physics*, 33:2019–2029, 1994.

[71] T. M. Cover and J. A. Thomas. *Elements of Information Theory*. John Wiley & Sons, New York, 1st edition, 1991.

[72] H. Cramér. *Mathematical Methods of Statistics*. Princeton University Press, Princeton, NJ, 1946.

[73] N. Cressie and T. R. C. Read. Multinomial goodness-of-fit tests. *Journal of the Royal Statistical Society, Series B (Methodological)*, 46(3):440–464, 1984.

[74] I. Csiszár. Generalized cutoff rates and Rényi's information measures. *IEEE Transactions on Information Theory*, 41(1):26–34, 1995.

[75] I. Csiszár. Axiomatic characterizations of information measures. *Entropy*, 10:261–273, 2008.

[76] I. Csiszár and P. C. Shields. Information theory and statistics: a tutorial. *Foundations and Trends in Communications and Information Theory*, 1(4):417–528, 2004.

[77] F. Dahlqvist, V. Danos, I. Garnier, and O. Kammar. Bayesian inversion by ω-complete cone duality. In J. Desharnais and R. Jagadeesan, editors, *27th International Conference on Concurrency Theory (CONCUR 2016)*, pages 1–15. Schloss Dagstuhl–Leibniz-Zentrum für Informatik, Saarbrücken, 2016.

[78] Z. Daróczy. Generalized information functions. *Information and Control*, 16:36–51, 1970.

[79] P.-T. De Boer, D. P. Kroese, S. Mannor, and R. Y. Rubinstein. A tutorial on the cross-entropy method. *Annals of Operations Research*, 134:19–67, 2005.

[80] C. Dellacherie and P.-A. Meyer. *Probabilities and Potential, C: Potential Theory for Discrete and Continuous Semigroups*, volume 151 of *North-Holland Mathematics Studies*. North-Holland, Amsterdam, 2011.

[81] L.-Y. Deng. The cross-entropy method: a unified approach to combinatorial optimization, Monte-Carlo simulation, and machine learning. *Technometrics*, 48:147–148, 2006.

[82] B. Dennis and G. P. Patil. Profiles of diversity. In S. Kotz and N. L. Johnson, editors, *Encyclopedia of Statistical Sciences*, volume 7, pages 292–296. John Wiley, New York, 1986.

[83] P. J. DeVries, D. Murray, and R. Lande. Species diversity in vertical, horizontal and temporal dimensions of a fruit-feeding butterfly community in an Ecuadorian rainforest. *Biological Journal of the Linnean Society*, 62:343–364, 1997.

[84] T. Downarowicz. *Entropy in Dynamical Systems*, volume 18 of *New Mathematical Monographs*. Cambridge University Press, Cambridge, 2011.

[85] R. M. Dudley. *Real Analysis and Probability*, volume 74 of *Cambridge Studies in Advanced Mathematics*. Cambridge University Press, Cambridge, 2002.

[86] S. Eguchi. A differential geometric approach to statistical inference on the basis of contrast functionals. *Hiroshima Mathematical Journal*, 15:341–391, 1985.

[87] S. Eguchi. Geometry of minimum contrast. *Hiroshima Mathematical Journal*, 22:641–647, 1992.

[88] P. Elbaz-Vincent and H. Gangl. On poly(ana)logs I. *Compositio Mathematica*, 130:161–214, 2002.

[89] P. Elbaz-Vincent and H. Gangl. Finite polylogarithms, their multiple analogues and the Shannon entropy. In F. Nielsen and F. Barbaresco, editors, *Geometric Science of Information 2015*, volume 9389 of *Lecture Notes in Computer Science*, pages 277–285. Springer, Cham, 2015.

[90] K. E. Ellingsen. Biodiversity of a continental shelf soft-sediment macrobenthos community. *Marine Ecology Progress Series*, 218:1–15, 2001.

[91] A. M. Ellison. Partitioning diversity. *Ecology*, 91:1962–1963, 2010.

[92] P. Erdős. On the distribution function of additive functions. *Annals of Mathematics*, 47:1–20, 1946.

[93] P. Erdős. On the distribution of additive arithmetical functions and on some related problems. *Rendiconti del Seminario Matematico e Fisico di Milano*, 27:45–49, 1957.

[94] T. Ernst. *A Comprehensive Treatment of q-Calculus*. Birkhäuser, Basel, 2012.

[95] G. Everest and T. Ward. *Heights of Polynomials and Entropy in Algebraic Dynamics*. Universitext. Springer, London, 1999.

[96] D. K. Faddeev. On the concept of entropy of a finite probabilistic scheme (in Russian). *Uspekhi Matematicheskikh Nauk*, 11:227–231, 1956.

[97] D. P. Faith. Conservation evaluation and phylogenetic diversity. *Biological Conservation*, 61:1–10, 1992.

[98] K. Falconer. *Fractal Geometry*. John Wiley & Sons, Chichester, 1990.

[99] A. Feinstein. *Foundations of Information Theory*. McGraw-Hill, New York, 1958.

[100] W. Fenchel. On conjugate convex functions. *Canadian Journal of Mathematics*, 1:73–77, 1949.

[101] E. Fermi. *Thermodynamics*. Dover Books on Physics. Dover, New York, 1956.

[102] C. Fernández-González, C. Palazuelos, and D. Pérez-García. The natural rearrangement invariant structure on tensor products. *Journal of Mathematical Analysis and Applications*, 343:40–47, 2008.

[103] R. P. Feynman. Negative probability. In B. Hiley and D. F. Peat, editors, *Quantum Implications: Essays in Honour of David Bohm*, pages 235–248. Routledge, London, 1987.

[104] M. Fiore and T. Leinster. An abstract characterization of Thompson's group F. *Semigroup Forum*, 80:325–340, 2010.

[105] T. M. Fiore, W. Lück, and R. Sauer. Euler characteristics of categories and homotopy colimits. *Documenta Mathematica*, 16:301–354, 2011.

[106] T. M. Fiore, W. Lück, and R. Sauer. Finiteness obstructions and Euler characteristics of categories. *Advances in Mathematics*, 226:2371–2469, 2011.

[107] J. C. Fodor and J.-L. Marichal. On nonstrict means. *Aequationes Mathematicae*, 54:308–327, 1997.

[108] G. B. Folland. *Real Analysis: Modern Techniques and Their Applications*. John Wiley & Sons, New York, 2nd edition, 1999.

[109] B. Forte and C. T. Ng. On a characterization of the entropies of degree β. *Utilitas Mathematica*, 4:193–205, 1973.

[110] M. Fréchet. Pri la funkcia ekvacio $f(x + y) = f(x) + f(y)$. *L'Enseignement Mathématique*, 15:390–393, 1913.

[111] P. Freyd. Algebraic real analysis. *Theory and Applications of Categories*, 20:215–306, 2008.

[112] B. Fruth, J. R. Hickey, C. André, T. Furuichi, J. Hart, T. Hart, H. Kuehl, F. Maisels, J. Nackoney, G. Reinartz, T. Sop, J. Thompson, and E. A. Williamson. *Pan paniscus* (errata version published in 2016). *IUCN Red List of Threatened Species* e.T15932A102331567, 2016.

[113] S. Furuichi. Uniqueness theorems for Tsallis entropy and Tsallis relative entropy. *IEEE Transactions on Information Theory*, 51:3638–3645, 2005.

[114] P. Gács. Quantum algorithmic entropy. *Journal of Physics A: Mathematical and General*, 34:6859–6880, 2001.

[115] R. Ghrist. Barcodes: the persistent topology of data. *Bulletin of the American Mathematical Society*, 45:61–75, 2008.

[116] R. Ghrist. *Elementary Applied Topology*. Createspace, 2014.

[117] H. Gimperlein and M. Goffeng. On the magnitude function of domains in Euclidean space. *American Journal of Mathematics*, to appear, 2021.

[118] C. Gini. Variabilità e mutabilità. Studi economico-giuridici della facoltà di Giurisprodenza della Regia Università di Cagliari, Anno III, Parte II, 1912.

[119] K. Gomi. Magnitude homology of geodesic space. Preprint arXiv:1902.07044, available at arXiv.org, 2019.

[120] K. Gomi. Smoothness filtration of the magnitude complex. *Forum Mathematicum*, 32(3):625–639, 2020.

[121] I. J. Good. Some terminology and notation in information theory. *Proceedings of the IEE, Part C: Monographs*, 103(3):200–204, 1956.

[122] I. J. Good. Maximum entropy for hypothesis formulation, especially for multidimensional contingency tables. *Annals of Mathematical Statistics*, 34(3):911–934, 1963.

[123] I. J. Good. Diversity as a concept and its measurement: comment. *Journal of the American Statistical Association*, 77(379):561–563, 1982.

[124] M. Gould. *Coherence for Categorified Operadic Theories*. PhD thesis, University of Glasgow, 2008.

[125] D. Govc and R. Hepworth. Persistent magnitude. *Journal of Pure and Applied Algebra*, 225(3):106517, 2021.

[126] M. Grabisch, J.-L. Marichal, R. Mesiar, and E. Pap. *Aggregation Functions*, volume 127 of *Encyclopedia of Mathematics and its Applications*. Cambridge University Press, Cambridge, 2009.

[127] A. Gray. *Tubes*, volume 221 of *Progress in Mathematics*. Springer, Basel, 2nd edition, 2004.

[128] R. Greco, A. Di Nardo, and G. Aantonastaso. Resilience and entropy as indices of robustness of water distribution networks. *Journal of Hydroinformatics*, 14:761–771, 2012.

[129] M. Grendár and R. K. Niven. The Pólya information divergence. *Information Sciences*, 180:4189–4194, 2010.

[130] G. Grimmett and D. Stirzaker. *Probability and Random Processes*. Oxford University Press, Oxford, 3rd edition, 2001.

[131] M. Gromov. *Metric Structures for Riemannian and Non-Riemannian Spaces*. Birkhäuser, Boston, MA, 2001.

[132] M. Gromov. In a search for a structure, part 1: On entropy. Preprint, 2012.

[133] Y. Gu. Graph magnitude homology via algebraic Morse theory. Preprint arXiv:1809.07240, available at arXiv.org, 2018.

[134] H. Hadwiger. *Vorlesungen über Inhalt, Oberfläche und Isoperimetrie*. Springer, Berlin, 1957.

[135] T. C. Hales. The NSA back door to NIST. *Notices of the American Mathematical Society*, 61(2):190–192, 2014.

[136] L. Hannah and J. Kay. *Concentration in the Modern Industry: Theory, Measurement, and the U.K. Experience*. MacMillan, London, 1977.

[137] G. Hardy, J. E. Littlewood, and G. Pólya. *Inequalities*. Cambridge University Press, Cambridge, 2nd edition, 1952.

[138] J. Harte. *Maximum Entropy and Ecology: A Theory of Abundance, Distribution, and Energetics.* Oxford Series in Ecology and Evolution. Oxford University Press, Oxford, 2011.

[139] T. Haszpra and T. Tél. Topological entropy: a Lagrangian measure of the state of the free atmosphere. *Journal of the Atmospheric Sciences*, 70:4030–4040, 2013.

[140] A. Hatcher. *Algebraic Topology*. Cambridge University Press, Cambridge, 2002.

[141] J. Havrda and F. Charvát. Quantification method of classification processes: concept of structural α-entropy. *Kybernetika*, 3:30–35, 1967.

[142] M. Hennessy and R. Milner. Algebraic laws for nondeterminism and concurrency. *Journal of the Association for Computing Machinery*, 32(1):137–161, 1985.

[143] R. Hepworth. Magnitude cohomology. Preprint arXiv:1807.06832, available at arXiv.org, 2018.

[144] R. Hepworth and S. Willerton. Categorifying the magnitude of a graph. *Homology, Homotopy and Applications*, 19(2):31–60, 2017.

[145] J. Hey. The mind of the species problem. *Trends in Ecology and Evolution*, 16(7):326–329, 2001.

[146] M. O. Hill. Diversity and evenness: a unifying notation and its consequences. *Ecology*, 54(2):427–432, 1973.

[147] A. Hobson. A new theorem of information theory. *Journal of Statistical Physics*, 1(3):383–391, 1969.

[148] D. A. Huffman. A method for the construction of minimum-redundancy codes. *Proceedings of the IRE*, 40:1098–1101, 1952.

[149] T. Humle, F. Maisels, J. F. Oates, A. Plumptre, and E. A. Williamson. *Pan troglodytes* (errata version published in 2016). *IUCN Red List of Threatened Species* e.T15933A102326672, 2016.

[150] S. H. Hurlbert. The nonconcept of species diversity: a critique and alternative parameters. *Ecology*, 52(4):577–586, 1971.

[151] A. R. Ives. Diversity and stability in ecological communities. In R. M. May and A. R. McLean, editors, *Theoretical Ecology: Principles and Applications.* Oxford University Press, Oxford, 2007.

[152] J. Izsák and L. Szeidl. Quadratic diversity: its maximization can reduce the richness of species. *Environmental and Ecological Statistics*, 9:423–430, 2002.

[153] P. Jaccard. Nouvelles recherches sur la distribution florale. *Bulletin de la Société Vaudoise des Sciences Naturelles*, 40:223–270, 1908.

[154] A. Jackson. *Comme appelé du néant* – as if summoned from the void: the life of Alexandre Grothendieck. *Notices of the American Mathematical Society*, 51(9):1038–1056, 2004.

[155] F. H. Jackson. On q-functions and a certain difference operator. *Transactions of the Royal Society of Edinburgh*, 46:253–281, 1908.

[156] E. T. Jaynes. Where do we stand on maximum entropy? In R. D. Levine and M. Tribus, editors, *The Maximum Entropy Formalism*, pages 15–118. MIT Press, Cambridge, MA, 1979.

[157] E. T. Jaynes. *Probability Theory: The Logic of Science.* Cambridge University Press, Cambridge, 2003.

[158] H. Jeffreys. *Theory of Probability.* Clarendon Press, Oxford, 1st edition, 1939.

[159] H. Jeffreys. An invariant form for the prior probability in estimation problems. *Proceedings of the Royal Society of London, Series A, Mathematical and Physical Sciences*, 186:453–461, 1946.

[160] A. Jeziorski, A. J. Tanentzap, N. D. Yan, A. M. Paterson, M. E. Palmer, J. B. Korosi, J. A. Rusak, M. T. Arts, W. Keller, R. Ingram, A. Cairns, and J. P. Smol. The jellification of north temperate lakes. *Proceedings of the Royal Society B*, 282:20142449, 2015.

[161] J. L. Johnson. Use of nucleic-acid homologies in taxonomy of anaerobic bacteria. *International Journal of Systematic Bacteriology*, 23(4):308–315, 1973.

[162] O. Johnson. *Information Theory and the Central Limit Theorem*. Imperial College Press, London, 2004.

[163] P. T. Johnstone. *Topos Theory*. Academic Press, London, 1977.

[164] G. A. Jones and J. M. Jones. *Information and Coding Theory*. Springer Undergraduate Mathematics Series. Springer, London, 2000.

[165] J. Jost. *Riemannian Geometry and Geometric Analysis*. Universitext. Springer, Berlin, 7th edition, 2017.

[166] L. Jost. Entropy and diversity. *Oikos*, 113:363–375, 2006.

[167] L. Jost. Partitioning diversity into independent alpha and beta components. *Ecology*, 88(10):2427–2439, 2007.

[168] L. Jost. G_{ST} and its relatives do not measure differentiation. *Molecular Ecology*, 17:4015–4026, 2008.

[169] L. Jost. Mismeasuring biological diversity: response to Hoffmann and Hoffmann (2008). *Ecological Economics*, 68:925–928, 2009.

[170] L. Jost. Independence of alpha and beta diversities. *Ecology*, 91:1969–1974, 2010.

[171] L. Jost, P. DeVries, T. Walla, H. Greeney, A. Chao, and C. Ricotta. Partitioning diversity for conservation analyses. *Diversity and Distributions*, 16:65–76, 2010.

[172] A. Joyal, M. Nielsen, and G. Winskel. Bisimulation from open maps. *Information and Computation*, 127:164–185, 1996.

[173] J. M. Joyce. Kullback–Leibler divergence. In M. Lovric, editor, *International Encyclopedia of Statistical Science*, pages 720–722. Springer, Berlin, 2011.

[174] B. Jubin. On the magnitude homology of metric spaces. Preprint arXiv:1803.05062, available at arXiv.org, 2018.

[175] V. Kac and P. Cheung. *Quantum Calculus*. Universitext. Springer, New York, 2002.

[176] R. Kaneta and M. Yoshinaga. Magnitude homology of metric spaces and order complexes. Preprint arXiv:1803.04247, available at arXiv.org, 2018.

[177] P. Kannappan and C. T. Ng. Measurable solutions of functional equations related to information theory. *Proceedings of the American Mathematical Society*, 38:303–310, 1973.

[178] P. Kannappan and C. T. Ng. On functional equations connected with directed divergence, inaccuracy and generalized directed divergence. *Pacific Journal of Mathematics*, 54(1):157–167, 1974.

[179] P. Kannappan and P. N. Rathie. On a characterization of directed divergence. *Information and Control*, 22:163–171, 1973.

[180] M. Kanter. Discrimination distance bounds and statistical applications. *Probability Theory and Related Fields*, 86:403–422, 1990.

[181] R. M. Karp. Reducibility among combinatorial problems. In R. E. Miller and J. W. Thatcher, editors, *Complexity of Computer Computations*, pages 85–103. Plenum Press, New York, 1972.

[182] R. E. Kass and L. Wasserman. The selection of prior distributions by formal rules. *Journal of the American Statistical Association*, 91:1343–1370, 1996.

[183] I. Kátai. A remark on additive arithmetical functions. *Annales Universitatis Scientiarum Budapestinensis de Rolando Eötvös Nominatae, Sectio Mathematica*, 10:81–83, 1967.

[184] G. M. Kelly. *Basic Concepts of Enriched Category Theory*, volume 64 of *London Mathematical Society Lecture Note Series*. Cambridge University Press, Cambridge, 1982. Also *Reprints in Theory and Applications of Categories* 10:1–136, 2005.

[185] G. M. Kelly. On the operads of J. P. May. *Reprints in Theory and Applications of Categories*, 13:1–13, 2005.

[186] D. F. Kerridge. Inaccuracy and inference. *Journal of the Royal Statistical Society, Series B (Methodological)*, 23:184–194, 1961.

[187] C. J. Keylock. Simpson diversity and the Shannon–Wiener index as special cases of a generalized entropy. *Oikos*, 109(1):203–207, 2005.

[188] A. I. Khinchin. *Mathematical Foundations of Information Theory*. Dover, New York, 1957.

[189] M. Khovanov. A categorification of the Jones polynomial. *Duke Mathematical Journal*, 101:359–426, 2000.

[190] D. A. Klain and G.-C. Rota. *Introduction to Geometric Probability*. Lezioni Lincee. Cambridge University Press, Cambridge, 1997.

[191] J. Kock. *Frobenius Algebras and 2D Topological Quantum Field Theories*, volume 59 of *London Mathematical Society Student Texts*. Cambridge University Press, Cambridge, 2003.

[192] A. N. Kolmogorov. Sur la notion de la moyenne. *Atti della Accademia Nazionale dei Lincei, Rendiconti, VI Serie*, 12:388–391, 1930.

[193] A. N. Kolmogorov. On certain asymptotic characteristics of completely bounded metric spaces. *Doklady Akademii Nauk SSSR*, 108:385–388, 1956.

[194] A. N. Kolmogorov. On the notion of mean. In V. M. Tikhomirov, editor, *Selected Works of A. N. Kolmogorov*, volume I: *Mathematics and Mechanics*, volume 25 of *Mathematics and Its Applications (Soviet Series)*, pages 144–146. Springer Science and Business Media, Dordrecht, 1991.

[195] M. Kontsevich. The $1\frac{1}{2}$-logarithm. Private note, 1995. Reprinted as appendix of [88].

[196] M. Kontsevich. Operads and motives in deformation quantization. *Letters in Mathematical Physics*, 48(1):35–72, 1999.

[197] M. Kovačević, I. Stanojević, and V. Šenk. On the entropy of couplings. *Information and Computation*, 242:369–382, 2015.

[198] S. Kullback. *Information Theory and Statistics*. John Wiley & Sons, New York, 1959.

[199] S. Kullback. The Kullback–Leibler distance. *American Statistician*, 41(4):340–341, 1987.

[200] M. Laakso and R. Taagepera. "Effective" number of parties: a measure with application to West Europe. *Comparative Political Studies*, 12:3–27, 1979.

[201] J. Lambek. Deductive systems and categories II: standard constructions and closed categories. In P. Hilton, editor, *Category Theory, Homology Theory and their Applications, I (Battelle Institute Conference, Seattle, 1968)*, volume 86 of *Lecture Notes in Mathematics*. Springer, Berlin, 1969.

[202] S. L. Lauritzen. Statistical manifolds. In S.-I. Amari, O. E. Barndorff-Nielsen, R. E. Kass, S. L. Lauritzen, and C. R. Rao, editors, *Differential Geometry in Statistical Inference*, volume 10 of *Lecture Notes – Monograph Series*, pages 163–216. Institute of Mathematical Statistics, Hayward, CA, 1987.

[203] F. W. Lawvere. *Functorial Semantics of Algebraic Theories*. PhD thesis, Columbia University, 1963. Also *Reprints in Theory and Applications of Categories* 5:1–121, 2004.

[204] F. W. Lawvere. Metric spaces, generalized logic and closed categories. *Rendiconti del Seminario Matematico e Fisico di Milano*, XLIII:135–166, 1973. Also *Reprints in Theory and Applications of Categories* 1:1–37, 2002.

[205] F. W. Lawvere. State categories, closed categories, and the existence semi-continuous entropy functions. IMA Preprint Series 86, Institute for Mathematics and Its Applications, University of Minnesota, Minneapolis, MN, 1984.

[206] P. M. Lee. On the axioms of information theory. *Annals of Mathematical Statistics*, 35:415–418, 1964.

[207] T. Leinster. *Higher Operads, Higher Categories*, volume 298 of *London Mathematical Society Lecture Notes Series*. Cambridge University Press, Cambridge, 2004.

[208] T. Leinster. Are operads algebraic theories? *Bulletin of the London Mathematical Society*, 38:233–238, 2006.

[209] T. Leinster. General self-similarity: an overview. In L. Paunescu, A. Harris, T. Fukui, and S. Koike, editors, *Real and Complex Singularities*. World Scientific, Singapore, 2007.

[210] T. Leinster. The Euler characteristic of a category. *Documenta Mathematica*, 13:21–49, 2008.

[211] T. Leinster. A maximum entropy theorem with applications to the measurement of biodiversity. Preprint arXiv:0910.0906, available at arXiv.org, 2009.

[212] T. Leinster. A general theory of self-similarity. *Advances in Mathematics*, 226:2935–3017, 2011.

[213] T. Leinster. Integral geometry for the 1-norm. *Advances in Applied Mathematics*, 49:81–96, 2012.

[214] T. Leinster. A multiplicative characterization of the power means. *Bulletin of the London Mathematical Society*, 44:106–112, 2012.

[215] T. Leinster. Notions of Möbius inversion. *Bulletin of the Belgian Mathematical Society*, 19:911–935, 2012.

[216] T. Leinster. The magnitude of metric spaces. *Documenta Mathematica*, 18:857–905, 2013.

[217] T. Leinster. The magnitude of a graph. *Mathematical Proceedings of the Cambridge Philosophical Society*, 166:247–264, 2019.

[218] T. Leinster. A short characterization of relative entropy. *Journal of Mathematical Physics*, 60(2):023302, 2019.

[219] T. Leinster. Entropy modulo a prime. *Communications in Number Theory and Physics*, to appear, 2021.

[220] T. Leinster and C. A. Cobbold. Measuring diversity: the importance of species similarity. *Ecology*, 93:477–489, 2012.

[221] T. Leinster and M. Meckes. Maximizing diversity in biology and beyond. *Entropy*, 18(88), 2016.

[222] T. Leinster and M. Meckes. The magnitude of a metric space: from category theory to geometric measure theory. In N. Gigli, editor, *Measure Theory in Non-Smooth Spaces*, pages 156–193. De Gruyter Open, Warsaw, 2017.

[223] T. Leinster and E. Roff. The maximum entropy of a metric space. *Quarterly Journal of Mathematics*, to appear, 2021.

[224] T. Leinster and M. Shulman. Magnitude homology of enriched categories and metric spaces. *Algebraic & Geometric Topology*, to appear, 2021.

[225] T. Leinster and S. Willerton. On the asymptotic magnitude of subsets of Euclidean space. *Geometriae Dedicata*, 164:287–310, 2013.

[226] P. Lemey, M. Salemi, and A.-M. Vandamme. *The Phylogenetic Handbook*. Cambridge University Press, Cambridge, 2nd edition, 2009.

[227] M. Lesnick. Studying the shape of data using topology. *The Institute Letter*, pages 10–11, Summer 2013. Institute for Advanced Study, Princeton, NJ.

[228] Z. Lin. Are groups algebras over an operad? Mathematics Stack Exchange, 2013. Available at https://math.stackexchange.com/q/366371.

[229] J. Lindhard and V. Nielsen. Studies in statistical dynamics. *Matematisk-Fysiske Meddelelser: Kongelige Danske Videnskabernes Selskab*, 38:1–42, 1971.

[230] D. V. Lindley. *Making Decisions*. Wiley, London, 2nd edition, 1985.

[231] D. V. Lindley. *Understanding Uncertainty*. Wiley, Hoboken, NJ, 2006.

[232] Y. V. Linnik. An information-theoretic proof of the central limit theorem with the Lindeberg condition. *Theory of Probability and its Applications*, 4(3):288–299, 1959.

[233] J. E. Littlewood. *Lectures on the Theory of Functions*. Oxford University Press, London, 1944.

[234] J.-L. Loday and B. Vallette. *Algebraic Operads*, volume 346 of *Grundlehren der Mathematischen Wissenschaften*. Springer, Berlin, 2012.

[235] N. Lusin. Sur les propriétés des fonctions mesurables. *Comptes Rendus Hebdomadaires des Séances de l'Académie des Sciences*, 154:1688–1690, 1912.

[236] S. Mac Lane. *Categories for the Working Mathematician*. Graduate Texts in Mathematics 5. Springer, New York, 1971.

[237] R. H. MacArthur. Patterns of species diversity. *Biological Reviews*, 40:510–533, 1965.

[238] D. J. C. MacKay. *Information Theory, Inference and Learning Algorithms*. Cambridge University Press, Cambridge, 2003.

[239] M. C. Mackey and P. K. Maini. What has mathematics done for biology? *Bulletin of Mathematical Biology*, 77:735–738, 2015.

[240] A. E. Magurran. *Measuring Biological Diversity*. Blackwell, Oxford, 2004.

[241] F. Maisels, R. A. Bergl, and E. A. Williamson. *Gorilla gorilla* (errata version published in 2016). *IUCN Red List of Threatened Species* e.T9404A102330408, 2016.

[242] E. Manes. *Algebraic Theories*, volume 26 of *Graduate Texts in Mathematics*. Springer, Berlin, 1976.

[243] T. Manke, L. Demetrius, and M. Vingron. An entropic characterization of protein interaction networks and cellular robustness. *Journal of the Royal Society Interface*, 3:843–850, 2006.

[244] M. Markl, S. Shnider, and J. Stasheff. *Operads in Algebra, Topology and Physics*, volume 96 of *Mathematical Surveys and Monographs*. American Mathematical Society, Providence, RI, 2002.

[245] A. Máté. A new proof of a theorem of P. Erdős. *Proceedings of the American Mathematical Society*, 18:159–162, 1967.

[246] A. E. Mather, L. Matthews, D. J. Mellor, R. Reeve, M. J. Denwood, P. Boerlin, R. J. Reid-Smith, D. J. Brown, J. E. Coia, L. M. Browning, D. T. Haydon, and S. W. J. Reid. An ecological approach to assessing the epidemiology of antimicrobial resistance in animal and human populations. *Proceedings of the Royal Society B*, 279:1630–1639, 2012.

[247] J. P. May. *The Geometry of Iterated Loop Spaces*, volume 271 of *Lecture Notes in Mathematics*. Springer, Berlin, 1972.

[248] J. P. May. Definitions: operads, algebras and modules. In J.-L. Loday, J. Stasheff, and A. Voronov, editors, *Operads: Proceedings of Renaissance Conferences*, volume 202 of *Contemporary Mathematics*. American Mathematical Society, Providence, RI, 1997.

[249] R. L. Mayden. A hierarchy of species concepts: the denouement in the saga of the species problem. In M. F. Claridge, H. A. Dawah, and M. R. Wilson, editors, *Species: The Units of Biodiversity*, pages 381–424. Chapman & Hall, London, 1997.

[250] M. W. Meckes. Positive definite metric spaces. *Positivity*, 17:733–757, 2013.

[251] M. W. Meckes. Magnitude, diversity, capacities, and dimensions of metric spaces. *Potential Analysis*, 42:549–572, 2015.

[252] M. W. Meckes. On the magnitude and intrinsic volumes of a convex body in Euclidean space. *Mathematika*, 66:343–355, 2020.

[253] R. S. Mendes, L. R. Evangelista, S. M. Thomaz, A. A. Agostinho, and L. C. Gomes. A unified index to measure ecological diversity and species rarity. *Ecography*, 31:450–456, 2008.

[254] R. A. Mittermeier, J. U. Ganzhorn, W. R. Konstant, K. Glander, I. Tattersall, C. P. Groves, A. B. Rylands, A. Hapke, J. Ratsimbazafy, M. I. Mayor, E. E. Louis, Jr., Y. Rumpler, C. Schwitzer, and R. M. Rasoloarison. Lemur diversity in Madagascar. *International Journal of Primatology*, 29:1607–1656, 2008.

[255] J.-M. Morvan. *Generalized Curvatures*. Springer, Berlin, 2008.

[256] T. S. Motzkin and E. G. Straus. Maxima for graphs and a new proof of a theorem of Turán. *Canadian Journal of Mathematics*, 17:533–540, 1965.

[257] H. J. Moulin. *Fair Division and Collective Welfare*. MIT Press, Cambridge, MA, 2003.

[258] M. Mureşan. *A Concrete Approach to Classical Analysis*. CMS Books in Mathematics. Springer, New York, 2009.

[259] H. Nagendra. Opposite trends in response for the Shannon and Simpson indices of landscape diversity. *Applied Geography*, 22:175–186, 2002.

[260] M. Nagumo. Über eine Klasse der Mittelwerte. *Japanese Journal of Mathematics*, 7:71–79, 1930.

[261] A. Nater, M. P. Mattle-Greminger, A. Nurcahyo, M. G. Nowak, M. de Manuel, T. Desai, C. Groves, M. Pybus, T. Bilgin Sonay, C. Roos, A. R. Lameira, S. A. Wich, J. Askew, M. Davila-Ross, G. Fredriksson, G. de Valles, F. Casals, J. Prado-Martinez, B. Goossens, E. J. Verschoor, K. S. Warren, I. Singleton, D. A. Marques, J. Pamungkas, D. Perwitasari-Farajallah, P. Rianti, A. Tuuga, I. G. Gut, M. Gut, P. Orozco-terWengel, C. P. van Schaik, J. Bertranpetit, M. Anisimova, A. Scally, T. Marques-Bonet, E. Meijaard, and M. Krützen. Morphometric, behavioral, and genomic evidence for a new orangutan species. *Current Biology*, 27(22):3487–3498, 2017.

[262] S. Nee. More than meets the eye. *Nature*, 429:804–805, 2004.

[263] M. Nicolau, A. J. Levine, and G. Carlsson. Topology based data analysis identifies a subgroup of breast cancers with a unique mutational profile and excellent survival. *Proceedings of the National Academy of Sciences of the USA*, 108(17):7265–7270, 2011.

[264] K. Noguchi. The Euler characteristic of acyclic categories. *Kyushu Journal of Mathematics*, 65:85–99, 2011.

[265] K. Noguchi. Euler characteristics of categories and barycentric subdivision. *Münster Journal of Mathematics*, 6:85–116, 2013.

[266] K. Noguchi. The zeta function of a finite category. *Documenta Mathematica*, 18:1243–1274, 2013.

[267] M. Nygaard and G. Winskel. Domain theory for concurrency. *Theoretical Computer Science*, 316:153–190, 2004.

[268] OECD. *Handbook of Biodiversity Valuation: A Guide for Policy Makers*. Organisation for Economic Co-operation and Development, Paris, 2002.

[269] N. Otter. Magnitude meets persistence. Homology theories for filtered simplicial sets. Preprint arXiv:1807.01540, available at arXiv.org, 2018.

[270] W. Parry. *Entropy and Generators in Ergodic Theory*. W. A. Benjamin, New York, 1969.

[271] G. P. Patil. Diversity profiles. In A. H. El-Shaarawi and W. W. Piegorsch, editors, *Encyclopedia of Environmetrics*. John Wiley & Sons, Chichester, 2002.

[272] G. P. Patil and C. Taillie. A study of diversity profiles and orderings for a bird community in the vicinity of Colstrip, Montana. In G. P. Patil and M. Rosenzweig, editors, *Contemporary Quantitative Ecology and Related Ecometrics*, pages 23–48. International Co-operative Publishing House, Fairland, MD, 1979.

[273] G. P. Patil and C. Taillie. Diversity as a concept and its measurement. *Journal of the American Statistical Association*, 77(379):548–561, 1982.

[274] D. Pavlovic. Quantitative concept analysis. In F. Domenach, D. I. Ignatov, and J. Poelmans, editors, *Formal Concept Analysis. ICFCA 2012*, volume 7278 of *Lecture Notes in Computer Science*, pages 260–277. Springer, Berlin, 2012.

[275] S. Pavoine and M. B. Bonsall. Biological diversity: distinct distributions can lead to the maximization of Rao's quadratic entropy. *Theoretical Population Biology*, 75:153–163, 2009.

[276] S. Pavoine, S. Ollier, and D. Pontier. Measuring diversity from dissimilarities with Rao's quadratic entropy: Are any dissimilarities suitable? *Theoretical Population Biology*, 67:231–239, 2005.

[277] R. K. Peet. The measurement of species diversity. *Annual Review of Ecology and Systematics*, 5:285–307, 1974.

[278] X. Pennec. Barycentric subspace analysis on manifolds. *Annals of Statistics*, 46:2711–2746, 2018.

[279] A. Peres, P. F. Scudo, and D. R. Terno. Quantum entropy and special relativity. *Physical Review Letters*, 88(23):230402, 2002.

[280] O. L. Petchey and K. J. Gaston. Functional diversity: back to basics and looking forward. *Ecology Letters*, 9:741–758, 2006.

[281] D. Petz. Characterization of the relative entropy of states of matrix algebras. *Acta Mathematica Hungarica*, 59:449–455, 1992.

[282] E. C. Pielou. *Ecological Diversity*. John Wiley & Sons, New York, 1975.

[283] E. C. Pielou. *Mathematical Ecology*. John Wiley & Sons, New York, 2nd edition, 1977.

[284] A. Plumptre, M. Robbins, and E. A. Williamson. *Gorilla beringei* (errata version published in 2016). *IUCN Red List of Threatened Species* e.T39994A102325702, 2016.

[285] C. R. Rao. Information and the accuracy attainable in the estimation of statistical parameters. *Bulletin of the Calcutta Mathematical Society*, 37:81–89, 1945.

[286] C. R. Rao. Diversity and dissimilarity coefficients: a unified approach. *Theoretical Population Biology*, 21:24–43, 1982.

[287] C. R. Rao. Diversity: its measurement, decomposition, apportionment and analysis. *Sankhyā: The Indian Journal of Statistics*, 44(1):1–22, 1982.

[288] C. R. Rao. Differential metrics in probability spaces. In *Differential Geometry in Statistical Inference*, volume 10 of *Lecture Notes – Monograph Series*, pages 217–240. Institute of Mathematical Statistics, Hayward, CA, 1987.

[289] P. N. Rathie and P. Kannappan. A directed-divergence function of type β. *Information and Control*, 20:38–45, 1972.

[290] A. Ratnaparkhi. Learning to parse natural language with maximum entropy models. *Machine Learning*, 34:151–175, 1999.

[291] M. C. Reed. Mathematical biology is good for mathematics. *Notices of the American Mathematical Society*, 62(10):1172–1176, 2015.

[292] D. Reem. Remarks on the Cauchy functional equation and variations of it. *Aequationes Mathematicae*, 91:237–264, 2017.

[293] R. Reeve, T. Leinster, C. A. Cobbold, J. Thompson, N. Brummitt, S. N. Mitchell, and L. Matthews. How to partition diversity. Preprint arXiv:1404.6520v3, available at arXiv.org, 2016.

[294] A. Rényi. On measures of entropy and information. In J. Neyman, editor, *Proceedings of the 4th Berkeley Symposium on Mathematical Statistics and Probability*, volume 1, pages 547–561. University of California Press, Berkeley, CA, 1961.

[295] A. Rényi. *Probability Theory*. North-Holland, Amsterdam, 1970.

[296] C. Ricotta and L. Szeidl. Towards a unifying approach to diversity measures: bridging the gap between the Shannon entropy and Rao's quadratic index. *Theoretical Population Biology*, 70:237–243, 2006.

[297] E. Riehl. *Category Theory in Context*. Dover, New York, 2016.

[298] C. P. Robert, N. Chopin, and J. Rousseau. Harold Jeffreys's *Theory of Probability* revisited. *Statistical Science*, 24:141–172, 2009.

[299] R. T. Rockafellar. *Convex Analysis*. Princeton Mathematical Series. Princeton University Press, Princeton, NJ, 1970.

[300] S. Roman. *Field Theory*, volume 158 of *Graduate Texts in Mathematics*. Springer, New York, 2nd edition, 2006.

[301] G.-C. Rota. On the foundations of combinatorial theory I: theory of Möbius functions. *Zeitschrift für Wahrscheinlichkeitstheorie und Verwandte Gebiete*, 2:340–368, 1964.

[302] R. D. Routledge. Diversity indices: which ones are admissible? *Journal of Theoretical Biology*, 76:503–515, 1979.

[303] I. Sason. Entropy bounds for discrete random variables via maximal coupling. *IEEE Transactions on Information Theory*, 59(11):7118–7131, 2013.

[304] S. H. Schanuel. Negative sets have Euler characteristic and dimension. In A. Carboni, M. C. Pedicchio, and G. Rosolini, editors, *Category Theory (Como, 1990)*, Lecture Notes in Mathematics 1488, pages 379–385. Springer, Berlin, 1991.

[305] M. J. Schervish. *Theory of Statistics*. Springer Series in Statistics. Springer, New York, 1995.

[306] R. Schneider. *Convex Bodies: the Brunn–Minkowski Theory*. Encyclopedia of Mathematics and its Applications 44. Cambridge University Press, Cambridge, 2nd edition, 2014.

[307] I. J. Schoenberg. Metric spaces and positive definite functions. *Transactions of the American Mathematical Society*, 44:522–536, 1938.

[308] G. Segal. Classifying spaces and spectral sequences. *Institut des Hautes Études Scientifiques Publications Mathématiques*, 34:105–112, 1968.

[309] C. E. Shannon. A mathematical theory of communication. *Bell System Technical Journal*, 27:379–423, 1948.

[310] C. E. Shannon. Prediction and entropy of printed English. *Bell System Technical Journal*, 30:50–64, 1951.

[311] M. Shiino. H-theorem with generalized relative entropies and the Tsallis statistics. *Journal of the Physical Society of Japan*, 67:3658–3660, 1998.

[312] K. Shimatani. The appearance of a different DNA sequence may decrease nucleotide diversity. *Journal of Molecular Evolution*, 49:810–813, 1999.

[313] J. E. Shore and R. W. Johnson. Axiomatic derivation of the principle of maximum entropy and the principle of minimum cross-entropy. *IEEE Transactions on Information Theory*, 26(1):26–37, 1980.

[314] E. H. Simpson. Measurement of diversity. *Nature*, 163:688, 1949.

[315] I. Singleton, S. A. Wich, M. Nowak, and G. Usher. *Pongo abelii* (errata version published in 2016). *IUCN Red List of Threatened Species* e.T39780A102329901, 2016.

[316] V. V. Skachkov, V. V. Chepkyi, H. D. Bratchenko, and A. N. Efymchykov. Entropy approach to the investigation of information capabilities of adaptive radio engineering system in conditions of intrasystem uncertainty. *Radioelectronics and Communications Systems*, 58:241–249, 2015.

[317] W. Smith and J. F. Grassle. Sampling properties of a family of diversity measures. *Biometrics*, 33(2):283–292, 1977.

[318] R. M. Solovay. A model of set theory in which every set of reals is Lebesgue measurable. *Annals of Mathematics*, 92:1–56, 1970.

[319] A. R. Solow and S. Polasky. Measuring biological diversity. *Environmental and Ecological Statistics*, 1:95–107, 1994.

[320] R. P. Stanley. *Enumerative Combinatorics, volume 1*, volume 49 of *Cambridge Studies in Advanced Mathematics*. Cambridge University Press, Cambridge, 1997.

[321] V. Summers. *Torsion in the Khovanov Homology of Links and the Magnitude Homology of Graphs*. PhD thesis, North Carolina State University, 2019.

[322] H. Suyari. On the most concise set of axioms and the uniqueness theorem for Tsallis entropy. *Journal of Physics A: Mathematical and General*, 35:10731–10738, 2002.

[323] K. Tanaka. The Euler characteristic of a bicategory and the product formula for fibered bicategories. Preprint arXiv:1410.0248, available at arXiv.org, 2014.

[324] I. J. Taneja and P. Kumar. Relative information of type s, Csiszár's f-divergence, and information inequalities. *Information Sciences*, 166:105–125, 2004.

[325] T. Tao. *Structure and Randomness: Pages from Year One of a Mathematical Blog*. American Mathematical Society, Providence, RI, 2008.

[326] N. Timme, W. Alford, B. Flecker, and J. M. Beggs. Synergy, redundancy, and multivariate information measures: an experimentalist's perspective. *Journal of Computational Neuroscience*, 36:119–140, 2014.

[327] B. Tóthmérész. Comparison of different methods for diversity ordering. *Journal of Vegetation Science*, 6:283–290, 1995.

[328] M. Tribus and E. C. McIrvine. Energy and information. *Scientific American*, 225(3):179–190, 1971.

[329] S. N. Tronin. Abstract clones and operads. *Siberian Mathematical Journal*, 43(4):746–755, 2002.

[330] S. N. Tronin. Operads and varieties of algebras defined by polylinear identities. *Siberian Mathematical Journal*, 47(3):555–573, 2006.

[331] C. Tsallis. Possible generalization of Boltzmann–Gibbs statistics. *Journal of Statistical Physics*, 52:479–487, 1988.

[332] C. Tsallis. What are the numbers that experiments provide? *Química Nova*, 17(6):468–471, 1994.

[333] C. Tsallis. Generalized entropy-based criterion for consistent testing. *Physical Review E*, 58:1442–1445, 1998.

[334] H. Tuomisto. A diversity of beta diversities: straightening up a concept gone awry. Part 1. Defining beta diversity as a function of alpha and gamma diversity. *Ecography*, 33:2–22, 2010.

[335] P. J. Turnbaugh, M. Hamady, T. Yatsunenko, B. L. Cantarel, A. Duncan, R. E. Ley, M. L. Sogin, W. J. Jones, B. A. Roe, J. P. Affourtit, M. Egholm, B. Henrissat, A. C. Heath, R. Knight, and J. I. Gordon. A core gut microbiome in obese and lean twins. *Nature*, 457:480–484, 2009.

[336] P. J. Turnbaugh, C. Quince, J. J. Faith, A. C. McHardy, T. Yatsunenko, F. Niazi, J. Affourtit, M. Egholm, B. Henrissat, R. Knight, and J. I. Gordon. Organismal, genetic, and transcriptional variation in the deeply sequenced gut microbiomes of identical twins. *Proceedings of the National Academy of Sciences of the USA*, 107:7503–7508, 2010.

[337] H. Tverberg. A new derivation of the information function. *Mathematica Scandinavica*, 6:297–298, 1958.

[338] R. Twarock, M. Valiunas, and E. Zappa. Orbits of crystallographic embedding of non-crystallographic groups and applications to virology. *Acta Crystallographica*, A71:569–582, 2015.

[339] United Nations, Department of Economic and Social Affairs, Population Division. World population prospects: the 2017 revision. Custom data acquired via website, 2017.

[340] I. Vajda. Axioms for a-entropy of a generalized probability scheme (in Czech). *Kybernetika*, 4(2):105–112, 1968.

[341] L. van den Dries. *Tame Topology and O-minimal Structures*, volume 248 of *London Mathematical Society Lecture Note Series*. Cambridge University Press, Cambridge, 1998.

[342] R. I. Vane-Wright, C. J. Humphries, and P. H. Williams. What to protect? – Systematics and the agony of choice. *Biological Conservation*, 55:235–254, 1991.

[343] R. S. Varga and R. Nabben. On symmetric ultrametric matrices. In L. Reichel, A. Ruttan, and R. S. Varga, editors, *Numerical Linear Algebra*, pages 193–199. Walter de Gruyter, Berlin, 1993.

[344] S. D. Veresoglou, J. R. Powell, J. Davison, Y. Lekberg, and M. C. Rillig. The Leinster and Cobbold indices improve inferences about microbial diversity. *Fungal Ecology*, 11:1–7, 2014.

[345] J. P. Vigneaux. *Topology of Statistical Systems: A Cohomological Approach to Information Theory*. PhD thesis, Université Paris Diderot, 2019.

[346] A. R. Wallace. *Tropical Nature, and Other Essays*. MacMillan and Co., London, 1878.

[347] L. Wang, M. Zhang, S. Jajodia, A. Singhal, and M. Albanese. Modeling network diversity for evaluating the robustness of networks against zero-day attacks. In M. Kutyłowski and J. Vaidya, editors, *Proceedings of the 19th European Symposium on Research in Computer Security (ESORICS 2014)*, pages 494–511. Springer, Cham, 2014.

[348] R. M. Warwick and K. R. Clarke. New 'biodiversity' measures reveal a decrease in taxonomic distinctness with increasing stress. *Marine Ecology Progress Series*, 129:301–305, 1995.

[349] M. G. Watve and R. M. Gangal. Problems in measuring bacterial diversity and a possible solution. *Applied and Environmental Microbiology*, 62(11):4299–4301, 1996.

[350] A. Wehrl. General properties of entropy. *Reviews of Modern Physics*, 50(2):221–260, 1978.

[351] R. H. Whittaker. Vegetation of the Siskiyou mountains, Oregon and California. *Ecological Monographs*, 30:279–338, 1960.

[352] R. H. Whittaker. Evolution and measurement of species diversity. *Taxon*, 21:213–251, 1972.

[353] S. Willerton. Heuristic and computer calculations for the magnitude of metric spaces. Preprint arXiv:0910.5500, available at arXiv.org, 2009.

[354] S. Willerton. Is this graph of reciprocal power means always convex? MathOverflow, 2014. Available at https://mathoverflow.net/q/176706.

[355] S. Willerton. On the magnitude of spheres, surfaces and other homogeneous spaces. *Geometriae Dedicata*, 168:291–310, 2014.

[356] S. Willerton. The Legendre–Fenchel transform from a category theoretic perspective. Preprint arXiv:1501.03791, available at arXiv.org, 2015.

[357] S. Willerton. Spread: a measure of the size of metric spaces. *International Journal of Computational Geometry and Applications*, 25(3):207–225, 2015.

[358] S. Willerton. On the magnitude of odd balls via potential functions. Preprint arXiv:1804.02174, available at arXiv.org, 2018.

[359] S. Willerton. The magnitude of odd balls via Hankel determinants of reverse Bessel polynomials. *Discrete Analysis*, 5:1–42, 2020.

[360] M. V. Wilson and A. Shmida. Measuring beta diversity with presence-absence data. *Journal of Ecology*, 72:1055–1064, 1984.

[361] G. U. Yule. *The Statistical Study of Literary Vocabulary*. Cambridge University Press, Cambridge, 1944.

[362] D. Zagier. The dilogarithm function. In P. Cartier, B. Julia, P. Moussa, and P. Vanhove, editors, *Frontiers in Number Theory, Physics and Geometry II*, pages 3–65. Springer, Berlin, 2007.

[363] A. Zygmund. *Trigonometric Series*, volume I. Cambridge University Press, Cambridge, 2nd edition, 1959.

Index of Notation

Standard Notation

$\mathbb{N}$	the set $\{0, 1, 2, \ldots\}$ of natural numbers
$\mathbb{Z}$	the set $\{\ldots, -1, 0, 1, \ldots\}$ of integers
$\mathbb{Q}, \mathbb{R}, \mathbb{C}$	the sets of rational, real and complex numbers
$[a, b]$	the interval $\{x \in \mathbb{R} : a \leq x \leq b\}$
$[a, b)$	the interval $\{x \in \mathbb{R} : a \leq x < b\}$
$(a, b]$	the interval $\{x \in \mathbb{R} : a < x \leq b\}$
(a, b)	the interval $\{x \in \mathbb{R} : a < x < b\}$
$\lfloor x \rfloor$	the greatest integer less than or equal to x
$\lceil x \rceil$	the least integer greater than or equal to x
$f\vert_A$	the restriction of a function $f : X \to Y$ to a subset $A \subseteq X$
M^{T}	the transpose of a matrix M
$\mathbb{P}(A)$	the probability of an event A
$\mathbb{E}(X)$	the expected value of a real random variable X
log	natural (base e) logarithm
$d\nu/d\mu$	Radon–Nikodym derivative
I	identity matrix (see also I below)
$\vert S \vert$	the cardinality of a finite set S (see also $\vert \cdot \vert$ below)

Notation Defined in the Text

Index

Printed in the United States
by Baker & Taylor Publisher Services